ANNUAL REVIEW OF NEUROSCIENCE

ANNUAL REVIEW OF NEUROSCIENCE

W. MAXWELL COWAN, *Editor*
Washington University School of Medicine

ZACH W. HALL, *Associate Editor*
University of California School of Medicine

ERIC R. KANDEL, *Associate Editor*
College of Physicians and Surgeons of Columbia University

VOLUME 2

1979

ANNUAL REVIEWS INC.   4139 EL CAMINO WAY   PALO ALTO, CALIFORNIA 94306

ANNUAL REVIEWS INC.
Palo Alto, California, USA

REPRINTS   The conspicuous number aligned in the margin with the title of each article in this volume is a key for use in ordering reprints. Available reprints are priced at the uniform rate of $1.00 each postpaid. The minimum acceptable reprint order is 5 reprints and/or $5.00 prepaid. A quantity discount is available.

International Standard Serial Number: 0147-006X
International Standard Book Number: 0-8243-2402-1

Annual Reviews Inc. and the Editors of its publications assume no responsibility for the statements expressed by the contributors to this Review.

PRINTED AND BOUND IN THE UNITED STATES OF AMERICA

Annual Review of Neuroscience
Volume 2, 1979

CONTENTS

SOME RELATED ARTICLES IN OTHER *ANNUAL REVIEWS*

From the *Annual Review of Biochemistry,* Volume 48 (1979)

Chemistry and Biology of the Neurophysins, E. Breslow
Epidermal Growth Factor, G. Carpenter and S. Cohen
Cell Surface Components Involved in Cell Recognition, W. Frazier and L. Glaser
Activation of Adenylate Cyclase by Choleragen, J. Moss and M. Vaughan
Peptide Neurotransmitters, S. H. Snyder and R. B. Innis

From the *Annual Review of Biophysics and Bioengineering,* Volume 8 (1979)

Endogenous Chemical Receptors: Some Physical Aspects, F. J. Barrantes
Electrical Properties of Egg Cell Membrane, S. Hagiwara and L. A. Jaffe
Neural Prostheses, F. T. Hambrecht
Three-Dimensional Computer Reconstruction of Neurons and Neuronal Assemblies, E. R. Macagno, C. Levinthal, and I. Sobel
Axoplasmic Transport of Proteins, D. L. Wilson and G. C. Stone

From the *Annual Review of Medicine,* Volume 30 (1979)

Efficacy of Biofeedback Therapy, M. T. Orne
Prenatal Sex Hormones and the Developing Brain: Effects on Psychosexual Differentiation and Cognitive Function, A. Erhardt and H. F. L. Meyer-Bahlburg
Computed Tomography in Neurologic Diagnosis, A. G. Osborn

From the *Annual Review of Pharmacology and Toxicology,* Volume 19 (1979)

Biology of Opioid Peptides, A. Beaumont and J. Hughes
Physical Mechanisms of Anesthesia, S. H. Roth
β-Adrenoceptor Blocking Drugs, A. Scriabine

From the *Annual Review of Physiology,* Volume 41 (1979)

Renal Nerves and Na^+ Secretion, C. Gottschalk
Biochemical Mechanisms of the Sodium Pump, A. Schwartz, E. Wallick, and L. Lane
Properties of Two Distinct Inward Membrane Currents in the Heart, H. Reuter
Electrogenesis of the Plateau and Pacemaker Potential, M. Vassalle

Ann. Rev. Neurosci. 1979. 2:1–15

THE METAMORPHOSIS OF A PSYCHOBIOLOGIST

❖11516

Seymour S. Kety

Ann. Rev. Neurosci. 1979. 2:1–15

THE METAMORPHOSIS   ❖11516
OF A PSYCHOBIOLOGIST

Seymour S. Kety

Harvard Medical School, Laboratories for Psychiatric Research,
Mailman Research Center, McLean Hospital, Belmont, Massachusetts 02178

My introduction to the challenge of the brain and human behavior came about, some fifty years ago, at the hands of Edwin Landis, a remarkable teacher of science at the Central High School in Philadelphia. Cognizant of the interest of several of us, he had organized a science and philosophy club for which we found an appropriate dedication from Virgil: "Felix, qui potuit rerum cognoscere causas." Through him I learned that to understand the causes of things was an extremely remote goal, particularly in the case of the brain, but that there was excitement enough in the process of trying to understand and in the individual small steps along the way. I also learned that there were many approaches to the understanding of a complex phenomenon and that the ultimate understanding was likely to involve most of them. Although fortune or reasoned choice might set us on one of these paths by virtue of idiosyncratic aptitudes, training and experience, that did not make it the best or the only path. The unparalleled complexity of the brain and of behavior, both individual and collective, calls for great humility. Yet, in its stead, we have often seen complacency and conviction, or the advocacy of one approach and the disparagement of others.

Neuroscience, whose very name represents a juxtaposition of disparate roots, is a discipline that grew out of the manifold approaches to the study of the nervous system. It may be unnecessary to remind ourselves that for the most part neuroscience deals with one important aspect of behavior—the machinery that mediates it. After the networks of the brain have been delineated to their uttermost detail, as well as the mechanisms for processing, storing, and acting upon information, and the physical and chemical mechanisms that modulate these processes, we will still need to define and

1

understand the informational content and its significance in terms of human experience. It is unlikely that either aspect can be reduced to the other, and neither alone will constitute a sufficient explanation of the diversity and versatility of human behavior.

In particular types or syndromes of behavior, however, it is possible that these two spheres of influence may have different weights, account for larger or lesser fractions of the variance, or be susceptible to more proximate or remote elucidation. Thus, there are few neurobiologists, indeed, who would feel that their techniques and competence would usefully be directed at uncovering differences between the brain of a Democrat and that of a Republican (Kety 1961), yet that proposition is no more absurd than the opposite presumption that the biological disciplines have little to contribute to an understanding of the major mental disorders, on the unproven assumption that these are essentially social and psychological problems.

Modern neurochemistry had its origins in the less restrictive and more constructive hypothesis that many forms of insanity were biochemical in nature. That was the basis on which, nearly one hundred years ago, Johann Thudichum received a grant from the Privy Council of England. It is particularly significant that, armed with that hypothesis, Thudichum did not then go to a mental hospital for the urine or blood of patients; instead, he went to the slaughterhouses for cattle brain and spent the next ten years isolating and characterizing the major components. I have recounted that glimpse of scientific history many times because it illustrates in a compelling way, the motivation of most scientists to contribute to the understanding and solution of a human problem, the value of enlightened public support that leaves the strategy of the effort to the scientist himself, and the wisdom of laying firm foundations of basic knowledge on which others may build, rather than directly but prematurely attacking the problem with inadequate tools, concepts, and information.

Thudichum did not discover the biochemical disturbances that occur in psychosis, but he developed one of the foundations on which modern neurochemistry was built (Thudichum 1884). It may turn out that in the balance and relationships among the minute quantities of dopamine, noradrenalin, serotonin, acetylcholine, GABA, or in neurotransmitters yet to be discovered, or the growing list of polypeptides that occur in the brain, lie some of the crucial disturbances of insanity. What would have been the chance that Thudichum would have discovered this at the time when those substances were still undreamed of? He would have frittered away the public funds and ten years of his career in futile research, which a more naive judgment would have called "relevant."

Forty years after Thudichum's death, renewed research on the chemical structure of the brain was undertaken by Jordi Folch-Pi, among whose major contributions were the characterization of three distinct phospholipids from Thudichum's "cephalin" and the discovery of the proteolipids, a new class of chemical compounds and the major components of myelin.

After the discovery that cells and tissues respire in vitro, brain slices were found to be the seat of active respiration. It was Ralph Gerard, the late Honorary President of the Society for Neuroscience, who coupled metabolism with neural function. As a post-doctoral fellow with A. V. Hill he measured the heat of axonal conduction, and with Otto Meyerhoff, the importance of oxidative metabolism in that process. Seven years later, Gerard and Hartline were successful in measuring the micro-quantities of oxygen utilized by optic nerve fibers during physiological conduction. At the same time physiologists were laying another foundation of neurochemistry. Dale, Loewi, Cannon, and, in time, many other scientists were elucidating the chemical processes involved in transmission at peripheral synapses.

Oxygen utilization and energy metabolism of the brain rapidly became the main focus of neurochemistry. Studies on the respiration and intermediary metabolism of brain slices and homogenates by Quastel, Elliott, and many others produced a wealth of information on the rates of oxygen and substrate utilization by different regions of the brain and in various species. The preference of brain for glucose as substrate was demonstrated as was the ability of neural tissue to engage in anerobic glycolysis. McIlwain produced a model of neuronal activation by electrical stimulation or with potassium ion, and Larrabee, in the isolated sympathetic ganglion, examined metabolism in relation to transsynaptic neuronal activation. An important conclusion that followed from his observations was the primary action of anesthetic agents on synaptic transmission rather than neuronal metabolism, the latter falling as a result of decreased bombardment. This was one of several early hints of the importance of synaptic transmission as a crucial site of action for a variety of chemical substances that act on the brain.

Valuable as in vitro techniques were in their ability to control most of the variables in the neuronal environment, they provided only elegant but highly simplified models of what the living brain was like and there was a need to study it directly.

STUDIES ON THE CIRCULATION AND ENERGY METABOLISM OF THE BRAIN

Nature has protected the brain, however, not only against gravitational and traumatic stresses, but also against the prying eyes and hands of scientists. The obvious way to measure the amounts of oxygen or other substances used or produced by an organ is by assaying their concentration difference

between blood entering and leaving the organ. The amount of blood entering or leaving per minute, multiplied by the arteriovenous difference of the
substance yields the amount utilized per minute. It was not, however, easy
to measure the cerebral blood flow in animals under reasonably physiological conditions. In most species the brain receives its arterial blood not only
from the internal carotid and vertebral arteries, but also from the external
carotid by way of a network of anastomoses—the *rete mirabile* of Galen.
It requires rather drastic surgery to restrict the arterial inflow to one or two
channels that can be monitored. In primates, however, the cerebral arterial
supply is much more discrete. Recognizing this, Dumke & Schmidt (1943)
made the first reliable quantitative measurements of blood flow in the
mammalian brain, using the rhesus monkey and inserting an ingenious
bubble flow meter into the arterial supply. I was in Boston at the time,
working on traumatic shock with Joseph Aub and his associates at the
Massachusetts General Hospital. Alfred Pope and I, the two young post-
doctoral fellows in the laboratory, had just written a review on the circula-
tory system in shock and had become impressed, as others had before us,
with the wealth of homeostatic mechanisms that had been evolved to pre-
serve the circulation of the brain.

I was so impressed with the paper by Dumke and Schmidt that I decided
to return to the University of Pennsylvania to work with Carl Schmidt on
the cerebral circulation. Shortly after I arrived, he invited me and Harry
Pennes, another newcomer to the Department of Pharmacology, to assist
in the studies he was about to undertake, of the oxygen consumption in the
brain of the monkey. Besides obtaining values for that function in the
mammalian brain under light anesthesia, that study also demonstrated a
striking increase in metabolism, exceeding the increase in cerebral blood
flow, during convulsions induced by metrazol or picrotoxin.

But just as the brain is unique among organs for its complexity, so is the
human brain unique in its capacity, its versatility and plasticity, its ability
to conceptualize and create, to experience ecstasy and deep grief, and to
describe to outside observers the results of its inner processes. It is also the
human brain that falls prey to serious disorders of these functions, for which
no comparable animal models exist. To study the metabolism of the human
brain while it was engaged in these functions and experiences might teach
us something about these processes, and its study in disease might be of
benefit to those suffering from neurological or mental disorders.

While in Boston, I had heard an inspiring lecture by Andre Cournand,
describing his early work on the output of the human heart in health and
disease, and was impressed with the possibility that clinical studies could
be more physiological, relevant, and fundamental, under certain circumstances, than studies in animals. Cournand had used the Fick principle,

calculating pulmonary blood flow, equivalent to cardiac output, from the oxygen taken up by the lung and the oxygen content in blood entering and leaving, all of which he could measure independently.

Much earlier, in 1927, the Boston psychiatrist Meyerson had described a simple technique for obtaining cerebral venous blood in man from the superior bulb of the internal jugular, making it possible to measure the arteriovenous difference across the brain for substances utilized or produced in significant amounts by that organ. Lennox and Gibbs then used that approach to infer cerebral circulation from the arteriovenous oxygen difference, assuming that cerebral oxygen consumption, which they could not measure, was unchanged. About the same time, Himwich used the arteriovenous oxygen difference to infer metabolic rate for the brain with the assumption that blood flow was constant. Interestingly enough, many of the findings of both of these groups were later found to be valid by virtue of a fortunate choice of problem to which each approach was applied. Lennox and Gibbs were able correctly to infer that carbon dioxide increased cerebral blood flow because metabolism is not, in fact, appreciably altered at the concentrations of carbon dioxide they used, and both they and Himwich discovered that the respiratory quotient for the human brain was close to unity since that calculation does not require a measurement of cerebral blood flow. However, they sometimes made inferences that were incorrect, i.e. that cerebral blood flow is markedly increased in coma (because of the narrowed arteriovenous difference, which is the result of diminished metabolism). What had limited the acceptance of that approach was that the arteriovenous oxygen difference, being a function of both blood flow and oxygen consumption, was not a valid measure of either. The oxygen consumption of the brain could not be measured independently and certainly it could not be assumed to be constant under a wide variety of functional states, since it would be expected to vary with the state of activity or disease that was the object of investigation.

By 1943 the Fick principle had been applied by Homer Smith to the kidney, and two years later by Stanley Bradley to the measurement of hepatic blood flow; in each case they took advantage of the ability of these organs specifically to excrete a foreign substance at a rate that could be independently measured. Unlike the kidney and liver, however, the brain was not known to remove selectively and specifically a foreign substance from the blood and excrete it for accurate measurement.

The brain does, however, absorb by physical solution an inert gas, which reaches it by way of the arterial blood. The accumulation of such a gas in the brain should be independent of the state of mental activity and determined instead by relatively simple physical principles such as diffusion and solubility, which I felt should be quite constant in the brain whether asleep or awake, working out a complex mathematical problem, or suffering from schizophrenia. In the brain of a monkey I found that an arteriovenous difference did exist for nitrous oxide as it was breathed, and that it narrowed as the brain approached equilibrium with the arterial partial pressure. The variable difference could be integrated to give the denominator of a Fick equation. The next problem was how to determine the numerator, representing the amount or concentration of the inert gas accumulated in the brain. One could, of course, do this by using a radioactive inert gas and external counting, but a simpler approach presented itself. The arteriovenous difference became progressively narrow with time because the brain was coming to equilibrium. If that was the case, there would be a time at which the venous blood draining the brain would be in equilibrium with the brain itself and could be used as a measure of the partial pressure of gas in the brain, which, multiplied by the ratio of solubilities between blood and brain, would yield the concentration in the brain.

In a series of determinations on the monkey, carried out with Carl Schmidt in 1944 and 1945, it appeared that the major assumptions were valid and that the indirect method was capable of yielding values that correlated well with direct measurement using the bubble flow meter. Then began a number of efforts to examine more rigorously these and a number of other assumptions on which the accuracy of the method depended.

The solubility of nitrous oxide in blood and brain was determined in vitro in collaboration with Harmel, Broomell, and Rhode. No significant difference was found in the solubility for blood, normal brain of animals and man, nor in the brains of patients dying with a variety of disorders. Moreover, the equilibration between blood and brain could be studied in animals and was found to be sufficiently complete at the end of ten minutes. Studies with radioactive krypton indicated that in man as well, ten minutes was sufficient for nearly complete equilibrium. Two other series of studies indicated that the venous blood was well mixed by the time it emerged and that contamination from the distribution of the external carotid was minor.

The first systematic study in man was carried out on 14 healthy young men who volunteered to serve as subjects (Kety & Schmidt 1948). The values for blood flow and oxygen consumption were in the same range as those we had previously found in the monkey, when both were reduced to unit weight of brain. I am still impressed, however, with how large a share of the body's economy is used in supporting the brain—about a fifth of the cardiac output and of the oxygen consumption at rest—but how small the utilization of energy by the human brain—a mere 20 watts—in comparison with what man-made computers require. The number of problems to which we wanted to apply the new technique were legion but we tried to select those that might contribute to fundamental knowledge about the brain and its physiological functions, or in the case of disease, where there was reason to think that an alteration in circulation or metabolism would be crucially involved. The first application was to the effects of altered concentrations of oxygen and carbon dioxide. Here we confirmed previous studies in animals and the remarkable insight of Roy and Sherrington, half a century earlier, that, by dilating cerebral vessels, the products of metabolism (diminished oxygen, increased carbon dioxide and H^+) played an important role in the homeostatic linkage of blood flow to metabolic requirements locally. We were not surprised to find that in deep anesthesia or coma, the cerebral oxygen consumption was reduced by 50%.

The study on sleep, however, produced some surprising results (Mangold et al 1955). Although it was not unusual for some of our subjects to fall asleep and have to be awakened during our previous studies, it was not easy to get them to sleep when we wished. I was a subject in a sleep study and although I was quite comfortable, the effort of trying to sleep kept me awake. In the course of more numerous unsuccessful trials we did have six subjects sleep for the ten minutes necessary to make a measurement. The results were unexpected. The utilization of oxygen by the brain was quite normal, in spite of the prevailing belief, which stemmed from Pavlov and Sherrington, that sleep was characterized by a suppression of neuronal activity. It was not until Evarts succeeded in recording from individual cortical neurons of sleeping animals, and the confirmation provided by our later studies of regional circulation that the observations in man showing sleep to be an active process became credible.

In disease states, the finding that cerebral circulation was normal in essential hypertension despite a perfusing pressure that was twice normal, was interesting. This occurred without intervention by the known sympathetic supply, suggesting a humoral vasoconstriction or a homeostatic autoregulation, both of which are now known to occur. We were unprepared for Sokoloff's finding that the brain did not share in the generalized increase in metabolism that occurs in hyperthyroidism. This led him to surmise, and eventually to demonstrate, that the important action of thyroxin was on protein rather than carbohydrate metabolism.

The studies on schizophrenia were begun because it had been proposed that a deficit in oxygen utilization occurred in the brain, but there was an equally cogent reason. The development of the nitrous oxide technique was supported in part by a small grant from the Scottish Rite Schizophrenia Program. Although the directors had never asked about the relevance of cerebral circulation to schizophrenia, when an application to that problem was possible we were eager to make it. The results were not illuminating: the brain of the schizophrenic patient has the same blood flow and utilizes oxygen at the same rate as that of normal individuals (Kety et al 1948).

From this we concluded that if there was a biochemical derangement in schizophrenia it was in processes more subtle and more specific than oxygen utilization. But as recently as thirty years ago we had no idea what those processes might be. We also suggested that alterations in oxygen utilization could occur in particular regions of the brain so small as not to be reflected in the overall measurements to which the nitrous oxide technique was limited. In the intervening time much more has been learned about both possibilities.

In 1948 I joined Julius Comroe in the Graduate School of Medicine at Penn where he had developed an exciting new approach to the teaching of the basic medical sciences to physicians. With his encouragement I decided to tackle the theoretical considerations involved in the uptake of inert gases by the lungs and their distribution in the tissues. Although the nitrous oxide technique was based upon such theory, its assumptions had been examined only empirically. Moreover from the arterial concentration curves for nitrous oxide obtained in a variety of clinical conditions it was obvious that the shape of that curve was altered by changes in ventilation or cardiac output. My rudimentary knowledge of mathematics did not go very far beyond linear differential equations, but using these it was possible to derive expressions that described the uptake and distribution of inert gases (Kety 1951), or for that matter, any diffusible and nonmetabolized substance. That description was a first approximation and has since been further elaborated by others, using more sophisticated approaches. The expression for the uptake at the lungs in terms of ventilation, cardiac output, and solubility was of some value to anesthesiology, and equations derived for capillary-tissue exchange facilitated the development of various techniques for the measurement of capillary permeability or circulation depending on the choice of tracer.

THE INTRAMURAL BASIC RESEARCH PROGRAM OF THE NIMH

In 1951 I had an unexpected visit from Robert Felix, the director of the newly established National Institute of Mental Health in Bethesda, who invited me to join him as its first scientific director. Although I was very happy with academic life, Dr. Felix had stimulated a receptive area. The studies on schizophrenia, which were what brought me to his attention, had served to impress on me the magnitude of the problem of mental illness and the depths of our ignorance about it. For one who was motivated to contribute to an understanding of the problem, yet did not feel that a quick breakthrough was likely, what better opportunity existed than to plan and develop a program of basic research that might help to provide the basis of an eventual understanding. After thinking about it for two months, visiting Bethesda, seeing the 200 laboratories being constructed for the new institute, meeting its small but dedicated staff, conferring with James Shannon and Harry Eagle, the scientific directors of the Heart and Cancer Institutes, I accepted the challenge and spent 16 years there.

Dr. Felix was the ideal director for the National Institute of Mental Health. He appreciated the need for substantially increased research and defended it valiantly. He did not presume to know in what directions our research program should go, or if he did, he did not permit that to influence me. For my part, these mysterious illnesses that had baffled the human race for centuries had not revealed any of their secrets to me. I could think of no better investment of these new and unprecedented resources than using them to establish a broad program of fundamental research, representing all of the disciplines concerned with the brain and behavior. Perhaps because my background was more deficient in the social sciences than in any of the others, I decided to establish the Laboratory of Socio-Environmental Studies first, with John Clausen as its chief. Shortly thereafter Wade Marshall became head of Neurophysiology. Alexander Rich was appointed to represent neurochemistry, and Giulio Cantoni, soon joined by Seymour Kaufman, to represent biochemistry. When the Clinical Center was completed Robert Cohen was asked to be Director of Clinical Research and together, we recruited David Shakow to direct a large laboratory of psychology, representing a wide spectrum of experimental, developmental, and clinical psychology.

I was also charged with organizing the Basic Research Program of the National Institute of Neurological Diseases and recruited several additional neuroscientists: William Windle, soon joined by Sanford Palay in neuroanatomy, Roscoe Brady in neurochemistry, Kenneth Cole in biophysics, Ichiji Tasaki in neurobiology, Karl Frank in neurophysiology, and, for a short time, Roger Sperry in developmental neurobiology, until he was wooed away by Cal Tech. One concern that some had expressed was rapidly put to rest. A government institution with the proper philosophy could attract a faculty as distinguished as that of any university. Of the initial group that joined me, six eventually became members of the National Academy of Science. The two programs were organized simultaneously and merged into a single basic research program as they should have been, since the neurosciences have as much pertinence to mental illness as to neurological disease. It is unfortunate that they were separated some years later for parochial reasons. Much more serious, however, has been the segregation of disciplines that has prevailed in the extramural research programs of the two institutes over the past decade or more, as the result of a tacit agreement that the Neurological Institute would support fundamental research in the neurobiological disciplines while the Mental Health Institute recognized the social and psychological sciences as its domain.

LOCAL CEREBRAL CIRCULATION
AND FUNCTIONAL ACTIVITY

I had accepted the position in Bethesda with the understanding that I would be able to continue some personal research, and this was made possible by the excellent administrative support that the National Institutes of Health provided. Most important, however, were the scientific collaborators with whom I became associated. Louis Sokoloff joined me as soon as a modest laboratory was available for our studies in cerebral metabolism and we were joined from time to time by a substantial number of post-doctoral fellows such as William Landau, Lewis Rowland, and Martin Reivich, and visiting scientists, including Niels Lassen from Denmark and Cesare Fieschi from Italy, all of whom have achieved well-deserved recognition in neurology, neurochemistry, and physiology.

The nitrous oxide technique that yielded average values for the whole brain was useful for the study of physiologic or pathologic changes that were generalized or affected a significant fraction of the brain. But that organ is remarkable for its complexity and heterogeneity, and for this, another approach was necessary. In the theoretical analysis of inert gas exchange at the tissues, I had developed an expression relating local tissue concentration of a freely diffusible, nonmetabolized substance, to the history of its concentration in arterial blood and the rate of blood flow through the region. Since the arterial and tissue concentrations of a radioactive tracer could be evaluated, it should be possible to measure regional blood flow. The theory was tested using trifluor-iodo-methane labeled with iodine-131, and measuring tissue concentrations by means of autoradiograms of frozen tissue sections (Landau et al 1955). Later, antipyrine-C^{14} was substituted in a simpler technique, which gave better resolution (Reivich et al 1968). Lassen and Ingvar applied the theory to measurement of regional blood flow in man, with techniques that have become widely used.

In the autoradiograms from animals or the visual display obtained in clinical studies, many of the structures of the brain were differentiated by means of their blood flow. The magnitude of the blood flow and the changes that could be induced by activation, suggested a linkage between perfusion, metabolism, and functional activity, although metabolism could not be measured at that time.

Recently, Sokoloff (1977) has succeeded in visualizing and measuring metabolic rate throughout the brain in terms of glucose uptake, using a nonmetabolized, radioactive congener (carbon-14 labeled deoxyglucose). So close is the coupling between glucose utilization and neuronal activity that the autoradiograms visualize, with a high degree of fidelity and resolution, small regions of the brain in terms of their functional activity. With the use of positron emission tomography and a suitably labeled deoxyglucose, application of the technique to man is feasible, so that one can look forward to visualizing levels of functional activity throughout the conscious human brain, in addition to the presently recognized value of the technique in fundamental neurophysiological, anatomical, and pharmacological research.

THE LABORATORY OF CLINICAL SCIENCE

By 1956 several investigators had joined the Intramural Research Program of the Mental Health Institute whose interests were in the interface between the basic neurobiological sciences and clinical psychiatric problems, and a new grouping was established designated the Laboratory of Clinical Science. By that time the Basic Research Program was fully staffed, I was eager to involve myself more in research and less in administration, and the implications of the new neurobiological knowledge to psychiatry attracted me. I asked to be allowed to step down from the scientific directorship to join the new laboratory as its nominal chief.

The initial group consisted of Edward Evarts, Julius Axelrod, Louis Sokoloff, Marian Kies, Roger McDonald who was succeeded by Irwin Kopin, Philippe Cardon, and Seymour Perlin who was succeeded by William Pollin. Although quite diverse in their scientific interests, they shared a commonality of motivation and a mutuality of spirit, which made the 11 years I spent in that laboratory one of the most exciting and rewarding periods of my life.

What were these new and promising implications that attracted many of us? It was not the numerous enthusiastic claims that were being made regarding abnormal proteins, metabolites, or toxic factors in the blood or urine of schizophrenic patients, which lacked plausibility and did not survive replication. Rather, it was a number of less spectacular but more credible observations with more remote relevance—observations that suggested that the synapses of the brain, like those in the periphery, were chemically mediated switches rather than electrical junctions. Acetylcholine by that time had achieved the status of a putative neurotransmitter in the brain, but there were other substances like serotonin, noradrenalin, and dopamine, which could conceivably serve in such a role, and which had only recently been identified in the brain. Lysergic acid diethylamide, a drug which had attracted wide attention because of its hallucinogenic properties, had also been found to block some of the pharmacological actions of serotonin. Three other psychotomimetic drugs, dimethyltryptamine, mescaline, and amphetamine, were substituted forms of serotonin or dopamine. Only a few years before, chlorpromazine had been found to be remarkably effective in the alleviation of psychotic behavior and reserpine was being used extensively as a major tranquilizer. Although it was to take ten years for the action of chlorpromazine on dopamine synapses to be discovered and substantiated, knowledge of the remarkable ability of reserpine to deplete the brain of serotonin was literally around the corner—in Bernard Brodie's laboratory at the Heart Institute.

If the synapses involved in the mental states and behaviors produced or ameliorated by such drugs were chemically mediated, that would offer a plausible site at which these drugs could act. Moreover, if central synapses in general were chemical switches, then a biochemistry of behavior was conceivable, and at the synapse, not only drugs, but genetic factors, dietary constituents, hormones, metabolic, immune, and infectious processes, could all be seen to act, altering the patterns of transsynaptic interaction and affecting behavior and mental processes. For the first time, plausible and heuristic approaches could now be opened and explored, that might some-day explain the biological disturbances of mental illness and the symptoms that depend on them.

The most productive way of exploring these new approaches was not by way of a crash program. The gap between the knowledge we had and the clinical problems was still too wide to be spanned all at once by any concerted effort. What was needed was to narrow the gap by an increase in knowledge on both sides, which is best done by relying on the creativity and judgment of individual scientists who know better than anyone else what their next step should be. The members of the laboratory pursued their own research goals, some studying the clinical problems in greater detail and in the light of new knowledge, most expanding the base of fundamental knowledge in areas that they perceived to be relevant. Where appropriate, collaborative efforts developed within the laboratory, and quite as often, outside of it. I believe that subsequent events have justified this approach.

Among the claims that were being made at that time was one postulating the formation of a toxic, hallucinogenic metabolite of circulating epineph-rine in schizophrenia, which was sufficiently provocative that some of us decided to examine it further. The difficulty was that in 1956 we knew little enough about the normal metabolism of epinephrine, let alone its metabo-lism in disease. One strategy would be to administer labeled epinephrine in pharmacologically insignificant amounts and compare the urinary chromatographic profiles of radioactivity. The carbon-14 labeled material that was available would not provide sufficient specific activity. It was possible that a tritium-labeled epinephrine could be prepared with the requisite stability and activity, and arrangements were made to have 7-H^3-epinephrine of high specific activity synthesized. By the time the labeled compound arrived, however, that strategy was no longer necessary. In the year that had elapsed, Axelrod, taking off from a brief report in the literature, had demonstrated the enzymatic O-methylation of catecholamines in vitro, characterized the enzyme responsible, predicted the major catecholamine metabolites, and then went on to extract and identify them in the urine of animals (Axelrod 1959). When the radioactive epinephrine became available, it was a simple matter to examine its metabolism in normal subjects (LaBrosse et al 1961) and in schizophrenics. No evidence was found for an abnormal metabolism of circulating epinephrine in that disorder.

There is not the space nor the necessity of indicating Axelrod's contributions to our present knowledge of catecholamine metabolism and inactivation at the synapse. They have not solved a major psychiatric problem as yet, but when the final chapter to our understanding of mental illness is written, his work will occupy a prominent place in it.

Studies by Axelrod, Glowinski, and Kopin, on the metabolism of norepinephrine in the brain and the effects of psychoactive drugs, further supported the role of that amine as a neurotransmitter in central synapses. It required evidence from many quarters before there was general agreement that the biogenic amines were important as neurotransmitters in the brain. Electron microscopy and fluorescence histology were the most direct and compelling, but these were reinforced by microinjection and electrophysiological studies, and most recently, by the demonstration in vitro of specific receptors for several of the transmitters.

The list of neurotransmitters and synaptic modulators has gotten longer, extending from the biogenic amines to include amino acids and polypeptides. The involvement of specific members of the list in the pharmacological action of most of the psychoative drugs has been reasonably well established. It is also clear that neurotransmitters and modulators play important roles in the mediation of certain mental states and types of behavior, although their precise action and interactions remain to be elucidated. It is not possible to state, at present, how they are involved in mediating the symptoms of mental illness or whether they play an etiological role. Those who choose to point out that no morphological or biochemical lesion has been identified in the major psychoses are still correct, although the inference they draw, that therefore none exists, is clearly a non sequitur.

GENETIC FACTORS IN SCHIZOPHRENIA

When I reviewed the field in 1959, I found no compelling evidence for a biochemical disturbance in schizophrenia, except for the observations on families and twins, which was compatible with the existence of forms of the illness with a genetic component. The inability to control or randomize environmental variables, however, made that conclusion less than rigorous, and suggested the desirability of studying a national sample of adopted individuals whose genetic endowment and environmental influences could be studied separately through their biological or adoptive relatives.

In 1962, David Rosenthal, Paul Wender, and I learned of our respective interests in such a strategy and began a collaboration with Fini Schulsinger in Copenhagen where it has been possible to compile such a national sample, identify adoptees or biological parents with schizophrenia or affective disorder, and to study their relatives or offspring, reared in another environment. In each of our studies and in those of others these mental illnesses continue to run in families, but in the biological and not in the adoptive families. The evidence permits the conclusion that genetic factors operate significantly in the transmission of these disorders, or at least in some of their major subgroups (Kinney & Matthysse 1978). Since the genes can express themselves only through biochemical mechanisms, this should constitute rather compelling justification for the relevance of biological research to mental illness.

The directions that such research can take are legion, limited only by what fraction of the staggering costs of mental illness we invest in its ultimate understanding and prevention. There is the development of a new neuropathology (Matthysse & Pope 1975), representing the application to the brain of powerful new morphological, histochemical, and molecular techniques. Studies, such as those by Nauta and his associates, of the interconnections of the mesolimbic system are now possible by autoradiographic and electron microscopic techniques (Cowan 1975). There is the further study of the manifold neurotransmitters—the systems in which they are involved, their interconnections, the properties of their receptors. It is possible to elucidate the neural basis of psychological functions, as Hubel & Wiesel, Mountcastle, Evarts, or Kandel have done in areas of perception, attention, voluntary behavior, or adaptation to experience. One could go on, through all of the neurosciences. The opportunities for significant research are many, and there is a new generation of eager and competent scientists to explore them. The prospects have never been more promising.

Those prospects have been recognized not only by neuroscientists, but by academic medicine, and many departments of psychiatry have been moving in the direction of their realization. On that basis I accepted the chair of psychiatry at Johns Hopkins in 1961, which would have been a most gratifying and important role if I had been a psychiatrist. When I saw the clinical responsibilities outstripping my involvement in research, I reluctantly resigned. In 1967 I moved to Harvard, not as a department chairman, but as a professor of psychiatry with my major responsibilities in research. At our laboratories at the McLean Hospital I have the good fortune to be associated with a number of scientists who represent a wide range of basic disciplines that sustain psychiatry—from Walle Nauta in neuroanatomy to Philip Holzman in psychology. They and their counterparts in genetics, pharmacology, molecular biology, and biochemistry exemplify the mature and multidisciplinary field that psychiatric research has become. The chemical nature of synaptic transmission, the efficacy of the new psychotherapeutic drugs, and the recognition of the significant genetic contribution to the etiology of the major psychoses, make clear the pertinence of the neurosciences, as well as the psychological and social sciences, to modern psychiatry and to further progress in the understanding, treatment, and prevention of mental illness.

Literature Cited

Axelrod, J. 1959. Metabolism of epinephrine and other sympathomimetic amines. *Physiol. Rev.* 39:751–76

Cowan, M. 1975. Recent advances in neuroanatomical methodology. In *The Nervous System,* ed. D. Tower, R. O. Brady, 1:59–70. New York: Raven. 685 pp.

Dumke, P. R., Schmidt, C. F. 1943. Quantitative measurements of cerebral blood flow in the macacque monkey. *Am. J. Physiol.* 138:421–28

Kety, S. S. 1951. The theory and applications of the exchange of inert gas at the lungs and tissues. *Pharmacol. Rev.* 3:1–41

Kety, S. S. 1961. A biologist examines the mind and behavior. Many disciplines contribute to understanding human behavior, each with peculiar virtues and limitations. *Science* 132:1861–70

Kety, S. S., Schmidt, C. F. 1948. Nitrous oxide method for the quantitative determination of cerebral blood flow in man: Theory, procedure, and normal values. *J. Clin. Invest.* 27:476–83

Kety, S. S., Woodford, R. B., Harmel, M. H., Freyhan, F. A., Appel, K. E., Schmidt, C. F. 1948. Cerebral blood flow and metabolism in schizophrenia. The effects of barbiturate semi-narcosis, insulin coma and electroshock. *Am. J. Psychiatry* 104:765–70

Kinney, D. K., Matthysse, S. 1978. Genetic transmission of schizophrenia. *Ann. Rev. Med.* 29:459–73

LaBrosse, E. H., Axelrod, J., Kopin, I. J., Kety, S. S. 1961. Metabolism of 7-H^3-epinephrine-d-bitartrate in normal young men. *J. Clin. Invest.* 40:253–60

Landau, W. M., Freygang, W. H., Rowland, L. P., Sokoloff, L., Kety, S. S. 1955. The local circulation of the living brain; values in the unanesthetized and anesthetized cat. *Trans. Am. Neurol. Assoc.* 80:125–29

Mangold, R., Sokoloff, L., Conner, E., Kleinerman, J., Therman, P. G., Kety, S. S. 1955. The effects of sleep and lack of sleep on the cerebral circulation and metabolism of normal young men. *J. Clin. Invest.* 34:1092–1100

Matthysse, S., Pope, A. 1975. The approach to schizophrenia through molecular pathology. In *Molecular Pathology,* ed. R. A. Good, S. B. Day, G. Yunis, pp. 744–68. Springfield, Ill.: Thomas

Reivich, M., Jehle, J., Sokoloff, L., Kety, S. S. 1968. The effect of slow wave sleep and REM sleep on regional cerebral blood flow in cats. *J. Neurochem.* 15:301–6

Sokoloff, L. 1977. Relation between physiological function and energy metabolism in the central nervous system. *J. Neurochem.* 29:13–26

Thudichum, J. W. L. 1884. "A Treatise on the Chemical Constitution of the Brain." London: Balliere, Tindall & Cox

Ann. Rev. Neurosci. 1979. 2:17–34

VISUAL TRANSDUCTION IN VERTEBRATE PHOTORECEPTORS

❖11517

Wayne L. Hubbell

Department of Chemistry, University of California, Berkeley, California 94720

M. Deric Bownds

Laboratory of Molecular Biology and Department of Zoology,
University of Wisconsin, Madison, Wisconsin 53706

INTRODUCTION

Studies on vertebrate and invertebrate photoreceptor cells have increased exponentially in the past ten years. Advances in both biochemical and physiological techniques have brought us to the threshold of understanding the pathways that regulate transduction and adaptation. This brief review is confined largely to studies on vertebrate rod photoreceptors, because biochemical correlates of transduction are being studied most actively in amphibian rod receptor cells. The site of visual transduction in these cells is the rod-shaped outer segment (ROS), which in the amphibian is 6–8 μm in diameter and 40–70 μm in length. The ROS is a modified cilium, composed of a plasma membrane that encloses a stacked series of 1000–2000 disc membranes. The disc membranes are formed by invagination of the plasma membrane at the base of the outer segment, but soon thereafter pinch off to become physically separate from the plasma membrane. When light strikes one of the approximately 3×10^9 rhodopsin molecules in the *disc* membranes, a transient suppression of the sodium permeability of the *plasma* membrane occurs, and the rod cell hyperpolarizes. The mechanisms that may link photon absorption and the permeability change are considered in this review. No attempt is made to provide an exhaustive coverage of the literature; instead we focus attention on recent and selected aspects of physiology, biochemistry, and structure relevant to the transduction

17

0147-006X/79/0315-0017$01.00

process. The citations emphasize relatively recent work published before May 1978, with bias towards the authors' laboratories. The reader is referred to recent reviews by Ostroy (1977), Ebrey & Honig (1975), Rosenkranz (1977), Daemen (1973), and Hagins (1972) for discussion of topics not mentioned here.

The membranes that comprise the ROS are a particularly favorable preparation for studying the molecular basis of a nerve signal, for they are the only nerve membranes known to contain an excitable component that is labeled with a chromophore. The retinyl group, which binds to opsin to form the visual pigment rhodopsin, can be used as a marker in isolating photoreceptor membranes or rhodopsin, and knowledge of the photochemistry of this system allows one to quantitate precisely sensory input. ROS can be detached easily by gentle shaking of a retina and freed from gross contamination. Their plasma membranes reseal, and the structures maintain their ability to perform transduction (Korenbrot & Cone 1972, Bownds & Brodie 1975). There is general agreement that the light-sensitive permeability of the ROS plasma membrane must be regulated by an internal transmitter (or transmitters) that mediates between the site of photon absorption in the disc membrane and the conductance mechanism of the plasma membrane (Baylor & Fuortes 1970, Hagins 1972, Cone 1973). Because many sodium channels are closed after one photon absorption, an amplification step must intervene.

Two general mechanisms are currently under consideration as explanations of the phenomena. The first involves amplification by enzymatic processes. In this scheme, a photolyzed rhodopsin molecule activates an enzyme, which in turn produces or degrades many transmitters. A specific version of this scheme, suggested by recent studies discussed below, is shown in Figure 1a. Sodium channels are considered to be kept in their open conformation in the dark by protein phosphorylation mediated by cyclic GMP. Illumination activates a phosphodiesterase enzyme, which lowers cyclic GMP levels, protein is dephosphorylated, and the channels close. The first part of this review, contributed by M. D. Bownds, deals briefly with the physiology of the ROS and considers several light-dependent enzymatic reactions that may participate in transduction.

A second mechanism by which amplification is achieved in biological systems is through relaxation of transmembrane gradients of ions or small molecules. Hagins & Yoshikami (1974) and Hagins (1972) have proposed a specific model based on this mechanism in which Ca^{++} is sequestered within the internal volume of the discs in the dark-adapted state and released into the cytoplasm when light strikes rhodopsin (see Fig. 1b). The sodium conductance mechanism is then inhibited as it binds the released calcium. One possibility is that rhodopsin itself forms a transmembrane

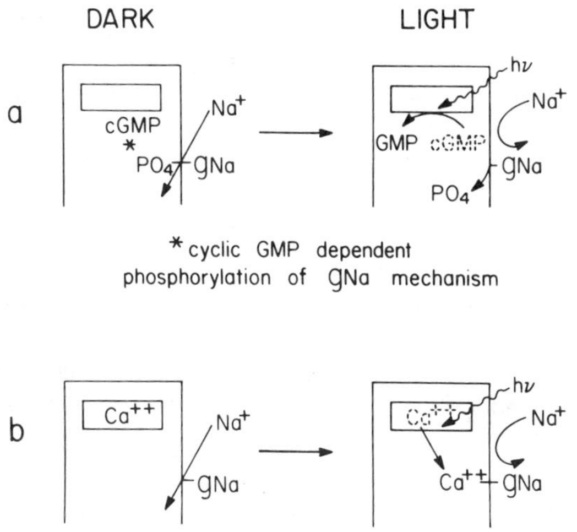

Figure 1 Two specific schemes for linking photon absorption to the sodium conductance decrease in the plasma membrane are shown in very simplified form. (*a*) Protein which comprises the sodium channel is kept in an open conformation in the dark by a cyclic GMP dependent phosphorylation. Light activates the enzymic degradation of cyclic GMP, protein is dephosphorylated, and sodium channels close. (*b*) Calcium sequestered in disc membranes is released by illumination and inhibits the sodium channels.

Ca^{++} channel upon photolysis. This suggestion has stimulated interest in the structure of rhodopsin, and this topic is considered in the second part of this review, contributed by W. L. Hubbell, along with a review of some recent data concerning light-stimulated transmembrane ion movements in rhodopsin-containing membranes.

PHYSIOLOGY AND BIOCHEMISTRY OF THE ROD OUTER SEGMENT

To monitor directly the process of visual transduction, one needs to specify the light-induced conductance changes that occur in individual outer segments, as distinct from responses that arise from the inner segment portion of the receptor cell, or from interactions between photoreceptor cells. Yau, Lamb & Baylor (1977) have approached this in a novel way by recording from single, living toad outer segments after drawing them into a close-fitting suction electrode. As in previous intracellular measurements, the response amplitude saturates with increasing light intensity in a hyperbolic manner. The conductance responses caused by single photon absorptions occur approximately within a 5 μm lateral distance of the rhodopsin molecule that is photoisomerized (Baylor, Lamb & Yau 1978). The time to peak response, after a dim flash of light, varies from 800 msec to 2–3 sec in different preparations, possibly because of their differing metabolic health. The single photon effects are more rapid at the base of the outer segment, near the cilium, than at the tip. Background illumination causes an approximately two-fold increase in their kinetics. In these experiments using the suction electrode technique, a more rapid transient response observed with intracellular microelectrode recording is absent. This response most likely originates from a voltage-dependent conductance in the inner segment portion of the receptor cell, and Fain et al (1978) have shown that it is reversibly abolished by 2 mM cesium.

These new electrophysiological techniques are extremely useful, for they now permit detailed studies of outer segment responses to flashes of light, extended illumination (light adaptation), and recovery of sensitivity in the dark after illumination (dark adaptation). These will be the relevant responses to compare with biochemical data obtained on outer segment, for previous intracellular studies of transduction and adaptation have also monitored the voltage-dependent conductances probably arising in the inner segment (Baylor & Hodgkin 1974, Kleinschmidt & Dowling 1975, Fain 1976). The voltage noise analysis of light-sensitive channel kinetics also can be pursued, to determine whether previous estimates of channel open time ($\sim$100–300 msec) and number of channels on an outer segment ($\sim$1000) are correct (Schwartz 1977, Lamb & Simon 1977).

The problem of establishing what mechanisms reside in the outer segment has been pursued in a different way in suspensions of isolated frog rod outer segments. Bownds & Brodie (1975), following the original observations of Korenbrot & Cone (1972), have used light suppression of outer segment swelling as an in vitro assay for the decrease in plasma membrane permeability caused by illumination. In this in vitro system, as in the intact retina, permeability is suppressed over 4–5 log units of light intensity, and varies with the logarithm of intensity at intermediate light levels. A sensitivity-controlling system operates, and several biochemical correlates of transduction and adaptation have been found (see below, and Brodie & Bownds 1976, Woodruff et al 1977). Thus, it appears that isolated rod outer segments contain the biochemical machinery for transduction and adaptation.

As mentioned above, there is general agreement that an internal transmitter mediates between rhodopsin excitation and the permeability mechanism. Baylor, Hodgkin & Lamb (1974) have argued that several simultaneous or sequential chemical reactions may intervene between photon absorption and the permeability decrease. At this juncture, one must leave open the possibilities that rate-limiting steps in transduction could reside in protein conformation changes, diffusion times, or small molecule transformations. Hardt & Cone (personal communication) have obtained evidence indicating that at least one diffusion-limited step is involved in the process. The presumed internal transmitter(s) might appear or disappear in response to illumination, be regulated by trans-disc membrane fluxes, or by enzyme synthesis and breakdown. Ions, small organic molecules, and protein are all potential transmitter candidates.

A problem in evaluating the transmitter model has been an absence of data describing ion or small molecule transformations that are triggered by the very low levels of illumination at which rod cells function (see Szuts & Cone 1977). However, some success has been encountered recently in establishing the potential relevance of cyclic GMP in transduction. Studies by Woodruff et al (1977, 1978a, b), prompted by growing evidence of a central role for cyclic GMP (Fletcher & Chader 1976, Orr et al 1976, Goridis 1977, and see below), have shown that bleaching one rhodopsin molecule can lead to the disappearance of 5×10^4 molecules of cyclic GMP. This represents a large amplification of the light signal, and this reaction becomes a candidate for the amplification step inferred from electrophysiological studies. The half-time for the cyclic GMP decrease is 125 msec, and its latency is less than 50 msec. The change is rapid enough to be involved in the transduction process. The decline in cyclic GMP concentration becomes larger as illumination increases, and varies with the logarithm of light intensity at levels that bleach between 5×10^2 and 5×10^5 rhodopsin molecules per outer segment-second. Light suppression of plasma membrane permeability assayed in vitro occurs over this same range of intensity (Brodie & Bownds 1976). The decrease in cyclic GMP is reversible, with dark levels being restored 30–60 seconds after illumination ceases. The rate of cyclic GMP restoration, like the membrane potential recovery after illumination, is a function of light intensity. Calcium ions suppress both the dark permeability of outer segments and their cyclic GMP levels (Woodruff & Bownds 1978a, b). Removal of calcium increases cyclic GMP levels (Cohen, Hall & Ferrendelli 1978, Woodruff & Bownds 1978a, b).

It would appear that these changes, at least among the nucleotides, may be unique to cyclic GMP. Biernbaum & Bownds (1978) have found that ATP levels are not changed by illumination, and that a light-induced decrease in GTP levels is slower than the cyclic GMP changes. Orr et al (1976) and Cohen, Hall & Ferrendelli (1978) have shown that cyclic AMP levels are not light sensitive. The effects of low levels of illumination on amino acids or other small molecules have not yet been studied.

In other systems, it has been speculated that cyclic nucleotides control ion permeability by regulating the covalent modification of proteins, most notably through phosphorylation/dephosphorylation sequences (Nathanson 1977). Polans, Hermolin & Bownds (1978) have recently demonstrated that two minor membrane-associated protein components of frog rod outer segments with apparent molecular weights of 12,000 and 13,000 daltons are phosphorylated in the dark by a cyclic GMP–dependent mechanism. They are rapidly dephosphorylated by illumination, and regain their phosphate after illumination ceases. Drugs that enhance cyclic GMP levels raise the phosphorylation level of the proteins. Decreases in cyclic GMP levels correlate with decreases in phosphorylation. Calcium ions mimic the effect of illumination by causing rapid dephosphorylation. It is not yet known whether these proteins are associated with the disc or plasma membranes of the ROS. Lolley, Brown & Farber (1977) have also noted a cyclic nucleotide-dependent phosphorylation in bovine rod outer segment preparations, which appear to involve a 30,000 molecular weight protein.

These studies, along with work on the physiology of isolated outer segment (Brodie & Bownds 1976) and intact toad rods (Lipton, Ostroy & Dowling 1977, Lipton, Rasmussen & Dowling 1977) suggest a possible role for cyclic GMP and the protein phosphorylations in regulating the permeability of the plasma membrane. Several electro-physiological studies have demonstrated also that calcium ions can suppress dark permeability (see references in next section), and thus mimic the effect of light. The observation that calcium addition suppresses guanylate cyclase activity (Troyer, Hall & Ferrendelli 1978), cyclic GMP levels (Woodruff & Bownds 1978a, b), and cyclic GMP–dependent protein phosphorylation (Polans, Hermolin & Bownds 1978), raises the possibility that the effect of calcium on permeability may be mediated by the cyclic GMP system. Further work will be required to determine how and whether calcium and cyclic GMP interact in controlling ROS permeability, as has been suggested for other systems (Rasmussen & Goodman 1977).

The direct studies of cyclic GMP levels were preceded by several years of work on the light-sensitive enzymes that regulate cyclic nucleotide metabolism in outer segments. Bitensky, Gorman & Miller (1971) first reported a light-activated adenylate cyclase in outer segments, but subsequent work from that laboratory and others (Keirns et al 1975, Miki et al 1975, Goridis & Weller 1976) demonstrated that the main action of light was to activate a cyclic GMP phosphodiesterase (PDE) by at least ten-fold. Guanylate cyclase activity may be slightly inhibited by light (Krishna et al 1976). Disc membranes contain approximately one molecule of PDE for every 1000 molecules of rhodopsin present. Light activation of the PDE requires GTP (Wheeler & Bitensky 1977), and is maximal when only 0.1% of the rhodopsin has been bleached. A GTPase activity has also been identified in ROS (Robinson & Hagins 1977, Biernbaum, Shedlovsky & Bownds 1978), which Wheeler & Bitensky (1977) and Wheeler, Matuo & Bitensky (1977) have suggested may play a role in both activation and inactivation of the PDE. While the relationship between the activities of guanylate cyclase, PDE, and cyclic GMP levels has not been adequately characterized, it seems likely that PDE activity is the main regulator of ROS cyclic GMP levels.

In contrast to the cyclic GMP transformations, another recently discovered reaction in ROS, the light-activated phosphorylation of rhodopsin and opsin (Kühn & Dreyer 1972, Bownds et al 1972, Frank & Buzney 1977, Weller, Virmaux & Mandell 1975a, McDowell & Kühn 1977), has not been as clearly associated with a physiological function. The cyclic GMP reactions occur on a time scale of milliseconds to seconds, while rhodopsin phosphorylation appears to require seconds to minutes. Rhodopsin phosphorylation measured in isolated frog ROS can involve an amplification step. In the presence of ATP or GTP, low levels of illumination trigger the incorporation of 10–40 molecules of phosphate for each rhodopsin molecule bleached (Miller, Paulsen & Bownds 1977). Since each rhodopsin molecule probably binds no more than 4–5 phosphate groups (Kühn & McDowell 1977) unbleached rhodopsin molecules can also be phosphorylated. Miller, Paulsen & Bownds (1977) have suggested that the amplification involves cooperative interactions in the plane of the disc membrane. It is not yet determined whether this amplification occurs in living receptor cells. There is general agreement that the phosphorylation reaction utilizes an extractable kinase that is not dependent on cyclic nucleotides and is not light sensitive.

Most evidence to date points to a role for rhodopsin phosphorylation in regulating sensitivity of the receptor after the initial transduction steps that have been discussed thus far (Miller, Brodie & Bownds 1975, Kühn 1974). One attractive possibility is that phosphorylation of opsin or rhodopsin permits it to inactivate the phosphodiesterase. Light adaptation, the loss of sensitivity that occurs on prolonged illumination in photoreceptors, might then be the consequence of accumulation of inactivated phosphodiesterase, which can no longer function in linking photon absorption to a cyclic GMP decrease.

RHODOPSIN STRUCTURE AND CONFORMATIONAL TRANSITIONS

The two general mechanisms for signal amplification mentioned in the introduction involve the rhodopsin molecule in fundamentally different roles. In enzymatic amplification schemes, the disc membrane itself plays no central role in the basic excitation mechanism. The hydrophobic domain of rhodopsin serves to anchor the protein to the membrane thus providing for proper orientation of the chromophore, but action of rhodopsin is mainly restricted to the membrane surface. In the Ca hypothesis, however, the membrane plays the central role of compartmentalizing the transmitter, and the action of rhodopsin is transmembrane [see Figure 1(b)]. Structural investigations of rhodopsin and direct measurement of ion movements can, in principle, provide some of the information necessary to distinguish between these two mechanisms, and we briefly review selected results from each area.

Considering structural aspects first, it is clear that if rhodopsin is a gated channel, it must span the thickness of the disc membrane, at least in the excited state. Neutron scattering (Saibil, Chabre & Worchester 1976), deuterium exchange (Osborne & Nabedryk-Viala 1977), and freeze-fracture microscopy (Clarke & Branton 1968, Chen & Hubbell 1973, Corless et al 1976) all indicate that a substantial fraction of the rhodopsin polypeptide is folded within the membrane interior. The freeze-fracture and neutron scattering experiments also directly indicate the intrinsic asymmetry of the molecule. Early freeze-fracture experiments (Chen & Hubbell 1973) and energy transfer measurements (Wu & Stryer 1972) provided suggestive evidence that rhodopsin spans the membrane thickness. Using papain proteolysis and lactoperoxidase catalyzed iodination in a comparative study of native and reconstituted ROS membranes, Hubbell et al (1977) and Hubbell & Fung (1978) obtained definitive evidence for the transmembrane organization of rhodopsin. In addition, these investigators have presented the partial model for the organization of the polypeptide in the membrane shown in Figure 2 (Hubbell & Fung 1978). An important aspect of the model is the intradiscal location of the rhodopsin carbohydrate, based on histochemical evidence (Röhlich 1976) and a study of the binding of native and reconstituted membranes to concanavalin A immobilized on Sepharose (Hubbell & Fung 1978). Very recent work by Adams & Shichi (1978) and J. M. Corless (personal communication) supports this assignment, which localizes the N-terminus at the intradiscal surface, since Hargrave (1977) has shown the sugars to be attached very close to the N-terminus. The C-terminus is located at the cytoplasmic surface of the disc since it is rapidly removed by papain or thermolysin proteolysis in topologically closed vesicles (Hargrave & Fong 1977, Hubbell & Fung 1978). The papain proteolytic cleavage pattern of rhodopsin suggests that the polypeptide crosses the membrane interface a *minimum* of 3 times, although the actual number of transmembrane excursions is likely to be larger (Hubbell & Fung 1978). Rhodopsin contains a substantial amount of α-helix (Shichi & Shelton 1974, Stubbs et al 1976), but little β-structure (Rothschild et al 1976). The transmembrane segments are believed to be oriented α-helices (Hubbell & Fung 1978, Saibil, Michel-Villaz & Chabre 1978), a conclusion strongly supported by the magnetic anisotropy of the rod outer segments (Chalazonitis, Chagneux & Arvanitaki 1970, Hong, Mauzerall & Mauro 1971;

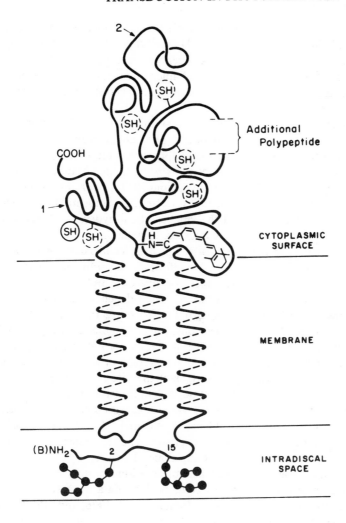

Figure 2 Proposed organization of the rhodopsin polypeptide in photoreceptor membranes. Only enough polypeptide is shown to illustrate principle features of the organization; that some polypeptide is not included is indicated by a break in the chain. Three transmembrane helical segments are shown, although it is likely that the complete polypeptide contains more. The filled circles (●) represent sugar monomers. The rhodopsin sulfhydryl groups are shown to indicate their position in the chain relative to the papain cleavage sites at the arrows marked 1 and 2; their absolute locations within the chain are not known. The 11-*cis* retinal chromophore is included only to indicate its parallel orientation with respect to the membrane surface. A more detailed discussion of the model is to be found in Hubbell & Fung (1978).

M. Chabre, personal communication). Possible functional implications of α-helical clusters are discussed by Hubbell & Fung (1978). It thus appears that the structure of the molecule is certainly compatible with a transmembrane function. Indeed, the rather complex folding pattern in the membrane interior suggests a function other than that of simply anchoring the molecule to the interface.

It is certainly reasonable to suggest that the electrical events in the photoreceptor are linked to rhodopsin photochemistry via conformational transitions in the protein, and many approaches too numerous to list here have directly detected such transitions in ROS membranes. Results from X-ray (Corless 1972, Chabre & Cavoggioni 1973, Chabre 1975) and neutron (Saibil, Chabre & Worcester 1976) scattering, lactoperoxidase catalyzed iodination (Fung 1977), and the kinetics of sulfhydryl reactions (Chen & Hubbell 1978) are particularly interesting, since they all indicate a conformational change in the region of the protein localized at the cytoplasmic surface of the disc membrane. A conformational transition in this region may activate the phosphodiesterase or convert the molecule into a substrate for the opsin kinase. A manifestation of light-dependent structural changes in rhodopsin has also been visualized in freeze-fracture images by a study of lateral phase separations in reconstituted membranes (Chen & Hubbell 1973). Results of these studies suggest that conformational changes modulate the organization of the hydrophobic region of the molecule as well as the region at the cytoplasmic surface of the disc membrane.

A very interesting class of light-dependent transformations in ROS membranes has recently been reported, which suggests cooperative interactions. Asai et al (1975) and Asai, Chiba & Watanabe (1977) report that light scattering from ROS suspensions decreases when rhodopsin is bleached, an effect interpreted in terms of disc shrinkage. The intensity of scattered light varies with the fraction of rhodopsin bleached in a nonlinear fashion. Their experiments were not time-resolved, but the data suggest a very rapid response to a bleaching stimulus. Hofmann et al (1976) and Uhl, Hofmann & Kreutz (1977) have also observed a rapid light scattering change from suspensions of ROS membranes upon bleaching, which is interpreted as a disc shrinkage. The scattering changes occur in the millisecond time range and saturate when about 15% of the rhodopsin is bleached. Wey & Cone (1978) have reported similar results, and find in addition that after saturation of the response, the sample "adapts" in the dark over a period of 10 min and further bleaching again elicits a scattering response. Kreutz, Hofmann & Uhl (1974) found that the quantum yield of metarhodopsin II production and proton uptake is higher at low bleach levels than at high bleach levels, and Miller, Paulsen & Bownds (1977) showed that the quantum yield of rhodopsin phosphorylation is highest at low bleach levels. The nonlinear dependence of these phenomena on the fraction of rhodopsin bleached suggests some form of cooperative interaction in ROS membranes. The apparent cooperativity observed in these experiments need not necessarily be a result of direct interactions between rhodopsin molecules, but could be a result of coupling through protein-phospholipid interactions. However, Reich & Emrich (1976) specifically postulate pairwise interactions of rhodopsins to explain the dependence of the metarhodopsin I–metarhodopsin II equilibrium on Ca^{++} and H^+. The relevance of cooperative interactions to excitation and adaptation responses has been discussed by Robinson (1975).

TRANSMEMBRANE ION MOVEMENTS IN RHODOPSIN-CONTAINING MEMBRANES

Hagins & Yoshikami (1977) demonstrated that buffering ROS cytoplasmic levels of free Ca^{++} with chelating agents produces effects on the photoresponse in quantitative agreement with predictions of the Ca hypothesis. Brown, Coles & Pinto (1977) and Pinto, Brown & Coles (1977) showed that an intracellular Ca^{++} injection mimics the action of light in producing hyperpolarization of the photoreceptor cell. The former experiments suggest that Ca^{++} is somehow involved in permeability control and the latter demonstrate that Ca^{++} alone is capable, directly or indirectly, of modulating the conductance of the plasma membrane. These experiments by themselves, although strongly suggestive, do not demonstrate the validity of the Ca hypothesis shown in Figure 1(b). In order to accept the hypothesis, a direct demonstration that Ca^{++} is released from the disc upon excitation is required. Moreover, it must also be shown that the release is amplified, i.e. each excited rhodopsin must yield at least 100–1000 Ca^{++} within about 100 msec (Hagins 1972, Cone 1973). Considering that there are only on the order of 0.2–0.7 Ca^{++} per rhodopsin in the rod outer segment (Liebman 1974, Yoshikami & Hagins 1976, Szuts and Cone, 1977), it is evident that Ca^{++} fluxes must be studied at very low bleach levels in order to observe hundreds to thousands of ions moving per photoinduced rhodopsin.

There are now numerous reports of experiments designed to detect Ca^{++} release from native membrane preparations, but the experiments of Szuts & Cone (1977) appear to be the most carefully conceived. Using atomic absorption analysis, these investigators examined the Ca^{++} content of freshly isolated frog rod outer segments in which the plasma membrane was ruptured immediately following a stimulus that bleached a small amount of rhodopsin. The average result of 26 independent experiments failed to demonstrate a significant Ca^{++} release. However, as pointed out by Szuts and Cone, the variability of the results at low bleach levels exceeds the release required by the Ca^{++} hypothesis and the negative result does not discount the idea. Sorbi & Cavaggioni (1975) also failed to find significant calcium release at low light levels from outer segment fragments loaded with $^{45}Ca^{++}$.

Smith, Fager & Litman (1977) recently reported light-induced release of 0.75 Ca^{++}/rhodopsin from sonicated vesicles derived from bovine rod outer segments. The stoichiometry of release was found to be independent of the amount of Ca^{++} trapped and the amount of rhodopsin bleached (in the range of 24–100% bleached). Although the release is not amplified as required by the Ca hypothesis, the result is of interest because of the near 1:1 stoichiometry and the fact that the released Ca^{++} appears to be derived from an internal pool. Kaupp & Junge (1977) have employed Ca^{++} detection by arsenazo III to observe a rapid (10–20 msec) ejection of about 1 Ca^{++} per 30 bleached rhodopsin from sonicated vesicles of bovine rod outer segment. Again, the ion release is not amplified, but the result is of interest because of the rapid kinetics, and may represent Ca^{++} release from bound sites on metarhodopsin I in the conversion to metarhodopsin II as discussed by Emrich & Reich (1976). Hendricks, Daeman & Bonting (1974), Hemminki (1975), Weller, Virmaux & Mandel (1975b), and Liebman (1974) also reported Ca^{++} release from ROS membranes, but experiments were not performed under conditions that allow the data to be interpreted as a critical test of the Ca hypothesis. In summary, it must be concluded that the light-activated release of a large number of Ca^{++} per bleached rhodopsin in native ROS membranes has yet to be demonstrated.

Hong & Hubbell (1972) showed that highly purified rhodopsin could be incorporated into phospholipid vesicles without denaturation. Subsequent experiments have shown that the reconstituted protein must have a conformation very similar to the native molecule (Hong & Hubbell 1973, Applebury et al 1974, O'Brien, Costa & Ott 1977). In studies of rhodopsin chemistry, such reconstituted systems offer the distinct advantage that light-dependent properties can be unequivocally attributed to rhodopsin itself, and several studies of light-dependent permeability in reconstituted vesicles have now been reported. Hubbell et al (1977) found that vesicles of phosphatidylcholine containing incorporated rhodopsin and internally trapped $^{45}Ca^{++}$ released the entire complement of isotope with the onset of illumination that bleached large amounts of rhodopsin. Trapped ^{14}C-insulin was not released, indicating that vesicle rupture was not responsible for the $^{45}Ca^{++}$ release. Similar experiments were carried out by Darszon, Montal & Zarco (1977) who showed that $^{22}Na^+$, $^{134}Cs^+$, $^{45}Ca^{++}$, glucose, and glycerol were released from the vesicles by a light flash bleaching that entire rhodopsin population. Insulin and sucrose were not released. O'Brien et al (1977) studied the permeability of reconstituted vesicles to Mn^{++}, Eu^{++}, and Co^{++} using NMR methods and found a light-dependent permeability increase to these ions. From all of the above results, it is clear that bleaching rhodopsin in reconstituted systems increases permeability. These experiments, however, cannot yet be interpreted to imply that rhodopsin acts as a light-gated channel. The ion fluxes in these experiments were all observed over long periods of time (2.5–120 min) at high bleach levels. The relatively small macroscopic permeability changes could simply result from a small change in the polarizability of the membrane interior or creation of molecular packing defects as a result of conformation changes of rhodopsin, which are known to occur (see above). Alternatively, the use of nonnative phospholipids in the reconstituted membrane might give rise to a host of physical phenomena, which could alter bulk membrane permeability by mechanisms that have little relevance to processes occurring in the native system (protein aggregation for example). This is not to imply that reconstituted systems are not valid models for study, nor that the results are uninteresting, but only to emphasize that caution must be exercised in interpreting slow, nonspecific fluxes obtained at high bleach levels. A rapid, amplified Ca^{++} release at low bleach levels, as expected for a Ca^{++} channel, has yet to be demonstrated in reconstituted membrane vesicles, but it is anticipated that continued investigation of these systems will more clearly define the intrinsic properties of the rhodopsin molecule.

Montal, Darszon & Trissl (1977) studied the electrical properties of single bilayer membranes ("black lipid membranes") containing rhodopsin, and find light- and voltage-dependent conductance changes occurring with a pronounced latency of 3–200 sec depending on the rhodopsin concentration in the membrane. The conductance is not ion-specific. These authors interpret the effect as the formation of discrete channels due to aggregation of the protein. If this mechanism of channel formation is to be invoked as an explanation of visual transduction, the aggregation must be imagined to occur between a single photoactivated rhodopsin and other native rhodopsin molecules, since a single photoactivated rhodopsin is sufficient to excite the receptor. However, Downer & Cone (1978) find no evidence for this type of aggregate formation in native membranes. Antanavage et al (1977) also investigated the electrical properties of single bilayer membranes containing rhodopsin and reported a transient, proton selective conductance change with flash illumination. At the present time, caution must be exercised in interpretation of results from the single bilayer experiments because of the rather harsh treatments of the protein involved in bilayer preparation;

one must entertain the possibility that the light-dependent changes are related to rhodopsin denaturation occurring upon loss of the additional stabilization conferred by the 11-*cis* retinal chromophore. Nevertheless, these studies provide information difficult to obtain in other ways and lead to definite hypotheses that can be tested in native systems.

In conclusion, it must be stated that studies of ion fluxes in both native and artificial systems have so far failed to provide definitive evidence for pore or channel formation by native rhodopsin and critical tests of the Ca hypothesis must await future experiments. On the other hand, it is now apparent that the cGMP changes stimulated by illumination are sufficiently rapid to be involved in membrane permeability changes, and cGMP must be seriously considered as a part of the excitation mechanism of the rod cell.

ACKNOWLEDGEMENTS

Works of the authors cited in the text were supported by NIH grants EY 00463 (to D. B.) and EY 00729 (to W. L. H.). We thank D. Baylor, G. Fain, A. Cohen, R. A. Cone, J. M. Corless, and M. Chabre for access to their unpublished work.

Literature Cited

Adams, A. J., Shichi, H. 1978. Conconavalin A binding sites of rod membranes. *Biophys. J.* 21:116a

Antanavage, J., Chien, P., Ching, G., Dunlop, C., Mueller, P. 1977. Rhodopsin mediated proton fluxes in lipid bilayers. *Biophys. J.* 17:182a

Applebury, M. L., Zuckerman, D. M., Lamola, A. A., Jovin, T. M. 1974. Rhodopsin purification and recombination with phospholipids assayed by the metarhodopsin I→metarhodopsin II transition. *Biochemistry* 13:3448–58

Asai, H., Chiba, T., Kimura, S., Takagi, M. 1975. A light-induced conformational change in rod photoreceptor disc membranes. *Exp. Eye Res.* 21:259–67

Asai, H., Chiba, T., Watanabe, M. 1977. Cooperative conformational change in rod photoreceptor disk membrane induced by bleaching. *Vision Res.* 17:983–84

Baylor, D. A., Fuortes, M. G. F. 1970. Electrical responses of single cones in the retina of the turtle. *J. Physiol. London* 207:77–92

Baylor, D. A., Hodgkin, A. L. 1974. Changes in time scale and sensitivity in turtle photoreceptors. *J. Physiol. London* 242:729–58

Baylor, D. A., Hodgkin, A. L., Lamb, T. D. 1974. The electical response to turtle cones to flashes and steps of light. *J. Physiol. London* 242:685–727

Baylor, D. A., Lamb, T. D., Yau, K. W. 1978. The membrane current of rod outer segments. *J. Physiol. London* In press

Biernbaum, M., Bownds, D. 1978. Light- and Ca^{++} dependent Guanosine 5'-triphosphate hydrolysis in frog retinal rod outer segments. *J. Gen. Physiol.* (submitted)

Bitensky, M. W., Gorman, R. E., Miller, W. H. 1971. Adenyl cyclase as a link between photon capture and changes in membrane permeability of frog photoreceptors. *Proc. Natl. Acad. Sci. USA* 68:561–62

Bownds, D., Brodie, A. E. 1975. Light-sensitive swelling of isolated frog rod outer segments as an *in vitro* assay for visual transduction and dark adaptation. *J. Gen. Physiol.* 66:407–25

Bownds, D., Dawes, J., Miller, J., Stahlman, M. 1972. Phosphorylation of frog photoreceptor membranes induced by light. *Nature New Biol.* 237:125–27

Brodie, A. E., Bownds, D. 1976. Biochemical correlates of adaptation processes in isolated frog photoreceptor membranes. *J. Gen. Physiol.* 68:1–11

Brown, J. E., Coles, J. A., Pinto, L. H. 1977. Effect of injection of calcium and EGTA into the outer segments of retinal rods of Bufo Marinus. *J. Physiol. London* 269:707–22

Chabre, M. 1975. X-ray diffraction of retinal rods I. Structure of disc membrane, effect of illumination. *Biochim. Biophys. Acta* 382:322–35

Chabre, M., Cavaggioni, A. 1973. Light induced changes of ionic flux in the retinal rod. *Nature New Biol.* 244:118–20

Chalazonitis, N., Chagneux, R., Arvanitaki, C. R. 1970. Rotation des segments externes des photorécepteurs dans le champ magnetique constant. *C. R. Acad. Sci. Ser. D* 271:130–33

Chen, Y. S., Hubbell, W. L. 1973. Temperature and light dependent structural changes in rhodopsin-lipid membranes. *Exp. Eye Res.* 17:517–32

Chen, Y. S., Hubbell, W. L. 1978. Reactions of the sulfhydryl groups of membrane-bound bovine rhodopsin. *Mem. Biochem.* 1:107–30

Clarke, A. W., Branton, D. 1968. Fracture faces in frozen outer segments from the guinea pig retina. *Z. Zellforsch. Mikrosk. Anat.* 91:586–603

Cohen, A. I., Hall, I. A., Ferrendelli, J. A. 1978. Calcium and cyclic nucleotide regulation in incubated mouse retinas. *J. Gen. Physiol.* In press

Cone, R. A. 1973. The internal transmitter model for visual excitation: some quantitative implications. In *Biochemistry and Physiology of Visual Pigments,* ed. H. Langer, pp. 275–82. New York: Springer. 366 pp.

Corless, J. M. 1972. Lamellar structure of bleached and unbleached rod photoreceptor membranes. *Nature* 237:229–31

Corless, J. M., Cobbs, W. H., Costello, M. J., Robertson, J. D. 1976. On the asymmetry of retinal rod outer segment disk membranes. *Exp. Eye Res.* 23:295–324

Daemen, F. J. M. 1973. Vertebrate rod outer segment membranes. *Biochim. Biophys. Acta* 300:255–88

Darszon, A., Montal, M., Zarco, J. 1977. Light increases the ion and non-electrolyte permeability of rhodopsin-phospholipid vesicles. *Biochem. Biophys. Res. Commun.* 76:820–27

Downer, N. W., Cone, R. A. 1978. Does rhodopsin change its state of aggregation during visual excitation? *Biophys. J.* 21:135a

Ebrey, T. G., Honig, B. 1975. Molecular aspects of photoreceptor function. *Q. Rev. Biophys.* 8:129–84

Emrich, H. M., Reich, R. 1976. The effect of Ca^{++} on the Metarhodopsin I–II transition 1. Experiments. *Pflügers Arch.* 364:17–21

Fain, G. L. 1976. Sensitivity of toad rods: dependence on wavelength and background illumination. *J. Physiol. London* 261:71–102

Fain, G. L., Quandt, F. N., Bastian, B. L., Gerschenfeld, H. M. 1978. Contribution of a cesium-sensitive conductance increase to the rod photoresponse. *Nature* 272:467–69

Fletcher, R. T., Chader, G. J. 1976. Cyclic GMP-control of concentration by light in retinal photoreceptors. *Biochem. Biophys. Res. Commun.* 70:1297–1302

Frank, R. N., Buzney, S. M. 1977. Rhodopsin phosphorylation and retinal rod outer segment cyclic nucleotide phosphodiesterase: lack of a casual relationship. *Exp. Eye Res.* 25:495–504

Fung, B. K.-K. 1977. *Molecular structure of rhodopsin: proteolytic cleavage and enzymatic iodination of disc membranes and reconstituted membranes.* PhD thesis. Univ. Calif. Berkeley. 239 pp.

Goridis, C. 1977. The effect of flash illumination of the endogenous cyclic GMP content of isolated frog retinae. *Exp. Eye Res.* 24:171–77

Goridis, C., Weller, M. 1976. A role for cyclic nucleotides and protein kinase in vertebrate photoreception. *Adv. Biochem. Psychopharmacol.* 15:391–412

Hagins, W. A. 1972. The visual process: excitatory mechanisms in the primary photoreceptor cells. *Ann. Rev. Biophys. Bioeng.* 1:131–158

Hagins, W. A., Yoshikami, S. 1974. A role for Ca^{++} in excitation of retinal rods and cones. *Exp. Eye Res.* 18:299–305

Hagins, W., Yoshikami, S. 1977. Intracellular transmission of visual excitation in photoreceptors: electrical effects of chelating agents introduced into rods by vesicle fusion. In *Vertebrate Photoreception,* ed. H. B. Barlow, P. Fatt, p. 97. London: Academic. 378 pp.

Hargrave, P. A. 1977. The amino-terminal tryptic peptide of bovine rhodopsin. A glycopeptide containing two sites of oligosaccharide attachment. *Biochim. Biophys. Acta* 492:83–94

Hargrave, P. A., Fong, S. 1977. The Amino- and Carboxyl-terminal sequence of bovine rhodopsin. *J. Supramol. Struct.* 6:559–70

Hemminki, K. 1975. Light-induced decrease

in calcium binding to isolated bovine photoreceptors. *Vision Res.* 15:69–72

Hendricks, T., Daeman, F. J. M., Bonting, S. L. 1974. Biochemical aspects of the visual process XXV. Light-induced movements in isolated frog rod outer segments. *Biochim. Biophys. Acta* 345: 468–73

Hofmann, K. P., Uhl, R., Hoffmann, W., Kreutz, W. 1976. Measurements of fast light-induced light scattering and absorption changes in outer segments of vertebrate light sensitive rod cells. *Biophys. Struct. Mechanism* 2:61–77

Hong, F. T., Mauzerall, D., Mauro, A. 1971. Magnetic anisotropy and the orientation of retinal rods in a homogeneous magnetic field. *Proc. Natl. Acad. Sci. USA* 68:1283–85

Hong, K., Hubbell, W. L. 1972. Preparation and properties of phospholipid bilayers containing rhodopsin. *Proc. Natl. Acad. Sci. USA* 69:2617–21

Hong, K., Hubbell, W. L. 1973. Lipid requirements for rhodopsin regenerability. *Biochemistry* 12:4517–23

Hubbell, W. L., Fung, B., Chen, Y., Hong, K. 1977. Molecular anatomy and light-dependent processes in photoreceptor membranes. See Hagins & Yoshikami 1977, pp. 41–59

Hubbell, W. L., Fung, B. K.-K. 1978. The structure and chemistry of rhodopsin: relationship to models of function. In *Membrane Transduction Mechanisms*, ed. R. A. Cone, J. Dowling. New York: Raven. In press

Kaupp, U. B., Junge, W. 1977. Rapid calcium release by passively loaded retinal discs on photoexcitation. *FEBS Lett.* 81:229–32

Keirns, J. J., Miki, N., Bitensky, M. W., Keirns, M. 1975. A link between rhodopsin and disc membrane cyclic nucleotide phosphodiesterase, action spectrum and sensitivity to illumination. *Biochemistry* 14:2760–66

Kleinschmidt, J., Dowling, J. E. 1975. Intracellular-recordings from Gecko photoreceptors during light and dark-adaptation. *J. Gen. Physiol.* 66:617–48

Korenbrot, J. I., Cone, R. A. 1972. Dark ionic flux and the effects of light in isolated rod outer segments. *J. Gen. Physiol.* 60:20–45

Kreutz, W., Hofmann, K., Uhl, R. 1974. On the significance of two-dimensional super-structures in biomembranes for energy-transfer and signal conversion. In *Biochemistry of Sensory Functions*, ed. L. Jaenicke, pp. 311–28. New York: Springer. 641 pp.

Krishna, G., Krishnan, N., Fletcher, R. T., Chader, G. 1976. Effects of light on cyclic GMP metabolism in retinal photoreceptors. *J. Neurochem.* 27: 717–22

Kühn, H. 1974. Light dependent phosphorylation of rhodopsin in living frogs. *Nature* 250:588–92

Kühn, H., Dreyer, W. J. 1972. Light dependent phosphorylation of rhodopsin by ATP. *FEBS Lett.* 20:1–6

Kühn, H., McDowell, J. H. 1977. Isoelectric focusing of phosphorylated cattle rhodopsin. *Biophys. Struct. Mech.* 3:199–203

Lamb, T. D., Simon, E. J. 1977. Analysis of electrical noise in turtle cones. *J. Physiol. London* 272:435–68

Liebman, P. A. 1974. Light-dependent Ca^{++} content of rod outer segment disk membranes. *Invest. Opthalmol.* 13: 700–1

Lipton, S. A., Ostroy, S. E., Dowling, J. E. 1977. Electrical and adaptive properties of rod photoreceptors in Bufo marinus I. Effects of altered extracellular Ca^{++} levels. *J. Gen. Physiol.* 70:747–70

Lipton, S. A., Rasmussen, H., Dowling, J. E. 1977. Electrical and adaptive properties of rod photoreceptors in Bufo marinus II. Effects of cyclic nucleotides and prostaglandins. *J. Gen. Physiol.* 70: 771–91

Lolley, R. N., Brown, B. M., Farber, D. B. 1977. Protein phosphorylation in rod outer segments from bovine retina: cyclic nucleotide-activated protein kinase and its endogenous substrate. *Biochem. Biophys. Res. Commun.* 78:572–78

McDowell, J. H., Kühn, H. 1977. Light-induced phosphorylation of rhodopsin in cattle photoreceptor membranes-substrate activation and inactivation. *Biochemistry* 16:4054–60

Miki, N., Baraban, J. M., Keirns, J. J., Boyce, J. J., Bitensky, M. W. 1975. Purification and properties of the light-activated cyclic nucleotide phosphodiesterase of rod outer segments. *J. Biol. Chem.* 250:6320–27

Miller, J. A., Brodie, A. E., Bownds, M. D. 1975. Light-activated rhodopsin phosphorylation may control light sensitivity in isolated rod outer segments. *FEBS Lett.* 59:20–23

Miller, J. A., Paulsen, R., Bownds, M. D. 1977. Control of light-activated phosphorylation in frog photoreceptor membranes. *Biochemistry* 16:2633–39

Montal, M., Darszon, A., Trissl, H. W. 1977. Transmembrane channel formation in rhodopsin-containing bilayer membranes. *Nature* 267:221–25

Nathanson, J. A. 1977. Cyclic nucleotides and nervous system function. *Physiol. Rev.* 57:157–256

O'Brien, D. F., Zumbulyadis, N., Michaels, F. M., Ott, R. 1977. Light-regulated permeability of rhodopsin:egg phosphatidylcholine recombinant membranes. *Proc. Natl. Acad. Sci. USA* 74:5222–26

O'Brien, D. F., Costa, L. F., Ott, R. A. 1977. Photochemical functionality of rhodopsin-phospholipid recombinant membranes. *Biochemistry* 16:1295–1303

Orr, H. T., Lowry, O. H., Cohen, A. I., Ferrendelli, J. A. 1976. Distribution of 3′:5′-cyclic AMP and 3′:5′-cyclic GMP in rabbit retina in vivo: selective effects of dark and light adaptation and ischemia. *Proc. Natl. Acad. Sci. USA* 73:4442–45

Osborne, H. B., Nabedryk-Viala, E. 1977. The hydrophobic heart of rhodopsin revealed by an infrared ^{1}H-^{2}H exchange study. *FEBS Lett.* 84:217–20

Ostroy, S. E. 1977. Rhodopsin and the visual process. *Biochim. Biophys. Acta* 463:91–175

Pinto, L. H., Brown, J. E., Coles, J. A. 1977. Mechanism for the generation of the receptor potential of rods of Bufo marinus. See Hagins & Yoshikami 1977, pp. 159–67

Polans, A. S., Hermolin, J., Bownds, D. 1978. Light induced dephosphorylation of frog rod outer segment proteins. *J. Gen. Physiol.* (Submitted)

Rasmussen, H., Goodman, D. B. 1977. Relationships between calcium and cyclic nucleotides in cell activation. *Physiol. Rev.* 57:421–509

Reich, R., Emrich, H. 1976. The effect of Ca^{++}on the metarhodopsin I-II transition II. Model calculations and hypothesis on a molecular mechanism of visual excitation. *Pfluegers Arch.* 364:23–28

Robinson, G. W. 1975. Rhodopsin domains and visual response. *Vision Res.* 15:35–48

Robinson, W. E., Hagins, W. A. 1977. A light-activated GTPase in retinal rod outer segments. *Biophys. J.* 17:196a

Röhlich, P. 1976. Photoreceptor membrane carbohydrate on the intradiscal surface of retinal rod disks. *Nature* 263:789–91

Rosenkranz, J. 1977. New aspects of the ultrastructure of frog rod outer segments. *Int. Rev. Cytol.* 50:26–154

Rothschild, K. J., Andrew, J. R., DeGrip, W. J., Stanley, H. E. 1976. *Science* 191:1176–78

Saibil, H., Chabre, M., Worcester, D. 1976. Neutron diffraction studies of retinal rod outer segment membranes. *Nature* 262:266–70

Saibil, H., Michel-Villaz, M., Chabre, M. 1978. Orientation of α-helical segments in frog rhodopsin studied by infra-red linear dichroism. *Biophys. J.* 21:172a

Schwartz, E. A. 1977. Voltage noise observed in rods of the turtle retina. *J. Physiol. London* 272:217–46

Schichi, H., Shelton, E. 1974. Assessment of physiological integrity of sonicated retinal rod membranes. *J. Supramol. Struct.* 2:7–16

Smith, H. G., Fager, R. S., Litman, B. 1977. Light-activated calcium release from sonicated bovine retinal rod outer segment disks. *Biochemistry* 16:1399–1405

Sorbi, R. T., Cavaggioni, A. 1975. Effect of strong illumination on the ion efflux from the isolated discs of frog photoreceptors. *Biochim. Biophys. Acta* 394:577–85

Stubbs, G. W., Smith, H. G., Litman, B. J. 1976. Alkyl glucoside as effective solubilizing agents for bovine rhodopsin. A comparison with several commonly used detergents. *Biochim. Biophys. Acta* 426:46–56

Szuts, E. Z., Cone, R. A. 1977. Calcium content of frog rod outer segments and discs. *Biochim. Biophys. Acta* 468:194–208

Troyer, E. W., Hall, I. A., Ferrendelli, J. A. 1978. Guanylate cyclase in CNS. Enzymatic characteristics of soluble and particulate enzymes from mouse cerebellum and retina. *J. Neurochem.* In press

Uhl, R., Hofmann, K. P., Kreutz, W. 1977. Measurement of fast light-induced disc shrikage within bovine rod outer segments by means of a light-scattering transient. *Biochim. Biophys. Acta* 469:113–22

Weller, M., Virmaux, N., Mandel, P. 1975a. Light-stimulated phosphorylation of rhodopsin in the retina: the presence of a protein kinase that is specific for photobleached rhodopsin. *Proc. Natl. Acad. Sci. USA* 74:4238–42

Weller, M., Virmaux, N., Mandel, P. 1975b. Role of light and rhodopsin phosphorylation in control of permeability of retinal rod outer segment discs to Ca^{++}. *Nature* 256:68

Wey, C. L., Cone, R. A. 1978. Light-induced light-scattering changes from rod outer segments (ROS). *Biophys. J.* 21:135a

Wheeler, G. L., Matuo, Y., Bitensky, M. W. 1977. Light-activated GTPase in vertebrate photoreceptors. *Nature* 269: 822–23

Wheeler, G. L., Bitensky, M. W. 1977. A light-activated GTPase in vertebrate photoreceptors: regulation of light-activated cyclic GMP phosphodiesterase. *Proc. Natl. Acad. Sci. USA* 74:4238–42

Woodruff, M. L., Bownds, D. 1978a. Characterization of the light-dependent decrease in guanosine 3', 5'-cyclic monophosphate in isolated frog photoreceptor membranes. *J. Gen. Physiol.* (submitted)

Woodruff, M. L., Bownds, M. D. 1978b. Amplitude, kinetics and reversibility of a light-induced decrease in guanosine 3', 5'-cyclic monophosphate in isolated

frog retinal rod outer segments. *J. Gen. Physiol.* (submitted)

Woodruff, M. L., Bownds, D., Green, S. H., Morrisey, J. L., Shedlovsky, A. 1977. Guanosine 3',5'-cyclic monophosphate and the in vitro physiology of frog photoreceptor membranes. *J. Gen. Physiol.* 69:667–79

Wu, C., Stryer, L. 1972. Proximity relationships in rhodopsin. *Proc. Natl. Acad. Sci. USA* 69:1104–8

Yau, K. W., Lamb, T. D., Baylor, D. A. 1977. Light-induced fluctuations in membrane current of single toad rod outer segments. *Nature* 269:78–80

Yoshikami, S., Hagins, W. A. 1976. Ionic composition of vertebrate photoreceptors by electron probe analysis. *Biophys. J.* 16:35a

Ann. Rev. Neurosci. 1979. 2:35–64
Copyright © 1979 by Annual Reviews Inc. All rights reserved

OPIATE RECEPTORS
AND OPIOID PEPTIDES

♦11518

Solomon H. Snyder and Steven R. Childers

Departments of Pharmacology and Experimental Therapeutics
and Psychiatry and Behavioral Sciences, Johns Hopkins University
School of Medicine, Baltimore, Maryland 21205

OPIATE RECEPTORS AND OPIATE-LIKE PEPTIDES

Among the central nervous system neurotransmitter candidates, the bio-genic amines, acetylcholine, norepinephrine, serotonin, and dopamine have been known for the longest period of time. During the past decade the role of amino acids such as glycine, glutamate, and γ-aminobutyric acid (GABA) as quantitatively major neurotransmitters has been verified. More recently numerous peptides have emerged as possible neurotransmitters or neuromodulators in the central nervous system. Among these peptides the greatest attention has been focused upon the opioid peptides, enkephalins, and endorphins. For most neurotransmitters, receptor sites were character-ized biochemically only long after identification of the transmitter itself. In the case of the opioid peptides, receptor sites were discovered first and their remarkable properties predicted the existence of a substrate that interacts uniquely with the receptors.

IDENTIFICATION OF OPIATE RECEPTORS

Several pharmacological properties of opiates portended the existence of specific receptor sites for them. While morphine is only moderately potent, certain opiates have extraordinary milligram potency. For instance, in some tests etorphine is 5,000–10,000 times more potent than morphine. Such an agent could act at extraordinarily low doses only if it combines with high affinity to a specific receptor site. The existence of opiate antagonists also favored the receptor concept. Pure antagonists such as naloxone elicit no euphoric or analgesic actions themselves, but block those of opiate agonists

35

0147-006X/79/0315-0035$01.00

such as morphine. Most opiate antagonists, such as nalorphine or levallorphan, seem "contaminated" with some agonist effects since they also elicit
analgesia. Some mixed agonist-antagonists such as pentazocine elicit
analgesia, yet are substantially less addicting than pure agonists. The existence of various drugs along the agonist-antagonist continuum was consistent with the selective receptor notion. Additionally, opiate actions are
generally stereospecific with the (−)-isomers being up to thousands of times
more potent than the (+)-isomers.

To identify the postulated opiate receptors by biochemical techniques,
numerous workers attempted to measure the binding of radioactive opiates
to brain membranes. As an initial screen for specific receptor binding the
relative potencies of (+)- and (−)-isomers in competing for binding were
compared. Goldstein et al (1971) described binding of ^{3}H-levorphanol to
mouse brain homogenates of which about 2 percent was displaced differentially by levorphanol isomers. When the extent of stereoselective binding
was amplified by solubilization and purification (Lowney et al 1974), it
became apparent that the stereoselective binding did not involve the pharmacologically relevant opiate receptor but cerebroside sulfate (Loh et al
1974).

Though some features of opiate binding to cerebroside sulfate mimic
those of the opiate receptors (Loh et al 1978), cerebroside sulfate is clearly
not identical with the opiate receptor; whether or not it constitutes a portion
of the opiate receptor is unclear. Opiate receptor binding is exquisitely
sensitive to degradation by proteolytic enzymes (Pasternak & Snyder
1975a) and protein modifying reagents (Pasternak et al 1975a). Opiates
have substantially less affinity for binding to cerebroside sulfate than to
opiate receptors. Moreover, some clones of neuroblastoma-glioma cells
display opiate receptor binding with essentially the same characteristics as
those of mammalian brain, yet lack detectable cerebroside sulfate (R.
Miller, personal communication). The regional distribution of opiate receptor binding does not parallel the distribution of cerebroside sulfate or other
lipids, or the binding of opiates to soluble fractions thought to be cerebroside sulfate (Lowney et al 1974). The stereoselective interactions of opiates
with cerebroside sulfate emphasizes the caution necessary before it is assumed that binding of a radioactive agent to membranes labels a particular
receptor site. Indeed, stereoselective binding of opiates to certain synthetic
filters has been identified in which the (−)-isomer of opiates was more potent
than the (+)-isomer (Snyder et al 1975).

The failure to identify specific opiate receptor binding in early studies was
probably related to the use of radioactive opiates of low specific activity. At
the high concentrations employed specific receptors were fully saturated
and most of the binding occurred to nonspecific sites of much lower affinity.

Also, early studies did not employ procedures designed to wash away ligands that were non-specifically bound. Utilizing radioactive opiates of high specific activity so that concentrations in the nanomolar range could be employed, as well as utilizing rapid and thorough washing procedures that would greatly reduce nonspecific binding while preserving specific receptor interactions, we (Pert & Snyder 1973a,) and others (Terenius 1973, Simon et al 1973) identified substantial levels of stereoselective binding to brain membranes. In the same initial investigations, Pert & Snyder (1973a,b) demonstrated stereospecific binding to the guinea pig intestine, a tissue in which opiates inhibit electrically induced contractions in close proportion to their pharmacological potencies (Kosterlitz & Waterfield 1975, Paton 1957). In the guinea pig intestine, stereospecific binding was confined to the nervous plexus since it was abolished by removal of the nervous plexus, which by weight constitutes only a minute proportion of the guinea pig intestinal strips (Pert & Snyder 1973b). The selectivity of localization of binding was apparent in that a large number of nonnervous tissues failed to demonstrate any receptor binding.

As already indicated, however, stereospecific binding, though necessary, is not a sufficient criterion for identification of a biologically meaningful receptor. Ideally, one would like to show a correlation in the same tissue between binding and pharmacological potency. Such a demonstration is not feasible in the brain where the pharmacological endpoint is analgesia. However, one can elicit inhibition in vitro of electrically induced contractions of the guinea pig intestine and measure binding in the same tissue. A close correlation was observed between the affinity of a large number of opiate agonists and antagonists for binding sites and their pharmacological potencies in the same strips of guinea pigs intestine (Creese & Snyder 1975).

Additional evidence for the specificity of opiate receptor binding derives from detailed comparisons of the analgesic and binding affinities of homologous series of opiates (Wilson et al 1975).

ION AND NUCLEOTIDE EFFECTS UPON OPIATE RECEPTOR BINDING

Differential pharmacological activities of agonists, antagonists, and mixed agonist-antagonists constitute one of the most striking and important pharmacological properties of opiates. Merely substituting an N-allyl for N-methyl transforms agonists into antagonists. While pure antagonists such as naloxome produce no analgesia or euphoria, the activity of many antagonists, such as nalorphine and levallorphan, is contaminated with some agonist activity. Certain mixed agonist-antagonist drugs such as pentazocine are clinically effective analgesics that are less addicting than pure agonists. It is as if the agonist component of the drug produces analgesia while the antagonist component prevents changes associated with addiction.

Clearly an early goal was to determine whether receptor binding interactions distinguished agonists and antagonists. In initial studies no differences were detected in the relative affinities for matching pairs of agonists and antagonists in competing for binding of ^{3}H-naloxone, the antagonist, to brain membranes in buffers lacking metal ions (Pert & Snyder 1973a). In these experiments addition of sodium failed to reduce ^{3}H-naloxone binding (Pert & Snyder 1973b) while in comparable experiments the binding of the agonist, ^{3}H-etorphine was reduced by sodium (Simon et al 1973). Addition of low concentrations of sodium in parallel experiments dramatically differentiated agonists and antagonists (Pert et al 1973, Pert & Snyder 1974). Sodium enhanced the binding of the antagonists ^{3}H-naloxone, ^{3}H-nalorphine, and ^{3}H-levallorphan and markedly decreased the binding of the agonists ^{3}H-dihydromorphine, ^{3}H-levorphanol, and ^{3}H-oxymorphone (Pert & Snyder 1974). Interestingly, the augmentation of antagonist binding was more marked with the pure antagonist naloxone than with nalorphine and levallorphan, both of which possess some agonist activity. The sodium effect was quite potent, being apparent with as little as 0.5 mM sodium chloride. The influence of sodium was also highly specific. Lithium, whose atomic radius and biological activity are similar to those of sodium, is the only other ion that can mimic the effects of sodium, though it does so to a lesser degree. By contrast potassium, rubidium, and cesium, do not discriminate between agonists and antagonists but merely decrease the binding of both agonists and antagonists at higher concentrations.

To evaluate the influence of sodium on receptor interactions of large numbers of opiates, the potencies of drugs in inhibiting binding of ^{3}H-naloxone was examined in the presence and absence of sodium (Pert et al 1973, Pert & Snyder 1974). Pure antagonists such as naloxone and diprenorphine are just as potent in the presence as in the absence of sodium. Antagonists with limited amounts of agonist activity are reduced in potency about two fold by sodium, while pure agonists become 12–60 times weaker when sodium is added to incubation media. The class of mixed agonist-antagonists with potential as relatively nonaddicting analgesics display a reduction in potency of about 3–6 fold in the presence of sodium. This "sodium index" of opiates has proved useful in screening large numbers of drugs for therapeutic potential as analgesics.

Besides its practical ramifications, the influence of sodium has theoretical importance. The reciprocal increase in antagonist and decrease in agonist binding elicited by sodium suggests that the opiate receptor may exist in two interconvertible conformations with selective affinites for agonist and antagonists respectively (Pert & Snyder 1974, Pasternak & Snyder 1975b, Snyder & Simantov 1977). According to this model, interconversion of receptor states is regulated by sodium, which suggests that sodium is the ion whose permeability is altered in eliciting the pharmacological effects of opiates and opioid peptides. Direct neurophysiological studies support such a possibility. In producing their inhibitory effects on neurons, opiates appear to block the excitatory effects of substances such as acetylcholine or glutamate by interfering at the sodium ion conductance modulator (Zieglgansberger & Bayerl 1976). Moreover, the opiate receptor-mediated inhibition of norepinephrine output and contractile response elicited by morphine is facilitated by low sodium concentrations, while the antagonist effects of naloxone are more apparent at high sodium concentrations (Enero 1977). These findings support the notion that the biological consequences of opiate receptors are mediated by changes in conductance of sodium ion.

Determining the molecular mechanism whereby neurotransmitter recognition is translated into a change in ion permeability is one of the central questions of the neurobiology of synaptic transmission. The ability to monitor these events biochemically at the opiate receptor provides a paradigm for other receptors. Influences of ions that may relate to synaptic transmission have also been described in the case of the glycine receptor (Young & Snyder 1974, Snyder & Young 1975, Muller & Snyder 1978). The postsynaptic hyperpolarization that mediates the inhibitory neurotransmitter actions of glycine in the spinal cord and brainstem involves an enhancement in chloride conductance. The relative potencies of chloride and other anions in producing these effects closely parallel their influence on the binding of strychnine to the glycine receptor. Similar selective effects of sodium ion differentiating agonists and antagonists have been described for α-noradrenergic (Greenberg et al 1978) and dopamine receptors (Creese, Usdin & Snyder, in preparation).

Detailed biochemical studies have examined the mechanism of the sodium effect on the opiate receptor. The enhancement of antagonist binding by sodium appears to be elicited at least in part by an accelerated dissociation of an endogenous opiate-like molecule, presumably enkephalin, from the opiate receptor (Pasternak et al 1975a, Pasternak et al 1976). Opiate receptor binding involves distinct high and low affinity sites (Pasternak & Snyder 1975b). The reduction in agonist binding produced by sodium is mediated by an abolition of high affinity agonist binding sites with negligible effect on the low affinity agonist sites. Thus depending on the range of ^{3}H-opiate concentrations examined, the effect of sodium on opiate agonist binding appears as either a change in the number (Pert & Snyder 1974) or affinity (Simon et al 1975) of binding sites. By examining displacement curves for the mutual inhibition of agonist and antagonist binding Birdsall et al (1976) have also emphasized the discrepancies between high and low affinity binding sites.

Divalent cations also distinguish agonist from antagonist binding (Pasternak et al 1975b, Simantov et al 1976a). Low concentrations of manganese and magnesium, but not calcium, selectively increase agonist binding, effects that are most marked in the presence of sodium. Endogenous divalent cations appear to have a physiological role in regulating opiate receptor binding, because removal of such ions by EDTA, which chelates magnesium and manganese as well as calcium, reduces agonist binding selectively while EGTA, which selectively chelates calcium, fails to affect receptor binding. The pharmacological relevance of the influence of divalent cations is demonstrated by the findings of Enero (1977) that manganese enhances the pharmacological actions of morphine upon the cat nictitating membrane. Moreover, both in binding studies and in pharmacological effects on the nictitating membrane, the effect of manganese is more pronounced in the presence than the absence of sodium. Studies in intact animals suggest some influences of calcium upon opiate actions since intraventricular injections of EDTA and calcium have reciprocal effects on the analgesic effects of morphine (Kakunga et al 1966).

Opiate agonist and antagonist interactions at receptors can also be differentiated by proteolytic enzymes (Pasternak & Snyder 1975a) and protein-modifying reagents (Pasternak et al 1975a). Both types of treatments selectively decrease binding of agonists to receptors and do so more effectively in the presence of sodium. The pharmacological relevance of such effects is demonstrated by the observation that protein-modifying reagents selectively diminish actions of morphine on the nictitating membrane in the presence of high rather than low sodium concentrations (Enero 1977).

Opiate receptor binding is also exquisitely sensitive to degradation by phospholipases, especially phospholipase A (Pasternak & Snyder 1975a). The possibility that endogenous phospholipids might constitute a functioning portion of the opiate receptor is suggested by the finding that several phospholipids can bind opiates stereoselectively (Abood & Hoss 1975). Moreover, addition of phospholipids, especially phosphatidylinositol, to brain membranes enhances opiate receptor binding (Abood & Takeda 1976).

Besides the selective effects of sodium and manganese on receptor functioning, there is evidence that cyclic nucleotides play a role as "second messengers" in the actions of opiates. Opiates inhibit prostaglandin E_1 stimulation of cyclic AMP formation in brain homogenates (Collier & Roy 1974) and neuroblastoma-glioma cells (Traber et al 1975). Opiates inhibit basal as well as prostaglandin-stimulated adenylate cyclase in neuroblastoma-glioma cells and the effect is reversed by naloxone (Sharma et al 1975). A role for cyclic GMP is suggested by the observation that analgesic doses of morphine administered in vivo increase the concentration of cyclic GMP in the corpus striatum (Racagni et al 1976). In vitro enkephalin and opiates increase levels of cyclic GMP in slices of rat corpus striatum; the increase is blocked by naloxone (Minneman & Iversen 1976).

Guanine nucleotides play prominent regulatory roles in the activity of adenylate cyclase as well as in binding interactions of various hormones with receptors (Lad et al 1977, Lefkowitz & Williams 1977, Mukherjee & Lefkowitz 1976) and at α-receptors in the brain (U'Prichard & Snyder 1978). In these systems guanyl nucleotides apparently maintain receptors in a sensitized state. Without guanyl nucleotides the receptors appear converted into a desensitized state with extremely high affinity for agonists; dissociation thus is slowed and the physiological effects of hormones or neurotransmitters impaired. Thus it is of interest that Blume (1978) has recently identified selective effects of guanine but not adenine nucleotides in decreasing the affinity of opiate receptors for ^{3}H-opiates.

ENDOGENOUS OPIOID PEPTIDES: NOMENCLATURE

The discovery of opiate receptor binding in brain led to a search for an endogenous morphine-like substance that would act as a natural opiate receptor ligand. The term enkephalin refers to two pentapeptides, methionine-enkephalin (met-enkephalin), H_2N-tyr-gly-gly-phe-met-OH, and leucine-enkephalin (leu-enkephalin), H_2N-try-gly-gly-phe-leu-OH. The amino acid sequence of met-enkephalin is contained within the sequence of the 91-amino acid polypeptide β-lipotropin (β-LPH) (Hughes et al 1975) isolated from pituitary (Li 1964) and represents amino acids 61–65. The 31-amino acid C-terminal fragment of β-LPH (amino acids 61–91) is a potent opioid peptide generally known as β-endorphin or C-fragment (Cox et al 1976, Li & Chung 1976, Bradbury et al 1976a). Other fragments of β-LPH that possess opioid activity include α-endorphin (β-LPH$_{61-76}$), γ-endorphin (β-LPH$_{61-77}$) and C'-fragment (β-LPH$_{61-87}$). A great deal of information about enkephalins and endorphins have been summarized in other reviews (Fredrickson 1977, Uhl et al 1978a, Miller & Cuatrecasas 1978, Snyder 1978).

IDENTIFICATION OF ENDOGENOUS OPIOID PEPTIDES

The striking properties of the opiate receptor suggested that it might not be merely an evolutionary vestige but could conceivably serve to interact with some normally occurring opiate-like substance. Direct evidence favoring such a possibility derives from neurophysiological studies. Electrical stimulation in the central grey area of the brainstem can elicit analgesia (Mayer & Liebeskind 1974). Strikingly, the apparent analgesia in rats elicited by central grey stimulation can be attenuated at least in part by naloxone (Akil et al 1976). Properties of opiate receptor binding also suggested the existence of an endogenous ligand for the opiate receptor. In early studies, injections of opiate agonists and antagonists in vivo elicited an increase in the apparent number of receptor sites measured in isolated membranes (Pert & Snyder 1976). Antagonists, which in the body's normal sodium environment have greater potency at the opiate receptor than corresponding agonists, were more potent in eliciting this effect. Such findings would be consistent with the displacement by these drugs of some endogenous substance. Similarly, incubation of receptor-containing brain membrane extracts produced an enhancement of binding accompanied by the release into the medium of a substance with considerable affinity for the opiate receptor (Pasternak et al 1975a).

How might one search for such a substance? Two approaches were taken. Hughes (1975) screened for the ability of brain extracts to mimic the influence of morphine upon electrically induced contractions of the mouse vas deferens and the guinea pig ileum. Brain extracts did contain morphine-like activity that was blocked by naloxone and had a regional distribution resembling that of the opiate receptor. Pasternak et al (1975c) and Terenius & Wahlstrom (1974) demonstrated that brain extracts contained a substance that competed for opiate receptor binding. Specificity of this effect was established by showing that the marked regional variations in opiate receptor density were paralleled by similar variations in the concentration of the morphine-like substance (Pasternak et al 1975c). With such straightforward biologically relevant assays for this substance, it was possible to purify it and attempt an isolation.

Hughes et al (1975) isolated the substance from pig brain and showed it to consist of two pentapeptides, met-enkephalin and leu-enkephalin, which differed only in their carboxyl amino acid. Using an assay based on competition for receptor binding, Simantov & Snyder (1976a) isolated the same two pentapeptides from calf brain confirming the findings of Hughes et al (1975). Whereas Hughes et al (1975) found four times more met-enkephalin than leu-enkephalin in pig brain, the ratio was reversed in calf brain. The greater concentration of leu-enkephalin and met-enkephalin in some parts of calf brain was confirmed by specific radioimmunoassay (Simantov et al 1977a). Once the amino acid sequence of enkephalin was known, it became evident that the five amino acids of met-enkephalin were contained within the 91-amino acids of the peptide β-lipotropin isolated ten years earlier from the pituitary by Li (1964). The question arose as to whether fragments of β-lipotropin might possess opiate-like activity. Even prior to the identification of the enkephalin structure, Cox et al (1975) had noted that crude pituitary extracts mimic the influence of morphine upon smooth muscle. Several groups showed that a variety of lipotropin fragments, all incorporating the met-enkephalin sequence, did possess opiate activity (Bradbury et al 1976, a, b, c, Cox et al 1976, Li & Chung 1976, Ling et al 1976, Chretien et al 1976). The most potent of these is β-endorphin, the terminal 31-amino acids of β-lipotropin.

Traditional assay procedures (i.e. radioreceptor binding and guinea pig ileum bioassays) detect all opioid peptides and cannot distinguish between individual peptides. The development of a radioimmunoassay for enkephalins (Simantov & Snyder 1976b, Weissman et al 1976, Yang et al 1977, Miller et al 1978b) provided an opportunity to make this distinction. The enkephalin radioimmunoassay not only distinguishes between met- and leu-enkephalin but also distinguishes enkephalins from higher molecular weight opioid peptides. For example, in our laboratory, antisera directed against either met- or leu-enkephalin cross-react with β-endorphin only at 10,000-fold molar excess of the latter peptide. When the radioimmunoassay is combined with the radioreceptor assay, estimates can be made concerning levels of different opioid peptides. In pituitary, large amounts of opioid activity are detected by radioreceptor assay (210,000 β-endorphin equivalents per gm tissue), while no significant amount of enkephalin is detected by radioimmunoassay (< 2 pmole enkephalin per gm tissue). Therefore, in pituitary the larger opioid peptides predominate.

More recently, radioimmunoassays for β-endorphin have also been developed (Guillemin et al 1977a, LaBella et al 1977, Li et al 1977, Snell et al 1977). Although these antisera readily distinguish β-endorphin from the enkephalins, they display considerable cross-reactivity with β-LPH. After separation of β-endorphin and β-LPH by gel filtration, radioimmunoassays have shown that levels of β-endorphin in brain are considerably lower than those of enkephalin (Rossier et al 1977b).

Spector and his colleagues (Gintzler et al 1976) have utilized a specific antiserum to morphine to identify a different endogenous opioid in brain. This morphine-like compound (MLC) is not a peptide and, although it has not yet been purified, presumably has a structure related to morphine.

Other methods have also been used to distinguish opioid peptides. High pressure liquid chromatography effectively separates met- and leu-enkephalin (Meek & Bohan 1978), while treatment of extracts with cyanogen bromide selectively destroys met-enkephalin activity (Smith et al 1976). Udenfriend and his associates have combined high pressure liquid chromatography and fluorescent derivatives to quantitate low levels of opioid peptides in brain and pituitary (Rubinstein et al 1977).

INTERACTIONS OF OPIOID PEPTIDES WITH OPIATE RECEPTORS

Studies on receptor binding have shown that the opioid peptides are potent ligands of the opiate receptor. In the bioassays, met-enkephalin is twenty times more potent than normorphine in the vas deferens and equipotent with normorphine in guinea pig ileum. Leu-enkephalin has half the potency of met-enkephalin in vas deferens but only one fifth the potency of met-enkephalin in guinea pig ileum (Smith et al 1976). Naloxone, at a concentration of 900 nM, completely blocks these effects. Studies of enkephalin competition for ^{3}H-opiate binding to brain opiate receptors (Simantov & Snyder 1976a, c) showed that both enkephalins compete for opiate receptor binding with affinities resembling that of morphine. Met-enkephalin is twice as potent as leu-enkephalin in reducing binding of ^{3}H-naloxone, while both peptides are equipotent in reducing binding of the agonist ^{3}H-dihydromorphine.

One problem in measuring enkephalin binding in brain is the rapid degradation of enkephalin that occurs at 25° and 37°. At 25°, 40 min of incubation with brain membranes results in 80% destruction of enkephalin. At 37°, 85% destruction occurs after 20 min incubation. No significant destruction, however, occurs at 0°. Degradation of enkephalin at the higher temperatures can be prevented by the addition of 50 μg/ml of the antibiotic bacitracin (Miller et al 1977, Simantov & Snyder 1976c). At this concentration, bacitracin does not affect ^{3}H-naloxone binding itself, but does protect enkephalin for 40 min at 25° and 20 min at 37°. Further studies of enkephalin degradation (Hambrook et al 1976, Meek et al 1977) showed that rapid destruction occurs in human plasma as well as rat brain homogenates. The half-life of enkephalins in plasma is about 2 min; in brain homogenates, enkephalin is completely destroyed within one minute at 37° at a brain protein concentration of 3 mg/ml. Examination of degradation products by thin layer chromatography showed that deactivation occurs by cleavage of the tyr–gly amide bond. Attempts were made by Pert et al (1976a) to stabilize this bond by the substitution of gly^2 by a D-amino acid that would presumably block the accessibility of this crucial bond to proteolytic enzymes. The introduction of D-Ala into position 2, thus creating D-Ala^2-met-enkephalin, produces an analogue with an equivalent potency to met-enkephalin in receptor binding, but which is not degraded. Therefore, D-Ala^2-met-enkephalin can be studied under conditions that completely destroy enkephalins.

Following these studies of purified, unlabeled enkephalins, ^{3}H-enkephalin became available for direct binding studies (Simantov & Snyder 1976b, Morin et al 1976, Audigier et al 1977, Lord et al 1977). In our studies

(Simantov et al 1978) ^{3}H-met-enkephalin binding to rat brain membranes was saturable, with half-maximal binding at a concentration of 1 nM. Scatchard analysis revealed two distinct linear components with a higher affinity dissociation constant of 0.64 nM and a lower affinity dissociation constant of 2.6 nM. Binding of ^{3}H-met-enkephalin is also stereospecific, with levorphanol displaying a 10,000 fold greater potency in displacing ^{3}H-met-enkephalin than dextrorphan.

Miller et al (1978a) have prepared ^{125}I-D-Ala-enkephalin and showed that the mono-iodo derivative possesses high affinity binding ($K_D = 0.8$ nM) while the di-iodo derivative does not. Since ^{125}I-labeled compounds have far greater specific activity than ^{3}H-labeled derivatives, they can be utilized in more sensitive receptor assays.

Detailed studies of the potencies of various opiates in competing for ^{3}H-enkephalin binding have revealed some interesting features of opiate receptor interactions. For some drugs, such as etorphine, levorphanol, phenazocine, and the opioid peptides, the affinity of binding is the same, whether measured by competiton with ^{3}H-met-enkephalin, ^{3}H-naloxone, or ^{3}H-dihydromorphine. Other drugs, like morphine, oxymorphone, normorphine, and fentanyl, are 19–55 times more potent in competing for ^{3}H-dihydromorphine and ^{3}H-naloxone than for ^{3}H-met-enkephalin binding (S. Childers, I. Creese, A. Snowman and S. Snyder, manuscript in preparation). Interestingly, agonists in the latter group display very high sodium ratios (i.e. ratio of potency in the presence of sodium to the potency in the absence of sodium), while agonists with similar affinities for various ^{3}H-opiate binding sides have relatively low sodium ratios. One explanation for the differential affinities of these two groups of agonists is two classes of opiate receptors. Drugs with similar affinities for ^{3}H-enkephalin and for ^{3}H-opiate binding sites would bind to a true "enkephalin receptor," while those drugs that are much less potent with ^{3}H-enkephalin binding would bind to a distinct "opiate receptor." Alternatively, one class of receptor may exist, and enkephalin may possess several points of attachment to the active site of the receptor. One such point may be the hydrophobic side chain of leucine or methionine in enkephalin, which may correspond to the C_6–C_7 region of ring C in opiates (Smith & Griffin 1978). Drugs that contain hydrophobic moieties in this region are as potent in displacing ^{3}H-enkephalin as ^{3}H-opiates, but drugs that possess an hydrophilic ring C are relatively weak in displacing ^{3}H-enkephalin. This model can be useful in predicting relative potencies of drugs by examining their ring C structure.

Other groups have interpreted enkephalin binding data as indicative of several classes of receptor. Multiple receptors were hypothesized by Martin (1976) in brain with physiological experiments: μ-opiate receptors have high affinity for morphine, and κ-receptors have high affinity for the mixed agonist-antagonist ketocyclazocine. Lord et al (1977) have compared drug potencies on biological activities of guinea pig ileum and mouse vas deferens with potencies based on competition for receptor binding in brain. The potency profile of ileum most closely resembles that of the κ- and μ-opiate receptors, while the potencies of opiates in vas deferens resembles the profile of δ-receptors, a class with high affinity for opioid peptides. In brain, high affinity ^{3}H-leu-enkephalin binding sites resemble δ-receptors while high-affinity ^{3}H-naloxone and ^{3}H-dihydromorphine sites resemble μ-receptors (Lord et al 1977).

The production of a number of synthetic analogues of enkephalin has led to extensive structure-activity studies that have determined which amino acid residues are required for receptor binding. Of primary importance is an unaltered N-terminal tyrosine: removal of the tyrosine amino or hydroxyl groups eliminates opioid activity (Fredrickson et al 1976, Morgan et al 1976, Terenius et al 1976, Chang et al 1976, Büscher et al 1976). Phenylalanine in position 4 is also crucial (Morgan et al 1976); replacing phenylalanine with tyrosine results in an almost total loss of activity (Morgan et al 1976, Trenius et al 1976). Loss of activity is also seen if the distance between the aromatic residues is altered (Terenius et al 1976). Alterations in the C-terminal portion also reduce activity, although this portion of the molecule is less crucial than the N-terminal moiety. In particular, activity depends on a hydrophobic C-terminal residue, since increasing hydrophilicity on this residue decreases activity (Miller & Cuatrecasas 1978).

A number of workers have devised conformations of enkephalin based on theoretical minimum energy calculations (Isogai et al 1977, Loew & Burt 1978), proton magnetic resonance experiments of enkephalin in solution (Jones et al 1976, Roques et al 1976), and X-ray diffraction of enkephalin in crystalline form (Smith & Griffin 1978). Some studies (Jones et al 1976, Roques et al 1976) suggest that enkephalin is folded in a β_1 turn involving the sequence gly-gly-phe-met and a hydrogen bond between the NH of met and the CO of gly$_2$. In the crystal structure, however, two hydrogen bonds are preferred: one between the NH of phe and the CO of tyr, the second between the NH of tyr and the CO of phe (Smith & Griffin 1978). Despite these variable interpretations of conformation, certain structural similarities between enkephalin and opiates have become obvious. The tyrosine phenol ring corresponds to the phenol ring A of opiates, the phenylalanine ring corresponds with ring F (present in phenazocine and other opiates but not morphine), and the N-terminal amino group corresponds to the tyramine amino group of opiates. In addition, the side chain of leu or met may correspond to C_6–C_7 on ring C, while the terminal carboxyl group may correspond to the C_6 hydroxyl group of morphine (Smith & Griffin 1978).

BIOSYNTHESIS OF OPIOID PEPTIDES

After the discovery that the sequence of met-enkephalin is identical to residues 61–65 of β-LPH and the isolation of β-endorphin as β-LPH $_{61-91}$ many workers reasoned that: (*a*) β-endorphin is synthesized in pituitary by breakdown of a large molecular weight precursor; (*b*) enkephalin in brain is probably synthesized from β-endorphin. Several studies have confirmed the first hypothesis. Unfortunately, the second hypothesis remains unproven and the mechanism of enkephalin synthesis remains unclear.

The first experiments to identify synthesis of β-endorphin utilized pituitary preparations to specifically hydrolyze β-LPH into β-endorphin. These studies have taken advantage of the fact that β-LPH itself has no opioid activity. Thus Lazarus et al (1976) have shown that incubation of β-LPH with rat brain extracts generates opioid activity, presumably by breaking β-LPH into opioid peptides. This activity disappears after 2 hour incubation as the peptides are degraded. These findings were extended by Bradbury et al (1976b) who demonstrated that the pituitary contains a trypsin-like enzyme that rapidly cleaves the bond between Arg and Tyr. Such a cleavage would result in the formation of β-endorphin from β-LPH. In addition, Rubinstein et al (1977) have shown that storage of fresh pituitaries increases levels of β-endorphin, suggesting endogenous formation from precursors. Pulse labeling of pituitary slices with radioactive amino acids produces labeled β-LPH and β-endorphin (Crine et al 1977), with the latter predominating over the former. These results suggest that β-LPH is a transient precursor that is readily broken down into β-endorphin.

A relationship of opioid peptides to ACTH was shown by Mains and Eipper, who observed that antisera to β-endorphin cross-reacted with "Big ACTH," the 31,000–molecular weight precursor of ACTH isolated from pituitary (Mains et al 1977). Sequential immunoprecipitation of "Big ACTH" with ACTH and β-endorphin antisera revealed that both antisera precipitated the same protein, and that trypsin digests of "Big ACTH" yielded peptides similar in structure to β-LPH and β-endorphin as well as to ACTH. Further support for the idea that "Big ACTH" is an endorphin precursor comes from Roberts & Herbert (1977) who synthesized "Big ACTH" in a cell-free system and showed that immunoreactive corticotropin and β-endorphin precursors are identical, and that β-LPH tryptic peptides are located on the C terminal side of ACTH tryptic peptides. These findings allow construction of the precursor arrangements of ACTH and β-endorphin on "Big ACTH" (Fig. 1).

Also consistent with the idea of common precursors of ACTH and β-endorphin is the observation of Guillemin et al (1977b; Rossier et al 1977a)

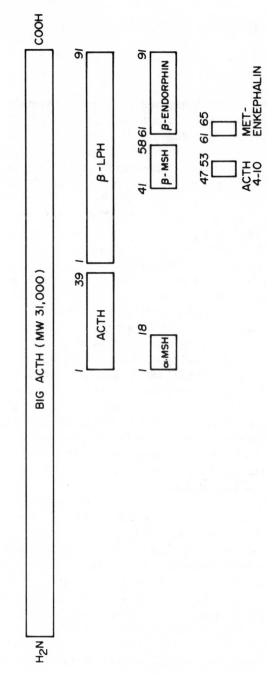

Figure 1   The 31,000 molecular weight peptide "big ACTH" contains within its sequence the entire ACTH and β-lipotropin (β-LPH) molecules, which appear to be located next to each other. Within the ACTH molecule lies the sequence of α-melanotropin (α-MSH), while the sequence of β-melanotropin (β-MSH), as well as the sequence of β-endorphin, is contained within the structure of β-LPH. The sequence of the fourth through tenth amino acids of ACTH is contained within the β-LPH sequence, so that this portion of the ACTH molecule is repeated twice within the big ACTH precursor. Though the sequence of met-enkephalin is contained within that of β-endorphin, free enkephalins have not been demonstrated in significant quantities in the pituitary gland. Moreover, there is no evidence that β-endorphin is the precursor of met-enkephalin in the brain.

that plasma levels of ACTH and β-endorphin vary in a parallel fashion following stress or glucocorticoid administration. Therefore secretions of ACTH and β-endorphin from pituitary are regulated in the same way and may arise from the same cells. Immunofluorescence studies (Bloom et al 1978) have shown that the same cells in pituitary stain for both ACTH and β-endorphin (see Anatomical Distribution section).

Whether β-endorphin serves as a precursor for enkephalin in the brain is the subject of much investigation. Incubation of brain slices with β-endorphin results in the formation of met-enkephalin (Austen et al 1977, Austen & Smyth 1977). Whether this conversion takes place physiologically is questionable. First, brain levels of β-endorphin are only 5–10% those of enkephalin (Simantov et al 1977a, Rossier et al 1977b). Moreover the regional localizations of β-endorphin and enkephalin throughout the brain differ considerably. Highest concentrations of enkephalin occur in parts of the brain in which only low levels of β-endorphin or β-lipotropin have been detected (Elde et al 1976, Simantov et al 1977a, Watson et al 1977b, Zimmerman et al 1978, Bloom et al 1978). Also β-endorphin could serve as the precursor of met-enkephalin, but no one has demonstrated the existence of a leucine-containing species of β-endorphin despite careful examination of pituitary extracts. It is of course possible that leucine-β-endorphin could exist in the brain but not in the pituitary. Synthesis of enkephalin in brain has been confirmed by Yang et al (1978), who isolated labeled enkephalin with an antibody affinity column after intraventricular infusion of ^{3}H-amino acids. Similarly, enkephalin synthesis in intestine has been detected by incubating ileum slices with labeled amino acids (Sosa et al 1978).

ANALGESIA, ADDICTION, AND BEHAVIOR

A number of studies have explored the possibility that opioid peptides are natural analgesics, i.e. neurotransmitters released by specific neurons in various pain-mediating pathways. A series of studies have explored the question by direct injection of enkephalin either intraventricularly or into the central grey of rat brain to produce analgesia that can be blocked by opiate antagonists (Belluzzi et al 1976, Chang et al 1976, Büscher et al 1976). These results showed that met-enkephalin is a weak analgesic, effective at doses of about 120 μg. Leu-enkephalin is even weaker, and actually failed to produce analgesia in many cases. However, the effective intracerebral dose of D-Ala-met-enkephalin is considerably lower and the duration of action is longer than that of met-enkephalin, suggesting that the weak potency of enkephalin is due to rapid degradation in the brain (Pert 1976, Wei et al 1977, Dutta et al 1977). In addition, D-Ala-enkephalin is effective

when administered intravenously (Dutta et al 1977). An even more effective
analgesic is the synthetic enkephalin analogue FK-33-824, which is 30,000
and 1000 times more potent than met-enkephalin and morphine respectively after intracerebral administration (Roemer et al 1977). In FK-33-824
D-alanine is substituted for glycine$_2$, phenylalanine is N-methylated, and in
methionine the sulfur is oxidized to a sulfoxide and the carboxyl is reduced
to a carbinol. It is also active by subcutaneous and oral administration.
Another effective enkephalin analogue is D-met$_2$, pro(NH$_2$)$_5$-enkephalin
(Bajusz et al 1977), which is as potent as morphine when given subcutane-
ously. β-endorphin itself is an effective intravenous analgesic and is over
1000 times more potent than met-enkephalin when given intracerebrally
(Bradbury et al 1976c, Tseng et al 1976, Graf et al 1976, Wei et al 1977).
However, β-endorphin is 23 times less active than FK-33-824 in intracere-
bral injections (Roemer et al 1977).

One persistent controversy in the area of analgesia has been the effect of
opiate antagonists on normal animals or patients. If opioid peptides are
endogenous analgesics, then opiate antagonists should alter pain perception
in normal subjects. A number of studies, however, have shown no effect of
naloxone on nociception in normal patients (El-Sobky et al 1976, Grevert
& Goldstein 1978), suggesting either that opioid peptides are not released
after painful stimuli or that the laboratory tests are not realistic pain models. Buchsbaum et al (1977) detected a significant hyperalgesic effect of
naloxone in pain-insensitive patients when subjects were divided into pain-
sensitive and insensitive groups. Levine et al (1978) have shown that naloxone increases pain sensation in patients already in acute pain from dental
extractions. Therefore, endogenous opioids may only be released under
relatively high pain situations.

Several studies (Chang et al 1976, Waterfield et al 1976) have shown that
repetitive administration of enkephalin results in development of tolerance
to its effects. Moreover, cross-tolerance between enkephalin and morphine
was demonstrated by successive treatments with both compounds. Van Ree
et al (1976) have demonstrated a similar tolerance phenomenon with repeated administration of β-endorphin to rats. In addition, Belluzzi et al
(1976) have observed sustained self-administration of enkephalin in rats
given the peptide intraventricularly, thus suggesting that dependence to
enkephalin does develop. Other studies (Pert 1976, Wei et al 1977, Loh et
al 1976) have shown that a number of withdrawal symptoms are produced
when naloxone is injected in animals after chronic administration of either
enkephalin or β-endorphin. The opioid peptides, particularly β-endorphin,
also cause profound behavioral changes when injected intracerebrally. Perhaps the most common behavioral change is the development of muscular
rigidity and immobility which lasts several hours (Segal et al 1977, Havlicek

et al 1976). Other behavioral changes include catalepsy (Bloom et al 1976, Jacquet & Marks 1976), hyperactivity (Segal et al 1977, Wei et al 1977), seizures (Urca et al 1977), and sedation (Jacquet & Marks 1976). All effects can be reversed by naloxone. β-endorphin is much more potent than met- or leu-enkephalin, and slightly more potent than D-ala-enkephalin (Wei et al 1977). There is some evidence that enkephalin analogues provide qualitatively different behavioral reactions from β-endorphin; for example, enkephalin analogues may produce hyperactivity more often than β-endorphin (Wei et al 1977).

ANATOMICAL DISTRIBUTION OF OPIOID PEPTIDES

Preliminary studies of the regional distribution of opioid peptides in brain relied on radioreceptor assay (Pasternak et al 1975c) or bioassay (Smith et al 1976) of enkephalin activity in discrete dissected brain areas. More recently, enkephalin radioimmunoassays have provided more precise anatomical localizations (Simantov et al 1977a, Yang et al 1977). In general, these results have shown: (a) in rat, mouse, guinea pig, human, and rabbit brains, met-enkephalin predominates over leu-enkephalin; the opposite is true in calf brains; (b) the ratio of met- to leu-enkephalin varies between brain regions; (c) the highest concentration of enkephalin in brain occurs in globus pallidus; other areas with high concentrations include caudate, hypothalamus, periaqueductal grey, amygdala, and spinal cord; cortex and cerebellum contain very little enkephalin. Radioimmunoassay of β-endorphin reveals a different distribution (Krieger et al 1977, Rossier et al 1977b, Bloom et al 1978). Whereas enkephalin has a wide distribution throughout many brain areas, β-endorphin is found in high concentrations only in hypothalamus. In particular, β-endorphin levels are low in globus pallidus, caudate, and spinal cord. Therefore, it appears that enkephalin and β-endorphin may be functionally part of different neuronal pathways.

More detailed localizations have been obtained with immunohistofluorescence utilizing specific antisera to met- and leu-enkephalin, and to α- and β-endorphin. These studies have complemented the earlier experiments with opiate receptor autoradiography (Pert et al 1976b, Atweh & Kuhar 1977a, b, c) and have generally shown that the distribution of enkephalin parallels that of opiate receptors, although some discrepancies exist (e.g. high concentration of opiate receptors in caudate with only moderate indication of enkephalin immunofluorescence). A number of groups (Elde et al 1976, Simantov et al 1977b, Watson et al 1977a) have reported similar distributions of enkephalin by immunofluorescence, with improvements in detail arising from visualization of enkephalin cell bodies by intracerebral injection of colchicine to inhibit axoplasmic transport (Hökfelt et al 1977b). As yet, no difference in distribution has been detected between staining by anti-met-enkephalin or anti-leu-enkephalin sera.

In spinal cord, enkephalin immunohistofluorescence is seen in much of the spinal grey matter as well as in the white matter adjacent to the lateral aspect of laminae I and II. In the grey matter, fluorescence is most concentrated in laminae I and II and in the area around the central canal. Long enkephalin-fluorescent fibers are evident in the ventral horn. At high cervical levels, the "substantia gelatinosa" of the nucleus caudalis of nerve V is also densely fluorescent. In the brainstem, enkephalin immunofluorescence is dense in the "substantia gelatinosa" of the Vth nerve and in the parabrachial nuclei. X nerve-associated structures with moderate to dense fluorescence include the nucleus of the tractus solitarius, nucleus commissuralis, and nucleus ambiguus. Cranial nerve nuclei VII and XII are also densely fluorescent as is the floor of the fourth ventricle. The area of the locus coeruleus and the pontine central grey show moderate fluorescence; the cerebellum shows relatively little. Enkephalin and opiate receptors in the brainstem could mediate a variety of opiate actions, including depression of central respiratory centers and of the chemo-receptor trigger zone in the floor of the fourth ventricle (Jaffe & Martin 1975, Lamberston 1974) and depression of cough reflexes in nuclei of the IX, X, and XII cranial nerves.

In the mesencephalon, studies have shown some enkephalin fluorescence in the periaqueductal grey, extending laterally into other regions of the midbrain tegmentum. Moderate fluorescence is seen in the dorsal portion of the inferior colliculus, and overlying the raphé nuclei. These enkephalin-rich regions in the midbrain may be important in mediating analgesic actions of opiates, since electrical stimulation of either the periaqueductal grey (Mayer & Liebeskind 1974) or of raphé nuclei (Oliveras et al 1975, Fields et al 1977) causes profound analgesia.

In the diencephalon, enkephalin is distributed unevenly. Enkephalin levels are high in the hypothalamus, with dense fluorescence in the areas of the periventricular, ventromedial, dorsomedial, and medial preoptic nuclei, in the floor of the hypothalamus, and in the infundibulum. Less flourescence is seen in the supraoptic, arcuate, mammillary, and suprachiasmatic nuclei. Immunofluorescence in the thalamus is most dense in medial dorsal regions and in the intralaminar nuclei, whereas more ventral regions display little enkephalin fluorescence. Thalamic intralaminar nuclei are thought to be associated with the more affective components of pain appreciation; conceivably, some of opiates' analgesia could be mediated at apparent enkephalinergic synapses here. Dorsomedial thalamic areas are strongly associated with frontal and limbic cortical regions thought to be more generally involved with affect; these thalamocortical systems might be involved in opiates' production of euphoria.

In the telencephalon, enkephalin immunohistofluorescence is dense in the rat central amygdaloid nucleus and somewhat less dense in corticomedial and basolateral areas. The hippocampus exhibits little enkephalin fluorescence. Moderate to sparse fluorescence is seen in the rat striatum, like the patchiness seen with the receptor distribution. The grey matter areas of the globus displays the densest fluorescence in the rat brain. Immunohistofluorescence in rat cerebral cortex is sparse, but is more concentrated in intermediate to deep cortical layers. The rat nucleus accumbens, lateral septal nucleus, and interstitial nucleus of the stria terminalis all possess fairly dense fluorescence.

The possible function of the large quantities of enkephalin in the globus pallidus is unclear. Since this region has been implicated in control of extrapyramidal motor activity, perhaps enkephalin in the globus pallidus mediates the actions of opiates on catalepsy and hypermotility (Bloom et al 1978).

Immunofluorescent studies of β-endorphin and β-LPH in the central nervous system (Bloom et al 1978, Watson et al 1977b) reveal cell bodies containing β-LPH and β-endorphin only in the basal hypothalamus. Fibers have been found in the paraventricular, supraoptic, and suprachiasmatic nuclei, in the median eminence, and especially in the stria terminalis of the anterior hypothalamus. In addition, some β-endorphin fluorescence is seen in the dorsal raphé nucleus, locus coeruleus, periaqueductal grey, and periventricular nucleus of the thalamus. One striking feature of the distribution within the hypothalamus is that the same cells which stain for β-endorphin and β-LPH also stain with ACTH antisera (Bloom et al 1978), consistent with the idea that these molecules are synthesized from a common precursor ("Big ACTH"). However, these cells do not overlap with those which stain for enkephalin, suggesting that enkephalin and β-endorphin arise from different precursors.

Studies of the pituitary have revealed high affinity opiate receptor binding in the posterior and intermediate lobes but not in the anterior lobe (Simantov & Snyder 1977). Pituitary and brain opiate receptors have similar affinities for many opiate drugs. Met- and leu-enkephalins, however, are about 20-fold less potent at displacing ^{3}H-naloxone from pituitary than from brain receptors, although morphine has a 10-fold lower affinity in the pituitary than in the brain. By contrast, β-endorphin is roughly equipotent at brain and pituitary receptors. The function of the pituitary opiate receptors is not known. It is possible that they bind opioid peptides either in the pituitary or those released from the hypothalamus and thereby mediate opiate effects on vasopressin release (Jaffe & Martin 1975). These opiate receptors in the pituitary and part of the hypothalamus may be physiological β-endorphin receptors while the majority of brain opiate receptors are enkephalin receptors.

Cells of the intermediate lobe of the pituitary do show α- and β-endorphin fluorescence (Bloom et al 1977). This fluorescence appears to be granular and to overlie the nonnuclear portions of virtually all pars intermedia parenchymal cells. The same cells stain with antisera directed against both α- and β-endorphins. The apparent intermediate lobe localization of α- and β-endorphins is interesting in light of the known presence of β-lipotropin in this lobe (Dessy et al 1973). Unfortunately, this picture is clouded somewhat by the high cross-reactivity of the β-endorphin antiserum used in these studies with β-lipotropin. Granular α- and β-endorphin fluorescence is also seen in certain cells of the anterior pituitary; positive cells are often noted to be in close apposition to blood vessels (Bloom et al 1977). By immunofluorescence, very little enkephalin can be seen in pituitary. Rubinstein et al (1977), using high pressure liquid chromatography, have confirmed the presence of low levels of met-enkephalin but not leu-enkephalin in pituitary.

Enkephalin has also been located in peripheral tissues. In particular, enkephalin is one of a series of neuroactive peptides (e.g., neurotensin, substance P) present in gut as well as brain and recently observed also in the adrenal medulla (T. Hökfelt, personal communication; L. Larsson, personal communication). Miller & Cuatrecasas (1978) have studied the distribution of enkephalin in the gastrointestinal tract and have detected immunoreactive fibers in the stomach, duodenum, and rectum of guinea pig. In the guinea pig ileum, enkephalin fibers appear mainly in Meissner's plexus and in the circular muscle layer. No immunoreactive cells or fibers have been detected in the gut with ACTH antisera, and one report (Polak et al 1977) shows that endocrine cells of the central mucosa that stain for gastrin also react with gastrin antisera.

ENKEPHALIN AS A NEUROTRANSMITTER

Following the discovery of enkephalin as an endogenous ligand of the opiate receptor, an important question has been whether enkephalin fulfills the criteria of a normal neurotransmitter. One such criterion is that enkephalin should be localized in nerve endings. Subcellular localization studies (Simantov et al 1976b) showed that opioid activity in rat brain is enriched in synaptosomal fractions. In bovine brain and pituitary, opioid activity is virtually entirely contained within rapidly sedimenting cytoplasmic granules (Queen et al 1976). These observations were recently confirmed by electron microscopic examination of met-enkephalin-immunoreactive structures in the locus coeruleus (Pickel et al 1978). In this study, met-enkephalin (and substance P) was localized in both large and small vesicles, exclusively in axons and axon terminals. In addition, enkephalin terminals appeared to form synapses on dendrites that strain for tyrosine hydroxylase, thus suggesting that enkephalin interacts with catecholaminergic systems. Other evidence of enkephalin-catecholamine interactions include a decrease in striatal opiate receptors following 6-hydroxydopamine lesion of the nigro-striatal pathway (Pollard et al 1977), and enkephalin-inhibited release of norepinephrine (discussed below).

A further requirement of a neurotransmitter role for enkephalin would be evidence demonstrating the release of enkephalin from presynaptic endings following nervous stimulation. In a preliminary study Smith et al (1976) showed that calcium-dependent release of enkephalin occurs from rat brain synaptosomes in the presence of a depolarizing concentration of potassium. Puig et al (1977) studied release of opiate-like activity in the guinea pig ileum. In this system, stimulation of ileum at a high frequency results in inhibition of electrically stimulated contraction, an effect similar to that of exogenous opiates or enkephalins. This effect is reversed by opiate antagonists in a stereospecific manner. Another study (Schulz 1977) utilized a specific radioimmunoassay for enkephalin to demonstrate that enkephalin itself is released from isolated strips of guinea pig ileum following electrical stimulation. More recent studies have demonstrated in vitro release of enkephalin from striatum (Henderson et al 1978) and globus pallidus slices (Iversen et al 1978). Release is stimulated by potassium and is calcium-dependent; release is also stimulated by veratridine and abolished by tetrodotoxin (Henderson et al 1978).

If enkephalin is a neurotransmitter, it should be present in identifiable neuronal pathways. A number of attempts have been made to locate specific enkephalinergic pathways in the brain by performing specific lesions:

1. hypophysectomy has no effect on CNS levels of enkephalin, indicating that enkephalin in brain is not derived from pituitary endorphins (Cheung & Goldstein 1976);
2. injection of kainic acid (a glutamic acid analogue that destroys cell bodies around the site of injection) into rat striatum reduces striatum enkephalin by 50% (Hong et al 1977, Childers et al 1978), suggesting that half of striatal enkephalin is present in small interneurons;
3. cutting the boundary between globus pallidus and caudate reduces enkephalin fluorescence in the globus (Cuello & Paxinos 1978), indicating that enkephalinergic cell bodies in the caudate project terminals into the globus pallidus;
4. an enkephalinergic pathway from the central amygdaloid nucleus to the stria terminalis has been revealed by a decrease of enkephalin fluorescence in the stria terminalis following lesions of the amygdala (Uhl et al 1978b);

5. lesions of the paraventricular nucleus of the hypothalamus decrease
   enkephalin immunofluorescence in the median eminence (Hökfelt et al
   1977b), indicating a pathway from the paraventricular nucleus to the
   median eminence;
6. in the spinal cord, enkephalin is largely confined to small interneurons
   in the dorsal grey, since dorsal rhizotomy and hemisections fail to
   change enkephalin histofluorescence (Hökfelt et al 1977a, b, Uhl et al
   1978b).

Other lesion experiments have localized opiate receptors. In the spinal
cord, section of the dorsal roots results in a decrease of receptor binding in
laminae I-III (Lamotte et al 1976), suggesting the presence of receptors on
nerve terminals of sensory neurons. Similarly, cutting the vagus nerve in the
neck abolishes opiate receptors in the vagal nuclei, indicating a presynaptic
localization of vagal opiate receptors (Atweh et al 1978). The presence of
opiate receptors on small interneurons in striatum has been demonstrated
by a loss of 40% of striatal receptor binding after injection of kainic acid
into striatum (Childers et al 1978).

The mechanism of action of enkephalins as neurotransmitters has been
studied by electrophysiologists. In general, the actions of microiontophoretically applied enkephalin are inhibitory; two exceptions are naloxone-reversible excitations in Renshaw cells in the spinal cord (Davies &
Dray 1976) and pyramidal cells in the hippocampus (Nicoll et al 1977).
Naloxone-reversible depression of neuronal firing by enkephalins has been
observed in cerebral cortex (Hill et al 1976, Zieglgansberger & Fry 1976,
Fredrickson & Norris 1976), thalamus (Hill et al 1976), caudate (Fredrickson & Norris 1976, Hill et al 1976, Zieglgansberger & Fry 1976), periaqueductal grey (Fredrickson & Norris 1976), brainstem (Bradley et al 1976),
and spinal cord (Zieglgansberger & Fry 1976, Duggan et al 1976). In
thalamus, neurons that are stimulated by noxious stimuli are also depressed
by enkephalin (Hill et al 1976). In addition, neurons in the myenteric plexus
of guinea pig ileum are depressed by enkephalin (North & Williams 1976).

The electrophysiological experiments have generated some debate concerning the mechanism of action of enkephalin. Since enkephalin abolishes
glutamate-induced activity in some regions, its major action may be on
postsynaptic receptors (Gent & Wolstencroft 1976, Zieglgansberger &
Bayerl 1976). Zieglgansberger & Fry (1976) have shown that in spinal cord,
enkephalin-induced depression occurs without hyperpolarization and suggest that enkephalin may block the actions of excitatory transmitters in
opening sodium channels. Barker et al (1978), studying the effects of enkephalin on cultured spinal cord neurons, have shown that enkephalin depresses glutamate responses in a noncompetitive manner, suggesting that

enkephalin may modulate other neurotransmitter actions rather than alter membrane conductance as do traditional neurotransmitters. Enkephalin may act presynaptically to alter neurotransmitter release. This conclusion follows especially from the existence of opiate receptors on terminals of sensory afferents in the spinal cord where enkephalin is located in small interneurons. Presynaptic inhibition of the release of the excitatory sensory transmitter by enkephalin could explain analgesic effects at the spinal level. A strong candidate for the "pain" sensory transmitter whose release would be inhibited by enkephalin is substance P. Opiates selectively block release of substance P from slices of the substantia gelatinosa of the trigeminal (Jessell & Iversen 1977). In mouse vas deferens, for example, enkephalin abolishes the potassium-stimulated release of norepinephrine (Henderson 1976). Enkephalin (Taube et al 1976) and β-endorphin (Arbilla & Langer 1978) inhibit the stimulated release of norepinephrine from cerebral cortex. In classical models of presynaptic inhibition partial depolarization of a sensory terminal prevents full depolarization by a subsequent nerve impulse. Since opiate inhibition of potassium-induced substance P release has been demonstrated in slices, the mechanism must involve instead hyperpolarization (E. Perl, personal communication).

In summary, enkephalin and other opioid peptides have been identified as natural ligands for the opiate receptor. The relationship between β-endorphin and enkephalin remain unclear, but available evidence points to the conclusion that the two peptides represent distinct systems. Enkephalin displays many characteristics of traditional neurotransmitters: (a) presence in synaptic terminals; (b) release by depolarizing agents; (c) rapid inactivation in brain; (d) synthesis in brain; (e) presence in discrete neuronal pathways; (f) selective microiontophoretic effects on neuronal firing. The actions of enkephalin on the neuronal level to regulate actions of other neurotransmitters may indicate a role as a "neuromodulator" rather than neurotransmitter. Whether such distinctions are meaningful is unclear.

Literature Cited

Abood, L. G., Hoss, W. 1975. Stereospecific morphine adsorption to phosphatidyl serine and other membranous components of brain. *Eur. J. Pharmacol.* 32:66–75

Abood, L. G., Takeda, F. 1976. Enhancement of stereospecific opiate binding to neural membranes by phosphatidyl serine. *Eur. J. Pharmacol.* 32:71–77

Akil, H., Mayer, D. J., Liebeskind, J. C. 1976. Antagonism of stimulation-produced analgesia by naloxone, a narcotic antagonist. *Science* 191:961–62

Arbilla, S., Langer, S. Z. 1978. Morphine and β-endorphin inhibit release of noradrenalin from cerebral cortex but not of dopamine from rat striatum. *Nature* 271:559–61

Atweh, S., Kuhar, M. J. 1977a. Autoradiographic localization of opiate receptors in rat brain. I. Spinal cord and lower medulla. *Brain Res.* 124:53–67

Atweh, S., Kuhar, M. J. 1977b. Autoradiographic localization of opiate receptors in rat brain. II. The brainstem. *Brain Res.* 129:1–12

Atweh, S., Kuhar, M. J. 1977c. Autoradio-graphic localization of opiate receptors in rat brain. III. The telencephalon. *Brain Res.* 134:393–405

Atweh, S. F., Murrin, L. C., Kuhar, M. J. 1978. Presynaptic localization of opiate receptors in the vagal and accessory optic systems: an autoradiographic study. *Neuropharmacology* 17:65–71

Audigier, Y., Malfroy-Camine, B. R., Schwartz, J. C. 1977. Binding of ³H-leu-enkephalin in rat striatum: partial inhibition by morphine or naloxone. *Eur. J. Pharmacol.* 41:247–48

Austen, B. M., Smyth, D. G. 1977. Specific cleavage of lipotropin C-fragment by endopeptidases: evidence for a preferred conformation. *Biochem. Biophys. Res. Commun.* 77:86–94

Austen, B. M., Smyth, D. G., Snell, C. R. 1977. γ-Endorphin, α-endorphin and met-enkephalin are formed extracellularly from lipotropin C-fragment. *Nature* 269:619–21

Bajusz, S., Ronai, A. Z., Szekely, J., Graf, L., Dunai-Kovacs, Z., Berzeti, L. 1977. A superactive antinociceptive pentapeptide: (D-met²-pro⁵)-enkephalinamide. *FEBS Lett.* 76:91–92

Barker, J. L., Neale, J. H., Smith, T. G. Jr. 1978. Opioid peptide modulation of amino acid responses on cultured neurons: evidence for a novel type of communication in the CNS. *Science.* 199:1451–53

Belluzzi, J. D., Grant, N., Garsky, V., Sarantakes, D., Wise, C. D., Stein, L. 1976. Analgesia induced *in vivo* by central administration of enkephalin in rat. *Nature* 260:625–26

Birdsall, N. J. M., Hulme, E. C., Bradbury, A. F., Smyth, D. G., Snell, C. R. 1976. The binding of the C-fragment of lipotropin and methionine-enkephalin to brain opiate receptors In *Opiates and Endogenous Opioid Peptides*, ed. H. W. Kosterlitz, pp. 19–26. Amsterdam: North Holland

Bloom, F., Segal, D., Ling, N., Guillemin, R. 1976. Endorphins: profound behavioral effects in rats suggest new etiological factors in mental illness. *Science* 194:630–32

Bloom, F. E., Battenberg, E., Rossier, J., Ling, N., Leppaluoto, J., Vargo, T., Guillemin, R. 1977. Endorphins are located in the intermediate and anterior lobes of the pituitary gland, not in the neurohypophysis. *Life Sci.* 20:43–48

Bloom, F. E., Rossier, J., Battenberg, E. L. F., Bayon, A., French, E., Henriksen, S. J., Siggins, G. R., Segal, D., Browne, R.,

Ling, N., Guillemin, R. 1978. β-endorphin: cellular localization, electrophysiological and behavioral effects. In *The Endorphins: Adv. Biochem. Psychopharmacol.,* ed. E. Costa, M. Trabucchi, 1:89–110. New York: Raven

Blume, A. J. 1978. Opiate binding to membrane preparations of neuroblastoma x glioma hybrid cells NG108–15: Effects of ions and nucleotides. *Life Sci.* 22:1843–52

Bradbury, A. F., Smyth, D. G., Snell, C. R. 1976a. The peptide hormones: molecular and cellular aspects. *Ciba Found. Symp.* 41:61–75

Bradbury, A. F., Smyth, D. G., Snell, C. R. 1976b. Lipotropin: precursor to two biologically active peptides. *Biochem. Biophys. Res. Commun.* 69:950–56

Bradbury, A. F., Feldberg, W. F., Smyth, D. G., Snell, C. R. 1976c. Lipotropin C-fragment: an endogenous peptide with potent analgesic activity. In *Opiates and Endogenous Opioid Peptides,* ed. H. W. Kosterlitz, pp. 9–17. Amsterdam: North Holland

Bradley, P. B., Briggs, I., Gayton, R. J., Lambert, L. A. 1976. Effects of microiontophoretically applied methionine-enkephalin on single neurones in rat brainstem. *Nature* 261:425–26

Buchsbaum, M. S., Davis, G. C., Bunney, W. E. Jr. 1977. Naloxone alters pain perception and somatosensory evoked potentials in normal subjects. *Nature* 270:620–22

Büscher, H. H., Hill, R. C. Römer, D., Cardinaux, F., Closse, A., Hauser, D., Pless, J. Jr. 1976. Evidence for analgesic activity of enkephalin in the mouse. *Nature* 261:423–24

Chang, J. K., Fong, B. T. W., Pert, A., Pert, C. B. 1976. Opiate receptor affinities and behavioral effects of enkephalin: structure-activity relationship of ten synthetic peptide analogues. *Life Sci.* 18:1473–82

Cheung, A., Goldstein, A. 1976. Failure of hypophysectomy to alter brain content of opioid peptides (endorphins). *Life Sci.* 19:1005–8

Childers, S. R., Schwarcz, R., Coyle, J. T., Snyder, S. H. 1978. Radioimmunoassay of enkephalins: levels of methionine-and leucine-enkephalin in morphine-dependent and kainic acid lesioned rat brains. In *The Endorphins: Adv. Biochem. Psychopharmacol.,* ed. E. Costa, M. Trabucchi, 18:161–174. New York: Raven

Chretien, M., Benjannet, S., Dragon, N., Seidah, N. G., Lis, M. 1976. Isolation of peptides with opiate activity from sheep and human pituitaries: relation to β-lipotropin. *Biochem. Biophys. Res. Commun.* 72:472–78

Collier, H. O. J., Roy, A. C. 1974. Morphine-like drugs inhibit stimulation by E prostaglandins of cyclic AMP formation by rat brain homogenate. *Nature* 248: 24–27

Cox, B. M., Opheim, K. E., Teschemacher, H., Goldstein, A. 1975. A peptide-like substance from pituitary that acts like morphine. 2. Purification and properties. *Life Sci.* 16:1777–82

Cox, B. M., Goldstein, A., Li, C. H. 1976. Opioid activity of a peptide, β-lipotropin-(61–91), derived from β-lipotropin. *Proc. Natl. Acad. Sci. USA* 73:1821–23

Creese, I., Snyder, S. H. 1975. Receptor binding and pharmacological activity of opiates in the guinea pig intestine. *J. Pharmacol. Exp. Ther.* 194:205–19

Crine, P., Benjannet, S., Seidah, N. G., Lis, M., Chrétien, M. 1977. In vitro biosynthesis of β-endorphin, γ-lipotropin, and β-lipotropin by pars intermedia of beef pituitary glands. *Proc. Natl. Acad. Sci. USA* 74:4276–80

Cuello, A. C., Paxinos, G. 1978. Evidence for a long leu-enkephalin striopallidal pathway in rat brain. *Nature* 271:178–80

Davies, J., Dray, A. 1976. Effects of enkephalin and morphine on Renshaw cells in feline spinal cord. *Nature* 262:603–4

Dessy, C., Herlant, M., Cretieu, M. 1973. Immunohistofluorescent detection of lipotropin synthesizing cells. *C.R. Acad. Sci. Ser. D.* 276:335–38

Duggan, A. W., Hall, J. G., Headley, P. R. 1976. Morphine, enkephalin and the substantia gelatinosa. *Nature* 264: 456–58

Dutta, A. S., Gormley, J. J., Hayward, C. F., Morley, J. S., Shaw, J. S., Stacey, G. J., Turnbull, M. T. 1977. Enkephalin analogues eliciting analgesia after intravenous injection. *Life Sci.* 21:559–62

Elde, R., Hökfelt, T., Johannson, O., Terenius, L. 1976. Immunohistochemical studies using antibodies to leu-enkephalin: initial observations on the nervous system of the rat. *Neurosci.* 1: 349–55

El-Sobky, A., Dostrovsky, J. O., Wall, P. D. 1976. Lack of effect of naloxone on pain perception in humans. *Nature* 263: 783–85

Enero, M. A. 1977. Properties of the peripheral opiate receptors in the cat nictitating membrane, *Eur. J. Pharmacol.* 45:349–56

Fields, H. L., Basbaum, A. I., Clanton, C. H., Anderson, S. D. 1977. Nucleus raphé magnus inhibition of spinal cord dorsal horn neurons. *Brain Res.* 126:441–53

Fredrickson, R. C. A. 1977. Enkephalin pentapeptides: a review of current evidence for a physiological role in vertebrate neurotransmission. *Life Sci.* 21:23–41

Fredrickson, R. C. A., Nickander, R., Smith-Wick, E. L., Shuman, R., Norris, F. H. 1976. Pharmacological activity of met-enkephalin and analogues *in vitro* and *in vivo*. In *Opiates and Endogenous Opioid Peptides*, ed. H. W. Kosterlitz, pp. 239–246. Amsterdam: North Holland

Fredrickson, R. C. A., Norris, F. H. 1976. Enkephalin-induced depression of single neurons in brain areas with opiate receptors: antagonism by naloxone. *Science* 194:440–42

Gent, J. P., Wolstencroft, J. H. 1976. Effects of methionine-enkephalin and leucine-enkephalin compared with those of morphine on brainstem neurones in cat. *Nature* 261:426–27

Gintzler, A. R., Levy, A., Spector, S. 1976. Antibodies as a means of isolating and characterizing biologically active substances: Presence of a non-peptide, morphine-like compound in the central nervous system. *Proc. Natl. Acad. Sci. USA* 73:2132–36

Goldstein, A., Lowney, L. I., Pal, B. K. 1971. Stereospecific and nonspecific interactions of the morphine congener levorphanol in subcellular fractions of the mouse brain. *Proc. Natl. Acad. Sci. USA* 68:1742–47

Graf, L., Szekely, J. I., Ronai, A. Z., Dunai-Kovacs, Z., Bajusz, S. 1976. Comparative study on analgesic effect of met[5]-enkephalin and related lipotropin fragments. *Nature* 263:240–41

Greenberg, D. A., U'Prichard, D. C., Sheehan, P., Snyder, S. H. 1978. α-Noradrenergic receptors in the brain: differential effects of sodium on binding of ^{3}H-agonists and ^{3}H-antagonists. *Brain Res.* 140:378–84

Grevert, P., Goldstein, A. 1978. Endorphins: naloxone fails to alter experimental pain or mood in humans. *Science* 199: 1093–95

Guillemin, R., Ling, N., Vargo, T. 1977a. Radioimmunoassays for α-endorphin and β-endorphin. *Biochem. Biophys. Res. Commun.* 77:361–66

Guillemin, R., Vargo, T., Rossier, J., Minick, S., Ling, N., Rivier, C., Vale, W., Bloom, F. 1977b. β-endorphin and adrenocorticotropin are secreted by the pituitary gland. *Science* 197:1367–69

Hambrook, J. M., Morgan, B. A., Rance, M. J., Smith, C. F. C. 1976. Mode of deactivation of the enkephalins by rat and human plasma and rat brain homogenates. *Nature* 262:782–83

Havlicek, V., Rezek, M., Friesen, H. G. 1976. Hexadecapeptide α-endorphin: central effects on motor function, EEG and sleep-waking cycle. *Neurosci. Abstr.* 2:568

Henderson, G. 1976. The effects of morphine on the release of noradrenaline from the mouse vas deferens. *Br. J. Pharmacol.* 57:551–57

Henderson, G., Hughes, J., Kosterlitz, H. W. 1978. *In vitro* release of leu- and met-enkephalin from the corpus striatum. *Nature* 271:677–79

Hill, R. G., Pepper, C. M., Mitchell, J. F. 1976. Depression of nociceptive and other neurones in the brain by iontophoretically applied met-enkephalin. *Nature* 262:604–6

Hökfelt, T., Ljungdahl, A., Terenius, L., Elde, R., Nilsson, G. 1977a. Immunohistochemical analysis of peptide pathways possibly related to pain and analgesia. *Proc. Natl. Acad. Sci. USA* 74:3081–85

Hökfelt, T., Elde, R., Johansson, O., Terenius, L., Stein, L. 1977b. The distribution of enkephalin immunoreactive cell bodies in the rat central nervous system. *Neurosci. Lett.* 5:25–31

Hong, J. S., Yang, H.-Y. T., Costa, E. 1977. On the location of methionine enkephalin neurons in rat striatum. *Neuropharmacol* 16:451–53

Hughes, J. T. 1975. Isolation of an endogenous compound from the brain with the pharmacological properties similar to morphine. *Brain Res.* 88:295–308

Hughes, J., Smith, T. W., Kosterlitz, H. W., Fothergill, L., Morgan, B. A., Morris, H. R. 1975. Identification of two related pentapeptides from the brain with potent opiate agonist activity. *Nature* 258:577–79

Isogai, Y., Nemethy, G., Scheraga, H. A. 1977. Enkephalin: conformation analysis by means of empirical energy calculations. *Proc. Natl. Acad. Sci. USA* 74:414–18

Iversen, L. L., Iversen, S. D., Bloom, F. E., Vargo, T., Guillemin, R. 1978. Release of enkephalin from rat globus pallidus *in vitro*. *Nature* 271:679–81

Jacquet, Y. F., Marks, N. 1976. The C-fragment of β-lipotropin: an endogenous neuroleptic or antipsychotogen? *Science* 194:632–35

Jaffe, J., Martin, W. 1975. Narcotic analgesics and antagonists. In *The Pharmacological Basis of Therapeutics*, ed. L. Goodman, A. Gilman, pp. 245–283. New York: Macmillan

Jessell, T. M., Iversen, L. L. 1977. Opiate analgesics inhibit substance P release from rat trigeminal nucleus. *Nature* 268:549–61

Jones, C. R., Gibbons, W. A., Garsky, V. 1976. Proton magnetic resonance studies of conformation and flexibility of enkephalin peptides. *Nature* 263:779–82

Kakunaga, T., Kaneto, H., Hano, K. 1966. Pharmacologic studies on analgesia: significance of the calcium ion in morphine analgesia. *J. Pharmacol. Exp. Ther.* 153:134–39

Kosterlitz, H. W., Waterfield, A. A. 1975. *In vitro* models in study of structure-activity relationships of narcotic agents. *Ann. Rev. Pharmacol.* 15:29–47

Krieger, D. T., Liotta, A., Brownstein, M. J. 1977. Presence of corticotropin in brain of normal and hypophysectomized rats. *Proc. Natl. Acad. Sci. USA* 74:648–52

LaBella, F., Queen, G., Senyshin, J., Lis, M., Chrétien, M. 1977. Lipotropin: localization by radioimmunoassay of endorphin precursor in pituitary and brain. *Biochem. Biophys. Res. Commun.* 75:350–57

Lad, P. M., Welton, A. F., Rodbell, M. 1977. Evidence for distinct guanine nucleotide sites in the regulation of the glucagon receptor and of adenylate cyclase activity. *J. Biol. Chem.* 252:5942–46

Lambertson, C. 1974. Neurogenic factors in control of respiration. In *Medical Physiology*, ed. V. B. Mountcastle, pp. 1447–1497. St. Louis: C. V. Mosby

Lamotte, C., Pert, C. B., Snyder, S. H. 1976. Opiate receptor binding in primate spinal cord: distribution and changes after dorsal root section. *Brain Res.* 112:407–12

Lazarus, L. H., Ling, N., Guillemin, R. 1976. β-endorphin as a prohormone for the morphinomimetic peptides endorphins and enkephalins. *Proc. Natl. Acad. Sci. USA* 73:2156–59

Lefkowitz, R. J., Williams, L. J. 1977. Catecholamine binding to the β-adrenergic receptor. *Proc. Natl. Acad. Sci. USA* 74:515–19

Levine, J. O., Gordon, N. C., Jones, R. T., Fields, H. L. 1978. The narcotic antagonist naloxone enhances clinical pain. *Nature* 272:826–27

Li, C. H. 1964. Lipotropin, a new active peptide from pituitary glands. *Nature* 201:924–25

Li, C. H., Chung, D. 1976. Isolation and structure of an untriakontapeptide with opiate activity from camel pituitary glands. *Proc. Natl. Acad. Sci. USA* 73:1145–48

Li, C. H., Rao, A. J., Doneen, B. A., Yamashiro, D. 1977. β-endorphin: lack of correlation between opiate activity and immunoreactivity by radioimmunoassay. *Biochem. Biophys. Res. Commun.* 74:576–80

Ling, N., Burgus, R., Guillemin, R. 1976. Isolation, primary structure and synthesis of α-endorphin and γ-endorphin, two peptides of hypothalamic-hypophysial origin with morphomimetic activity. *Proc. Natl. Acad. Sci. USA* 73:3942–46

Loew, G. H., Burt, S. K. 1978. Energy conformation study of met-enkephalin and its D-Ala analogue and their resemblance to rigid opiates. *Proc. Natl. Acad. Sci. USA* 75:7–11

Loh, H. H., Cho, T. M., Wu, Y.-C., Way, E. L. 1974. Stereospecific binding of narcotics to brain cerebrosides. *Life Sci.* 14:2231–45

Loh, H. H., Law, P. Y., Ostwald, T., Cho, T. M., Way, E. L. 1978. Possible involvement of cerebroside sulfate in opiate receptor binding. *Fed. Proc.* 37:147–52

Loh, H. H., Tseng, L. F., Wei, E., Li, C. H. 1976. β-endorphin is a potent analgesic agent. *Proc. Natl. Acad. Sci. USA* 83:2895–98

Lord, J. A. H., Waterfield, A. A., Hughes, J., Kosterlitz, H. W. 1977. Endogenous opioid peptides: multiple agonists and receptors. *Nature* 267:495–99

Lowney, L. I., Schulz, K., Lowery, P. J., Goldstein, A. 1974. Partial purification of an opiate receptor from mouse brain. *Science* 183:749–53

Mains, R. E., Eipper, B. A., Ling, N. 1977. Common precursor to corticotropins and endorphins. *Proc. Natl. Acad. Sci. USA* 74:3014–18

Martin, W. R., Eades, C. G., Thompson, J. A., Huppler, R. E., Gilbert, P. E. 1976. The effects of morphine- and nalorphine-like drugs in the nondependent and morphine-dependent chronic spinal dog. *J. Pharmacol. Exp. Ther.* 197:517–22

Mayer, D., Liebeskind, J. 1974. Pain reduction by focal electrical stimulation of the brain: an anatomical and behavioral analysis. *Brain Res.* 68:73–93

Meek, J., Yang, H.-Y., Costa, E. 1977. Enkephalin catabolism *in vitro* and *in vivo*. *Neuropharmacol.*, 16:151–54

Meek, J. L., Bohan, T. P. 1978. Use of high pressure liquid chromatography to study enkephalins. In *The Endorphins: Adv. Biochem. Psychopharmacol.*, ed. E. Costa, M. Trabucchi, 18:141–148. New York: Raven

Miller, R. J., Chang, K.-J., Cuatrecasas, P., Wilkinson, S. 1977. The metabolic stability of the enkephalins. *Biochem. Biophys. Res. Commun.* 74:1311–18

Miller, R. J., Cuatrecasas, P. 1978. Enkephalins and endorphins. In *Vitamins and Hormones.* In press

Miller, R. J., Chang, K.-J., Cuatrecasas, P., Leighton, J. 1978a. Interaction of iodinated enkephalin analogues with opiate receptors. *Life Sci.* 22:379–88

Miller, R. J., Chang, K.-J., Cooper, B., Cuatrecasas, P. 1978b. Radioimmunoassay and characterization of enkephalins in rat tissues. *J. Biol. Chem.* 253:531–38

Minneman, K. P., Iversen, L. L. 1976. Enkephalin and opiate narcotics increase cyclic GMP accumulation in slices of rat neostriatum. *Nature* 262:313–14

Morgan, B. A., Smith, C. F. C., Waterfield, A. A., Hughes, J., Kosterlitz, H. W. 1976. Structure-activity relationships of methionine-enkephalin. *J. Pharm. Pharmacol.* 28:660–61

Morin, O., Caron, M. G., DeLean, A., LaBrie, F. 1976. Binding of opiate-like pentapeptide methionine-enkephaplin to a particulate fraction from rat brain. *Biochem. Biophys. Res. Commun.* 73:940–46

Mukherjee, C., Lefkowitz, R. J. 1976. Desensitization of β-adrenergic receptors by β-adrenergic agonists in cell-free system: Resensitization by other purine nucleotides. *Proc. Natl. Acad. Sci. USA* 73:494–98

Muller, W. E., Snyder, S. H. 1978. Strychnine binding associated with synaptic glycine receptors in rat spinal cord membranes: ionic influences. *Brain Res.* 147:107–16

Nicoll, R., Siggins, G., Bloom, F. 1977. Neuronal actions of endorphins and enkephalin among brain regions: a comparative microiontophoretic study. *Proc. Natl. Acad. Sci. USA* 74:2584–88

North, R. A., Williams, J. T. 1976. Enkephalin inhibits firing of myenteric neurons. *Nature* 264:460–61

Oliveras, J. L., Redjemi, F., Guibaud, G., Besson, J. M. 1975. Analgesia induced by electrical stimulation of the inferior centralis nucleus of the raphe in the cat. *Pain* 1:139–45

Pasternak, G. W., Snyder, S. H. 1975a. Opiate receptor binding: enzymatic treatments discriminate between agonist and antagonist interactions. *Mol. Pharmacol.* 11:474–78

Pasternak, G. W., Snyder, S. H. 1975b. Identification of novel high-affinity opiate receptor binding in rat brain. *Nature* 253:563–65

Pasternak, G. W., Wilson, H. A., Snyder, S. H. 1975a. Differential effects of protein modifying reagents on receptor binding of opiate agonists and antagonists. *Mol. Pharmacol.* 11:478–84

Pasternak, G. W., Snowman, A. M., Snyder, S. H. 1975b. Selective enhancement of ^{3}H-opiate agonist binding by divalent cations. *Mol. Pharmacol.* 11:735–44

Pasternak, G. W., Goodman, R., Snyder, S. H. 1975c. An endogenous morphine-like factor in mammalian brain. *Life Sci.* 16:1765–69

Pasternak, G. W., Simantov, R., Snyder, S. H. 1976. Characterization of an endogenous morphine-like factor (enkephalin) in mammalian brain. *Mol. Pharmacol.* 11:735–44

Paton, W. D. M. 1957. Action of morphine and related substances on contractions and on acetylcholine output of coaxially-stimulated guinea pig ileum. *Br. J. Pharmacol. Chemother.* 12:119–27

Pert, A. 1976. Behavioral pharmacology of D-alanine²-methionine enkephalin amide and other long-acting opiate peptides. In *Opiates and Endogenous Opioid Peptides*, ed. H. W. Kosterlitz, pp. 87–94. Amsterdam: North Holland

Pert, C. B., Pasternak, G. W., Snyder, S. H. 1973. Opiate agonists and antagonists discriminated by receptor binding in brain. *Science* 182:1359–61

Pert, C. B., Bowie, D. L., Fong, B. T. W., Change, J. K. 1976a. Synthetic analogues of met-enkephalin which resist enzymatic destruction. In *Opiates and Endogenous Opioid Peptides*, ed. H. W. Kosterlitz, pp. 79–86. Amsterdam: North Holland

Pert, C. B., Kuhar, M. J., Snyder, S. H. 1976b. Autoradiographic localization of opiate receptor in rat brain. *Proc. Natl. Acad. Sci. USA* 73:3729–33

Pert, C. B., Snyder, S. H. 1973a. Opiate receptor: demonstration in nervous tissue. *Science* 179:1011–14

Pert, C. B., Snyder, S. H. 1973b. Properties of opiate receptor binding in rat brain. *Proc. Natl. Acad. Sci. USA*, 70:2243–47

Pert, C. B., Snyder, S. H. 1974. Opiate receptor binding of agonists and antagonists affected differentially by sodium. *Mol. Pharmacol.* 10:868–79

Pert, C. B., Snyder, S. H. 1976. Opiate receptor binding: enhancement by opiate administration *in vivo*. *Biochem. Pharmacol.* 25:847–53

Pickel, V. M., Joh, T. H., Reis, D. J., Leeman, S. E., Miller, R. J. 1978. Axon terminals related to dendrites of catecholaminergic neurons. *Brain Res.* In press

Polak, J. M., Sullivan, S. N., Bloom, S. R., Facer, R., Pearse, A. G. E. 1977. Enkephalin-like immunoreactivity in the human gastrointestinal tract. *Lancet* i:972–74

Pollard, H., Llorens-Cortes, C., Schwartz, J. C. 1977. Enkephalin receptors on dopaminergic neurones in rat striatum. *Nature* 268:745–46

Puig, M. M., Gascon, P., Craviso, G. L., Musacchio, J. M. 1977. Endogenous opiate receptor ligand: electrically induced release in the guinea pig ileum. *Science* 195:419–20

Queen, G., Pinsky, C., LaBella, F. 1976. Subcellular localization of endorphine activity in bovine pituitary and brain. *Biochem. Biophys. Res. Commun.* 72:1021–27

Racagni, G., Zsilla, G., Guidotti, A., Costa E. 1976. Accumulation of cGMP in striatum of rats injected with narcotic analgesics: antagonism by naltrexone. *J. Pharm. Pharmacol.* 28:258–60

Roberts, J. L., Herbert, E. 1977. Characterization of a common precursor to corticotropin and β-lipotropin: identification of β-lipotropin peptides and their arrangement relative to corticotropin in the precursor synthesized in a cell-free system. *Proc. Natl. Acad. Sci. USA* 74:5300–4

Roemer, D., Buescher, H. H., Hill, R. C., Pless, J., Bauer, W., Cardinaux, F., Closse, A., Hauser, D., Huguenin, R. 1977. A synthetic enkephalin analogue with prolonged parenteral and oral analgesic activity. *Nature* 268:547–49

Roques, B. P., Barbay-Jaurequiberry, C., Oberlin, R., Anteunis, M., Lala, A. K. 1976. Conformation of met-enkephalin determined by high field PMR spectroscopy. *Nature* 262:778–79

Rossier, J., French, E. D., Rivier, C., Ling, N., Guillemin, R., Bloom, F. 1977a. Foot shock induced stress increases β-endorphin levels in blood but not brain. *Nature* 270:618–20

Rossier, J., Vargo, T. M., Minick, S., Ling, N., Bloom, F. E., Guillemin, R. 1977b. Regional dissociation of β-endorphin and enkephalin contents in rat brain and pituitary. *Proc. Natl. Acad. Sci. USA* 74:5162–65

Rubinstein, M., Stein, S., Gerber, L. D., Udenfriend, S. 1977. Isolation and characterization of the opioid peptides from rat pituitary: β-lipotropin. *Proc. Natl. Acad. Sci. USA* 74:3052–55

Schulz, R., Wuster, M., Simantov, R., Snyder, S. H., Herz, A. 1977. Electrically stimulated release of opiate-like material from myenteric plexus of the guinea pig ileum. *Eur. J. Pharmacol.* 41:347–48

Segal, D. S., Browne, R. G., Bloom, F., Ling, N., Guillemin, R. 1977. β-endorphin: endogenous opiate or neuroleptic? *Science* 198:411–14

Sharma, S. K., Nirenberg, M., Klee, W. A. 1975. Morphine receptors as regulators of adenylate cyclase activity. *Proc. Natl. Acad. Sci. USA* 72:590–94

Simantov, R., Snowman, A. M., Snyder, S. H. 1976a. Temperature and ionic influences on opiate receptor binding. *Mol. Pharmacol.* 12:977–86

Simantov, R., Snowman, A. M., Snyder, S. H. 1976b. A morphine-like factor in rat brain: subcellular localization. *Brain Res.* 107:650–57

Simantov, R., Snyder, S. H. 1976a. Morphine-like factors in mammalian brain: structure elucidation and interactions with opiate receptor. *Proc. Natl. Acad. Sci. USA* 73:2515–19

Simantov, R., Snyder, S. H. 1976b. Brain-pituitary opiate mechanisms: pituitary opiate receptor binding, radioimmunoassays for methionine enkephalin and leucine enkephalin and ^{3}H-enkephalin interactions with the opiate receptor. In *Endogenous Opioid Peptides*, ed. H. W. Kosterlitz, pp. 41–48. Amsterdam: North Holland

Simantov, R., Snyder, S. H. 1976c. Morphine-like factors, leucine-enkephalin and methionine-enkephalin: interactions with opiate receptor. *Mol. Pharmacol.* 12:987–98

Simantov, R., Snyder, S. H. 1977. Opiate receptor binding in the pituitary gland. *Brain Res.* 124:178–84

Simantov, R., Childers, S. R., Snyder, S. H. 1977a. Opioid peptides: differentiation by radioimmunoassay and radioreceptor assay. *Brain Res.* 135:358–67

Simantov, R., Kuhar, M. J., Uhl, G., Snyder, S. H. 1977b. Opioid peptide enkephalin: immunohistochemical mapping in the rat central nervous system. *Proc. Natl. Acad. Sci. USA* 74:467–71

Simantov, R., Childers, S. R., Snyder, S. H. 1978. The opiate receptor binding in-teractions of ^{3}H-methionine enkephalin, an opioid peptide. *Eur. J. Pharmacol.* 47:319–31

Simon, E. J., Hiller, J. M., Edelman, I. 1973. Stereospecific binding of the potent narcotic analgesic ^{3}H-etorphine to rat brain homogenate. *Proc. Natl. Acad. Sci. USA* 70:1947–49

Simon, E. J., Hiller, J. M., Groth, J., Edelman, I. 1975. Further properties of ste-reospecific opiate binding sites in rat brain: on the nature of the sodium effect. *J. Pharmacol. Exp. Ther.* 192:531–37

Smith, G. D., Griffin, J. F. 1978. Conforma-tion of [Leu^5] enkephalin from X-ray diffraction: features important for rec-ognition at opiate receptor. *Science* 199:1214–16

Smith, T. W., Hughes, J., Kosterlitz, H. W., Sosa, R. P. 1976. Enkephalins: isola-tion, distribution, and function. In *Opi-ates and Endogenous Opioid Peptides*, ed. H. W. Kosterlitz, pp. 57–62. Am-sterdam: North Holland

Snell, C. R., Jeffcoate, W., Lowry, P. J., Reese, L. H., Smyth, D. G. 1977. Prepa-ration and characterization of a specific antiserum to the C-fragment of lipotro-pin. *FEBS Lett.* 81:427–30

Snyder, S. H. 1978. The opiate receptor and morphine-like peptides in the brain. *Am. J. Psychiatry.* 135:645–52

Snyder, S. H., Pasternak, G. W., Pert, C. B. 1975. Opiate receptor mechanisms. In *Handbook of Psychopharmacology*, ed. L. L. Iversen, S. D. Iversen, S. H. Sny-der, 5:329–360. New York: Plenum

Snyder, S. H., Simantov, R. 1977. The opiate receptor and opioid peptides. *J. Neuro-chem.* 29:13–20

Snyder, S. H., Young, A. B. 1975. The gly-cine synaptic receptor in the mam-malian central nervous system. *Br. J. Pharmacol.* 53:473–84

Sosa, R. P., McKnight, A. T., Hughes, J., Kosterlitz, H. W. 1978. Incorporation of labelled amino acids into the enke-phalins. *FEBS Lett.* In press

Taube, H. D., Borowski, E., Endo, T., Starke, K. 1976. Enkephalin: a potential modulator of noradrenaline release in rat brain. *Eur. J. Pharmacol.* 38:377–80

Terenius, L. 1973. Characteristics of the "receptor" for narcotic analgesics in synaptic plasma membrane fractions from rat brain. *Acta Pharmacol. Tox-icol.* 33:377–84

Terenius, L., Wahlstrom A. 1974. Inhibi-tor(s) of narcotic receptor binding in brain extracts and in cerebrospinal fluid. *Acta Pharmacol.* (Kbh.) [Suppl. 1] 33:55

Terenius, L., Wahlstrom, A., Lindeberg, G., Karlsson, S., Ragnarsson, U. 1976. Opiate receptor affinities of peptides related to leu-enkephalin. *Biochem. Biophys. Res. Commun.* 71:175–79

Traber, J., Fischer, K., Latzin, S., Hamprecht, B. 1975. Morphine antagonizes action of prostaglandin in neuroblastoma and neuroblastoma x glioma hybrid cells. *Nature* 253:120–22

Tseng, L. F., Loh, H. H., Li, C. H. 1976. β-endorphin as a potent analgesic by intravenous injection. *Nature* 263:239–40

Uhl, G. R., Childers, S. R., Snyder, 1978a. Opioid peptides and the opiate receptor. In *Frontiers in Neuroendocrinology,* ed. W. F. Ganong, L. Martini, 5:289–328. New York: Raven

Uhl, G. R., Goodman, R. R., Kuhar, M. J., Snyder, S. H. 1978b. Enkephalin and neurotensin: immunohistochemical localization and identification of an amygdalofugal pathway. In *The Endorphins, Adv. Biochem. Psychopharmacol.,* ed. E. Costa, M. Trabucchi, 18:71–88. New York: Raven

U'Prichard, D. C., Snyder, S. H. 1978. Guanyl nucleotide influences on ³H-ligand binding to α-noradrenergic receptors in calf brain membranes. *J. Biol. Chem.* 253:3444–52

Urca, G., Frenk, H., Liebeskind, J. C., Taylor, A. N. 1977. Morphine and enkephalin: analgesic and epileptic properties *Science* 197:83–86

Van Ree, J. M., DeWeid, D., Bradbury, A. F., Hulme, E. C., Smyth, D. G., Snell, C. R. 1976. Induction of tolerance to the analgesic action of lipotropin C-fragment. *Nature* 264:792–94

Waterfield, A. A., Hughes, J., Kosterlitz, H. W. 1976. Cross-tolerance between morphine and methionine enkephalin. *Nature* 260:624–25

Watson, S. J., Akil, H., Sullivan, S., Barchas, J. D. 1977a. Immunocytochemical localization of methionine enkephalin: preliminary observations. *Life Sci.* 21:733–38

Watson, S. J., Barchas, J. D., Li, C. H. 1977b. β-lipotropin localization of cells and axons in rat brain by immunocytochemistry. *Proc. Natl. Acad. Sci. USA* 74:5155–58

Wei, E. T., Tseng, L. F., Loh, H. H., Li, C. H. 1977. Comparison of the behavioral effects of β-endorphin and enkephalin analogues. *Life Sci.* 21:321–28

Weissman, B. A., Gershon, H., Pert, C. B. 1976. Specific antiserum to leu-enkephalin and its use in a radioimmunoassay. *FEBS Lett.* 70:245–48

Wilson, R. S., Rogers, M. E., Pert, C. B., Snyder, S. H. 1975. Homologous N-alkylnorketobemidones. Correlation of receptor binding with analgesic potency. *J. Med. Chem.* 18:240–42

Yang, H.-Y., Hong, J.-S., Costa, E. 1977. Regional distribution of leu- and met-enkephalin in rat brain. *Neuropharmacology* 16:303–7

Yang, H.-Y. T., Hong, J. S., Fratta, W., Costa, E. 1978. Rat brain enkephalins: distribution and biosynthesis. In *The Endorphins, Adv. Biochem. Psychopharmacol.,* ed. E. Costa, M. Trabucchi, 18:149–160. New York: Raven

Young, A. B., Snyder, S. H. 1974. The glycine synaptic receptor: evidence that strychnine binding is associated with the ionic conductance mechanism. *Proc. Natl. Acad. Sci. USA* 71:4002–5

Zieglgansberger, W., Bayerl, H. 1976. The mechanism of inhibition of neuronal activity by opiates in the spinal cord of the cat. *Brain Res.* 115:111–28

Zieglgansberger, W., Fry, J. P. 1976. Actions of enkephalin on cortical and striatal neurones of naive and morphine tolerant/dependent rats. In *Opiates and Endogenous Opioid Peptides,* ed. H. W. Kosterlitz, pp. 231–238. Amsterdam: North Holland

Zimmerman, E. A., Liotta, A., Krieger, D. T. 1978. β-Lipotropin in brain: localization in hypothalamic neurons by immunoperoxidase technique. *Cell Tissue Res.* In press

Ann. Rev. Neurosci. 1979. 2:65–112

THE BRAIN AS A TARGET FOR STEROID HORMONE ACTION

♦11519

Bruce S. McEwen, Paula G. Davis, Bruce Parsons, and Donald W. Pfaff

The Rockefeller University, New York, New York 10021

INTRODUCTION

The study of steroid hormone effects on brain is as old as the field of endocrinology itself, for it was A. A. Berthold (1849) who is often credited as having performed the first formal experiment in endocrinology. Berthold transplanted testes into castrated roosters and observed the restoration of crowing, sexual behavior, and aggression, which were lost after gonadal ablation. The investigation of hormone-brain interactions received an important boost with the publication of the book *Hormones and Behavior* by Beach (1948), and more recently a journal of the same name has come into existence. Study of the cellular aspects of hormone effects on the brain has profited greatly from advances in neuroanatomy, neuroendocrinology, neurophysiology, and molecular endocrinology in the past two decades. Especially in the last area, the introduction of tritiated steroid hormones has produced knowledge about receptor mechanisms of steroid hormone action in many tissues, and has opened the way for studies of the steroid sensitive cells of the brain that are directed toward their anatomical mapping and their identification with specific neural pathways and neurotransmitter substances. The eventual goal of these studies is to describe the cellular events by which steroid hormones activate particular behaviors. The present review is a progress report of these efforts and is intended to update the last joint review article from our two laboratories (McEwen and Pfaff 1973). We focus in greatest detail on the best-studied system, the action of estrogens to facilitate mating behavior in female rats, after first considering some more general aspects of steroid hormone receptors and steroid hormone metabolism in the rodent brain.

65

0147-006X/79/0315-0065$01.00

GENERAL ASPECTS OF STEROID HORMONE ACTION IN BRAIN AND PITUITARY

Cellular Sites and Modes of Action

Steroid hormones circulating in the blood gain rapid and relatively unrestricted access to all parts of the nervous system (see McEwen et al 1972, Morrell et al 1975), whereupon the hormone may be metabolised and/or interact with receptor sites to produce an effect. Thus the biochemical approach to steroid-brain interactions has focused on three aspects: (*a*) recognition of putative receptor sites and mapping of their distribution; (*b*) determination of the extent and functional importance of steroid transformations by neural tissue; (*c*) delineation of cellular effects of steroid hormones and their metabolites. These topics are dealt with in order in this section of the chapter.

Underlying the search for receptor sites and the delineation of steroid effects are certain notions about the cellular mode of action of these hormones. Progress in the field of steroid hormone action over the past decade has emphasized the role of intracellular receptor sites that translocate hormone into the cell nucleus and trigger the expression of genetic information in the form of altered synthesis of ribonucleic acid, leading to alterations in the synthesis of specific proteins (e.g. estrogens acting in this manner stimulate synthesis of the messenger RNA for ovalbumin in the chick oviduct: O'Malley & Means 1974). There is ample evidence, reviewed below, for the existence of such a mechanism in the brain for the five major classes of steroid hormones. At the same time there is some preliminary evidence for the direct action of steroids on neural activity and synaptic function. For example, glucocorticoids alter the uptake, in vitro, of the amino acid tryptophan by synaptosomes (see Sze 1976) and inhibit acetylcholine-induced release of corticotropin releasing factor by hypothalamic fragments (Jones et al 1977). And estradiol-17β (the natural isomer), but not its stereoisomer estradiol-17α, exerts rapid inhibitory effects on septal-preoptic neurons when iontophoresed directly as the hemisuccinate ester (Kelley et al 1977). Thus far there is no biochemical evidence for specific membrane "receptors" that might mediate these effects.

The two modes of steroid action are depicted diagrammatically in Figure 1. The best way at present to recognize direct effects of steroids (see McEwen et al 1978, for discussion) is by their latency of onset, which should be very short (e.g. seconds or minutes), in contrast to genomic effects of steroids, which have a latency of onset of minutes to hours. (A case in point is the lordosis reflex, discussed in the section on Temporal Correlations, which has an onset latency of 18–24 hr after estradiol administration.)

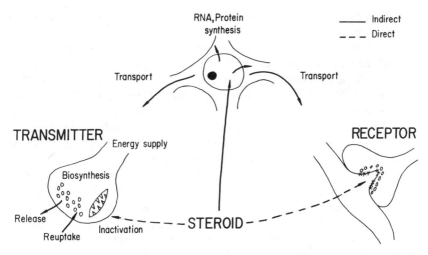

Figure 1 Genomic and nongenomic effects of steroid hormones in pre- and postsynaptic events. Nongenomic effects (dashed line) may involve the action of the hormone on the pre- or postsynaptic membrane to alter permeability to neurotransmitters or their precursors and/or functioning of neurotransmitter receptors. Genomic action of the steroid (unbroken line) leads to altered synthesis of proteins, which, after axonal or dendritic transport, may participate in pre- or postsynaptic events. From McEwen et al 1978. Used by permission of Raven Press.

Properties and Distribution of Putative Steroid Hormone Receptors

Prompted by the success of the genomic model of steroid hormone action in the oviduct, uterus, and other non-neural tissues, the investigation of neural steroid receptor mechanisms has focused on identifying, characterizing, and mapping intracellular steroid-binding macromolecules that translocate steroid to the cell nucleus. The fact that tritium-labeled steroids, infused in vivo to animals lacking endogenous steroid-producing tissue, can be localized to the nuclear compartment of neurons (with little indication of glial cell labeling) has facilitated the histological mapping of hormone-sensitive cells by means of autoradiography (Morrell et al 1975) as well as biochemical studies of ^{3}H-hormone distribution (see McEwen 1976).

Like those of non-neural target tissues, the intracellular steroid-binding macromolecules of the brain and pituitary are proteins that are thermolabile and contain sulfhydryl groups, oxidation or alkylation of which destroys steroid binding activity (see McEwen & Luine 1978). As shown in Table 1, these receptors are reasonably large macromolecules, with sedimentation coefficients in the range of 6–9 S (molecular weights of several hundred thousand units). Each class of receptor displays a high degree of stereoselectivity toward the appropriate hormones, synthetic analogs, and specific antagonists (Table 1). Very little "cross-talk" exists between these different binding systems.

The neuroanatomical distribution of steroid receptor systems has revealed distinct, though sometimes overlapping, patterns for each class of steroid (Table 2). Thus the pattern of estrogen-sensitive neurons (see Pfaff & Keiner 1973) is similar to, but recognizably different from, that of androgen-sensitive neurons (see Sar & Stumpf 1977). Both systems are concentrated primarily in the hypophysiotrophic area (hypothalamus, preoptic area, septum) and amygdala. Progestin receptors in the rat brain can be detected biochemically in most areas, but only those receptors in the hypothalamus, preoptic area, and pituitary are induced by estradiol, which suggests that certain cells in these tissues may contain both estrogen and progestin receptors (see MacLusky et al 1978). Glucocorticoid receptors are found by autoradiography (Warembourg 1975a, b, McEwen et al 1975a) and by in vivo cell nuclear uptake measurements (McEwen et al 1976) in the hippocampus, septum, and amygdala, and very few sensitive cells are de-

Table 1 Properties of brain and pituitary steroid hormone receptors[a]

Class of steroid	Molecular size	Specificity
Estrogen	$\approx 8S$[b]	$DES > E_2\beta > 2OHE_2 >$[i] $E_3 \equiv E_1 \equiv E_2\alpha > 3\beta Aol$ $3\alpha\,Aol > T$
Androgen	$6\text{–}9S$[c,d,e]	$T \equiv DHT > CypAC > P >$[j] $E_2 = F$
Progestin	$\approx 7S$[f]	$R5020 > dNorgestrol >$[k] $P \geqslant E_2\beta > B > T > Dex$
Glucocorticoid	$\approx 7S$[g,h]	$\beta \equiv Dex \equiv F > DOC \equiv P >$[l,m] $ALD > T \equiv E_2$
Mineralocorticoid	?	Binds ALD, Dex[n] Spironolactone

[a] Abbreviations are DES, diethylstilbestrol; E_2B, estradiol 17β; 2OHE$_2$, 2hydroxyestradiol 17β; E_3, estriol; E_1, estrone; $E_2\alpha$, estradiol 17α; 3βAol, 3β5α androstanediol; 3αAol, 3α5α androstanediol; T, testosterone; DHT, 5α dihydrotestosterone; R5020, dimethyl 19 norpregna-4,9-dione 3,20 dione; B, corticosterone; Dex, dexamethasone; F, cortisol; DOC, deoxycorticosterone; P, progesterone; ALD, aldosterone.

[b] McEwen 1978a (Review).
[c] Naess et al 1975a.
[d] Mercier et al 1976.
[e] Kato & Onouchi 1973a, b.
[f] Kato & Onouchi 1977.
[g] Chytil & Toft 1972.
[h] Watanabe et al 1973.

[i] Ginsburg et al 1977.
[j] Naess et al 1975b.
[k] MacLusky et al 1978.
[l] McEwen et al 1976.
[m] Grosser et al 1973.
[n] Anderson & Fanestil 1976.

tected in the hypothalamus, preoptic area, or pituitary. Thus the glucocorticoid pattern is distinctly different from those for estrogen, androgen, and progestin sensitive neurons.

Full understanding of the glucocorticoid pattern is complicated by three problems. First, there is the surprising similarity in the in vivo distribution of ^{3}H-aldosterone uptake to that of ^{3}H-corticosterone (Table 2), which has been obtained using tracer hormone doses at which one would expect to find the least cross-talk between the two systems and would expect to achieve a resolution of mineralocorticoid and glucocorticoid sites. It is conceivable that mineralocorticoid and glucocorticoid sites coexist in the same cells or at least in the same brain regions, but it must be emphasized that the putative mineralocorticoid receptors of neural tissue are very poorly defined (see Table 1) so no unequivocal conclusion can be drawn at this time. Second, the pattern of in vivo uptake and cell nuclear retention of ^{3}H-dexamethasone, a synthetic glucocorticoid with potent ACTH-release inhibiting properties, is totally different from that of ^{3}H-corticosterone, the natural glucocorticoid in the rat. Low uptake is seen in the hippocampus; moderate neuronal labeling in the basomedial hypothalamus (Warembourg 1975b); and high uptake is found in the pituitary (DeKloet et al 1975). A tentative explanation for this has been advanced, namely, that there may be at least two classes of glucocorticoid receptor sites differing in regional distribution and steroid specificity (see DeKloet & McEwen 1976a, b, MacLusky et al 1977, McEwen 1978b for discussion). The third aspect of the complexity of neural glucocorticoid receptor systems is that while autoradiography and in vivo cell nuclear uptake studies with ^{3}H-corticosterone give little indication of glial, as opposed to neuronal, labeling, in vitro studies of a glial cell tumor line maintained in culture (DeVellis et al 1974) and of the optic nerve in enucleated rats (Meyer et al 1978) indicate the

Table 2 Topography of steroid receptor systems in rat brain[a]

^{3}H-Steroid	Septum			Preoptic	Hypothalamus				Amygdala		Hippocampus Ammons Horn	Anterior pituitary
	nsl	ndb	nst	pom	pv	vm	arc	vpm	aco	am		
Estradiol[b]	+	+	+	++	++	++	++	++	++	++	−	++
5αDHT[c]	++	−	+	+	++	+	+	++	−	+	+	+
Corticosterone[d]	++	−	−	−	−	−	−	−	++	−	++	+/−
Aldosterone[e]	++	−	−	−	−	−	−	−	++	−	++	+

[a] Structures are nsl, nucleus septi lateralis; ndb, nucleus of the diagonal band of Broca; nst, nucleus of the stria terminalis; pom, nucleus preopticus medialis; pv, nucleus periventricularis hypothalami; vpm, nucleus premammillaris ventralis; arc, nucleus arcuatus; vm, nucleus ventromedialis hypothalami; aco, cortical nucleus of the amygdala; am, medial nucleus of the amygdala.
[b] Pfaff & Keiner 1973.
[c] Sar & Stumpf 1977.
[d] Warembourg 1975a, b.
[e] Ermisch, personal communication; MacLusky, Ermisch & McEwen, unpublished.

presence of considerable glucocorticoid receptor as well as the induction of an oligodendroglial-specific enzyme, glycerol phosphate dehydrogenase (DeVellis & Inglish 1973, Leveille et al 1977, Meyer et al 1978).

Steroid Metabolism by Brain and Pituitary Tissue

Brain and pituitary tissue both contain enzymes capable of converting steroid hormones to a variety of metabolites by reversible and irreversible reactions. Besides 20-keto reduction of progesterone and acetylation of corticosterone (Table 3), reversible transformations include the formation of estrone from estradiol, androstenedione from testosterone, and 11 dehydrocorticosterone from corticosterone. With the possible exception of 20α-hydroxyprogesterone, which has been reported to substitute for progesterone and to rapidly facilitate estrogen-dependent lordosis behavior (Kubli-Garfias & Whalen 1977), none of these reversibly generated steroid metabolites appear to have an unique role in the neuroendocrine system and are less potent than the parent hormone.

The irreversible transformations are summarized in Figure 2. Of the irreversibly generated metabolites, 5α dihydroprogesterone (5α DHP) and the 3α and 3β tetrahydrometabolites of 5α DHP (not shown in Figure 2)

Table 3 Steroid metabolism in brain and pituitary of rat

Class of steroid	Transformation	Tissues	Functional involvement
Estradiol	2 hydroxylation[a]	pituitary[c] hypothalamus	see text
Testosterone	5 α reduction[a]	pituitary[d] brain	gonadotropin feedback[h]
	aromatization[a]	limbic[e] brain and hypothalamus	male sexual behavior[h,i,j]
Progesterone	20 keto reduction[b]	pituitary[f] brain	lordosis behavior, LH and FSH secretion[f]
	5 α reduction[a]	pituitary[f]	lordosis behavior, LH and FSH secretion[f]
Corticosterone	acetylation[b]	brain[g]	unknown

[a] Essentially irreversible transformation.
[b] Reversible reaction.
[c] Fishman & Norton 1975.
[d] Denef et al 1973, 1974.
[e] Naftolin et al 1975b.
[f] Karavolas & Nuti 1976.
[g] Purdy & Axelrod 1968.
[h] McEwen 1978a.
[i] Davis & Barfield, in prep.
[j] Morali et al 1977.

appear to be no more effective than progesterone itself in promoting lordosis behavior and LH or FSH secretion (see Feder & Marrone 1977, Karavolas & Nuti 1976, Gorzalka & Whalen 1977, Kubli-Garfias & Whalen 1977). Thus there is considerable uncertainty at this point whether the A-ring reduction of progesterone plays a critical role in the action of this steroid. Yet because progesterone appears to be a preferential substrate for 5α reductase (which also acts upon testosterone to produce active metabolites —see below), this interaction has been proposed as a possible route for anti-androgenicity of progestins (Massa & Martini 1971/72). The A-ring reduction of testosterone to 5α DHT has been known for some time to be an important reaction for the action of androgens upon peripheral target organs (see Liao & Fang 1969) as well as the neuroendocrine system (see McEwen 1978a for review). The pituitary and all regions of the brain contain 5α reductase, and ^3H 5α DHT has been recovered from cell nuclear receptor sites in brain and pituitary after the infusion of ^3H-testosterone. The regional distribution of ^3H DHT, as a metabolite of ^3H T, is identical to that of ^3H DHT infused directly into the animal (Lieberburg & McEwen 1977), which indicates that 5α reduction is not a limiting factor in receptor occupation. The other major transformation of T is aromatization to estradiol (E_2) (see Figure 2 and Table 3), and ^3H E_2, as a metabolite of ^3H T, is found to be distributed differently than if ^3H E_2 were infused directly. Specifically, the pituitary is not labeled by T-derived ^3H E_2 which indicates that it lacks aromatase activity, and the amygdala shows a disproportionately high level of T-derived ^3H E_2 (Lieberburg & McEwen 1977). Since DHT is largely without effect in castrated male rats in restoring copulatory behavior, while T is effective (as is the combined injection of DHT plus E_2) the notion has arisen that aromatization of T may be an obligatory step in the action of T in male sexual behavior (see McEwen 1978a, for references). Recent evidence that a steroid inhibitor of aromatization, androsta-1,4,6-triene-3,17-dione (ATD), can block T induction of male copulatory behavior (Christensen & Clemens 1975, Morali et al 1977) supports this notion.

The phylogenetically stable pattern of neural estrogen receptor sites (see above) together with the existence of paradoxical, in contrast to sex-specific, effects of estrogens on mating behavior in various lower and higher vertebrates (see Young 1961) raise questions regarding the generality of aromatization of testosterone among vertebrates. Indeed aromatization is found in the brains of various mammals (rat, mouse, cow, rabbit, cat, and rhesus monkey) and in amphibian (Kelley et al 1978) and reptilian (Callard et al 1977) brains (for discussion and references see Callard et al 1977). The occurrence of 5α reductase and the presence of 5α DHT is less well studied across species. Tentative identification of 5α DHT has been made in an

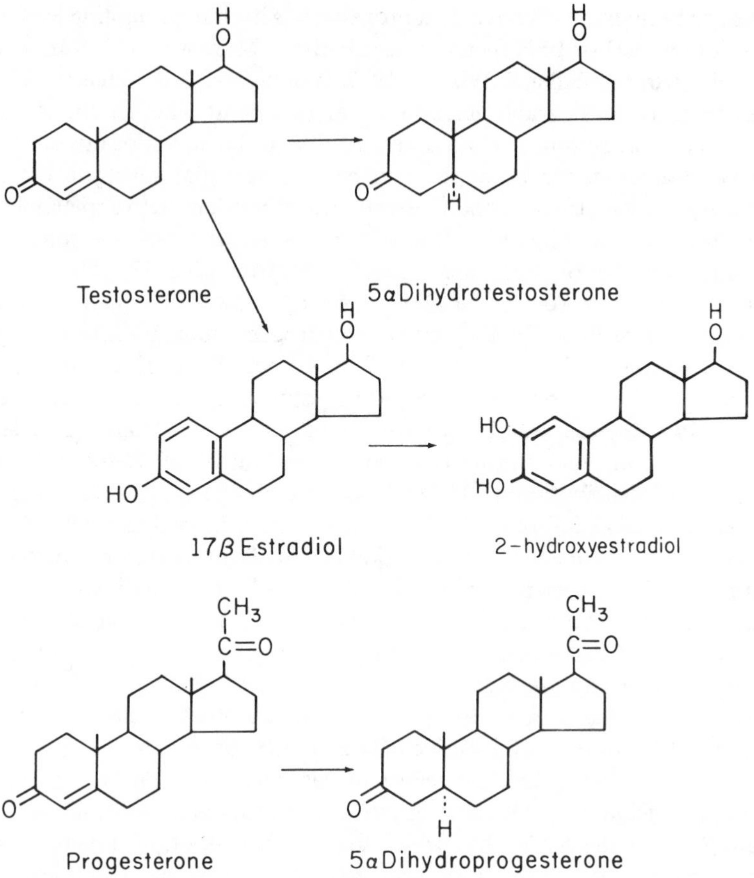

Figure 2 Some of the major steroid transformations that occur in brain and/or pituitary tissue of the rat. From McEwen et al 1978. Used by permission of Raven Press.

amphibian (Kelley et al 1978) and 5α reductase has been measured in certain birds (see Massa et al 1977).

Recent observations on the formation of another class of steroid metabolites, the 2 hydroxylated estrogens (catecholestrogens), in rat brain tissue has focused attention on the possible role of these estrogens as mediators of some estrogen effects in the brain (Fishman & Norton 1975, Paul and Axelrod 1977, see Fishman 1976 for review). Possibly the most interesting and unique actions of catecholestrogens are their stimulatory effects on LH which occur within 6 hr (Table 4). It should be noted that catecholestradiol, unlike estradiol itself, does not suppress LH and FSH levels in ovariectomized rats (cited by Ball et al 1978), although it appears to do so in

Table 4 Neuroendocrine effects of catecholestrogens

Effect	Reference
Implants in amygdala of miniature pig reduces plasma LH levels	Parvizi & Ellendorf 1975
Systemic injection in 35d ♂ rats increases plasma LH in 6 hr	Naftolin et al 1975a
Systemic injection in adult ovariectomized ♀ rats increases plasma LH in 6 hr	Gethmann & Knuppen 1976
Systemic and intracranial administration reveals weak estrogenic action on female sexual behavior in ovariectomized rats	Luttge & Jasper 1977; Marrone et al 1977
Neonatal administration produces less anovulatory sterility than estradiol itself	Parvizi & Naftolin 1977
Antagonism of estrogen effect on hypothalamic cAMP levels	Paul & Skolnick 1977

miniature pigs (see Table 4). Among the neuroendocrine effects summarized in Table 4, two indicate that catecholestrogens may have weak estrogenic effects on development and sexual behavior. This evidence of weak estrogenicity is supported by studies showing weak estrogenicity of catecholestrogens on the uterus (Martucci & Fishman 1977). And in spite of evidence for an antagonistic effect of catecholestrogens on the induction by high doses (20 μM) of estrogen of increases in cAMP in hypothalamus (Table 4), studies on the uterus (Martucci & Fishman 1977) and on the induction of lordosis behavior (Luttge & Jasper 1977) reveal no such "anti-estrogenicity." Catecholestradiol does have a moderate affinity for estrogen receptors in the brain and uterus (Davies et al 1975, Martucci & Fishman 1976), but nothing is known about its efficacy in translocating receptor to the cell nuclei.

Because the above-mentioned effects of estrogens on hypothalamic cAMP levels (Table 4) may be mediated by catecholamines (see Weissman & Johnson 1976) it is important to mention that catecholestrogens are extremely effective inhibitors of the O-methylation of catecholamines (Ball et al 1972) as well as various liver enzymes of drug metabolism (Bolt & Kappus 1976). Thus the interaction of catecholestrogens with enzymes of catecholamine metabolism, and possibly even with catecholamine receptors, must be considered as a possible mode of action of these steroids.

Steroid Hormone Effects on Cell Chemistry and Function

One goal of the biochemical approach to steroid action in brain is the elucidation of the cellular and chemical events regulated by the hormone which are essential for its physiological and behavioral actions. While we do not yet understand how a steroid hormone, acting at the genomic level via RNA and protein synthesis, can alter the electrical activity of neurons, we do have an ever-increasing catalog of data regarding the cellular and biochemical consequences of hormone action in the brain. These consequences, which are summarized in Tables 5 and 6 for glucocorticoids and in Tables 7 and 8 for estrogens, include effects on protein synthesis as well as effects on neurotransmitter metabolism.

It should be noted that many of the reported effects of glucocorticoids listed in Tables 5 and 6 have been produced by high doses and prolonged treatment with corticoids such as do not occur naturally in the rat or mouse,

Table 5 Glucocorticoid effects in adult brain[c]

Effect	Tissue		Steroid dose	Time	Reference
1. Inhibits CRF release	Hypothalamic fragments in vitro[a]	B	10–10–10–7M	10′	Jones et al 1977
2. Increases brain tyrosine level	Whole brain, in vivo[a]	F	20 mg/kg	15′	Diez et al 1977
3. Increases P/O ratio	Whole brain, in vivo[a]	F	3–5 mg/kg	60′	Roosevelt et al 1973
4. Increases single unit activity	Hippocampus, in vivo[a]	B	5 mg/kg	30–60′	Pfaff et al 1971
5. Increases labeling of proteins	Hippocampus, in vitro, slices[a]	B	10^{-10} M	60′	Lee et al 1977
6. Decreases high affinity GABA transport	Hippocampus, in vivo[a]	B	solid implant	4–7 days	Miller et al 1978
7. Increases GPDH activity	Whole brain, nerve, in vivo[a]	B B	endogenous solid implant	7–14 days	DeVellis & Inglish 1968 Meyer & McEwen, unpubl.
8. Increases DBH activity	Hypothalamus, in vivo[a]	B	100 mg/kg	4 hr	Shen & Ganong 1976
9. Increases NA, DA turnover	Whole brain, in vivo[b]	B	5–15 mg/kg	60′	Iuvone et al 1977
10. Decreases NA turnover	Whole brain, in vivo[a]	B F	endogenous 25 mg/kg	2 days	Javoy et al 1968 Fuxe et al 1970
11. Increases TH activity	Superior cervical ganglion, in vivo[a]	Dex	3 mg/kg	48 hr	Hanbauer et al 1975a
12. Potentiates reserpine induction of TH	Superior cervical ganglion, in vivo[a]	B	endogenous		Hanbauer et al 1975b
13. Potentiates NGF effect on TH activity	Superior cervical ganglion, culture[a]	B	5×10^{-6} M	24–48 hr	Otten & Thoenen 1976, 1977
14. Increases TH activity	Median eminence, in vivo[a]	Dex	0.3 mg/kg/day	7 days	Kizer et al 1974
15. Induces PNMT	Hypothalamus, medulla, in vivo[a]	Dex	1 mg/kg/day	7–13 days	Moore & Phillipson 1975
	Adrenal medulla, in vivo[a]	Dex	3–5 mg/kg/day	3 days	Pohorecky & Wurtman 1971
16. Increases NA, DA levels and size of large and small vesicles	Carotid body, in vivo[a]	Dex	1 mg/kg/day	10 days	Hellstrom & Koslow 1976

[a] Rat.
[b] Mouse.
[c] Abbreviations are CRF, corticotropin releasing factor; P/O ratio, ratio of ATP generated to oxygen used; NA, noradrenaline; DA, dopamine; DBH, dopamine β hydroxylase; GPDH, glycerol phosphate dehydrogenase; TH, tyrosine hydroxylase; NGF, nerve growth factor; PNMT, phenylethanolamine N methyl transferase; B, corticosterone; F, cortisol; Dex, dexamethasone.

and that few of these effects have been systematically studied with respect to adrenalectomy or hormone replacement therapy. Thus the physiological relevance of many of these changes remains uncertain. Exceptions to this generalization include the effects of physiological levels of corticosterone on high affinity GABA uptake in synaptosomes from the hippocampus (#6, Table 5) and on GPDH activity in nerve and brain (#7, Table 5), which reverses the changes in these activities that follow adrenalectomy. In addition, not only do glucocorticoids alter the turnover of serotonin (Table 6), possibly through a variety of effects on tryptophan and serotonin uptake and monoamine oxidase activity, but they also must be present for a variety of agents to elevate tryptophan hydroxylase activity and to promote brain seizure activity. It is unclear whether any of these glucocorticoid effects require protein synthesis, because the effects are very rapid (e.g. see Sze 1976, Azmitia & McEwen 1974) and because protein synthesis inhibitors suppress tryptophan hydroxylase activity in vivo, independently of glucocorticoid administration, thus precluding the definitive use of these inhibitors in blocking a hormone effect (Azmitia & McEwen 1976). The postnatal evaluation of tryptophan hydroxylase activity is also dependent on glucocorticoids and may involve protein synthesis (see Table 6 and Sze 1976).

Table 6   Effects of glucocorticoids on the serotonergic system[a]

Effect	Reference
Elevates 5HT synthesis	Azmitia et al 1970a, Gal et al 1968, Millard et al 1972
Acute elevation of 5HT levels	Telegdy & Vermes 1975
Depression of 5HT levels	Curzon & Green 1971
Enhances uptake of Tp by synaptosomes; by brain	Hillier et al 1975, Sze 1976
Enhances uptake and release of 5HT by synaptosomes	Vermes et al 1976
Inhibition of MAO activity	Parvez & Parvez 1973; Petrovic & Janic 1974
Required for elevation of tryptophan hydroxylase activity by	Sze 1976
Ethanol intoxication	Kuriyama et al 1971
Morphine	Azmitia et al 1970b
Reserpine	Sze et al 1976
Foot shock, cold, ether stress	Azmitia & McEwen 1974
Required for appearance of	
Ethanol withdrawal seizures	Sze et al 1974
Acoustic reduction of susceptibility to audiogenic seizures	Sze & Maxson 1975

[a] Abbreviations are 5HT, serotonin; Tp, tryptophan; MAO, monoamine oxidase.

Table 7 Biochemical effects of estradiol in hypothalamus

I. "Short Term" (within 48 hrs. post-steroid)

Parameter	Time first observed (hour)	Reference
Increase in RNA-polymerase II	3	Peck 1978
Facilitation of mating behavior in female rats	17	Green et al 1970
Increase in cystine aminopep-tidase	18	Heil et al 1971
Increase in choline acetylase	24	Luine & McEwen 1977

II. "Long Term" (in excess of 48 hrs. post-steroid)

Parameter	Time observed	Reference
Increase in spontaneous activity of MBH neurons[a]	Day 10	Bueno & Pfaff 1976
Decrease in spontaneous activity of POA neurons[b]		
Decrease in norepinephrine synthesis in anterior hypothalamus	Hour 60	Bapna et al 1971
Increase in reuptake of dopamine, serotonin and norepinephrine in anterior hypothalamus	Hour 60	Cardinali & Gomez 1977
Increased incorporation of lysine into protein	Week 5 following ovariectomy	Litteria & Thorner 1974
Increase in spontaneous activity of MPO[c] neurons	Hour 48	Kubo et al 1975
Decrease in LH-RH in[d] anterior hypothalamus	Day 7 following ovariectomy	Araki et al 1975
Increase in LH-RH[d] in MBH[a]	Day 2	Kalra 1976
Increase in G6PDH[e], MDH[f], ICDH[g] in MBH[a]	Day 7	Luine et al 1974, 1975, 1977
Decrease in MAO[h] and TH[i]		

[a] basal medial hypothalamus.
[b] preoptic area.
[c] medial preoptic area.
[d] luteinizing hormone releasing hormone.
[e] glucose-6-phosphate dehydrogenase.
[f] malate dehydrogenase.
[g] isocitrate dehydrogenase.
[h] monoamine oxidase.
[i] tyrosine hydroxylase.

Table 8 Biochemical effects of estradiol in pituitary

I. "Short Term" (within 48 hrs. post-steroid)

Parameter	Time first observed (hour)	Reference
Decrease in sensitivity to LH-RH[a] (in culture)	4	Tang & Spies 1975
Decrease in sensitivity to LH-RH[a] in vivo	2	Vilchez-Martinez et al 1974
Increased incorporation of ^{3}H-alanine and ^{3}H-glucos-amine into LH[b]	5.5	Liu & Jackson 1977
Decrease in FSH[c] (in culture)	6	Miller et al 1977
Increase in sensitivity to LH-RH[a] (in culture)	10	Drouin et al 1976
Increase in sensitivity to LH-RH[a] in vivo	14	Vilchez-Martinez et al 1974
Increased incorporation of ^{3}H-thymidine into DNA	30	Jacobi et al 1977

II. "Long Term" (in excess of 48 hrs.)

Parameter	Time observed (day)	Reference
Increase in sensitivity to T-RH[d]	2–4	DeLean et al 1977
Increase in DNA polymerase	7	Mastro & Hymer 1973
Increase in net RNA and in RNA/DNA ratio	16	Robinson & Leavitt 1971

[a]luteinizing hormone releasing hormone
[b]Luteinizing hormone
[c]follicle stimulating hormone
[d]thyroid releasing hormone

The estradiol effects, summarized in Table 7 for the hypothalamus and preoptic area, and in Table 8 for the pituitary, have been categorized according to their time course; this temporal aspect of estrogen action forms an important part of the discussion below in the section on Temporal Correlations. The hypothesis that most, if not all, of these effects of estradiol involve intracellular receptors acting via the genome is based on: (a) the fact that they occur in brain regions that contain estrogen receptor sites, (b) the finding that they have a time course compatible with a genomic effect, and (c) the observation that they can be blocked (or in some cases mimicked) by nonsteroidal estrogen antagonists (which are at the same time weak agonists) that are known to interact with intracellular estrogen receptors (for discussion, see Luine & McEwen 1977, McEwen et al 1978). Further proof that these effects are the result of the direct action of estradiol on neuronal or pituitary estrogen receptors is a subject of active investigation in various laboratories.

In this connection, it is important to note that some effects of estradiol on the brain may be indirect and mediated by pituitary hormones, whose release is influenced by estradiol. This may be the case, for example, in the short-term alterations in catecholamine turnover. Noradrenaline turnover is increased following gonadectomy and is decreased by estrogen replacement therapy (see Coppola 1969, Bapna et al 1971, Löfström et al 1977). These changes parallel alterations in gonadotropin secretion, and it has been shown that the administration of FSH to normal rats with low levels of this hormone increases noradrenaline turnover (Anton-Tay et al 1970). Dopamine turnover is increased by estrogen treatment and these changes parallel increases in prolactin secretion (Löfström et al 1977, Jiminez et al 1977). Prolactin administration to ovariectomized rats enhances dopamine turnover in the median eminence and anterior hypothalamus after a latent period of 10–26 hr (Gudelsky et al 1976). It remains to be seen whether FSH and prolactin action might explain estradiol effects on hypothalamic tyrosine hydroxylase and Type A monoamine oxidase activity (Table 7). The definitive experiments would have to be performed on hypophysectomized rats. Against the view that estrogen acts only indirectly via the pituitary is the report that dopamine-secreting neurons of the arcuate nucleus region have the ability to concentrate ^{3}H-estradiol (Grant & Stumpf 1973, 1975) and the possibility that some noradrenergic neurons of the brainstem may also have limited numbers of estrogen receptors (Heritage et al 1977).

Hormones, Brain Function, and Behavior

The actions of steroid hormones such as corticosterone and estradiol upon the brain and pituitary are measurable in terms of neuroendocrine and behavioral parameters. Estradiol, testosterone, and progesterone, for example, are required in many vertebrate species for the expression of mating behavior (see Young 1961), and estradiol and testosterone also exert regulatory influences over gonadotropin secretion. Testosterone is also an important regulatory hormone in the expression of certain forms of aggressive behavior (see Bronson & Desjardins 1971). Glucocorticoids have been implicated not only in the regulation of ACTH secretion but also in the detection and recognition of sensory stimuli, in the regulation of paradoxical sleep, and in the extinction of conditioned avoidance behaviors (see McEwen 1978b for references). And mineralocorticoids have been shown to produce a suppression of salt appetite, which increases after adrenalectomy in rats (Fregly & Waters 1966).

In order to examine further the important interrelationships between hormones, specific brain regions, and specific behaviors, we have chosen to examine in greater detail the interaction of estrogens with the female rat brain. This choice is dictated in part by the stereotypy of the lordosis response in the rat, which facilitates quantitation and analysis of its components, as well as by the extensive information on the neural sites that control this behavior.

ESTROGENS IN BRAIN: RECEPTORS AND FUNCTIONS

Background

The net result of the action of estrogen on the brain is a sequence of orderly changes in pituitary function and behavior. For instance, in the case of estrogen effects on female reproductive behavior, the application of a variety of physiological methods (electrical stimulation, electrical recording, transections, etc) has led to the discovery of a neural circuit that includes not only hypothalamic cells, but obligatory links in the brainstem and spinal cord (reviewed by Pfaff & Modianos 1978). At hypothalamic levels, there is a clear separation between the cells responsible for the activation of male, as opposed to female, behavior: cells facilitating masculine sex responses are located in the medial preoptic area, while those responsible for activating female sex behavior are located in and around the ventromedial nucleus of the hypothalamus (reviewed by Pfaff & Modianos 1978). Although the study of extrahypothalamic circuitry is required for the explanation of complete behavioral mechanisms, the most detailed attempts to relate the cell biology of hormone action to neuronal function have come from experiments with hypothalamic tissue.

The ultimate purpose of these experiments is to show a causal sequence of events, beginning with estrogen entry into the brain and its accumulation by specific receptors, and ending in functionally relevant alterations in the activity of individual nerve cells. However, the information necessary for the construction of the sequence is still lacking. To understand how estrogen receptors are related to behavioral mechanisms we presently are limited to the study of correlations between receptor function and behavioral function. Three types of correlations have been shown: those based on neuroanatomical studies of estrogen receptors and behavioral mechanisms (correlations in space), see section on Spatial Correlations; correlations based on the temporal properties of estrogen action, see section on Temporal Correlations; and correlations based on the fact that anti-estrogens can block both estrogen accumulation and estrogen effects on mating behavior, see section on Anti-Estrogens.

Spatial Correlations

Autoradiographic studies of the location of estrogen and androgen receptors in the brain have allowed a correlation with the functional aspects of steroid hormone action. For instance, in the female rat (Pfaff 1968, Stumpf 1968, Pfaff & Keiner 1973) a system of limbic, medial preoptic, and medial hypothalamic cells shows the strongest concentration of radioactive estrogen. These same cells have been strongly implicated in the control of pituitary gonadotropin release and female mating behavior. In fact, many of the regions of highest estrogen concentration are interconnected and this anatomical feature provides the strongest basis for the inference that there is an estrogen concentrating system. In the female hamster (Krieger, Morrell & Pfaff 1977) estrogen is also accumulated in the same limbic-hypothalamic system, but binding in the medial anterior hypothalamus is elevated compared to that in the medial preoptic area. This cross-species difference appears to be correlated with the fact that estrogen stimulation of female mating behavior in the hamster is accomplished by cells located in the medial anterior hypothalamus (Ciaccio & Lisk 1973/74). In the rhesus monkey (Gerlach et al 1976, Pfaff et al 1976) estrogen is strongly accumulated in the same limbic-hypothalamic system, with impressive concentrations in the arcuate (infundibular) nucleus. Estrogen-accumulating nerve cell groups are prominent in basomedial hypothalamic tissue, which has been shown to be adequate for the estrogen-stimulated release of luteinizing hormone (Krey et al 1975).

In birds, radioactive androgen is accumulated by nerve cell groups that are crucial for the control of gonadotropic hormone release (tuberal hypothalamic cell groups), courtship and copulatory behavior (medial preoptic area) and hormone-dependent bird song (nucleus intercollicularis and the motoneurons of the nucleus tracheosyringalis) (chaffinch, Zigmond et al 1973; zebra finch, Arnold et al 1976; chicken, Barfield et al 1978). In fact, in all the major vertebrate classes, there is a close correlation between estrogen accumulation (defined autoradiographically) and estrogen action (reviewed by Morrell et al 1975 and Pfaff 1976).

The functional data that provide the most direct comparison to the autoradiographic studies are those based on local implants of steroid hormones in the brain. The wide distribution of hormone-sensitive cells in the brain corresponds to sites implicated in the control of a variety of hormone dependent endpoints including sexual differentiation, gonadotropin secretion, and masculine and feminine sexual behavior, as well as maternal behavior, aggression, food intake, and activity levels. Results from lesion studies support the view that the neural substrates of some of these endpoints are at least partially distinct. For example, lesions of the anterior hypothalamic-preoptic area (AHPOA) of both males (Brookhart & Dey 1941, Heimer & Larsson 1966/67, Hart 1974 for additional references) and females (Dörner et al 1969, Singer 1968) abolish masculine mating behavior, whereas lesions of the ventromedial hypothalamic region disrupt feminine sexual behavior (for references and review, Malsbury, Kow & Pfaff, 1977). However, lesions of the ventromedial-arcuate region also produce alterations in food intake, aggressive behavior, and gonadotropin secretion (e.g. Mogenson & Calaresu 1975, Moyer 1976, Martini et al 1970). Thus, the possibility exists that the same or different steroid-sensitive cells in the area are involved in more than one behavior; or that the effects of the lesions are the result of damage to nonsteroidal cells, or to fibers of passage. The hormone implant method affords a direct means of assessing the functional concomitants of steroid uptake in neural tissue. This experimental approach has been most widely applied to sexual behavior and gonadotropin secretion and the following discussion focuses primarily upon these endpoints.

One of the assumptions underlying the application of small quantities of steroids to the central nervous system is that the substance is acting primarily, if not solely, at the site of implantation. If this were true, then (a) neural sites of action could be distinguished from possible pituitary and peripheral target organ sites, and (b) the specific locus of action within the brain could be determined. For these reasons, other steroid target tissues such as the vaginal tract and uterus in the female and the seminal vesicles, penis, and prostate in the male are evaluated for signs of hormonal stimulation. However, the absence of such stimulation does not mean that the intracranial implant is acting only at the site of implantation.

FEMININE SEXUAL BEHAVIOR   Experiments in which activation of feminine sexual behavior was accomplished by intracranial hormone implants provide conflicting evidence concerning the sites of action in the stimulation of estrous behavior (see Table 9). The results of the early studies by Harris, Michael & Scott (1958) indicated that implants of the synthetic estrogen, stilbestrol, into the posterior hypothalamus of spayed cats would activate estrous behavior. Later, it was shown that mating in cats could be elicited from a wide range of ventral hypothalamic sites extending from the POA rostrally to the mammillary region caudally (Michael 1965). It had previously been shown that systemic estrogen treatment could produce changes in the vaginal tract prior to the restoration of behavior; but the vaginal tracts of cats responding to intrahypothalamic implants remained atrophic and thus a central nervous site of action was presumed.

In other species, the site of hormone action has been ascribed to the AHPOA or the ventromedial region (VMN) or both (for review, Barfield & Chen 1977). Even within a single species, the rat, detailed comparisons

Table 9 Studies on the sites of action of sex hormones in the activation of feminine sexual behavior in rats[a]

Type of implant	Most effective location(s)	Other sites investigated	Reference
27 & 30g E$_2$-filled tubes, unilateral	AH-POA	Scattered hypothalamic sites, sparing region of VMH	Lisk 1962
1 μg EB in capillary tubes	VMH	POA	Dörner, Döcke & Moustafa 1968b
32g Hexestrol-filled tubes, bilateral	Habenula, septum, AH, amygdala (No localization observed)	—	Rodgers & Law 1968
27g, EB-filled tubes	AH-POA	None	Chambers & Howe 1968
27g EB-filled tubes, unilateral	DBB-POA	MRF, CPU	Yanase & Gorski 1976
27g EB-filled tubes, unilateral	POA, anterior-lateral hypothalamic area, VMH	None	Barfield & Chen 1977
30g EB-filled tubes, unilateral	VMH & secondarily, POA	Cortex, midbrain, septum, DBB, bed nucleus of ST, anterior-lateral hypothalamus, PVN, dorsal hypothalamus, posterior mammillary region	Barfield & Chen 1977

[a] Abbreviations are AH, anterior hypothalamus; DBB, diagonal bands of Broca; CPU, caudate-putamen; MRF, midbrain reticular formation; POA, preoptic area; PVN, paraventricular nucleus of the hypothalamus; ST, stria terminalis; VMH, ventromedial hypothalamus; EB, estradiol benzoate; g, gauge of tubing.

are difficult to make because of: (*a*) variability in the type of implant preparation employed (see Table 9); (*b*) failure to provide sufficient information on the quantification of the behavioral measures (e.g. Dörner et al 1968b, Lisk 1962); (*c*) failure to adequately sample within each of a variety of implant sites (e.g. Chambers & Howe 1968); and (*d*) the possibility of true species or strain differences.

Perhaps the single most important variable in determining if a discrete locus of action exists is the size of implant employed. For example, diffusion from a 27 gauge cannula may amount to as much as 2 mm in a 4–5 day period and radioactivity is also detectable in plasma and uterus (Palka, Ramirez & Sawyer 1966). Similar EB-filled cannulae induce lordotic behavior from sites in the AHPOA, VMN, and lateral hypothalamus (Barfield

& Chen 1977). When smaller, 30 gauge implants were employed, ventro-medial hypothalamic implant sites were significantly more effective than AHPOA; however, positive responses were still obtained from both implant locations. Sites in between these two were ineffective. These findings are similar to those obtained for the guinea pig (Morin & Feder 1974a), which suggests the possibility that the AHPOA and the VMN are both sites of estrogen action in the induction of lordosis behavior. This would be consistent with the observations that lesions of the POA may enhance lordotic responsiveness (Powers & Valenstein 1972), and that estrogen treatment produces opposite effects on electrophysiological responses of POA and VMN neurons (Bueno & Pfaff 1976).

We have reexamined the effectiveness of implants in the VMN using a refined technique in which tritium labeled estradiol was diluted with cholesterol in a ratio of approximately 1 to 300 (Davis, McEwen and Pfaff, unpublished observations). In this way we could correlate behavioral effectiveness with the spread of the hormone from the region of implantation. Thirty gauge cannulae prepared by this method contain approximately 10–14 ng of estradiol and deliver about 30% of the total hormone to the brain in a 8-day period. In ovariectomized, progestrone-treated rats, both unilateral and bilateral implants produce significant and substantial increases in lordotic behavior accompanied by solicitation behaviors. As determined by scintillation counting methods, there is no evidence for a significant spread of hormone to the contralateral side of the hypothalamus (from unilateral implants) or to the POA, cortex, amygdala, pituitary, or uterus with either bilateral or unilateral implants. Moreover, similarly prepared estradiol/cholesterol implants applied to the POA are not effective in stimulating feminine sexual behavior (Rubin and Barfield, unpublished observations). Taken together, these preliminary findings suggest that estrogenic stimulation of the VMN is sufficient to activate feminine sexual behavior.

MASCULINE SEXUAL BEHAVIOR   In contrast to feminine sexual behavior, there is substantial agreement concerning the principal locus of action of sex hormones in the activation of masculine copulatory behavior. In birds (Barfield 1969, 1971, Hutchison 1970), as well as mammals (see Table 10), this site is considered to be the AHPOA. Hormone action at the AHPOA would be consistent with demonstrations that lesions of the same area abolish the behavior (Heimer & Larsson 1966/67), whereas electrical stimulation activates copulation (e.g. Malsbury 1971). Even so, there is some evidence for the existence of multiple sites of action including the posterior hypothalamus (Johnston & Davidson 1972) and several extrahypothalamic sites (Kierniesky & Gerall 1973). However, because the implants employed

Table 10 Sites of action of gonadal steroids in activation of masculine sexual behavior in male rats[a]

Type of implant	Effective sites	Other sites investigated	Reference
TP pellet or TP in 20g tube	Medial POA & secondarily in posterior hypothalamus	supra & lateral POA, thalamus, cortex-corpus callosum, hippocampus, midbrain, CPU, posterior colliculus, zona incerta	Davidson 1966a
Bilateral blobs of T Bilateral blobs of TP	None region of suprachiasmatic and paraventricular nuclei	Scattered throughout hypothalamus	Lisk 1967
TP in 20g tubes, bilateral	Medial POA, CPU, substantia nigra, RF (No localization observed)	Not specified	Kierniesky & Gerall 1973
TP in 20g tubes, unilateral	Medial POA with some positive responses from posterior hypothalamic sites	None	Johnston & Davidson 1972
.2mg TP	POA	VMH	Dörner, Döcke & Moustafa 1968a
E$_2$ pellets, 3 X 10μg, bilateral	POA	Posterior hypothalamus	Christensen & Clemens 1974
T pellets 3 X 15μg, bilateral			
27 & 30g EB-filled tubes, unilateral	Medial AH-POA some positive responses throughout hypothalamus	DBB through posterior mammillary region	Davis & Barfield in prep.

[a] Abbreviations are TP, testosterone propionate; other abbreviations, see Table 9.

in these studies were quite large, there is some question that the results may be due to the spread of hormone from the site of implantation to the AHPOA.

There is now considerable evidence that the induction of masculine copulatory behavior may depend upon the aromatization of testosterone to estradiol (see the section on Steroid Metabolism.) If this were true, then it might be expected that estrogenic intracranial implants would be at least as effective as similarly placed androgenic implants in restoring copulatory behavior of castrated males. Indeed, unilateral 27 and 30 gauge EB-filled cannulae restore ejaculatory responding in castrated male rats treated systemically with dihydrotestosterone (Davis and Barfield, in preparation). Under these circumstances, the AHPOA was the most effective implant site, but positive responses were also obtained from animals with implants located more posteriorly in the hypothalamus. At present it is difficult to know if the restoration of behavior following posterior hypothalamic implants is related to diffusion of the steroid or to the variability of masculine copulatory behavior after castration (Davidson 1966a, b).

GONADOTROPIN SECRETION Evidence provided by hormone implants indicates that at least part of the positive and negative feedback actions of steroids upon gonadotropin secretion are mediated by hypothalamic (and perhaps extrahypothalamic mechanisms) in addition to possible direct effects upon the pituitary gland. From lesion, knife cut, and electrical stimulation experiments, it has been proposed that the medial basal hypothalamus (MBH) controls tonic gonadotropin secretion, whereas the POA provides the neural signal required for the ovulatory surge of LH (for review see Arai 1973, Martini et al 1970).

Implantation of hormones into these regions of the hypothalamus and preoptic area produces effects consistent with this hypothesis. Thus, implantation of steroid into the MBH results in decreases in LH secretion that do not appear to be mediated by transport of the steroid to the pituitary gland (e.g. Smith & Davidson 1974). Conversely, implantation of estradiol into the POA, but not into the MBH, initiates increases of LH (Goodman 1978). Although both implant sites resulted in detectable hormone levels in the pituitary gland, ineffective MBH implants produced pituitary estradiol levels well in excess of those normally associated with LH release. Therefore, it does not appear that estrogenic stimulation of the pituitary is sufficient for stimulating surges of LH. These results are consistent with the hypothesis that the preoptic region is the site of estrogen positive feedback, but because of the possible diffusion of the hormone we cannot rule out the participation of other neural regions.

Temporal Correlations

Studies of estrogen effects in the uterus, pituitary, and brain have been pursued to determine such temporal properties of these effects as the minimal latency, minimal necessary time of estrogen application, and duration. These parameters allow for some deductions about the possible relationships between estrogen receptors and estrogen action. In general, very fast effects may not depend upon some kinds of estrogen receptors; slower actions of estrogen may generally be thought to allow time for the hormone to enter the nucleus and alter genomic expression, with subsequent changes in biosynthetic events.

UTERUS Following a systemic injection of estradiol-17β, two distinct biochemical changes may be detected in the uterus within 15 minutes (Table 11). One of these is an increase in the activity of RNA-polymerase II (DNA-like RNA synthesis; Glasser et al 1972). The other is the appearance of a mRNA that codes for a protein in the uterus which is specifically induced by estrogen (DeAngelo & Gorski 1970, Notides & Gorski 1966).

Table 11   Biochemical effects of estradiol in uterus

Parameter	Time first observed (min.)	Reference
Increase in RNA-polymerase II	15	Glasser et al 1972
Appearance of RNA for "induced protein"	15	DeAngelo & Gorski 1970, Notides & Gorski 1966
Synthesis of "induced protein"	40	Barnea & Gorski 1970
Increase in RNA polymerase I	60	Gorski 1964
Increase in de novo fatty acid synthesis	60	Spooner & Gorski 1972
Increased conversion of glucose to glucose-6-phosphate	60	Smith & Gorski 1968
Altered template capacity of chromatin	60	Barker & Warren 1966, Glasser & Spelsberg 1973
Increase in acidic content of chromatin	60	Glasser & Spelsberg 1973
Increase in rate of peptide elongation	60	Whelly & Barker 1974
Increase in H_2O retention	120	MacLeod & Reynolds 1938
Increase in number of active ribosomes	240	Whelly & Barker 1974
Increase in net protein synthesis	240	Hamilton 1963, Noteboom & Gorski 1963
Increase in net RNA	480	Billing et al 1969b, Mueller et al 1958
Increase in DNA synthesis	1,080	Kaye et al 1972
Increase in mitotic activity	1,440	Kaye et al 1972

Shortly thereafter (40–60 minutes), the synthesis of this "induced protein" is observed in uterine cells (Barnea & Gorski 1970).

Within 60 minutes of estradiol-17β administration, distinct increases in the levels of RNA polymerase I (ribosomal RNA synthesis; Gorski 1964), in de novo fatty acid synthesis (Spooner & Gorski 1972), and in the conversion of glucose to glucose-6-phosphate (Smith & Gorski 1968), are seen in the uterus. At this time, laboratories have reported an alteration in the template capacity of uterine chromatin, as assayed by measuring transcribed RNA from purified DNA and an E. coli RNA polymerase in vitro (Barker & Warren 1966; Glasser et al 1972). In addition, an increase in the acidic, but not basic, content of uterine chromatin is observed one hour after estradiol injection (Glasser et al 1972).

Within 4 hours following an injection of estradiol, there is an increase in the overall rate of protein synthesis in the uterus (Hamilton 1963, Noteboom & Gorski 1963). In a cell free system containing uterine polyribosomes, estradiol has been shown to first affect the rate of peptide elongation at each active ribosome (1 hour), and to subsequently increase the number of active ribosomes (marked at 4 hours, maximal at 12 hours; Whelly & Barker 1974). Quantitative work has revealed almost a five-fold increase in the number of polysomes in the rat uterus, during this 4–12 hour post estradiol period (Eilon & Gorski 1972).

Maximal water retention by uterine cells occurs 6 hours after the administration of estradiol (Szego & Roberts 1953), although increased retention may be seen as early as 2 hours (MacLeod & Reynolds 1938). An increase in net RNA is observed in the uterus at 8–12 hours after estradiol administration (Billing et al 1969b, Mueller et al 1958). A two-fold increase in uterine DNA synthesis is seen 18–24 hours after estradiol injections (Kaye et al 1972); this precedes the change in mitotic activity at 24–48 hours (Kaye et al 1972).

New species of RNA are synthesized in the uterus rather shortly following estradiol treatment, as shown by the use of the rate of precursor incorporation into newly-synthesized RNA. For example, Means & Hamilton (1966) noted a change in the rate of ^{3}H-uridine incorporation into rapidly labeled nuclear RNA within 2 minutes following estradiol-17β in vivo. Likewise, Knowler & Smellie (1973) reported an increase in the rate of ^{3}H-uridine and ^{3}H-guanosine incorporation into 45S RNA, 2 hours following estradiol-17β in vivo. As noted by Billings et al (1969a, b), an increase in the specific activity of RNA precursor pools, i.e. precursor transport, could produce a change in the levels of precursor incorporation into RNA, without altering the in vivo rate of RNA synthesis (for an excellent review of this problem, see Katzenellenbogen & Gorski 1975). Indeed, Billings and co-workers (1969b) found that estradiol-17β greatly increases the uterine transport of RNA precursors during the first eight hours after its administration; furthermore, they conclude that this increase is sufficient to account for the previously observed increase in the incorporation of labeled uridine into RNA without an increase in in vivo synthesis. Net increases in RNA do not begin until 8 hours post estradiol (Billings et al 1969b).

Considerable interest has been focused upon the minimal length of time during which estradiol-17β must be retained in uterine nuclei to promote the described changes in cellular metabolism. In a series of experiments, Anderson and co-workers characterized two 'types' of uterine nuclear binding sites: (a) a large number (80–90%) of nonspecific sites, which exhibit short-term retention of estradiol; and, (b) a small number (10–20%) of specific acceptor sites, which exhibit long-term retention of estradiol (for references, see Clark & Peck 1976). The authors hypothesized that the retention of estradiol by the specific receptor sites for a minimum of 4–6 hours is necessary for uterine growth. Anderson et al (1975) found that estradiol and estriol accumulated in uterine nuclei rapidly and at the same rate; but differed with respect to their rate of retention. The nuclear concentrations of these two steroids were identical at 1–3 hours; however, nuclear concentrations of estriol fell to control levels by hour 6. Estradiol and estriol were equally effective in stimulating uterotrophic responses (RNA polymerase activity, and metabolism of ^{14}C-glucose) at 0–3 hours, but only estradiol stimulated an increase in uterine weight at 24 hours. However, multiple injections of estriol, which produced high nuclear concentrations of the receptor-hormone complex at all times, were effective in stimulating uterine growth at 24 hours (Anderson et al 1975). Lan & Katzenellenbogen (1976) chemically modified estriol, producing several compounds that differed in their nuclear retention times. Although ethinyl estriol, estradiol, and estriol all show maximal accumulation by 1 hour, ethinyl estriol and estradiol show a two-fold increase in nuclear retention 24 hours after drug treatment, while estriol has fallen to control levels by this time. All three compounds were effective in eliciting early uterotrophic responses (increased uterine wet weight and 2-deoxyglucose phosphorylation at 2–6 hours); however, only ethinyl estriol and estradiol were effective in stimulating the late uterotrophic responses (increased rates of 2-deoxyglucose phosphorylation at 24 hours, and increased DNA synthesis and wet weight at 72 hours; Lan & Katzenellenbogen 1976). Taken as a whole, these results suggest that the receptor-hormone complex must be retained for a minimum of 6 hours to promote the late changes in uterine metabolism.

Following a physiological dose of estradiol, the uterine cytosol receptor concentration is restored to 1.5 times its original value within 24 hours (Jensen et al 1969, Sarff & Gorski 1971). Thus, in the uterus, estrogen facilitates the synthesis of its own protein receptor. Several workers have investigated the mechanisms and possible physiological importance of this phenomenon. Mester & Baulieu (1975) characterized two stages of estradiol cytosol receptor replenishment: (a) there is an initial increase of receptors, which is insensitive to cycloheximide, at 0.5–6 hours post estradiol; and, (b) there is an additional increase of receptors at 6–11 hours, which is abolished by cycloheximide. In other words, two separate processes, one of them dependent upon protein synthesis, contribute to the restoration of uterine cytoplasmic estradiol receptors. This concept is substantiated by the work of Hsueh et al (1976), who demonstrated that progesterone interferes with the second, but not the first, phase of estradiol receptor replenishment in the uterus. Furthermore, this inhibition appears to be of physiological importance. Subsequent nuclear uptake of estradiol is depressed because of the decreased quantity of cytosol receptor available for translocation (Hsueh et al 1976).

BRAIN AND PITUITARY   To date, three principal paradigms have been used to characterize the changes of brain and pituitary tissue following estradiol. The first of these paradigms is the study of the changes in the cellular biochemistry of the hypothalamus and the pituitary, as a function of the estrous cycle. Many of these observed changes may be partially correlated with fluctuations in serum estradiol (see McEwen & Luine 1978 for summary).

The second of these paradigms is the study of the changes in hypothalamic and pituitary biochemistry, hypothalamic neural activity, and sexual behavior, which occur in the ovariectomized animal treated with estradiol (Tables 9 and 10). Vilchez-Martinez et al (1974) found that estradiol benzoate (EB) decreased the sensitivity of the pituitary to LH-RH, 2–6 hours after its administration in vivo. However, the authors noted that EB increased the sensitivity of the pituitary to LH-RH at 14–24 hours. Estradiol has elicited similar effects in cells cultured from pituitary. Tang & Spies (1975) demonstrated that 10^{-8} M estradiol decreased the sensitivity of cultured cells to LH-RH at 4 hours. Drouin et al (1976) found that 10^{-9} M estradiol increased the response of cultured cells to LH-RH at 10–24 hours.

Estrogen also stimulates the incorporation of protein and nucleic acid precursors into pituitary LH and DNA. Jacobi et al (1977) found that DES stimulated the in vivo incorporation of ^{3}H-thymidine into pituitary DNA, beginning at 30 hours. Liu & Jackson (1977) showed that estradiol stimulated the in vitro incorporation of ^{3}H-alanine and ^{3}H-glucosamine into LH at hour 4, in cells cultured from pituitary.

An effect of estradiol on the genomic expression of pituitary cells is implied by the work from two laboratories. Miller et al (1977) discovered that 10^{-9} M estradiol decreased the synthesis of FSH in cells cultured from pituitary, following 18 hours of treatment. Luine & McEwen (1977) found that 24 hours of EB treatment increased the levels of G6PDH, as well as the total protein content of the pituitary in vivo. Furthermore, when nuclear uptake of estrogen was inhibited by an antiestrogen, CI-628 (see below), no increase in pituitary G6PDH was seen (Luine & McEwen 1977).

In the hypothalamus, only four events have been characterized with respect to their exact time of induction by estrogen. The first of these is an increase in RNA polymerase II, which occurs in vivo within several hours of estradiol treatment (Peck 1978). Secondly, Heil and co-workers (1971) observed an increase in the levels of cystine aminopeptidase in the hypothalamus of the female rat, 18 hours after a subcutaneous dose of ethinyl-estradiol or estradiol-17β. Thirdly, Luine & McEwen (1977) reported an increase in the in vivo levels of choline acetylase (CAT) in the preoptic area of the female rat 24 hours following EB injection. Fourthly, Green et al (1970) demonstrated that an intravenous injection of estradiol-17β induces the onset of sexual receptivity in the female rat. Although a facilitative effect of estrogen on sexual behavior was seen as early as hour 17, the maximal increase in receptivity occurred at 24 hours (Green et al 1970). Thus, the time course of estrogen-induced sexual receptivity is strikingly similar to those reported for at least two hypothalamic proteins.

Many studies involving ovariectomized animals have focused on changes elicited by estradiol over a period of days or weeks. Hence, the time courses of these changes are unknown. Tables 9 (hypothalamus) and 10 (pituitary) include a summary of the events that have been shown to be affected by exposures to estradiol in excess of 48 hours.

The third paradigm is the study of estrogen-mediated events in the pituitary and the hypothalamus that are blocked by RNA and protein synthesis inhibitors. These events are most likely mediated by estrogen-induced changes in genomic expression. The use of biosynthetic inhibitors at various times illustrates the time-course of these changes in genomic expression. Debeljuk et al (1975) investigated the blockade of the estrogen-induced sensitization of the pituitary to LH-RH. The authors found that 50 μg of estradiol, followed by a systemic injection of LH-RH at 18 hours, increased serum LH levels in the female rat within 20 minutes. However, if actinomycin D was given simultaneously with estradiol, these changes were not observed (Debeljuk et al 1975). This observation suggests that the estrogen-induced sensitization of the pituitary to LH-RH involves a change in pituitary RNA synthesis within 18 hours.

Several authors have been successful in blocking the positive and negative feedback effects of estradiol on LH release. It should be noted that estradiol may effect LH release by changing molecular events in the hypothalamus and/or in the pituitary. Schally et al (1969) found that 50 μg of EB (+ 25 mg P) decreased serum levels in the female rat at 72 hours. However, actinomycin D administered at hour 52 abolished this decrease. Jackson (1972) reported that treatment of female rats with 0.5 μg of EB for 6 days, followed by 50 μg of EB on day 7, increased serum LH levels 29 hours later. Actinomycin D blocked this increase, if given simultaneously with EB. In a similar set of experiments, Jackson (1973) reported that cycloheximide blocked estrogen-induced increases in serum LH, between −5 and +24 hours of steroid administration. These experiments suggest that both the positive and negative feedback effects of estradiol on gonadotropin release depend on RNA and protein synthesis within 24 hours of the observed response.

The genomic origin of the lordosis response has been implicated by the work of two laboratories. Terkel et al (1973) demonstrated that actinomycin D applied directly to the preoptic area was effective in depressing lordosis quotients if given within 12 hours of EB. Quadagno & Ho (1975) showed that cycloheximide applied directly to the preoptic region depressed lordosis quotients if given within 12 hours of EB. Thus, it appears that hypothalamic RNA synthesis at 6 hours, and hypothalamic protein synthesis at 12 hours, are essential for the induction of sexual receptivity by estradiol.

Questions have also arisen concerning the minimal length of time that estradiol must be retained by hypothalamic nuclei, in order to promote changes in sexual behavior. Green et al (1970) reported that no facilitation of sexual behavior is seen within 16 hours of an intravenous estradiol injection, but thereafter a linear increase in receptivity is seen, until maximal receptivity is attained at hour 24. McEwen et al (1975b) investigated the occupation of nuclear estradiol receptors in the hypothalamus and pituitary during the time of sexual receptivity. Ovariectomized rats given a behaviorally effective intravenous injection of ^{3}H-estradiol-17β retained little labeled hormone in brain and pituitary nuclei 12 to 24 hours later. Furthermore, maximal nuclear occupation was seen between 1 and 2 hours of steroid administration. However, considerable nuclear occupation was seen 4 hours after estradiol administration (50% maximal occupation in pituitary; 30% maximal occupation in hypothalamus; McEwen et al 1975b). While this study clearly demonstrates that estradiol need not be present in the brain during the time of induced sexual receptivity, it does not define the minimal length of time that estradiol must be retained by hypothalamic nuclei to potentiate mating behavior.

The effects of estradiol on mating behavior are not restricted to changes within 16–48 hours of steroid administration. Several investigations have indicated that estradiol may affect the onset of sexual receptivity many days following administration. Whalen & Nakayama (1965) first reported such an effect in rats which had been ovariectomized for 1–4 weeks. Rats which received 3 μg of EB on days 1–3 of each week for 4 weeks came into estrogen-induced receptivity with shorter latencies than control animals (Whalen & Nakayama 1965). Gerall & Dunlap (1973) demonstrated that ovariectomized rats that received 3.3 μg EB + 0.5 mg progesterone (P) on day 1 of weeks 1–3, were more receptive to EB + P four weeks later than were control rats. Beach & Orndoff (1974) showed that rats which had been ovariectomized for 30 days and which had received 6 μg EB + 0.4 mg P 7, 14, 21, and 81 days later were more receptive to EB + P on day 88, than were animals which had received only P on days 7, 14, 21, and 81. These studies indicate that EB may potentiate behavior for many days. However, they are subject to the criticism that the route of hormonal administration (subcutaneous injections in oil) did not insure that residual estrogen was not present at the time of testing in the experimental animals.

A frequent topic of investigation is the decline in target tissue sensitivity that occurs during hormonal deprivation. Damassa & Davidson (1973) investigated the effect of 'long term' ovariectomy on estrogen-potentiated lordosis. Rats which had been ovariectomized for 6 weeks showed fewer daily lordoses during 2 weeks of daily estrogen treatment (0.1 or 0.4 μg EB/100 g body weight) than rats which had been ovariectomized for one day (Damassa & Davidson 1973).

While these results suggest that estrogen may potentiate sexual behavior over a period of many days, the molecular basis for this potentiated receptivity is unknown. We have investigated this phenomenon in rats that have been ovariectomized for a minimum of 21 days. Each animal received a 5 mm silastic implant of crystalline estradiol or of cholesterol for one week. The implants were then removed for 5 days, to permit serum levels of estradiol in experimental animals to return to control levels. At this time, animals were either sacrificed for chemical analyses, or were reimplanted with 5 mm estradiol for behavioral testing. Females were tested with an experienced male 44–46 hours after reimplantation. After testing, females received 0.5 mg P, and were retested 4 hours later. Animals pretreated with estradiol showed significantly higher lordosis quotients than animals pretreated with cholesterol, both when tested with estradiol alone and with estradiol + P. An increase in either brain or pituitary cytoplasmic estradiol levels does not appear to be responsible for the observed potentiation of mating behavior by estradiol (Parsons, unpublished results). Thus, the potentiation of lordosis does not seem to be the result of a change in steroid receptor dynamics, but may be due to the long-term regulation of enzymes and other cell constituents (see McEwen et al 1978).

Anti-Estrogens

UTERUS Estrogen antagonists have been used to study estradiol-induced events in target tissues. These synthetic compounds, such as MER-25, CI-628, clomiphene, tamoxifen, and nafoxidine, share three properties: (a) they inhibit many of the estradiol-induced events in target tissues when administered concurrently with estradiol; (b) they mimic some, but not all, of the estrogen-induced events in target tissues when administered in the absence of estradiol; and, (c) they compete with estradiol for the cytoplasmic protein receptor. Thus, they provide direct correlations between estrogen actions and estrogen receptors. For example, CI-628 competes with estradiol for binding to the uterine cytoplasmic receptor; however, CI-628 has the lower binding affinity (K_d-CI: 1.7×10^{-9}M, E_2: 1×10^{-10}M; Katzenellenbogen et al 1978). CI-628 and estradiol have similar rates of association with the receptor; however, CI-628 has the faster rate of dissociation

(Katzenellenbogen et al 1978). Nuclear receptor complexes of CI-628 and estradiol sediment at the same rate on linear sucrose density gradients (Katzenellenbogen et al 1978). If CI-628 is administered 15 minutes prior to ^{3}H-estradiol, the amount of radioactivity recovered from uterine nuclei 3 hours post steroid is a function of the dose of CI-628 (Chazal et al 1975).

Given that estrogen antagonists compete with estradiol for the cytoplasmic receptor, are there any other mechanisms (other than competitive inhibition) by which they attenuate the actions of estradiol? Clark et al (1973) reported that a single injection of nafoxidine promotes retention of the estrogen receptor by uterine nuclei for as long as 19 days. During this period, uterine wet weight was elevated, and vaginal epithelial cells were cornified. The authors conclude that nafoxidine is a partial agonist because it initiates a chain of events that is similar to those initiated by estradiol— binding to the cytoplasmic protein and translocation to the nucleus. However, a single injection of estradiol produces a two-fold increase in cytoplasmic estrogen receptors 24 hours later, while an injection of nafoxidine produces a three- to four-fold decrease at hour 24 (Clark et al 1973). The authors suggest that nafoxidine causes translocation of the receptor, which does not result in the replenishment and/or the synthesis of cytoplasmic receptors within 24 hours. CI-628 and clomiphene were later shown to have similar mechanisms of action (Clark et al 1974). From these data, Clark et al (1974) advanced the hypothesis that estrogen antagonists render the uterus refractory to a subsequent dose of estradiol, because they interfere with cytoplasmic replenishment and synthesis.

The work of Katzenellenbogen & Ferguson (1975) supports the above hypothesis. These workers demonstrated that both CI-628 and U-11,100A (a non-steroid clomiphene-like compound) depleted cytoplasmic estradiol receptors in the uterus for 24–42 hours. During this period, estradiol was ineffective in eliciting wet weight increases, or induced protein synthesis. Furthermore, induced protein sensitivity returned as the levels of cytoplasmic receptors increased (Katzenellenbogen & Ferguson 1975).

Taken as a whole, the above work suggests that estrogen antagonists not only compete directly with estradiol for cytosol receptors, but also interfere with the replenishment and/or synthesis of cytosol estradiol receptors. The former immediately limits estradiol uptake into uterine nuclei, while the latter subsequently limits steroid uptake. Although these are not the only two mechanisms that have been proposed for the action of estrogen antagonists in the uterus (for a complete discussion of these mechanisms, see Roy 1978), they are the most widely documented at the present time.

PITUITARY AND BRAIN   Estrogen antagonists have been used to investigate steroid-receptor interactions and estrogen-mediated responses both in the pituitary and in the brain. Estrogen antagonists have been shown to have 'anti-estrogenic' effects on sexual behavior and gonadotropin release (see Roy 1978, for references), and 'weakly estrogenic' effects on eating and body weight (Roy & Wade 1976 a, b, 1977).

Chazal et al (1975) reported that CI-628 affects the hypothalamus and the pituitary in the same manner as it affects the uterus—that is, when administered 15 minutes to 24 hours prior to ^{3}H-estradiol, CI-628 blocked nuclear uptake of labeled estradiol. Although the degree of antagonism by CI-628 was different in peripheral (90% in the uterus and pituitary) and central (39% in anterior hypothalamus, 22% in medial posterior hypothalamus) structures, the nature of the antagonism was similar among tissues (Chazal et al 1975).

Estradiol may either increase or decrease gonadotropin secretion, depending on dose and schedule of treatment. The effects of estrogen antagonists on gonadotropin secretion are therefore rather difficult to evaluate, particularly if one wishes to explain the effect of these compounds on serum gonadotropin levels. In spite of such difficulties, two different antagonists have been shown to inhibit the positive feedback effects of estradiol on gonadotropin secretion. Single injections of MER-25 and ICI 46472 (Tamoxifen) may delay or completely inhibit ovulation in intact animals by inhibiting the LH surge; however, this blockade of ovulation may be overridden by single injections of estradiol or of LH (Shirley et al 1968, Labhsetwar 1970). Direct implantation of MER-25 and Tamoxifen into the pituitary and/or hypothalamus, has shown that these target tissues are the sites of action of these compounds on the blockade of ovulation (Bainbridge & Labhsetwar, 1971, Billard & McDonald 1973). Namely, a different action of CI-628 on gonadotropin secretion feedback has been reported. Callantine et al (1966) found that daily treatment of ovariectomized rats with CI-628 for 14 days produced increased serum levels of FSH and LH. It is not clear whether this represents an interaction of CI-628 with positive or negative feedback. It is somewhat clearer that CI-628, given as a single dose, does block inhibitory effects of estradiol on pulsatile LH release (Krey, unpublished results).

Arai & Gorski (1968) were the first workers to report an effect of an estrogen antagonist on sexual behavior in rats. Arai & Gorski (1968) induced sexual receptivity in ovariectomized rats with subcutaneous injections of EB, followed 24 hours later by P. When CI-628 was administered by gavage within 24 hours of EB, the lordosis quotients at 52 hours post steroid were much lower than control scores (Arai & Gorski 1968). Sufficiently large doses of CI-628 have been shown to completely abolish EB-induced sexual behavior, if the two drugs are given simultaneously (Powers 1975). Furthermore, CI-628 completely abolishes estradiol-induced sexual behavior, if given up to 6 hours after the steroid; and reduces estradiol-induced behavior if given up to 18 hours post steroid, or 5 hours before testing (Whalen & Gorzalka 1973). Additionally, if MER-25 is administered to ovariectomized rats simultaneously with EB, lordosis quotients are greatly reduced over the ensuing 5 day period (Komisaruk & Beyer 1972).

Roy & Wade (1977) were the first workers to report correlations between antagonists that depress estradiol-induced sexual behavior and their capacity to inhibit the uptake of ^{3}H-estradiol in both whole homogenates and nuclear fractions of brain regions (hypothalamus, preoptic area-septum, and cerebral cortex) in female rats. In one set of experiments, MER-25, CI-628, and nafoxidine were injected intraperitoneally 2 hours before a systemic injection of ^{3}H-estradiol. CI-628 and nafoxidine completely inhibited lordosis 24 or 36 hours post steroid. These two compounds also greatly reduced the amount of ^{3}H-estradiol present in whole homogenates and nuclear fractions of the brain, 2 and 12 hours post steroid. However, MER-25 had no effect on lordosis 24 or 36 hours following estrogen treatment. Neither did MER-25 depress whole homogenate uptake in brain at 2 or 12 hours, nor nuclear uptake in the brain at 2 hours. But, when MER-25 was injected 12 hours prior to ^{3}H-estradiol, lordosis was inhibited at 24 and 36 hours. Furthermore, both whole homogenate and nuclear uptake of ^{3}H-estradiol in brain were depressed by 2 hours (Roy & Wade 1977). Additional correlations between uptake of ^{3}H-estrogen and the induction of sexual behavior following treatment with estrogen antagonists, have been provided by Walker & Feder (1977a, b) in guinea pigs. Treatment of ovariectomized guinea pigs with enclomiphene 48, 24, 12, and 0 hours prior to EB decreased the percentage of animals displaying lordosis 39 hours later. Furthermore, this regimen of enclomiphene reduced ^{3}H-estradiol uptake in whole homogenates of the hypothalamus-preoptic area 2 and 39 hours following EB. Similar pretreatment with CI-628 had no effect on lordosis, nor on whole homogenate uptake of the hypothalamus-preoptic area at 2 or 39 hours (Walker & Feder 1977a, b).

Progesterone Action

Progesterone administration to estrogen-primed females exerts a number of effects on sexual receptivity including both facilitation and inhibition. The specific behavioral response depends on the timing and the dose of progesterone relative to estrogen priming, as well as the time and duration of the behavioral testing (Feder & Marone 1977). Moreover, there are considerable species differences in the role of these progesterone actions in the estrous cycle of the gonadally intact female (Morin 1977). For example, guinea pigs enter a period of post-estrus refractoriness, which is not as easily demonstrated in rats.

Four important aspects of progesterone action in the rat are considered in this section: timing, hormonal specificity, intracranial localization, and cellular mechanism of action. With respect to timing, the normal interval for progesterone facilitation of lordosis in estrogen treated rats is 4–5 hr, following subcutaneous administration of the steroid in oil. But it has been reported that intravenous administration of progestins leads to more rapid elevations of lordosis behavior, within as little as 10–30 min for progestins with little anesthetic effect (Meyerson 1972). Even in the case of progesterone itself, a steroid with considerable anesthetic effect when given in large doses, intravenous administration leads to elevation of lordosis quotients in 1–2 hr (Kubli-Garfias & Whelan 1977).

Anesthetic progestogenic potency is not a good index of potency in facilitating lordosis (Meyerson 1967) and, in general, it is difficult to specify an exact structural model of a steroid that facilitates lordosis or LH release in the rat. Both deoxycorticosterone and progesterone facilitate lordosis, while aldosterone and corticosterone do not (Gorzalka & Whalen 1977). The 5α progestins, including those with further reduction of the 3 keto group, facilitate lordosis (Zucker 1967, Meyerson 1972, Whalen & Gorzalka 1972, Kubli-Garfias & Whalen 1977) and LH release (Brown-Grant 1974), but less well than progesterone. There are, however, many disagreements in the cited references regarding the exact order of potencies of the various 5α progestins. Another progesterone metabolite, the 20α hydroxy progesterone, facilitates lordosis (Zucker 1967, Meyerson 1972, Whalen & Gorzalka 1972, Kubli-Garfias & Whalen 1977) and LH release (Swerdloff et al 1972, Brown-Grant 1974), but less well than progesterone. Thus, there does not appear to be an obligatory role of progesterone metabolism in either of these two effects in the rat. Finally, it should be pointed out that large doses of estradiol will effectively facilitate lordosis in estrogen primed ovariectomized rats with the same time course as progesterone (Kow & Pfaff 1975). The possible significance of this observation is considered below.

Intracranial implants of progesterone presumably could mimic any of the effects observed following systemic administration of the hormone; both inhibitory and facilitative effects of progesterone have been demonstrated following implantation of the steroid into hypothalamic and midbrain loci (see Table 12). In guinea pigs, in which both inhibitory and facilitative effects are readily demonstrable, the regions mediating the two responses appear to be independent, inhibition following midbrain implants and facilitation following hypothalamic implants (Morin & Feder 1974a, b). Similar inhibitory effects occur after mesencephalic progesterone implants in hamsters (DeBold, Martin & Whalen 1976).

Table 12 Effects of progesterone implants on feminine sexual behavior in several species[a]

Species	Reference	Type of implant	Testing paradigm (intervals post P-implantation)	Site(s)	Main effect
Guinea Pigs					
(ovx, EB-primed)	Morin & Feder 1974b	27g, unilateral, P, 17 α OHP	hourly × 8	zona compacta of mid-brain substantia nigra	inhibition
	Morin & Feder 1974c	27g, unilateral, P, 17 α OHP	hourly × 8	basal hypothalamus (VMH, ARC, premam-millary)	facilitation
Hamsters					
(ovx, EB-primed)	DeBold, Martin & Whalen 1976	27g, unilateral, P	30, 60, 90, 120 & 240 min	mesencephalon in region of IP;	inhibition
				basal hypothalamus	no effect
Rats					
(ovx, EB-primed)	Ross et al 1971	27g, unilateral, P	15, 30, 60, 90 120 min	medial & lateral POA, AH, VMH, ARC;	no effect
				MRF, lateral to IP	facilitation
(ovx, EB- and P-primed)	Powers 1972	27g, bilateral, P	6 & 12 hr	MBH	facilitation
				MRF	no effect
ovx, intracranial EB-primed	Yanase & Gorski 1976	27g, unilateral, P	30–90 min & 2–3 hr	MRF, CPU	facilitation
ovx, EB-primed	Luttge & Hughes 1976	27g, unilateral, P	30, 60, 180 min	IP, VTA;	facilitation
				pons, RF, POA, lateral mesencephalon, mid and ant. hypothalamus, frontal cortex	no effect

[a] Abbreviations are ARC, arcuate nucleus; IP, interpeduncular nucleus; NBH, medial-basal hypothalamus; VTA, ventral tegmental area; P, progesterone; 17αOHP — 17αhydroxyprogesterone. Others as in previous tables.

In rats, the site of the facilitative actions of progesterone remains unclear. Both hypothalamic and ventral midbrain sites have been suggested and the discrepancy between these findings may be due to the specific experimental procedures employed. For example, the time of testing may be especially important. Powers (1972) tested at 6 hr post implantation whereas others have tested at earlier times. Since the actions of progesterone upon the brain may be extremely rapid even when the hormone is administered systemically, early testing times may be more appropriate for evaluating the effects of progesterone implants (Luttge & Hughes 1976). The method of hormone priming prior to hormone implantation may also contribute to the variable results. Specifically, Powers (1972) gave all his animals a subthreshold dose of progesterone, which clearly could have influenced the effectiveness of particular implant sites.

There is some indication that effective progesterone implants correspond with regions in which uptake of the hormone may occur such as the ventral mesencephalon (see Luttge & Hughes 1976). But not all sites suspected of accumulating progesterone are unequivocally sites of action in affecting behavior (e.g., basal hypothalamus, Sar & Stumpf 1973).

Initial attempts to identify a progesterone receptor in the brain and pituitary using ^{3}H-progesterone led to largely negative results. Large accumulations of labeled steroid were observed in vivo by these tissues, but there was no evidence either for selective cell nuclear binding or for saturable binding components in any part of the brain (see Feder & Marrone 1977, McEwen 1978a for review). An exception was the autoradiographic study of Sar & Stumpf (1973), which showed some cells in the basal hypothalamus and pituitary of estrogen-primed guinea pigs that accumulated radioactivity injected as ^{3}H-progesterone. The introduction of a synthetic progestin, ^{3}H R 5020 (see Table 1) permitted the identification of a cytosol-binding component with the properties expected of a progestin receptor including a high degree of stereospecificity for progestins (MacLusky et al 1978), and the ability to be induced by estrogen in the pituitary, hypothalamus, and preoptic area as well as in the uterus of the rat (Kato & Onouchi 1977, MacLusky & McEwen 1978, Moguilewsky & Raynaud 1978). In the rat this induction of receptor occurs against a background level of receptor that exists in virtually all regions of the brain and that can be demonstrated in adrenalectomized, gonadectomized animals; both estrogen-inducible and noninducible receptors show the same physicochemical properties and steroid specificity (MacLusky et al 1978). In contrast to the rat, the bonnet monkey (*M. radiata*) lacks progestin binding outside of the hypothalamus, but, like the rat, shows a strong estrogen induction of progestin receptors in the uterus, pituitary, and hypothalamus. This difference may be related to species differences in the neural control of ovulation.

The progestin receptors of uterus, pituitary, and hypothalamus appear to be similar to each other in physical properties and specificity toward various steroids and display a markedly higher affinity toward R 5020. R 5020 is also many times more potent than progesterone in promoting lordosis behavior in estrogen-primed rats and in inhibiting lordosis when administered concurrently with estradiol (Blaustein & Wade 1978). The progestin receptor also shows a moderate affinity for estradiol, which might explain the ability of this steroid to substitute for progesterone (see above). Taken together, the binding properties of the progestin receptor and its inducibility by estrogen suggest that the receptor may participate in the behavioral and neuroendocrine effects of progestins. Currently lacking are definitive mapping studies of the localization of progestin receptors in the preoptic area and hypothalamus and thorough structure-activity studies with the various steroids that have binding affinity for the receptor so as to ascertain the relative potencies in behavioral and neuroendocrine events.

Based on the available information we cannot be sure that the progestin receptor participates in any of these events, and even how it works. It has been shown that actinomycin D in the arcuate-ventromedial hypothalamus blocks progesterone action in facilitating LH release (Jackson 1975) and that cycloheximide attenuates progesterone-dependent post estrus refractoriness in guinea pigs (Wallen et al 1972). Despite this hint of genomic involvement, the cell nuclear retention in vivo of H R 5020 in the brain and pituitary is very limited, though detectable (Blaustein & Wade 1978), and the latency for progesterone to exert at least one of its effects, namely, the facilitation of lordosis, can be so short (10–30 min, see Meyerson 1972) as to suggest the possibility of a more direct action of progestins, perhaps at the synaptic level.

DISCUSSION OF SOME KEY TERMS AND CONCEPTS

Several key terms and concepts have emerged from the study of estrogen-mediated responses in the uterus and in the brain. These terms and concepts have primarily referred to the temporal properties of either steroid-receptor interactions, or of steroid-induced events. A brief discussion of these ideas may serve to clarify the mechanism(s) of steroid action in target tissues.

Following a systemic injection of ^{3}H-estradiol, peak levels of radioactivity appear in uterine nuclei within one hour, and decline to less than one third of these values by 6 hrs (Anderson et al 1975). However, the estradiol that is present in nuclear fractions at 6 hr appears to be of physiological significance. Clark & Peck (1976) hypothesized that "long term retention" (4–6 hrs) of estradiol by specific nuclear receptor sites is a prerequisite for uterine growth. In other words, there is a minimal temporal requirement for nuclear estradiol retention, which is necessary to induce late uterotrophic responses. This concept has been supported by the work of two laboratories. Estriol is neither retained for 6 hours following steroid administration, nor is it effective in eliciting late uterotrophic responses (Anderson et al 1975). However, ethinyl estriol, a synthetic derivative, shows a two-fold increase in nuclear retention at 24 hours, and is effective in stimulating late uterotrophic responses (Lan & Katzenellenbogen 1976).

McEwen et al (1976) investigated the time course of estradiol retention in hypothalamic and pituitary nuclei following a systemic injection of ^{3}H-estradiol. Peak levels of radioactivity appear in brain and pituitary nuclei within 2 hrs; and decline to less than one tenth of these values by 12 hrs. However, no facilitation of sexual behavior is seen within 16 hrs of an intravenous estradiol injection (Green et al 1970). McEwen et al (1975b) advanced the notion that estradiol need not be present in brain nuclei during the time of induced sexual receptivity. Thus, while there is a minimal temporal requirement for nuclear estradiol retention to induce late uterotrophic responses, it is possible for maximal nuclear estradiol retention to temporally precede changes in mating behavior.

It should be noted that the above two concepts have often been referred to as "maintenance" versus "triggering" mechanisms of estradiol action. It has been suggested that a maintenance mechanism depends upon the continued presence of nuclear bound estradiol to elicit the physiological response, while a triggering mechanism depends only upon the initial nuclear uptake of estradiol to promote the observed change (for references, see Moreines & Powers 1977). However, the above two mechanisms are not mutually exclusive. Taken together, they suggest that: (a) in the uterus, there is a minimal temporal requirement for nuclear retention of estradiol, which is necessary to induce late uterotrophic responses. The minimal time period(s) need not be identical for all estrogen-mediated responses; a short exposure to estradiol may be necessary to stimulate the synthesis of "induced protein," while longer exposures may be necessary to stimulate uterine weight; (b) in the brain, there is a period of maximal nuclear retention of estradiol, which is sufficient to induce sexual receptivity at a later time. Thus, the period of maximal nuclear retention is not necessarily conterminous with the facilitation of lordosis—the latency between estradiol retention and the appearance of late responses may be attributed to a change in genomic activity, and subsequent biosynthetic events. Therefore, these two concepts address the periods of time of nuclear estradiol retention, which are *necessary* to induce late uterotrophic responses, and which are *sufficient* to induce sexual receptivity.

Several laboratories have reported that estradiol may affect the onset of sexual receptivity many days following administration. The molecular basis for this potentiated receptivity is unknown. Although a variety of mechanisms has been suggested to explain the effects of estradiol on mating behavior many days following steroid administration, none of these mechanisms has received experimental verification. Work in our laboratory indicates that an increase in either brain or pituitary estradiol receptor levels does not appear to be responsible for the 'long-term' potentiation of mating behavior by estradiol. Likewise, an increase in brain progestin receptors does not appear to be responsible for "long-term' facilitation of lordosis by estradiol. Although estradiol induces progestin receptor synthesis in the brain (MacLusky et al 1978), these receptors fall to control levels by 120 hours following steroid administration (Parsons and MacLusky, unpublished results). Therefore, the long-term potentiation of lordosis after estrogen treatment cannot be the result of a change in steroid receptor dynamics, but may be due to a change in genomic expression over time. In turn, the relationship of the cellular changes following long-term estradiol treatment to those following the final estrogen injection (24–48 hrs before behavioral test) remains to be determined.

CONCLUSION AND PROSPECTUS

As chemically defined substances which are secreted within the body and which enter the brain from the blood, steroid hormones are unique tools for investigating brain function from the molecular level up to the level of behavior. Few other tools in neuroscience permit such breadth of analysis and at the same time offer the possibility of an in-depth analysis of mechanisms at the cellular and subcellular level. This review has summarized recent progress in our understanding of how steroid hormones affect functions of the mature brain.

Much progress has occurred in the past 5 years, especially in the study of estrogen action, and many more important revelations appear to be just over the horizon. Of particular interest will be the further elucidation of the neuroanatomical networks and circuits responsible for lordosis behavior, and furthur analysis of the chemical changes brought about by estradiol and progesterone in the limited area of the basal hypothalamus within which local estrogen stimulation is sufficient to activate lordosis behavior.

The activation of lordosis behavior in the rat is a reversible event triggered by estradiol and it is representative of a number of reversible effects of steroid hormones on the mature brain (e.g. male sexual behavior, aggressive behavior). Estrogens and testosterone are also capable of exerting permanent effects on the brain during a critical period of early development (see Plapinger & McEwen 1978, for review). These so-called organizational effects include the permanent attenuation of the ability of the rat to display lordosis behavior to activation by estradiol plus progesterone in adulthood. The cellular mechanism of these organizational effects may be one involving altered neuronal growth and formation of synaptic contacts (see Raisman & Field 1973) and it appears to involve, at least in the rat brain, an important participation of estrogen receptors like those found in the adult brain (see McEwen et al 1977). It is hoped that future experiments will address themselves to the important questions of what factors of cellular differentiation enable the same hormone (i.e. estradiol) to exert permanent effects on the brain at one stage of development and reversible, activational effects on the brain at another.

ACKNOWLEDGMENTS

Research in the authors' laboratories is supported by research grants NS 07080 to Dr. McEwen and HD 05751 to Dr. Pfaff and by an institutional grant for research in reproductive biology, RF 70095, from the Rockefeller Foundation. Dr. Davis is a recipient of a USPHS Postdoctoral Fellowship MH 05781 and Mr. Parsons is a graduate fellow of The Rockefeller University. The authors wish to thank Ms. Gabriele Zummer for editorial assistance and typing of the manuscript.

Literature Cited

Anderson, N. S. III, Fanestil, D. D. 1976. Corticoid receptors in rat brain: evidence for an aldosterone receptor. *Endocrinology* 98:676–84

Anderson, J. N., Peck, E. J., Clark, J. H. 1975. Estrogen-induced uterine responses and growth: relationship to receptor estrogen binding by uterine nuclei. *Endocrinology* 96:160–67

Anton-Tay, F., Anton, S. M., Wurtman, R. J. 1970. Mechanism of changes in brain norepinephrine metabolism after ovariectomy. *Neuroendocrinology* 6:265–73

Arai, Y. 1973. Sexual differentiation and development of the hypothalamus and steroid-induced sterility. In *Neuroendocrine Control,* ed. K. Yagi, S. Yoskida, pp. 27–55. New York: Wiley. 394 pp.

Arai, Y., Gorski, R. A. 1968. Effect of an anti-estrogen on steroid induced sexual receptivity in ovariectomized rats. *Physiol. Behav.* 3:351–53

Araki, S., Ferin, M., Zimmerman, E. A., Vande Wiele, R. L. 1975. Ovarian modulation of immunoreactive gonadotropins-releasing-hormone (GnRH) in the rat brain: evidence for a differential effect on the anterior and mid-hypothalamus. *Endocrinology* 96:644–50

Arnold, A., Nottebohm, F., Pfaff, D. W. 1976. Hormone-concentrating cells in vocal control and other areas of the brain of the zebra finch (Poephila guttata). *J. Comp. Neurol.* 165:487–512

Azmitia, E. C. Jr., Algeri, S., and Costa, E. 1970a. Turnover rate of *in vivo* conversion of tryptophan into serotonin in brain areas of adrenalectomized rats. *Science* 169:201–3

Azmitia, E. C. Jr., Hess, P., Reis, D. 1970b. Tryptophan hydroxylase changes in midbrain of the rat after chronic morphine administration. *Life Sci.* 9:633–37

Azmitia, E. C. Jr., McEwen, B. S. 1974. Adrenocortical influence on rat brain tryptophan hydroxylase activity. *Brain Res.* 78:291–302

Azmitia, E. C. Jr., McEwen, B. S. 1976. Early response of rat brain tryptophan hydroxylase activity to cycloheximide, puromycin and corticosterone. *J. Neurochem.* 27:773–78

Bainbridge, J. G., Labhsetwar, A. P. 1971. The role of estrogens in spontaneous ovulation: location of the site of action of positive feedback of estrogen by intracranial implantation of the antiestrogen I.C.I. 46474. *J. Endocrinol.* 50:321–27

Ball, P., Haupt, M., Knuppen, R. 1978. Comparative studies on the metabolism of oestradiol in the brain, the pituitary and the liver of the rat. *Acta Endocrinol. Copenhagen* 87:1–11

Ball, P., Knuppen, R., Haupt, M., Breuer, H. 1972. Interactions between estrogens and catecholamines, III. Studies on the methylation of catecholestrogens, catecholamines and other catechols by the catechol-O-methyltransferase of human liver. *J. Clin. Endocrinol. Metab.* 34:736–46

Bapna, J., Neff, N. H., Costa, E. 1971. A method for studying norepinephrine and serotonin metabolism in small regions of rat brain: effect of ovariectomy on amine metabolism in anterior and posterior hypothalamus. *Endocrinology* 89:1345–49

Barfield, R. J. 1969. Activation of copulatory behavior by androgen implanted into the preoptic area of the male fowl. *Horm. Behav.* 1:37–52

Barfield, R. J. 1971. Activation of sexual and aggressive behavior by androgen implanted into the male ring dove brain. *Endocrinology* 89:1470–76

Barfield, R. J., Chen, J. J. 1977. Activation of estrous behavior in ovariectomized rats by intracerebral implants of estradiol benzoate. *Endocrinology* 101:1716–25

Barfield, R., Ronay, G., Pfaff, D. W. 1978. Autoradiographic localization of androgen-concentrating cells in the chicken brain. *Neuroendocrinology* In press

Barker, K. L., Warren, J. D. 1966. Template capacity of uterine chromatin: control by estradiol. *Proc. Natl. Acad. Sci. US* 56:1298–1302

Barnea, A., Gorski, J. 1970. Estrogen-induced protein. Time course of synthesis. *Biochemistry* 9:1899–1904

Beach, F. 1948. *Hormones and Behavior,* New York: Hoeber

Beach, F. A., Orndoff, R. K. 1974. Variation in the responsiveness of female rates to ovarian hormones as a function of preceding hormonal deprivation. *Horm. Behav.* 5:202–5

Berthold, A. A. 1849. Transplantation der Hoden. *Arch. Anat. Physiol. Wiss. Med.* 16:42–46

Billard, R., McDonald, P. G. 1973. Inhibition of ovulation in the rat by intrahypothalamic implants of an antiestrogen. *J. Endocrinol.* 56:585–90

Billing, R. J., Barbiroli, B., Smellie, R. 1969a. Mode of action of estradiol: I. The transport of RNA precursors into the uterus. *Biochim. Biophys. Acta* 190:52–59

Billing, R. J., Barbiroli, B., Smellie, R. 1969b. Mode of action of estradiol: II. The synthesis of RNA. *Biochim. Biophys. Acta* 190:60–65

Blaustein, J. D., Wade, G. N. 1978. Progestin binding by brain and pituitary cell nuclei and female rat sexual behavior. *Brain Res.* 140:360–67

Bolt, H. M., Kappus, H. 1976. Interaction by 2-hydroxyestrogens with enzymes of drug metabolism. *J. Steroid Biochem.* 7:311–13

Bronson, F. H., Desjardins, C. 1971. Steroid hormones and aggressive behavior in mammals. In *The Physiology of Aggression and Defeat*, ed. B. E. Eleftheriou, J. P. Scott. New York: Plenum. 312 pp.

Brookhart, J. M., Dey, F. L. 1941. Reduction of sexual behavior in male guinea pigs by hypothalamic lesions. *Amer. J. Physiol.* 133:551–54

Brown-Grant, K. 1974. Steroid hormone administration and gonadotrophin secretion in the gonadectomized rat. *J. Endocrinol.* 62:319–32

Bueno, J., Pfaff, D. W. 1976. Single unit recording in hypothalamus and preoptic area of estrogen-treated and untreated ovariectomized female rats. *Brain Res.* 101:67–78

Callantine, M. R., Humphrey, R. R., Lee, S. L., Windsor, B. L., Schottin, N. H., O'Brien, O. P. 1966. Action of an estrogen antagonist on reproductive mechanisms in the rat. *Endocrinology* 79:153–67

Callard, G. V., Petro, Z., Ryan, K. J. 1977. Identification of aromatase in the reptilian brain. *Endocrinology* 100:1214–18

Cardinali, D. P., Gomez, E. 1977. Changes in hypothalamic noradrenaline, dopamine and serotonin uptake after estradiol administration to rats. *J. Endocrinol.* 73:181–82

Chambers, W. F., Howe, G. 1968. A study of estrogen-sensitive hypothalamic centers using a technique for rapid application and removal of estradiol. *Proc. Soc. Exp. Biol. Med.* 128:292–94

Chazal, G., Faudon, M., Gogan, F., Rotsztejn, W. 1975. Effects of two estradiol antagonists upon the estradiol uptake in the rat brain and peripheral tissues. *Brain Res.* 89:245–54

Christensen, L. W., Clemens, L. G. 1974. Intrahypothalamic implants of testosterone or estradiol and resumption of masculine sexual behavior in long-term castrated male rats. *Endocrinology* 95:984–90

Christensen, L. W., Clemens, L. G. 1975. Blockade of testosterone-induced behavior in the male rat with intracranial application of the aromatization inhibitor, androst-1,4,6-triene-3,17-dione. *Endocrinology* 97:1545–51

Chytil, F., Toft, D. 1972. Corticoid binding component in brain. *J. Neurochem.* 19:2877–80

Ciaccio, L. A., Lisk, R. D. 1973/74. Central control of estrous behavior in the female golden hamster. *Neuroendocrinology* 13:21–28

Clark, J. H., Anderson, J. N., Peck, E. J. 1973. Estrogen receptor-antiestrogen complex: atypical binding by uterine nuclei and effects on uterine growth. *Steroids* 22:707–18

Clark, J. H., Peck, E. J. 1976. Nuclear retention of receptor-estrogen complexes and nuclear acceptor sites. *Nature* 260:635–37

Clark, J. H., Peck, E. J., Anderson, J. N. 1974. Estrogen receptors and antagonism of steroid hormone action. *Nature* 251:446–48

Coppola, J. A. 1969. Turnover of hypothalamic catecholamines during various states of gonodotropin secretion. *Neuroendocrinology* 5:75–80

Curzon, G., Green, A. R. 1971. Regional and subcellular changes in the concentration of 5-hydroxytryptamine and 5-hydroxyindoleacetic acid in the rat brain caused by hydrocortisone, DL-α methyltryptophan, L-kynurenine and immobilization. *J. Pharm. Pharmacol.* 43:39–52

Damassa, D., Davidson, J. M. 1973. Effects of ovariectomy and constant light on responsiveness to estrogen in the rat. *Horm. Behav.* 4:269–79

Davidson, J. M. 1966a. Activation of the male rat's sexual behavior by intracerebral implantation of androgen. *Endocrinology* 79:783–94

Davidson, J. M. 1966b. Characteristics of sex behaviour in male rats following castration. *Anim. Behav.* 14:266–72

Davies, I. J., Naftolin, F., Ryan, K. J., Fishman, J., Siu, J. 1975. The affinity of catechol estrogens for estrogen receptors in the pituitary and anterior hypothalamus of the rat. *Endocrinology* 97:554–57

DeAngelo, A. B., Gorski, J. 1970. Role of RNA synthesis in the estrogen induction of a specific uterine protein. *Proc. Natl. Acad. Sci. USA* 66:693–700

DeBold, J. F., Martin, J. V., Whalen, R. E. 1976. The excitation and inhibition of sexual receptivity in female hamsters by progesterone: time and dose relationships, neural localization and mechanisms of action. *Endocrinology* 99: 1519–27

Debeljuk, L., Rettori, V., Rozados, R. V., Velez, C. V. 1975. Effect of actinomycin D and estradiol on the response to LH-releasing hormone in neonatally androgenized female rats. *Proc. Soc. Exp. Biol. Med.* 150:229–31

DeKloet, E. R., McEwen, B. S. 1976a. A putative glucocorticoid receptor and a transcortin-like macromolecule in pituitary cytosol. *Biochim. Biophys. Acta* 421:115–23

DeKloet, E. R., McEwen, B. S. 1976b. Differences between cytosol receptor complexes with corticosterone and dexamethasone in hippocampal tissue from rat brain. *Biochim. Biophys. Acta* 421: 124–32

DeKloet, E. R., Wallach, G., McEwen, B. S. 1975. Differences in corticosterone and dexamethasone binding to rat brain and pituitary. *Endocrinology* 96:598–609

DeLean, A., Goron, M., Kelly, P. A., Labrie, F. 1977. Changes of pituitary thyrotropin releasing hormone (TRH) receptor level and prolactin response to TRH during the rat estrous cycle. *Endocrinology* 100:1505–10

Denef, C., Magnus, C., McEwen, B. S. 1974. Sex-dependent changes in pituitary 5α dihydrotestosterone and 3α androstanediol formation during postnatal development and puberty in the rat. *Endocrinology* 94:1265–74

Denef, C., Magnus, C., McEwen, B. S. 1973. Sex differences and hormonal control of testosterone metabolism in rat pituitary and brain. *J. Endocrinol.* 59:605–21

DeVellis, J., Inglish, D. 1968. Hormonal control of glycerol phosphate dehydrogenase in the rat brain. *J. Neurochem.* 15:1061–70

DeVellis, J., Inglish, D. 1973. Age-dependent changes in the regulation of glycerolphosphate dehydrogenase in the rat brain and in a glial cell line. *Prog. Brain Res.* 40:321–30

DeVellis, J., McEwen, B. S., Cole, R., Inglish, D. 1974. Relations between glucocorticoid nuclear binding, cytosol receptor activity and enzyme induction in a rat glial cell line. *J. Steroid Biochem.* 5:392–93

Diez, J. A., Sze, P. Y., Ginsburg, B. E. 1977. Effects of hydrocortisone and electric footshock on mouse brain tyrosine hydroxylase activity and tyrosine levels. *Neurochem. Res.* 2:161–70

Dörner, G., Döcke, F., Hinz, G. 1969. Homo- and hypersexuality in rats with hypothalamic lesions. *Neuroendocrinology* 4:20–24

Dörner, G., Döcke, F., Moustafa, S. 1968a. Homosexuality in female rats following testosterone implantation in the anterior hypothalamus. *J. Reprod. Fertil.* 17:173–75

Dörner, G., Döcke, F., Moustafa, S. 1968b. Differential localization of a male and a female hypothalamic mating centre. *J. Reprod. Fertil.* 17:583–86

Drouin, J., LaGace, L., Labrie, F. 1976. Estradiol-induced increase of the LH responsiveness to LH releasing hormone (LHRH) in rat anterior pituitary cells in culture. *Endocrinology* 99:1477–81

Eilon, G., Gorski, J. 1972. Nuclear estrogen receptors: accumulation and retention as influenced by exogenous and endogenous estrogens. *Fed. Proc.* 31:245

Feder, H. H., Marrone, B. L. 1977. Progesterone: its role in the central nervous system as a facilitator and inhibitor of sexual behavior and gonadotropin release. *Ann. NY Acad. Sci.* 286:331–52

Fishman, J. 1976. The catechol estrogens. *Neuroendocrinology* 22:363–74

Fishman, J., Norton, B. 1975. Catechol estrogen formation in the central nervous system of the rat. *Endocrinology* 96: 1054–59

Fregly, M. J., Waters, I. W. 1966. Effect of mineralocorticoids on spontaneous sodium chloride appetite of adrenalectomized rats. *Physiol. Behav.* 1:65–74

Fuxe, K., Corrodi, H., Hökfelt, T., Jonsson, G. 1970. Central monoamine neurons and pituitary adrenal activity. *Prog. Brain Res.* 32:42–56

Gal, E. M., Heater, R. D., Millard, S. A. 1968. Studies on the metabolism of 5-hydroxytryptamine (serotonin) VI. hydroxylation and amines in cold-stressed reserpinized rats. *Proc. Soc. Exp. Biol. Med.* 128:412–15

Gerall, A. A., Dunlap, J. L. 1973. The effect of experience and hormones on the initial receptivity in female and male rats. *Physiol. Behav.* 10:851–54

Gerlach, J., McEwen, B., Pfaff, D., Moskovitz, S., Ferin, M., Carmel, P., Zimmerman, E. 1976. Cells in regions of rhesus monkey brain and pituitary retain radioactive estradiol, corticosterone and cortisol differentially. *Brain Res.* 103:603–12

Gethmann, U., Knuppen, R. 1976. Effect of 2-hydroxyestrone on lutropin (LH) and follitropin (FSH) secretion in the ovariectomized primed rat. *Z. Physiol. Chem.* 357:1011–13

Ginsburg, M., MacLusky, N., Morris, I. D., Thomas, P. J. 1977. The specificity of oestrogen receptor in brain, pituitary and uterus. *Br. J. Pharmacol.* 59:397–402

Glasser, S. R., Chytil, F., Spelsberg, T. C. 1972. Early effects of estradiol 17-β on the chromatin and activity of deoxyribonucleic acid-dependent ribonuclease acid polymerases in the rat uterus. *Biochem. J.* 130:947–57

Glasser, S. R., Spelsberg, T. C. 1973. Differential modulation of estrogen induced activity of uterine nuclear RNA polymerases (I and II) by cycloheximide and actinomycin D. *Meet. Am. Endocrinol. Soc., 55th* Abstr. No. 80, p. A-88

Goodman, R. L. 1978. The site of the positive feedback action of estradiol in the rat. *Endocrinology* 102:151–59

Gorski, J. 1964. Early estrogen effects on the activity of uterine ribonucleic acid polymerase. *J. Biol. Chem.* 239:889–92

Gorzalka, B. B., Whalen, R. E. 1977. The effects of progestins, mineralocorticoids, glucocorticoids, and steroid solubility on the induction of sexual receptivity in rats. *Horm. Behav.* 8:94–99

Grant, L. D., Stumpf, W. E. 1973. Localization of ³H-estradiol and catecholamines in identical neurons in the hypothalamus. *J. Histochem. Cytochem.* 21:404

Grant, L. D., Stumpf, W. E. 1975. Hormone uptake sites in relation to CNS biogenic amine systems. In *Anatomical Neuroendocrinology*, ed. W. E. Stumpf, L. D. Grant, p. 445. Basel: Karger. 472 pp.

Green, R., Luttge, W. G., Whalen, R. E. 1970. Induction of receptivity in ovariectomized female rats by a single intravenous injection of estradiol 17-β. *Physiol. Behav.* 5:137–41

Grosser, B. I., Stevens, W., Reed, D. J. 1973. Properties of corticosterone-binding macromolecules from rat brain cytosol. *Brain Res.* 57:387–95

Gudelsky, G. A., Simpkins, J., Mueller, G. P., Meites, J., Moore, K. E. 1976. Selective actions of prolactin on catecholamine turnover in the hypothalamus and on serum LH and FSH. *Neuroendocrinology* 22:206–15

Hamilton, T. H. 1963. Isotopic studies on estrogen-induced accelerations of ribonucleic acid and protein synthesis. *Proc. Natl. Acad. Sci. USA* 49:373–79

Hanbauer, I., Guidotti, A., Costa, E. 1975a. Dexamethasone induces tyrosine hy-

droxylase in sympathetic ganglia but not in adrenal medulla. *Brain Res.* 85:527–31

Hanbauer, I., Lovenberg, W., Guidotti, A., Costa, E. 1975b. Role of cholinergic and glucocorticosteroid receptors in the tyrosine hydroxylase induction elicited by reserpine in superior cervical ganglion. *Brain Res.* 96:197–200

Harris, G. W., Michael, R. P., Scott, P. P. 1958. In *Neurological Basis of Behavior*, ed. G. E. W. Wolstenholme, C. M. O'Connor. London: Churchill. 400 pp.

Hart, B. L. 1974. Medial preoptic-anterior hypothalamic area and sociosexual behavior of male dogs. A comparative neuropsychological analysis. *J. Comp. Physiol. Psychol.* 86:328–49

Heil, H., Meltzer, V., Kuhl, H., Abraham, R., Taubert, H. D. 1971. Stimulation of L-cystine-aminopeptidase activity by hormonal steroids and steroid-analogs in the hypothalamus and other tissues of the female rat. *Fertil. Steril.* 22:181–87

Heimer, L., Larsson, K. 1966/67. Impairment of mating behavior in male rats following lesions in the preoptic-anterior hypothalamic continuum. *Brain Res.* 3:248–63

Hellstrom, S., Koslow, S. H. 1976. Effects of glucocorticoid treatment on catecholamine content and ultrastructure of adult rat carotid body. *Brain Res.* 102:245–54

Heritage, A. S., Grant, L. D., Stumpf, W. E. 1977. ³H-Estradiol in catecholamine neurons of rat brain stem: combined localization by autoradiography and formaldehyde-induced fluorescence. *J. Comp. Neurol.* 176:607–30

Hillier, J., Hillier, J. G., Redfern, P. H. 1975. Liver tryptophan pyrrolase activity and metabolism of brain 5-HT in rat. *Nature* 253:566–67

Hsueh, A. J., Peck, E. J., Clark, J. H. 1976. Control of uterine estrogen receptor levels by progesterone. *Endocrinology* 98:438–44

Hutchison, J. B. 1970. Influence of gonadal hormones on the hypothalamic integration of courtship behavior in the Barbary Dove. *J. Reprod. Fertil. Suppl.* 11:15–41

Iuvone, P. M., Morasco, J., Dunn, A. J. 1977. Effect of corticosterone on the synthesis of ³H-catecholamines in the brains of CD-1 mice. *Brain Res.* 120:571–76

Jackson, G. L. 1972. Effect of actinomycin D on estrogen-induced release of lutenizing hormone in ovariectomized rats. *Endocrinology* 91:1284–87

Jackson, G. L. 1973. Time interval between injection of estradiol benzoate and LH release in the rat and effect of actinomycin D or cycloheximide. *Endocrinology* 93:887–91

Jackson, G. L. 1975. Blockage of progesterone-induced release of LH by intrabrain implants of actinomycin D. *Neuroendocrinology* 17:236–44

Jacobi, J., Lloyd, H. M., Mearnes, J. D. 1977. Onset of estrogen-induced prolactin secretion and DNA synthesis by the rat pituitary gland. *J. Endocrinol.* 72: 35–39

Javoy, F., Glowinski, J., Kordon, C. 1968. Effects of adrenalectomy on the turnover of norepinephrine in the rat brain. *Eur. J. Pharmacol.* 4:103–4

Jensen, E. V., Suzuki, T., Numata, M., Smith, S., DeSombre, E. R. 1969. Estrogen binding substances of target tissues. *Steroids* 13:417–27

Jiminez, A. E., Voogt, J. L., Carr, L. A. 1977. Plasma luteinizing hormone and prolactin levels and hypothalamic catecholamine synthesis in steroid-treated ovariectomized rats. *Neuroendocrinology* 23:341–51

Johnston, P., Davidson, J. M. 1972. Intracerebral androgens and sexual behavior in the male rat. *Horm. Behav.* 3:345–57

Jones, M. T., Hillhouse, E. W., Burden, J. L. 1977. Dynamics and mechanics of corticosteroid feedback at the hypothalamus and anterior pituitary gland. *J. Endocrinol.* 73:405–17

Kalra, S. P. 1976. Tissue levels of luteinizing hormone-releasing hormone in the preoptic area and hypothalamus, and serum concentrations of gonadotropins following anterior hypothalamic deafferentation and estrogen treatment of the female rat. *Endocrinology* 99:101–7

Karavolas, H. J., Nuti, K. M. 1976. Progesterone metabolism by neuroendocrine tissues. In *Subcellular Mechanisms in Reproductive Neuroendocrinology*, ed. F. Naftolin, K. J. Ryan, J. Davies, pp. 305–326. Amsterdam: Elsevier. 529 pp.

Kato, J., Onouchi, T. 1973a. 5-Alpha-dihydrotestosterone "receptor" in the rat hypothalamus. *Endocrinol. Jpn.* 20: 429–32

Kato, J., Onouchi, T. 1973b. 5-Alpha-dihydrotestosterone "receptor" in the rat hypophysis. *Endocrinol. Jpn.* 20: 641–44

Kato, J., Onouchi, T. 1977. Specific progesterone receptors in the hypothalamus and anterior hypophysis of the rat. *Endocrinology* 101:920–28

Katzenellenbogen, B. S., Ferguson, E. R. 1975. Antiestrogen action in the uterus: biological ineffectiveness of nuclear bound estradiol after antiestrogen. *Endocrinology* 97:1–12

Katzenellenbogen, B. S., Gorski, J. 1975. Estrogen actions on synthesis of macromolecules in target cells. In *Biochemical Actions of Hormones*, ed. G. Litwack, pp. 187–243. New York: Academic. 415 pp.

Katzenellenbogen, B. S., Katzenellenbogen, J. A., Ferguson, E. R., Krautchammer, N. 1978. Antiestrogen interaction with uterine estrogen receptors: studies with a radiolabeled antiestrogen (CI 628). *J. Biol. Chem.* 253:697–707

Kaye, A. M., Sheratzky, D., Lindner, H. R. 1972. Kinetics of DNA synthesis in immature rat uterus: age dependence and estradiol stimulation. *Biochim. Biophys. Acta* 261:475–86

Kelley, D. B., Lieberburg, I., McEwen, B. S., Pfaff, D. W. 1978. Autoradiographic and biochemical studies of steroid hormone-concentrating cells in the brain of *rana pipiens. Brain Res.* 140:287–305

Kelly, M. J., Moss, R. L., Dudley, C. A., Fawcett, C. P. 1977. The specificity of the response of preoptic-septal area neurons to estrogen: 17α-estradiol versus 17β-estradiol and the response of extrahypothalamic neurons. *Exp. Brain Res.* 30:43–52

Kierniesky, N. C., Gerall, A. A. 1973. Effects of testosterone propionate implants in the brain on the sexual behavior and peripheral tissue of the male rat. *Physiol. Behav.* 11:633–40

Kizer, J. S., Palkovits, M., Zivin, J., Brownstein, M., Saavedra, J. M., Kopin, I. J. 1974. The effect of endocrinological manipulations on tyrosine hydroxylase and dopamine β hydroxylase activities in individual hypothalamic nuclei of the adult male rat. *Endocrinology* 95:799–812

Knowler, J. T., Smellie, R. M. 1973. The estrogen-stimulated synthesis of heterogenous nuclear ribonucleic acid in the uterus of immature rats. *Biochem. J.* 131:689–97

Komisaruk, B. R., Beyer, C. 1972. Differential antagonism, by MER-25, of behavioral and morphological effects of estradiol benzoate in rats. *Horm. Behav.* 3:63–70

Kow, L.-M., Pfaff, D. W. 1975. Induction of lordosis in female rats: two modes of estrogen action and the effect of adrenalectomy. *Horm. Behav.* 6:259–76

Krey, L. C., Butler, W. R., Knobil, E. 1975. Surgical disconnection of the medial basal hypothalamus and pituitary function in the rhesus monkey. I. Gonadotropin secretion. *Endocrinology* 96: 1073–87

Krieger, M. S., Morrell, J. I., Pfaff, D. W. 1977. Autoradiographic localization of estradiol-concentrating cells in the female hamster brain. *Neuroendocrinology* 22:193–205

Kubli-Garfias, C., Whalen, R. E. 1977. Induction of lordosis behavior in female rats by intravenous administration of progestins. *Horm. Behav.* 9:380–86

Kubo, K., Gorski, R. A., Kawakami, M. 1975. Effects of estrogen on neuronal excitability in the hippocampal-septal-hypothalamic system. *Neuroendocrinology* 18:176–91

Kuriyama, K., Rauscher, G. E., Sze, P. Y. 1971. Effect of acute and chronic administration of ethanol on the 5-hydroxytryptamine turnover and tryptophan hydroxylase activity of the mouse brain. *Brain Res.* 26:450–54

Labhsetwar, A. P. 1970. The role of estrogens in spontaneous ovulation: evidence for positive estrogen feedback in the 4-day estrous cycle. *J. Endocrinol.* 50:321–27

Lan, N. C., Katzenellenbogen, B. S. 1976. Temporal relationships between hormone receptor binding and biological responses in the uterus: studies with short- and long-acting derivatives of estriol. *Endocrinology* 98:220–27

Lee, K. S., Etgen, A. M., Lynch, G. S. 1977. Corticosterone modification of the synthesis of specific proteins in the central nervous system. *Ann. Meet. Soc. Neurosci., Anaheim*, Abstr. 1120

Leveille, P. J., de Vellis, J., Maxwell, D. S. 1977. Immunocytochemical localization of glycerol-3-phosphate dehydrogenase in rat brain: are oligodendrocytes target cells for glucocorticoids? *Ann. Meet. Soc. Neurosci., Anaheim*, Abstr. 1066

Liao, S., Fang, S. 1969. Receptor proteins for androgens and the mode of action of androgens on gene transcription in ventral prostate. *Vitam. Horm. NY* 27: 17–90

Lieberburg, I., McEwen, B. S. 1977. Brain cell nuclear retention of testosterone metabolites, 5α-dihydrotestosterone and estradiol-17β, in adult rats. *Endocrinology* 100:588–97

Lisk, R. D. 1962. Diencephalic placement of estradiol and sexual receptivity in female rats. *Am. J. Physiol.* 203:493–97

Lisk, R. D. 1967. Neural localization for androgen activation of copulatory behavior in the male rat. *Endocrinology* 80:754–61

Litteria, M., Thorner, M. W. 1974. Alterations in the incorporation of ^{3}H-lysine into proteins of the medial preoptic area and specific hypothalamic nuclei after ovariectomy in the adult female rat. *J. Endocrinol.* 60:377–78

Liu, T. C., Jackson, G. L. 1977. Effect of *in vivo* treatment of estrogen on luteinizing hormone synthesis and release by rat pituitaries *in vitro*. *Endocrinology* 100:1294–1302

Löfström, A., Eneroth, P. Gustafsson, J.-A., Skett, P. 1977. Effects of estradiol benzoate on catecholamine levels and turnover in discrete areas of the median eminence and the limbic forebrain, and on serum luteinizing hormone, follicle stimulating hormone and prolactin concentrations in the ovariectomized female rat. *Endocrinology* 101:1559–69

Luine, V. N., Khylchevskaya, R. I., McEwen, B. S. 1974. Estrogen effects on brain and pituitary enzyme activities. *J. Neurochem.* 23:925–34

Luine, V. N., Khylchevskaya, R. I., McEwen, B. S. 1975. Effect of gonadal hormones on enzyme activities in brain and pituitaries of male and female rats. *Brain Res.* 86:283–92

Luine, V. N., McEwen, B. S. 1977. Effects of an estrogen antagonist on enzyme activities and ^{3}H-estradiol nuclear binding in uterus, pituitary and brain. *Endocrinology* 100:903–10

Luine, V. N., McEwen, B. S., Black, I. B. 1977. Effect of 17 β-estradiol on hypothalamic tyrosine hydroxylase activity. *Brain Res.* 120:188–92

Luttge, W. G., Hughes, J. R. 1976. Intracerebral implantation of progesterone: reexamination of the brain sites responsible for facilitation of sexual receptivity in estrogen-primed ovariectomized rats. *Physiol. Behav.* 17:771–75

Luttge, W. G., Jasper, T. W. 1977. Studies on the possible role of 2-OH-estradiol in the control of sexual behavior in female rats. *Life Sci.* 20:419–26.

MacLeod, J., Reynolds, S. R. 1938. Vascular, metabolic and motility responses of uterine tissue following administration of oestrin. *Proc. Soc. Exp. Biol. Med.* 37:666–68

MacLusky, N. J., Krey, L., Lieberburg, I., McEwen, B. S. 1978. Estrogen modulation of progestin receptors in the bonnet monkey (m. radiata) and the rat. *Endocrinol. Soc. Ann. Meet. Miami, 1978*

MacLusky, N. J., McEwen, B. S. 1978. Progestin receptors in the rat brain: oestrogen modulates progestin receptor concentrations in some brain regions, but not in others. *Nature.* In press

MacLusky, N. J., Turner, B. B., McEwen, B. S. 1977. Corticosterone binding in rat brain and pituitary cytosols: resolution of multiple binding components by polyacrylamide gel based isoelectric focusing. *Brain Res.* 130:564–71

Malsbury, C. W. 1971. Facilitation of male rat copulatory behavior by electrical stimulation of the medial preoptic area. *Physiol. Behav.* 7:797–805

Malsbury, C. W., Kow, L.-M., Pfaff, D. W. 1977. Effects of medial hypothalamic lesions on the lordosis response and other behaviors in female golden hamsters. *Physiol. Behav.* 19:223–38

Marrone, B. L., Rodriguez-Sierra, J. F., Feder, H. H. 1977. Role of catechol estrogens in activation of lordosis in female rats and guinea pigs. *Pharmacol. Biochem. Behav.* 7:13–17

Martini, L., Motta, M., Fraschini, F. 1970. *The Hypothalamus.* New York: Academic. 705 pp.

Martucci, C., Fishman, J. 1976. Uterine estrogen receptor binding of catecholestrogens and of estetrol (1,3,5(10)-estratriene-3,15α,16α,17β-tetral). *Steroids* 23:325–33

Martucci, C., Fishman, J. 1977. Direction of estradiol metabolism as a control of its hormonal action-uterotrophic activity of estradiol metabolites. *Endocrinology* 101:1709–15

Massa, R., Cresti, L., Martini, L. 1977. Metabolism of testosterone in the anterior pituitary gland and the central nervous system of the European starling (*sturnus vulgaris*). *J. Endocrinol.* 75:347–54

Massa, R., Martini, L. 1971/72. Interference with the 5α-reductase system: a new approach for developing antiandrogens. *Gynecol. Invest.* 2:253–70

Mastro, A., Hymer, W. C. 1973. The effects of age and estrone treatment on DNA polymerase activity in anterior pituitary glands of male rats. *J. Endocrinol.* 59:107–19

McEwen, B. S. 1976. Steroid receptors in neuroendocrine tissues: topography, subcellular distribution and functional implications. In *International Symposium on Subcellular Mechanisms in Reproductive Neuroendocrinology,* ed. F. Naftolin, K. J. Ryan, J. Davies, pp. 277–304. Amsterdam: Elsevier. 529 pp.

McEwen, B. S. 1978a. Gonadal steroid receptors in neuroendocrine tissues. In *Hormone Receptors, Vol. I: Steroid Hormones,* ed. B. O'Malley, L. Birnbaumer, pp. 353–400. New York: Academic. 602 pp.

McEwen, B. S. 1978b. Influences of adrenocortical hormones on pituitary and brain function. In *Mechanisms of Glucocorticoid Action,* ed. G. Rousseau and J. Baxter. New York: Springer. In press

McEwen, B. S., de Kloet, R., Wallach, G. 1976. Interactions *in vivo* and *in vitro* of corticoids and progesterone with cell nuclei and soluble macromolecules from rat brain regions and pituitary. *Brain Res.* 105:129–36

McEwen, B. S., Gerlach, J. L., Micco, D. J. Jr. 1975a. Putative glucocorticoid receptors in hippocampus and other regions of the rat brain. In *The Hippocampus: A Comprehensive Treatise,* ed. R. Isaacson, K. Pribram, pp. 285–322. New York: Plenum. 418 pp.

McEwen, B. S., Krey, L. C., Luine, V. N. 1978. Steroid hormone action in the neuroendocrine system: when is the genome involved? In *The Hypothalamus,* ed. R. J. Baldessarini, J. B. Martini, pp. 255–268. New York: Raven. 490 pp.

McEwen, B. S., Lieberburg, I., Chaptal, C., Krey, L. C. 1977. Aromatization: important for sexual differentiation of the neonatal rat brain. *Horm. Behav.* 9:249–63

McEwen, B. S., Luine, V. N. 1978. Specificity, mechanisms and functional significance of steroid-receptor interactions in the brain and pituitary. *Colloq. Int. CNRS. Biol. Cell. Process. Neurosec. Hypothal.* In press

McEwen, B. S., Pfaff, D. W. 1973. Chemical and physiological approaches to neuroendocrine mechanisms: attempts at integration. In *Frontiers in Neuroendocrinology,* ed. W. F. Ganong, L. Martini, pp. 267–335. New York: Oxford Univ. Press

McEwen, B. S., Pfaff, D. W., Chaptal, C., Luine, V. N. 1975b. Brain cell nuclear retention of ^{3}H-estradiol doses able to promote lordosis: temporal and regional aspects. *Brain Res.* 86:155–61

McEwen, B. S., Zigmond, R. E., Gerlach, J. L. 1972. Sites of steroid binding and action in the brain. In *Structure and Function of Nervous Tissue,* Vol. 5, ed. G. H. Bourne, pp. 205–291. New York: Academic.

Means, A. R., Hamilton, T. H. 1966. Early estrogen action: concomitant synthesis within two minutes of nuclear RN synthesis and uptake of RNA precursor by the uterus. *Proc. Natl. Acad. Sci. US* 56:1594–98

Mercier, L., Le Guellec, C., Thieulant, M.-L., Samperez, S., Jouan, P. 1976. Androgen and estrogen receptors in the cytosol from male rat anterior hypophysis: further characteristics and differentiation between androgen and estrogen receptors. *J. Steroid Biochem.* 7:779–85

Mester, J., Baulieu, E. E. 1975. Dynamics of estrogen-receptor distribution between the cytosol and nuclear fractions of immature rat uterus after estradiol administration. *Biochem. J.* 146:617–23

Meyer, J. S., Leveille, P. J., McEwen, B. S., de Vellis, J. 1978. Corticoids and glial cells: glycerolphosphate dehydrogenase induction and cytosol binding in normal and degenerated rat optic nerve. *Endocr. Soc. Ann. Meet. Miami, 1978.*

Meyerson, B. J. 1967. Relationship between the anesthetic and gestagenic action and estrous behavior-inducing activity of different progestins. *Endocrinology* 81:369–74

Meyerson, B. J. 1972. Latency between intravenous injection of progestins and the appearance of estrous behavior in estrogen-treated ovariectomized rats. *Horm. Behav.* 3:1–9

Michael, R. P. 1965. Oestrogen in the central nervous system. *Br. Med. Bull.* 21:87–90

Millard, S. A., Costa, E., Gal, E. M. 1972. On the control of brain serotonin turnover rate by end product inhibition. *Brain Res.* 40:545–51

Miller, A. L., Chaptal, C., McEwen, B. S., Peck, E. J. 1978. Modulation of high affinity GABA uptake into hippocampal synaptosomes by glucocorticoids. *Psychoneuroendocrinology* In press

Miller, W. L., Knight, M. M., Grimek, H. J., Gorski, J. 1977. Estrogen regulation of follicle stimulating hormone in cell cultures of sheep pituitaries. *Endocrinology* 100:1306–16

Mogenson, G. J., Calaresu, F. R. 1975. *Neural Integration of Physiological Mechanisms and Behaviour.* Toronto: Univ. Toronto Press. 442 pp.

Moguilewsky, M., Raynaud, J.-P. 1978. Progestin binding sites in the rat hypothalamus pituitary and uterus. *Steroids.* In press

Moore, K. E., Phillipson, O. T. 1975. Effects of dexamethasone on phenylethanolamine N-Methyl-transferase and adrenaline in the brains and superior cervical

ganglia of adult and neonatal rats. *J. Neurochem.* 25:289–94

Morali, G., Larsson, K., Beyer, C. 1977. Inhibition of testosterone-induced sexual behavior in the castrated male rat by aromatase blockers. *Horm. Behav.* 9:203–13

Moreines, J. K., Powers, J. B. 1977. Effects of acute ovariectomy on the lordosis response of female rats. *Physiol. Behav.* 19:277–83

Morin, L. P. 1977. Progesterone: inhibition of rodent sexual behavior. *Physiol. Behav.* 18:701–15

Morin, L. P., Feder, H. H. 1974a. Intracranial estradiol benzoate implants and lordosis behaviour of ovariectomized guinea pigs. *Brain Res.* 70:95–102

Morin, L. P., Feder, H. H. 1974b. Inhibition of lordosis behavior in ovariectomized guinea pigs by mesencephalic implants of progesterone. *Brain Res.* 70:71–80

Morin, L. P., Feder, H. H. 1974c. Hypothalamic progesterone implants and facilitation of lordosis behavior in estrogen-primed ovariectomized guinea pigs. *Brain Res.* 70:81–93

Morrell, J. I., Kelley, D. B., Pfaff, D. W. 1975. Sex steroid binding in the brains of vertebrates: studies with light microscopic autoradiography. In *Brain Endocrine Interactions: II. The Ventricular System,* ed. K. Knigge, D. S. Scott, K. Kobayashi, S. Ishi, pp. 230–256. Basel: Karger.

Moyer, K. E. 1976. *The Psychobiology of Aggression.* New York: Harper & Row 402 pp.

Mueller, G. C., Herranen, A. M., Jervell, K. F. 1958. Studies on the mechanism of action of estrogens. *Recent Prog. Horm. Res.* 14:95–139

Naess, O., Attramadal, A., Aakvaag, A. 1975a. Androgen binding proteins in the anterior pituitary, hypothalamus, preoptic area and brain cortex of the rat. *Endocrinology* 96:1–9

Naess, O., Hansson, V., Djoeseland, O., Attramadal, A. 1975b. Characterization of the androgen receptor in the anterior pituitary of the rat. *Endocrinology* 97:1355–63

Naftolin, F., Morishita, H., Davies, I. J., Todd, R., Ryan, K. J., Fishman, J. 1975a. 2-Hydroxyestrone induced rise in serum luteinizing hormone in the immature male rat. *Biochem. Biophys. Res. Commun.* 64:905–10

Naftolin, F., Ryan, K. J., Davies, I. J., Reddy, V. V., Flores, F., Petro, Z., Kuhn, M. 1975b. The formation of estrogens by central neuroendocrine tissues. *Recent Prog. Horm. Res.* 31:295–315

Noteboom, W. D., Gorski, J. 1963. An early effect of estrogen on protein synthesis. *Proc. Natl. Acad. Sci. USA* 50:250–55

Notides, A. C., Gorski, J. 1966. Estrogen-induced synthesis of a specific uterine protein. *Proc. Natl. Acad. Sci. USA* 56:230–35

O'Malley, B. W., Means, A. R. 1974. Female steroid hormones and target cell nuclei. *Science* 183:610–20

Otten, U., Thoenen, H. 1976. Modulatory role of glucocorticoids on NGF-mediated enzyme induction in organ cultures of sympathetic ganglia. *Brain Res.* 111:438–41

Otten, U., Thoenen, H. 1977. Effect of glucocorticoids on nerve growth factor-mediated enzyme induction in organ cultures of rat sympathetic ganglia: enhanced response and reduced time requirement to initiate enzyme induction. *J. Neurochem.* 29:69–75

Palka, Y. S., Ramirez, V. D., Sawyer, C. H. 1966. Distribution and biological effects of tritiated estradiol implanted in the hypothalamo-hypophyseal region of female rats. *Endocrinology* 78:487–99

Parvez, H., Parvez, S. 1973. The effects of metopirone and adrenalectomy on the regulation of the enzymes monoamine oxidase and catechol-o-methyl transferase in different brain regions. *J. Neurochem.* 20:1011–20

Parvizi, N., Ellendorff, F. 1975. 2-Hydroxy-oestradiol-17β as a possible link in steroid brain interaction. *Nature* 256: 59–60

Parvizi, N., Naftolin, F. 1977. Effects of catechol estrogens on sexual differentiation in neonatal female rats. *Psychoneuroendocrinology* 2:409–11

Paul, S. M., Axelrod, J. 1977. Catechol estrogens: presence in brain and endocrine tissues. *Science* 197:657–59

Paul, S. M., Skolnick, P. 1977. Catechol oestrogens inhibit oestrogen elicited accumulation of hypothalamic cyclic AMP suggesting role as endogenous anti-oestrogens. *Nature* 226:559–61

Peck, E. J. 1978. Dynamics of receptor-estrogen complexes in the hypothalamus. In *Ontogeny of Receptors and Molecular Mechanisms of Reproductive Hormone Action,* ed. T. Hamilton. New York, Raven. In press

Petrovic, V. M., Janic, V. 1974. Adrenocortical control of monoamine oxidase activity in the ground squirrel (citellus citellus) during the winter. *J. Endocrinol.* 62:407–8

Pfaff, D. W. 1968. Uptake of estradiol-17-H[3] in the female rat brain. An autoradiographic study. *Endocrinology* 82: 1149–55

Pfaff, D. W. 1976. The neuroanatomy of sex hormone receptors in the vertebrate brain. In *Neuroendocrine Regulation of Fertility,* ed. T. C. A. Kumar, pp. 30–45. Basel: Karger. 322 pp.

Pfaff, D. W., Gerlach, J., McEwen, B. S., Ferin, M., Carmel, P., Zimmerman, E. 1976. Autographic localization of hormone-concentrating cells in the brain of the female rhesus monkey. *J. Comp. Neurol.* 170:279–94

Pfaff, D. W., Keiner, M. 1973. Atlas of estradiol-concentrating cells in the central nervous system of the female rat. *J. Comp. Neurol.* 151:121–58

Pfaff, D. W., Modianos, D. 1978. Neural mechanisms of female reproductive behavior. In *Neurobiology of Reproduction (Handbook of Behavioral Neurobiology),* ed. R. Goy, D. W. Pfaff. New York: Plenum. In press

Pfaff, D. W., Silva, M. T. A., Weiss, J. M. 1971. Telemetered recording of hormone effects on hippocampal neurons. *Science* 172:394–95

Plapinger, L., McEwen, B. S. 1978. Gonadal steroid-brain interactions in sexual differentiation. In *Biological Determinants of Sexual Behavior,* pp. 153–218. New York: Wiley. 822 pp.

Pohorecky, L. A., Wurtman, R. J. 1971. Adrenocortical control of epinephrine synthesis. *Pharmacol. Rev.* 23:1–35

Powers, J. B. 1972. Facilitation of lordosis in ovariectomized rats by intracerebral progesterone implants. *Brain Res.* 48:311–25

Powers, J. B. 1975. Anti-estrogenic suppression of the lordosis response in female rats. *Horm. Behav.* 6:379–92

Powers, J. B., Valenstein, E. S. 1972. Sexual receptivity: facilitation by medial preoptic lesions in female rats. *Science* 175:1003–5

Purdy, R. H., Axelrod, L. R. 1968. Properties of corticosteroid-21-O-acetyltransferase from the baboon brain. *Steroids* 11:851–62

Quadagno, D. M., Ho, G. K. 1975. The reversible inhibition of steroid-induced sexual behavior by intracranial cycloheximide. *Horm. Behav.* 6:19–26

Raisman, G., Field, P. M. 1973. Sexual dimorphism in the neuropil of the preoptic area of the rat and its dependence on neonatal androgen. *Brain Res.* 54:1–29

Robinson, J. A., Leavitt, W. W. 1971. Estrogen-related changes in anterior pituitary RNA levels. *Proc. Soc. Exp. Biol. Med.* 139:471–75

Rodgers, C. H., Law, O. T. 1968. Effects of chemical stimulation of the "limbic system" on lordosis in female rats. *Physiol. Behav.* 3:241–46

Roosevelt, T. S., Ruhmann-Wennhold, A., Nelson, D. H. 1973. Adrenal corticosteroid effects upon rat brain mitochondrial metabolism. *Endocrinology* 53:619–25

Ross, J., Claybaugh, C., Clemens, L. G., Gorski, R. A. 1971. Short latency induction of estrous behavior with intracerebral gonadal hormones in ovariectomized rats. *Endocrinology* 89:32–38

Roy, E. J. 1978. Antiestrogens and nuclear estrogen receptors in the brain. In *Current Studies of Hypothalamic Function,* ed. K. Lederis. Basal: Karger, In press

Roy, E. J., Wade, G. N. 1976a. Central action and a species comparison of the estrogenic effects of an antiestrogen on eating and body weight. *Physiol. Behav.* 18:137–40

Roy, E. J., Wade, G. N. 1976b. Estrogen effects of an antiestrogen, MER-25, on eating and body weight in rats. *J. Comp. Physiol. Psychol.* 90:156–66

Roy, E. J., Wade, G. N. 1977. Binding of ³H-estradiol by brain cell nuclei and female rat sexual behavior: inhibition by antiestrogens. *Brain Res.* 126:73–87

Sar, M., Stumpf, W. E. 1973. Neurons of the hypothalamus concentrate ³H-progesterone or its metabolites. *Science* 182:1266–68

Sar, M., Stumpf, W. E. 1977. Distribution of androgen target cells in rat forebrain and pituitary after ³H-dihydrotestosterone administration. *J. Steroid Biochem.* 8:1131–35

Sarff, M., Gorski, J. 1971. Control of estrogen binding protein concentration under basal conditions and after estradiol administration. *Biochemistry* 10:2557–63

Schally, A. V., Bowers, C. Y., Carter, W. H., Arimura, A., Redding, T. W., Saito, M. 1969. Effect of actinomycin D on the inhibitory response of estrogen on LH release. *Endocrinology* 85:290–99

Shen, J.-T., Ganong, W. F. 1976. Effect of variations in pituitary-adrenal activity on dopamine-β-hydroxylase activity in various regions of rat brain. *Neuroendocrinology* 20:311–18

Shirley, B., Wolinsky, J., Schwartz, N. B. 1968. Effects of a single injection of an estrogen antagonist in the estrous cycle of the rat. *Endocrinology* 82:959–68

Singer, J. J. 1968. Hypothalamic control of male and female sexual behavior female

rats. *J. Comp. Physiol. Psych.* 66:738–42

Smith, D. E., Gorski, J. 1968. Estrogen control of uterine glucose metabolism: an analysis based on the transport and phosphorylation of 2-deoxyglucose. *J. Biol. Chem.* 243:4169–74

Smith, E. R., Davidson, J. M. 1974. Localization of feedback receptors: effects of intracranially implanted steroids on plasma LH and LRF release. *Endocrinology* 95:1566–73

Spooner, P., J. Gorski, J. 1972. Early estrogen effects on lipid metabolism in the rat uterus. *Endocrinology* 91:1273–84

Stumpf, W. E. 1968. Estradiol-concentrating neurons: topography in the hypothalamus by dry mount autoradiography. *Science* 162:1001–3

Swerdloff, R. S., Jacobs, H. S., Odell, W. D. 1972. Synergistic role of progestogens in estrogen induction of LH and FSH surge. *Endocrinology* 90:1529–36

Sze, P. Y. 1976. Glucocorticoid regulation of the serotonergic system of the brain. In *Advances in Biochemical Psychopharmacology, Vol. 15,* ed. E. Costa, E. Giacobini, R. Paoletti, pp. 251–65. New York: Raven. 498 pp.

Sze, P. Y., Maxson, S. C. 1975. Involvement of corticosteroids in acoustic induction of audiogenic seizure susceptibility in mice. *Psychopharmacologia* 45:79–82

Sze, P. Y., Neckers, L., Towle, A. C. 1976. Glucocorticoids as a regulatory factor for brain tryptophan hydroxylase. *J. Neurochem.* 26:169–73

Sze, P. Y., Yanai, J., Ginsburg, B. E. 1974. Adrenal glucocorticoids as a required factor in the development of ethanol withdrawal seizures in mice. *Brain Res.* 80:155–59

Szego, C. M., Roberts, S. 1953. Steroid action and interaction in uterine metabolism. *REcent Prog. Horm. Res.* 8:419–69

Tang, T. K. L., Spies, H. G. 1975. Effects of gonadal steroids on the basal and LRF-induced gonadotropin secretion by cultures of rat-pituitary. *Endocrinology* 96:349–86

Telegdy, G., Vermes, I. 1975. Effect of Adrenocortical hormones on activity of the serotoninergic system in limbic structures in rats. *Neuroendocrinology* 18:16–26

Terkel, A. S., Shryne, J., Gorski, R. A. 1973. Inhibition of estrogen facilitation of sexual behavior by the intracerebral infusion of actinomycin D. *Horm. Behav.* 4:377–86

Vermes, I., Smelik, P. G., Mulder, A. H. 1976. Effects of hypophysectomy, adrenalectomy, and corticosterone

112   MCEWEN, DAVIS, PARSONS & PFAFF

treatment on uptake and release of putative central neurotransmitters by rat hypothalamic tissue *in vitro. Life Sci.* 19:1719–26

Vilchez-Martinez, J. A., Arimura, A., Debeljuk, L., Schally, A. V. 1974. Biphasic effect of estradiol benzoate on the pituitary responsiveness to LH-RH. *Endocrinology* 94:1300–3

Walker, W. W., Feder, H. H. 1977a. Inhibitory and facilitatory effects of various anti-estrogens on the induction of female sexual behavior by estradiol benzoate in guinea pigs. *Brain Res.* 134:445–65

Walker, W. W., Feder, H. H. 1977b. Antiestrogen effects on estrogen accumulation in brain cell nuclei: neurochemical correlates of estrogen action on female sexual behavior in guinea pigs. *Brain Res.* 134:467–78

Wallen, K., Goldfoot, D. A., Joslyn, W. D., Paris, C. A. 1972. Modification of behavioral estrus in the guinea pig following intracranial cycloheximide. *Physiol. Behav.* 8:221–23

Warembourg, M. 1975a. Radioautographic study of the rat brain after injection of 1,2,³H-corticosterone. *Brain Res.* 89: 61–70

Warembourg, M. 1975b. Radiographic study of the rat brain and pituitary after injection of ³H dexamethasone. *Cell Tissue Res.* 161:183–91

Watanabe, H., Orth, D. N., Toft, D. O. 1973. Glucocorticoid receptors in pituitary tumor cells. I. Cytosol receptors. *J. Biol. Chem.* 248:7625–30

Weissman, B. A., Johnson, D. F. 1976. Possible role of dopamine in diethylstilbestrol-elicited accumulation of cyclic AMP in incubated male rat hypothalamus. *Neuroendocrinology* 21:1–9

Whalen, R. E., Gorzalka, B. B. 1972. The effects of progesterone and its metabolites on the induction of sexual receptivity in rats. *Horm. Behav.* 3:221–26

Whalen, R. E., Gorzalka, B. B. 1973. Effects of an estrogen antagonist on behavior and on estrogen retention in neural and peripheral target tissues. *Physiol. Behav.* 10:35–40

Whalen R. E., Nakayama, K. 1965. Induction of estrous behavior: facilitation by repeated hormone treatment. *J. Endocrinology* 33:525–26

Whelly, S. M., Barker, K. L. 1974. Early effect of estradiol on the peptide elongation rate by uterine polyribosomes. *Biochemistry* 12:341–46

Yanase, M., Gorski, R. A. 1976. Sites of estrogen and progesterone facilitation of lordosis behavior in the spayed rat. *Biol. Reprod.* 15:536–43

Young, W. C. 1961. The hormones and mating behavior. In *Sex and Internal Secretions* Vol. II, ed. W. C. Young, pp. 1173–1239. Baltimore: Williams & Wilkins. 1609 pp.

Zigmond, R. E., Nottebohm, F., Pfaff, D. W. 1973. Androgen-concentrating cells in the midbrain of a songbird. *Science* 179:1005–7

Zucker, I. 1967. Actions of progesterone in the control of sexual receptivity of the spayed female rat. *J. Comp. Physiol. Psych.* 63:313–16

Ann. Rev. Neurosci. 1979. 2:113–68

CENTRAL CATECHOLAMINE NEURON SYSTEMS: ANATOMY AND PHYSIOLOGY OF THE NOREPINEPHRINE AND EPINEPHRINE SYSTEMS

♦11520

R. Y. Moore

Department of Neurosciences, University of California, San Diego, La Jolla, California 92093

F. E. Bloom

Arthur V. Davis Center for Behavioral Neurobiology, The Salk Institute, La Jolla, California 92037

INTRODUCTION

The intensity of research on the central catecholaminergic neuron systems has continued unabated since last year's review of the dopaminergic systems (Moore & Bloom 1978). Although our original intent was to attempt a review of the structure and function of all these systems within one chapter, the richness of important details deserving of inclusion resulted in a contribution exceeding the pages available. As a result, we have had an additional year in which to acquire even more material pertaining to the central adrenergic and noradrenergic neuron systems. It is the noradrenergic systems in particular that present the more flagrant violations (Dismukes 1977) of the structural and functional archetypes normally attributed to central neuronal systems. Their widespread efferent trajectories with high degrees of collateral arborization, their marked propensity for the expression of post-lesion regrowth, and their unique combination of functional actions, enriched in some cases by association with adenylate cyclase, all combine

113

to make these noradrenergic neurons of great interest. At the same time, the development of immunocytochemical methodologies has revealed the existence of separate and distinctive adrenergic neuron systems whose anatomical and physiological characteristics may have relevance to a variety of functions not subserved by the noradrenergic systems. Finally, we conclude this review of the catecholaminergic systems by addressing some of the questions raised by this enormous collection of research data. The literature search upon which this review is based was concluded on April 1, 1978. As with the review of the dopamine neuron systems, there is much more literature on the noradrenergic systems than could reasonably be cited. Consequently, we again selected references that we feel will guide readers to the most inclusive material. We very much regret being unable to cite all of the references that constitute the background for this review.

NOREPINEPHRINE SYSTEMS

Overview

The anatomical and functional analysis of norepinephrine (NE)-containing neurons in the brain has advanced steadily and less controversially than the DA circuits. In particular, the improved methodologies for neuronal circuit analysis and the more widespread use of fluorescence and immunohisto-chemistry have all added immeasurably to our understanding of the NE neuron projection systems. As detailed in the following sections, the extensive ascending projections of NE neuron systems have been studied very effectively in the recent past. These studies have brought forth evidence that NE systems contrast with DA systems in having remarkably extensive projection systems that distribute connections widely through the neuraxis (Table 1). Similarly, physiological analysis of the NE systems, especially

Table 1   Norepinephrine neuron systems in the mammalian brain

System	Nucleus of origin	Site(s) of termination
Locus coeruleus	Locus coeruleus	Spinal cord, brainstem, cerebellum, hypothalamus, thalamus, basal telencephalon, and the entire isocortex
Lateral tegmental[a]	Dorsal motor vagus, nucleus tractus solitarius, and adjacent tegmentum, lateral tegmentum	Spinal cord, brainstem, hypothalamus, basal telencephalon

[a]The brainstem norepinephrine neurons located in the dorsal, caudal, medulla medially and laterally are considered, along with the pontine and isthmic cell groups, to constitute one system.

analysis of their qualitative actions and mechanisms of effect, have also progressed rapidly of late, largely as a result of the application of the iontophoretic method. To corroborate the inference that iontophoretic effects of NE could be equated with NE synapses has required the demonstration that selective stimulation of the afferent NE axons can reproduce precisely the effects produced by microiontophoresis of NE (see Bloom 1974, 1975a, b, 1978). In fact, the major conceptual and experimental advantage of studies on the NE system is the ability now to examine discretely NE actions on cells known to be the targets of NE circuits. Due to space restrictions on references here, other recent reviews on this topic should also be consulted (Bloom 1968, 1975a, b, 1978, Bloom & Hoffer 1973, Hoffer & Bloom 1976, Salmoiraghi & Stefanis 1971), especially for references before 1967.

The Locus Coeruleus Norepinephrine System

THE NUCLEUS   The locus coeruleus is a prominent nucleus located in the brainstem reticular formation at the level of the isthmus. Its existence has been recognized for many years and Russell (1955) extensively reviewed its appearance in a number of mammalian species. In man it is also a prominent nucleus and forms one of the pigmented nuclei of the brainstem (Olszewski & Baxter 1954, Braak 1975). Until the development of monoamine histochemistry, however, very little was known of the connections or significance of the locus coeruleus. The early work of Dahlström & Fuxe (1964) demonstrated that the locus coeruleus is a nucleus composed entirely of norepinephrine (NE)-producing neurons. And, as such, it is the largest NE nucleus within the mammalian brain. The quantitative analysis of Swanson & Hartman (1975) indicates that it includes 43% of all the NE-producing neurons in the rat brain. The early studies of the Stockholm group demonstrated that the locus coeruleus has widespread projections through the neuraxis (Andén et al 1966, Fuxe et al 1969, Olson & Fuxe 1971, Ungerstedt 1971) but, it was the work of Ungerstedt in particular that first demonstrated the extensive innervation of telencephalic structures by locus coeruleus neurons. This work has subsequently been confirmed and extended using a variety of techniques and the following review of locus coeruleus anatomy and projections is based on observations from a number of laboratories. The neurons of the locus coeruleus (LC) form a distinct, compact cell group largely contained within the central gray of the isthmus, medial to the mesencephalic nucleus of the trigeminal nerve (Russel 1955, Swanson 1976a). There are a few exceptions to this but it is worth noting that, in the cat, the nucleus is much more widely dispersed than in other mammalian species that have been described (Chu & Bloom 1974a, Jones & Moore 1974). The locus coeruleus has been examined using the fluorescence histochemical method in several primate species (Hubbard & DiCarlo 1973, DiCarlo et al 1973, Felten et al 1974, Garver & Sladek 1975). In the primate, some LC cells may contain serotonin (Sladek & Walker 1977). The nucleus is quite homogenous in size and location, except in man where it is more widely dispersed than in lower species, particularly in that there are a large number of cells distributed over the anterior medullary velum of the fourth ventricle (Braak 1975).

Since nearly all of the work on the projections of the locus coeruleus has been carried out on the rat, the remainder of this review deals with that species. In the rat, the locus coeruleus is comprised of approximately 1,500 neurons on each side of the brainstem (Swanson 1976a). The rat locus coeruleus arises early in gestation (Nicholson & Bloom 1974) and is innervated soon thereafter (Lauder & Bloom 1975). In Nissl preparations of the adult rat the cells are deeply staining, medium-size neurons, which appear fusiform or bipolar in shape. The Nissl substance is prominent throughout the cytoplasm of the cell. In Golgi preparations (Swanson 1976a) locus coeruleus neurons appear to be predominantly multipolar with 3–5 large, rather thin dendrites that radiate from the soma and typically branch at least once or twice. Many of the secondary and tertiary branches of the dendrites extend well beyond the limits of the nucleus into the surrounding neuropil. The axon of locus coeruleus neurons typically gives off two or three fine collaterals within the nucleus and further collaterals beyond it.

The locus coeruleus has been the subject of few ultrastructural studies (Lenn 1965, Hökfelt 1967). In osmium-fixed material the neurons of the locus exhibit no distinguishing features; they show abundant rough endoplasmic reticulum, a Golgi apparatus, and the usual cytoplasmic inclusions (Lenn 1965). In some instances large granular vesicles are evident in the cytoplasm but these are infrequent. After permanganate fixation large numbers of small (500 Å), dense-core vesicles are seen in the cytoplasm of the LC neurons and in a large number of the terminals in the nucleus (Hökfelt 1967). The latter could arise from collaterals of NE neurons of the nucleus (Swanson 1976a, Aghajanian et al 1977) innervating neurons of the nucleus (Sakai et al 1977) or the epinephrine (E) input from the medulla (Hökfelt et al 1974).

As emphasized subsequently, extensive collateralization appears to be a prominent feature of the axons of the locus coeruleus neurons and this is evident not only in Golgi material (Swanson 1976a) but also in histochemical fluorescence material (Lindvall & Björklund 1974b, Hökfelt et al 1977, Fuxe et al 1978, Levitt & Moore 1978, Jacobowitz 1978).

PATHWAYS FROM THE LOCUS COERULEUS   The following description of the projection pathways of locus coeruleus neurons is derived from studies using fluorescence histochemical techniques (Olson & Fuxe 1971, Ungerstedt 1971, Maeda & Shimizu 1972, Lindvall & Björklund 1974b, 1978), immunofluorescent techniques (Swanson & Hartman 1975), and autoradiographic tracing (Segal et al 1974, Pickel et al 1974a, Jones et al 1977, Jones & Moore 1977). The latter has proven to be extremely useful in demonstrating pathways from the locus coeruleus. Jones et al (1977) demonstrated that tritiated proline injected into the locus coeruleus is taken up into neurons of the nucleus and incorporated into protein. This is transported along the axons of locus coeruleus neurons and, using pharmacologic manipulation, specific transport can be shown. That is, a comparison of regional distribution of labeled protein in animals with control injections, animals with locus coeruleus injections, and animals with locus coeruleus injections following pretreatment with intraventricular 6-hydroxydopamine permits the pattern of transport specific to locus coeruleus neurons to be estimated. Injection of tritiated L-DOPA gives a pattern of distribution of transported material identical to that with the tritiated amino acid. The transport of labeled protein in locus coeruleus neurons is blocked by pretreatment of animals with 6-hydroxydopamine. In the untreated animal, following injection of tritiated amino acid into locus coeruleus, there is transport of tritiated protein to the ipsilateral midbrain, hypothalamus, thalamus, and nearly all telencephalic areas. In particular, there is transport to all neocortical areas. Contralateral transport is evident in some basal forebrain areas such as the septum, amygdala, hippocampus, and in the frontal and cingulate cortex. Most neocortical areas, however, show relatively little contralateral transport.

In material prepared either by the glyoxylic acid method (Lindvall & Björklund 1974a) or by the autoradiographic tracing method (Jones & Moore 1977) three major ascending pathways can be seen to originate from the locus coeruleus (Figure 1). The first is the projection into the mesencephalic tegmentum, which was referred to by Ungerstedt (1971) as the dorsal catecholamine bundle. This is by far the largest ascending projection. A second projection enters the central gray and ascends as a component of the dorsal longitudinal fasciculus. A third ascending component turns ventrally from the locus coeruleus to traverse the mesencephalic tegmentum in the central tegmental tract and then ascends through the ventral tegmental area into the medial forebrain bundle (Figure 1). Several pathways arise from the dorsal pathway, which appears as a compact group of fibers lying ventral and lateral to the cerebral aqueduct just outside the periaqueductal gray. The fibers are maintained in this position until they reach the level of the fasciculus retroflexus at which point the major bundle turns ventrally to traverse the prerubral field of Forel, medial to the zona incerta, before joining the dorsal portion of the medial forebrain bundle complex at the

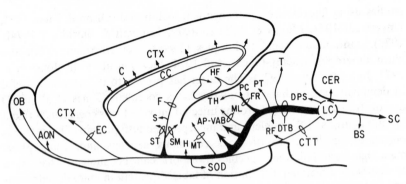

Figure 1 Diagram of the projections of the locus coeruleus viewed in the sagittal plane. See text for description. Abbreviations: AON, anterior olfactory nucleus; AP-VAB, ansa pedun-cularis-ventral amygdaloid bundle system; BS, brainstem nuclei; C, cingulum; CC, corpus callosum; CER, cerebellum; CTT, central tegmental tract; CTX, cerebral neocortex; DPS, dorsal periventricular system; DTB, dorsal catecholamine bundle; EC, external capsule; F, fornix; FR, fasciculus retroflexus; H, hypothalamus; HF, hippocampal formation; LC, locus coeruleus; ML, medial lemniscus; MT, mammillothalamic tract; OB, olfactory bulb; PC, posterior commissure; PT, pretectal area; RF, reticular formation; S, septal area; SC, spinal cord; SM, stria terminalis; T, tectum; TH, thalamus. (From observations of Lindvall & Bjorklund 1974b, Jones & Moore 1977).

level of the caudal hypothalamus. Although most of the dorsal bundle fibers turn ventrally at the level of the fasciculus retroflexus, a large group turns dorsally and runs along the fasciculus retroflex toward the habenular complex. A major component of this group enters the internal medullary lamina of the thalamus and ascends within it. A second group of fibers leaving the dorsal bundle at the level of the fasciculus retroflexus ascends within the external medullary lamina of the thalamus and a few fibers continue into the superficial zone of the thalamus. These groups of fibers provide most of the locus coeruleus NE input to thalamus.

The ascending fibers of the locus coeruleus projection that enter the medial forebrain bundle give rise to several distinct groups of fibers. The largest of these is made up of fascicles of fibers that leave the medial forebrain bundle complex laterally along its course to enter the ansa pedun-cularis-ventral amygdaloid bundle system. Part of these enter basal telence-phalic areas such as the amygdala, whereas others continue into the external capsule. Another group enters the mammillothalamic tract to ascend to the anterior nuclei of the thalamus. As the fibers in the medial forebrain bundle reach the level of the caudal septum, they break up into five major groups. One group turns medially into the diagonal band of Broca to innervate the septum and then enters the fornix. The second enters the stria medullaris turning caudally and continues along its length to the habenular nuclei. The third enters the stria terminalis and follows this path to the amygdaloid complex. The fourth group of fibers continues in the medial forebrain bundle as it enters the basal telencephalon. A portion of this continues rostrally into the external capsule, whereas other fibers are given off to deep layers of the olfactory tubercle and anterior olfactory nucleus. Lastly, the fifth group of fibers traverses the diagonal band and Zuckerkandl's bundle to turn around the genu of the corpus callosum and then run caudally in the cingulum; these fibers appear to correspond to that portion of the cingulum projection that was viewed as a component of the external sagittal stratum by Domesick (1970).

In addition to these ipsilateral projections, there are a number of commissures in the locus coeruleus system that give rise to contralateral projections. The first of these occurs just rostral to the locus coeruleus in the isthmic tegmentum. The fibers cross ventral to the medial longitudinal fasciculus adjacent to the commissure of Probst and then join the contralateral dorsal pathway. A second component crosses in the posterior commissure; fibers leaving the dorsal pathway and ascending along the fasciculus retroflexus course over the periaqueductal gray and enter the posterior commissure (Figure 2). These fibers then form a component of the contralateral dorsal pathway with a distribution similar to those arising ipsilaterally. The third commissural pathway arises from dorsal pathway fibers on the ventral lateral surface of the internal capsule adjacent to the optic tract. At rostral, tuberal hypothalamic levels, these turn medially to cross in the dorsal supraoptic commissure and enter the contralateral medial forebrain bundle. A fourth group of fibers crosses in the anterior commissure; and there is a small, fifth commissure that traverses the genu of the corpus callosum.

Two other major groups of fibers leave the locus coeruleus. The first is a group that ascends in the superior cerebellar peduncle to innervate the cerebellum. The second group of fibers descends in the central tegmental bundle through the brainstem to enter the ventral portion of the lateral column of the spinal cord.

TERMINAL PROJECTIONS OF THE LOCUS COERULEUS   The terminal projections of the locus coeruleus are best viewed in fluorescence histochemical material, particularly that prepared by the glyoxylic acid method (Lindvall & Björklund 1974a). In such material locus coeruleus axons have a distinctive morphology. Preterminal axons are thin with fusiform varicosities that do not have intense fluorescence. When a terminal field is reached, however, the preterminal fibers break up into a highly collateralized network with many features common to all areas innervated (Lindvall & Björklund 1974b). The primary feature is the axonal morphology. In

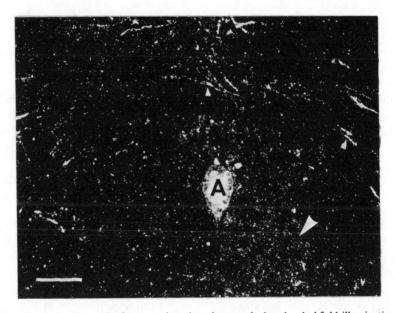

Figure 2 Autoradiograph of a coronal section photographed under darkfield illumination. The animal received an injection of tritiated proline into the right locus coeruleus. A, cerebral aqueduct. The photograph shows numerous silver grains in a pattern of terminal field innervation in the periaqueductal gray (large arrow). In addition, clusters of silver grains in linear arrays (small arrows) designate fibers of passage entering and decussating in the posterior commissure. Marker bar = 120 μm. (B. E. Jones and R. Y. Moore, unpublished).

terminal areas, the axons are typically fine, with regularly spaced, round, intensely fluorescent varicosities approximately 1–2 μm in diameter. The most frequent form of terminal innervation is a network or plexus of fibers, but the pattern varies from area to area. Some specific examples will be enumerated to illustrate this. It should be noted, however, that whereas there are some areas in which the locus forms the sole catecholamine neuron innervation, there are others in which it is mixed with NE-containing axons from other brainstem cell groups and areas where it is mixed with this group and with DA neurons originating from the mesencephalic DA neuron cell groups. There is sufficient evidence from a variety of techniques that the locus coeruleus innervates the entire neuraxis.

Spinal Cord   The recent work of Kuypers & Maisky (1975) and of Nygren & Olson (1977) indicates that the locus coeruleus innervates all segments of the spinal cord. Locus fibers were thought originally to contribute only a minor component of the plexiform network of catecholamine-containing fibers present throughout the spinal cord gray (Dahlström & Fuxe 1965), but the work of Nygren & Olson (1977) indicates that the locus NE innervation of spinal cord gray is a major component of the total innervation, distributing widely to both dorsal and ventral horns.

Brainstem   Until recently no major study had been carried out to delineate the distribution of locus coeruleus NE axons (as opposed to other brainstem NE axons) to brainstem nuclei and areas. Some studies such as that of Kromer & Moore (1977) showing that the NE innervation of the cochlear nuclei arises from locus coeruleus have been reported, but the first extensive study of the NE innervation of the brainstem is that of Levitt and Moore (unpublished observations, Table 4). The locus projects predominantly, but not exclusively, to sensory nuclei of the brainstem. Some sensory nuclei, particularly the vestibular complex, receive little or no catecholamine innervation. In a few areas, for example, the periaqueductal gray and the pontine nuclei, there is overlap of locus NE innervation with that arising from other brainstem NE neuron groups.

Cerebellum   A third area in which the innervation from the locus coeruleus has been described in detail is the cerebellum (Olson & Fuxe 1971, Bloom et al 1971, Mugnaini & Dahl 1975). As with the cochlear nucleus, this is an innervation that derives exclusively from the locus coeruleus. In the work of Bloom and collaborators (1971) a fairly sparse innervation of the cerebellar cortex was observed, although subsequent reexamination of this innervation with GA methods (Landis et al 1975, Bloom & Battenberg 1976) (Figure 3) reveals significantly more. The pattern of innervation is as follows. Nearly the entire innervation is to the molecular layer of the cerebellar cortex in the vicinity of the proximal dendrites of Purkinje cells. Some fibers appear to run in a longitudinal direction along the Purkinje cell layer and others appear to run vertically toward the surface of the folium, along the Purkinje cell dendrites. In outer molecular layer the fibers bifurcate to run parallel to the pial surface. Even the innervation in the rat and mouse cerebellum (Landis et al 1975, Bloom & Battenberg 1976) seen with GA methods appears sparse compared to the extremely rich innervation of cerebellar cortex reported for the chicken (Mugnaini & Dahl 1975). Fibers within the white matter of the cerebellar folia have the typical appearance of preterminal locus fibers. These turn into the granule cell layer where they form a plexus of fibers highly collateralized and distributed diffusely throughout the granule cell layer. Branches from this plexus then turn vertically into the Purkinje cell layer and pass through it into the molecular layer. In the molecular layer the fibers also give rise to a T-branch with each component turning radially in the direction of the parallel fibers. This branching is quite dense in the chicken and Magnaini & Dahl (1975) interpret these observations to indicate that the NE innervation terminates on more than one class of cerebellar neurons. This important interpretation requires additional data before it can be accepted. An extensive comparative study of cerebellar noradrenergic innervation has been reported by Tohyama (1976). While all mammals and birds appear to derive their cerebellar NE fibers from locus coeruleus, this does not seem to be the case for amphibians or teleosts.

Figure 3 Cerebellar cortex of normal rat, prepared by the GA-cryostat method of Bloom & Battenberg (1976). Fine varicose fibers can be seen coursing through the granule cell layer, sweeping around the Purkinje cell perikarya (P), to cross the molecular layer with their main axis more-or-less perpendicular to the surface of the folia. In the middle to outer third of the molecular layer, the fibers bifurcate in the typical cortical pattern of T-bifurcations (see text) to give off branches that run parallel to the outer pial surface. Marker bar = 100 μm. (Unpublished micrograph by L. Koda, E. L. Battenberg and F. E. Bloom).

Hypothalamus   As noted in the description of pathways, one group of locus coeruleus fibers appears to enter the periaqueductal gray directly to form a component of the dorsal longitudinal fasciculus. This group of fibers forms a plexus within the periaqueductal gray, which continues into the diencephalon as a component of the CA innervation of the periventricular nucleus of the hypothalamus. It continues rostrally to the most rostral portion of that nucleus (Lindvall & Björklund 1974b, 1978, Jones & Moore 1977). In the view of Lindvall & Björklund (1974b), parts of the periventricular system give rise to fibers innervating the dorsomedial nucleus, the paraventricular nucleus, and the supraoptic nucleus, which has a dense, pericellular innervation. This constitutes the major hypothalamic innervation from the locus coeruleus. As described below, the major hypothalamic innervation arises from brainstem NE cell groups (Table 2).

Thalamus and telencephalon   In addition to the projections of the hypothalamus there are three other ascending locus coeruleus terminal innervations that should be considered. The first of these is to the basal forebrain, the second to the thalamus, and the third to the cerebral neocortex. With the possible exceptions of the basal ganglia, the olfactory tubercle, and the nucleus accumbens, the entire telencephalon appears to receive some input from the locus coeruleus. This varies greatly in extent and appearance and no attempt is made to describe the entire terminal system. Some points can be emphasized, however, in regard to the relationship of locus coeruleus projections to other catecholamine innervation. For example, there is a locus coeruleus projection to virtually all components of the amygdaloid complex (Fallon et al 1978) and to nearly all major components of the olfactory system including the olfactory bulb, anterior olfactory nucleus, and piriform cortex (Fallon & Moore 1978). The innervation to the amygdala includes the central amygdaloid nucleus, which also contains an innervation from caudal brainstem NE cell groups and extremely dense dopamine (DA) innervation. The septal area shows similar patterns (Moore et al 1971, Moore 1975, 1978). There is a locus coeruleus innervation to the medial septal nucleus and to the nucleus septofimbrialis, which has the typical plexiform appearance of locus innervation. There is a locus innervation to the nucleus of the diagonal band but this, too, is mixed with an innervation from the caudal brainstem NE cell groups. In the interstitial nucleus of the stria terminalis and in the lateral septal nucleus, the locus coeruleus innervation is also mixed with a NE innervation from caudal brainstem cell groups and with an innervation from mesencephalic DA cell groups. The same appears to be true in the anterior olfactory nucleus and in the olfactory bulb, although the total catecholamine innervation to these latter areas is sparse.

Table 2   The distribution of norepinephrine axons to the hypothalamus in the rat[a]

Hypothalamic nucleus or area	Density of innervation[b]
Anterior hypothalamus	
Preoptic periventricular nucleus	4–5+
Medial preoptic area	3–4+
Lateral Preoptic area	2–4+
Anterior hypothalamic nucleus	2+
Anterior hypothalamic area	2–5+
Suprachiasmatic nucleus	1+
Supraoptic nucleus	4+
Paraventricular nucleus	5+
Tuberal hypothalamus	
Retrochiasmatic area	4–5+
Periventricular nucleus	5+
Arcuate nucleus	3+
Ventromedial nucleus	1+
Dorsomedial nucleus	5+
Ventral tuberal area	4–5+
Dorsal tuberal area	1–2+
Lateral hypothalmic area	2–3+
Mammillary (posterior) hypothalamus	
Premammillary nucleus	3+
Posterior hypothalamic nucleus	3+
Medial mammillary nucleus, medial part	2–3+
Medial mammillary nucleus, lateral part	2+
Posterior mammillary nucleus	1+
Lateral mammillary nucleus	2+
Supramammillary nucleus	2+
Tuberomammillary nucleus	4+
Lateral hypothalamic area	2–3+

[a] Original observations from Fuxe (1965) and S. S. Mosko and R. Y. Moore (unpublished observations). The innervation of hypothalamic nuclei and fields is almost exclusively from lateral tegmental NE and E neuron groups. Exceptions to this are locus coeruleus neuron projections to the supraoptic nucleus, dorsomedial nucleus, paraventricular nucleus, and periventricular nucleus.
[b] 5+, very dense innervation; 4+, dense innervation; 3+, moderate innervation; 2+, light innervation; 1+, scattered varicosities.

A further example of an area which receives a CA innervation exclusively from the locus coeruleus is the hippocampal formation (Figure 4). This innervation reaches the hippocampal formation from three sources, the ansa peduncularis-ventral amygdaloid bundle system, the fornix, and the cingulum. Fibers entering from the ansa peduncularis-ventral amygdaloid bundle terminate predominantly in the ventral portion of the hippocampal formation whereas those from the cingulum terminate in the dorsal part of Ammon's horn. The fornix fibers terminate predominantly in the stratum

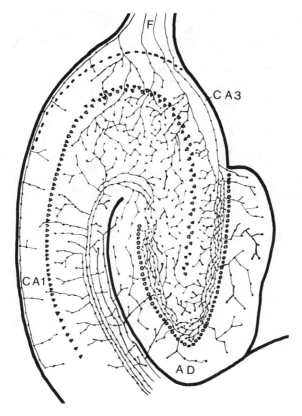

Figure 4 Diagram of a horizontal section of the hippocampal formation showing the pattern of locus coeruleus NE axon innervation. See text for description. F, fornix; CA1, CA3, fields of Ammon's horn; AD, area dentata. (From observations of Moore 1975; R. Loy and R. Y. Moore, unpublished).

radiatum of CA3 and in the area dentata. The innervation of the hippocampal formation is plexiform, as is true of other areas, but its organization is also determined by the laminar structure of the hippocampal formation. There is a dense plexus of fibers that runs through the stratum lacunosummoleculare of CA1 with fibers turning deeply along the apical dentrites of the hippocampal pyramidal cells. These may run to the apex of the dendrite and then branch laterally along the pyramidal cell layer. There is a dense plexus of fibers in the stratum radiatum of CA3 and a very dense plexus within the hilus of the area dentata. This appears to be oriented predominantly as radial fibers running beneath the granule cell layer, many of them in the zona limitans of Cajal (1911). The molecular layer of the area dentata has a typical, loose plexiform appearance (Figure 5). As shown below, this is a variant of a general cortical pattern of innervation.

Figure 5   Molecular layer of area dentata with typical plexiform innervation by locus coeruleus NE axons. The pial surface is toward the top of the photomicrograph. Marker bar = 20 μm. (From R. Y. Moore, unpublished).

The two major components of the ascending locus coeruleus innervation are to the dorsal thalamus and to the cerebral cortex. The innervation to the dorsal thalamus has been described in detail by Lindvall et al (1974). Virtually the entire dorsal thalamus is innervated (Table 3). The most dense innervation is evident in the anterior nuclei, and particularly, in the antero-ventral nucleus. The remainder of the dorsal thalamic nuclei show a plexiform arrangement of varying density, with fibers throughout all of the nuclei. These include the geniculate nuclei. In particular, the dorsal lateral geniculate nucleus has a quite dense plexiform innervation that appears to be predominantly axodendritic in arrangement.

The entire neocortex receives innervation from the locus coeruleus (Fuxe et al 1968, Ungerstedt 1971, Levitt & Moore 1978). The pattern of this innervation is similar from area to area and, again, it is generally the typical plexiform pattern of locus coeruleus innervation (Figure 6). Preterminal locus coeruleus fibers running in the external capsule turn vertically into the cortex and undergo an extensive collateralization. The collaterals in the fifth and sixth layers are oriented predominantly radially along the layers and

Table 3 Locus coeruleus NE neuron innervation of the dorsal thalamus in the rat[a]

Dorsal thalamic nucleus	Density of innervation
Anterior nuclear group	
Anterodorsal nucleus	3+
Anteromedial nucleus	4+
Anteroventral nucleus	5+
Lateral nuclear group	
Lateral dorsal nucleus	3+
Lateral posterior nucleus	2+
Posterior nucleus	2+
Ventral nuclear group	
Ventral anterior nucleus	3+
Ventral lateral nucleus	3+
Ventromedial nucleus	3+
Ventrobasal complex	3+
Lateral geniculate nucleus, dorsal	4+
Lateral geniculate nucleus, ventral	3+
Medial geniculate nucleus	3+
Midlind intralaminar nuclear group	
Central medial nucleus	2+
Central lateral nucleus	2+
Parataenial nucleus	2+
Nucleus reunieus	1+
Rhomboid nucleus	2+
Medial nuclear group	
Dorsal medial nucleus	3+

[a]The locus coeruleus NE innervation of the dorsal thalamic nuclei was graded subjectively as follows: very dense innervation, 5+; dense innervation, 4+; moderate innervation, 3+; light innervation, 2+; scattered varicosities, 1+. Original observations from Lindvall et al (1974) and Moore & Kromer (unpublished observations). The paraventricular nucleus is probably epithalamic in origin. It has a moderate locus coeruleus NE innervation and a dense brainstem CA innervation, which is, at least in part, from E neurons (Hökfelt et al 1974).

these give rise to ascending and descending branches. The fibers are typically of the terminal type with regularly spaced varicosities along the fine intervaricose axonal segment. Other fibers continue vertically through the fourth and third layers where there is less extensive collateralization into the layers as radial fibers. This again becomes more extensive within the second layer, and as the vertical fibers reach the molecular layer they form T-branches or turn horizontally to run parallel to the pial surface. The fibers give off frequent short branches that contain varicosities of the terminal type. The innervation of the molecular layer is much greater than that of other layers of cortex and this is particularly true in the cingulate and in the frontal cortex.

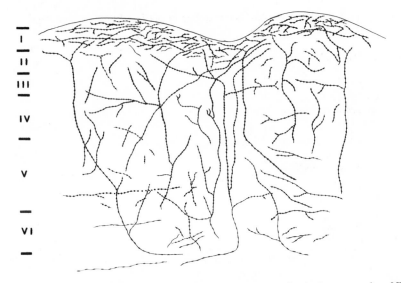

Figure 6   Diagram of the typical pattern of neocortical innervation by locus coeruleus NE axons. See text for description. (From Levitt & Moore 1978).

ULTRASTRUCTURAL PROPERTIES OF LOCUS COERULEUS BOUTONS

The ultrastructural identification of the terminals of locus coeruleus neurons, and other catecholamine-producing neuron terminals, was not accomplished as easily as the identification of NE-containing terminals of the sympathetic autonomic ground plexus. In the periphery, after a variety of fixation procedures, the terminals of sympathetic neurons exhibit small, densely granular synaptic vesicles about 500 Å in diameter (Grillo 1966). In the central nervous system, NE-containing terminals do not show small, granular vesicles with routine fixation and staining. With special treatment, however, these terminals can be identified. Three techniques are in current use: potassium permanganate fixation with or without exposure to catecholamine analogs (Hökfelt 1967), pretreatment with 5-hydroxydopamine followed by aldehyde-osmium or permanganate fixation (Tranzer & Thoenen 1967), and superfusion or CSF injection of tritiated NE followed by electron microscopic autoradiography (Descarries & Droz 1970). Immunohistochemistry may also eventually be usable (Pickel et al 1977). Each has its special advantages and disadvantages, but these are beyond the scope of the present review and the reader is referred to the original papers and to Bloom (1973) and Hökfelt (1968) for reviews. By combining permanga-

nate postfixation with a brief perfusion with paraformaldehyde and glyoxylic acid, direct visualization of small granular vesicles is obtained with no
pretreatment in most brain regions (Koda & Bloom 1977).

In areas known to be innervated by locus coeruleus neuron axons, and
in which there is no other CA neuron innervation, such as the cerebellum
(Bloom et al 1971, Landis & Bloom 1975), hippocampal formation (Koda
& Bloom 1977, Figure 7), and cerebral cortex (Descarries & Lapierre 1973,
Lapierre et al 1973, Descarries et al 1977) NE terminals have been demonstrated ultrastructurally by one or more of these methods. These terminals
typically contain pleomorphic, small vesicles approximately 500 Å in diameter but often also large, dense-core vesicles. When the material is prepared
by the permanganate or 5-hydroxy-dopamine methods, the small vesicles
are also dense cored.

The most exhaustive study of the ultrastructure of neocortical NE terminals has been carried out by Descarries and his associates (Descarries &
Lapierre 1973, Lapierre et al 1973, Descarries et al 1977). In this series of
studies these workers demonstrated that NE axons and terminals can be
selectively labeled by superfusion of the cerebral cortex with tritiated NE
followed by fixation with glutaraldehyde and electron microscopic autoradiography. The labeling of axons and terminals is blocked by pretreatment with 6-hydroxydopamine or by desmethylimipramine, a blocker of
NE uptake. Uptake was not altered by pretreatment with 5,6-dihydroxy-
tryptamine, a selective serotonin neuron neurotoxin. The labeled axons
appear oriented parallel to the pial surface in layer I and perpendicularly
or obliquely in layers II–IV. This agrees with the fluorescent histochemical
studies of Fuxe et al (1969) and Levitt & Moore (1978). Terminals were
identified in the first study (Descarries & Lapierre 1973) making contacts
with dendrites, but no quantitative analysis of the number of terminals or
other parameters was attempted. Subsequently, Lapierre et al (1973) demonstrated that the mean diameter of locus coeruleus NE axon terminals in
cerebral cortex is 1.15 μm, more than 35% of labeled terminals are present
in the molecular layer, the total number of NE nerve endings in cortex is
approximately 96,000 per cu. mm. of cortex, which represents a mean
incidence of NE terminals of approximately 1 per 8800–14,500 cortical
synapses. Finally, in an exhaustive topometric ultrastructural analysis of
serial sections through NE terminals, Descarries et al (1977) concluded that
only 5% of the labeled terminals make synaptic complexes with postsynaptic elements. Of these, approximately half make asymmetrical (Gray type
I) and half symmetrical (Gray type II) contacts. The remaining labeled
terminals were not observed to make identifiable synaptic complexes. This
is in contrast to other, unlabeled terminals in the sections, 50% of which
make typical synaptic complexes.

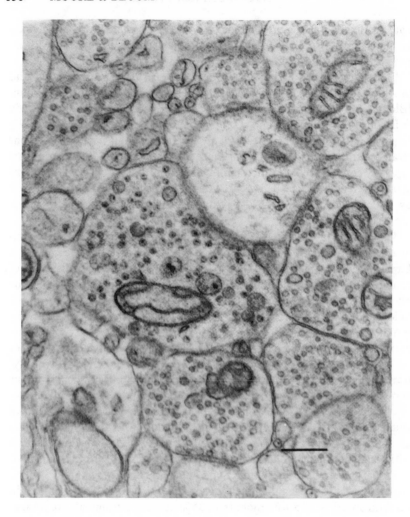

Figure 7 Electron micrograph of hilus of dentate gyrus of normal untreated rat, prepared by the permanganate/glyoxylic method of Koda & Bloom (1977). A medium sized bouton containing small and large synaptic vesicles with electron dense granular material can be seen making contact with the perimeter of a dendrite around which several boutons containing agranular vesicles can also be seen to make contact. The bouton, which has granular synaptic vesicles, exhibits a discrete area of contact over which the synaptic vesicles are gathered close to the membrane surface abutting the dendrite at which point both the presynaptic and postsynaptic surfaces of the contact exhibit some signs of paramembranous material suggestive of synaptic specializations. Marker bar = 0.5 μm. (From Koda & Bloom 1977).

The interpretation of this difference is not evident from the data currently available. Several alternatives are available. First, it may simply imply a lack of specificity of locus coeruleus axons for postsynaptic elements. In this case, stimulation of the locus coeruleus would presumably result in the release of NE from all of the terminals and the effects of this stimulation would depend upon the presence of appropriate receptors in nearby neural elements. This organization would not be unlike that of the peripheral sympathetic axon terminals and the structures they innervate (Hökfelt 1968). Second, the data could imply, as Descarries et al (1977) have suggested, that these are highly plastic elements, continually making new synaptic contacts (cf. Sotelo & Palay 1971, Moore et al 1971, for discussion), so that the small number of terminals apparent in a synaptic complex only reflects the extreme plasticity of the system. A third alternative is that our morphological definition of synaptic complex as equivalent to synapse, in the functional sense, may be too restrictive and requires further study (see Ramon-Moliner 1977).

In addition, it is not known whether the pattern observed in cerebral cortex will hold for other areas whose catecholamine innervation is solely from locus coeruleus axons. In the hippocampus of the rat, Koda & Bloom (1977) and their colleagues (Koda et al 1978a, b) have made a detailed examination of the endogenous NE-containing boutons as revealed by permanganate fixation without the necessity of superfused or topical norepinephrine. With this fixation procedure, NE terminals contain the characteristic small granular vesicles and become relatively easily identified ultrastructurally. The conclusion that the boutons exhibiting small granular vesicles are equivalent to the NE-containing boutons is derived from several pieces of data. The distribution of the small granular vesicle boutons can be correlated quite closely with the topographic distribution of glyoxylic acid-induced fluorescent boutons in the dentate gyrus and adjacent hippocampal lamina (Koda & Bloom 1977); both the fluorescent boutons and the small granular vesicle boutons disappear following treatments that decrease brain NE, such as reserpine, intracisternal injections of 6-hydroxydopamine (Koda & Bloom 1977), surgical transection of the ascending dorsal CA bundle, or electrical stimulation of the locus coeruleus (Koda et al 1978a). Discrete lesions of the locus coeruleus following microinjections of 6-hydroxydopamine (Koda et al 1978b) produce dramatic, but incomplete loss of glyoxylic acid-induced fluorescence in axon varicosities and permanganate reactive small granular vesicle-containing boutons, suggesting that there may well be significant crossed innervation between the locus coeruleus and the hippocampus (see above). In the dentate gyrus, when the incidence of specialized contacts for small granular vesicle-containing boutons was compared with all other boutons in the same fields, there were no differences; both groups of boutons showed specialized junctional contacts in about 18–20% of the boutons tabulated (Koda et al 1978b). In the cerebral cortex of the immature rat, after parenteral injection of 5-hydroxydopamine (while the blood-brain barrier is permeable to catecholamine analogs), those presumptive NE terminals identified as small granular vesicle-containing also show a very high incidence of specialized synapse-like contacts (Coyle & Molliver 1977). These synaptic-like contacts (Coyle & Molliver 1977) were seen mainly in the deeper cortical layers, rather than in the superficial layers, which were those most densely labeled in the studies by Descarries et al (1977) on the adult rat cortex.

As in all other cases of attempted ultrastructural-physiological correlation, the assumption that the specialized contact zone is *the* site of synaptic transmission remains to be documented with certainty. It is clear that sympathetic fibers of the peripheral autonomic nervous system transmit to smooth muscle without such specializations, but do show them at intraganglionic nerve-nerve contacts (see Grillo 1966). In this respect it is of interest that Swanson et al (1977) recently have reported LC NE neuron terminals innervating blood vessels within the brain parenchyma. In view of the observations reviewed by Bevan (1977) which suggest that receptiveness and receptor distribution and sensitivity bear close correlations with the separation between release and response sites, accurate determination of central adrenoreceptors with reference to identified adrenergic boutons seems to be the only ultimate solution to this dilemma.

One final point concerning locus coeruleus NE neuron terminal fields requires discussion. From the work of Lapierre et al (1973) it is suggested that each cu. mm. of cerebral cortex contains approximately 96,000 locus NE terminals. Since the mean diameter of the terminals is approximately 1 μm, the interterminal axon averages about 2 μm, and the mean number of locus coeruleus neurons per rat brain is about 3000 (Swanson 1976a), one can calculate the value for the mean number of terminals and the total axon length of each locus neuron if the volume of the areas known to receive a locus projection and their mean NE content is known. These latter figures have been estimated (Moore, unpublished observations) and, using them, each locus coeruleus neuron can be calculated to have an axon at least 30 cm in length, in terminal areas alone, with 100,000 terminals. Considering the DA content of the neostriatum, the density of DA axon innervation of the neostriatum, and the number of substantia nigra neurons (Andén et al 1966), these figures for locus coeruleus neurons appear appropriate as the density of NE innervation in any area is much less, the neurons are fewer in number by only a relatively small amount, but the area innervated is much greater. The question unanswered at the present time is whether the locus coeruleus NE neurons project in any topographic manner. The evidence available suggests that, if any topography exists in the projection, it is minor and that locus coeruleus NE neurons all project over wide areas of the neuroaxis. As with other problems outlined above, this requires a definitive experimental resolution.

PHYSIOLOGY

The Coeruleo-Cerebellar Noradrenergic Projection

CHARACTERIZATION OF THE PURKINJE CELL ADRENERGIC RE-CEPTOR   When NE is applied to Purkinje cells by iontophoresis, a uniform and powerful depression of spontaneous discharge results (Hoffer et al 1969, Siggins et al 1971a). Interspike interval histograms show that NE produces no effect on climbing fiber bursts or on the most probable single spike interval, but rather, that NE specifically augments the population of long pauses seen during normal Purkinje cell firing (Hoffer et al 1969, 1971b).

Several lines of evidence suggest that a beta receptor is involved in the Purkinje cell response to NE. For example, NE and isoproterenol produce changes in mean rate and in the interspike interval histogram analogous to those of NE (Siggins et al 1971d). Moreover, iontophoretic administration of MJ-1999 (Hoffer et al 1971a), or dichloroisoproterenol (Freedman et al 1975), specific beta-adrenergic antagonists, will block the NE responses. Iontophoresis of NE slows Purkinje cell discharge more effectively than adrenergic synaptic terminals are selectively destroyed by prior injection of 6-OHDA (Hoffer et al 1971c); NE inhibitory responses are seen in the early postnatal development of cerebellum (Woodward et al 1971). Furthermore, NE inhibits Purkinje cells in mice in which there is a congenital absence of most inhibitory interneurons (Siggins et al 1971d). These data all suggest that the NE effect is mediated postsynaptically on the Purkinje cell.

MECHANISM OF THE NORADRENERGIC RECEPTOR   By recording intracellularly from Purkinje cells during extracellular application of NE, or stimulation of locus coeruleus (Siggins et al 1971d, Hoffer et al 1973), a unique hyperpolarization is observed associated with either no change or with an increase in membrane resistance (Figure 8). Similar transmembrane changes after iontophoresis of NE have also been described in cat movements by Engberg & Marshall (1971) and in the hippocampus by Segal & Oliver (1975). The hyperpolarizing effect of NE that occurs with increased resistance is in direct contrast to changes seen with classical inhibitory postsynaptic potentials (IPSPs) or with iontophoresis of GABA (see Siggins et al 1971c). The classical inhibitory pathways and inhibitory amino acid transmitters are thought to operate exclusively through mechanisms that increase conductance to ionic species whose equilibrium potentials are more negative than the resting membrane potential. In such cases, the hyperpo-

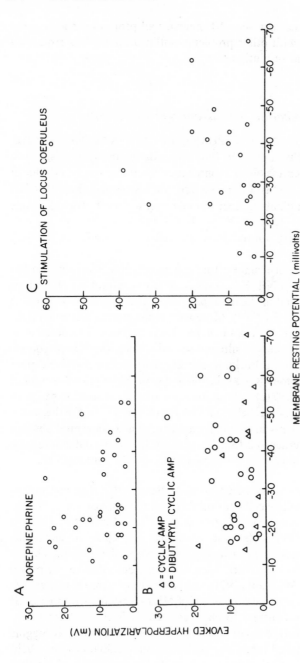

Figure 8 Relationship of the magnitude of drug or stimulus-evoked hyperpolarizations with initial (resting) membrane potential of individual Purkinje cells. Each point represents the peak hyperpolarizing response for a single test; displacement upward indicates a greater hyperpolarization. (*A*) Purkinje cell responses to iontophoresis of norepinephrine. (*B*) Responses to iontophoresis of the cyclic nucleotides, cyclic AMP, and dibutyryl cyclic AMP. (*C*) Responses to stimulation of the locus coeruleus with trains of 100 to 120 pulses at 10 sec. (From Hoffer et al 1973).

larization is associated with a decrease in membrane resistance. Hyperpolarization produced by NE, on the other hand, may be due to a decrease in conductance to some ion such as sodium or calcium, or to activation of an electrogenic pump (Phillis et al 1973). However, the depressant actions of NE do not require extracellular calcium (Geller & Hoffer 1977), nor do inhibitory substances, which also "antagonize" calcium actions, interfere selectively with NE inhibitory effects (Freedman et al 1975).

ACTIVATION OF THE COERULEO-CEREBELLAR NE PATHWAY   Purkinje cells show remarkably uniform inhibitory responses to stimulation of the locus coeruleus (LC); with trains of pulses, a complete cessation of discharge outlasts the stimulation period by 4–65 seconds (Hoffer et al 1973, Siggins et al 1971d; Figure 9). The latency of this inhibition is approximately 125 msec (Hoffer et al 1973). These inhibitory effects of LC in cerebellum are blocked by antipsychotic phenothiazines (Freedman & Hoffer 1975), by cobalt or lead (Freedman et al 1975), and by lithium (Henriksen, Siggins & Bloom, unpublished), as well as by prostaglandins (see Hoffer et al 1973). The effects of LC stimulation are also abolished when the synthesis and storage of NE are blocked pharmacologically (Hoffer et al 1973) and when the pathway is destroyed with 6-OHDA (Hoffer et al 1973).

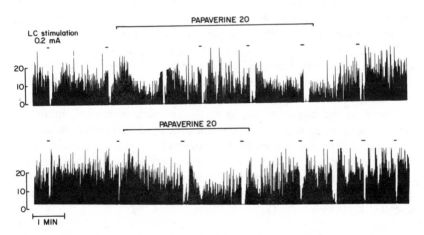

Figure 9   Papaverine, applied iontophoretically, potentiates the inhibitory actions of the noradrenergic locus coeruleus-hippocampus synaptic pathway. *A* and *B* indicate the responses during two different tests. Periods of LC stimulation (100 msec periods of 10 Hz at 0.2 mA) are indicated by the short horizontal bars; periods of papaverine iontophoresis are indicated by the brackets. In both cases, the effects of LC stimulation were potentiated by a 2–3 fold increase in duration with almost immediate recovery. Some direct slowing of the cell by papaverine can also be seen (From Segal & Bloom 1974b).

MECHANISM OF RESPONSES TO NE PATHWAY STIMULATION   Intracellular recording from some Purkinje cells during stimulation of LC with single shocks reveals long latency (around 150 msec) and relatively small hyperpolarizations. With trains of pulses to LC, large hyperpolarizations were observed that outlasted the stimulation period. One index of membrane resistance was obtained by measuring the size of climbing fiber excitatory postsynaptic potentials (EPSPs), and the potential deflections produced by hyperpolarizing currents passed through the recording micropipette by means of a Wheatstone bridge circuit. In all cases, input resistance, as measured by these two parameters, either increased or did not change during the LC evoked hyperpolarizations. Thus, LC stimulation exactly mimics the action of exogenous NE; both produce hyperpolarization with no decrease and frequent increases in membrane resistance (Hoffer et al 1973, Siggins et al 1971c, d).

THE COERULEO-HIPPOCAMPAL PROJECTION   The hippocampal cortex receives an extensive input of NE-containing fibers that contact pyramidal cells and cells of the dentate gyrus (Blackstad et al 1967, Moore 1975, Koda & Bloom 1977). Physiological studies indicate that the hippocampal NE projection from LC produces cellular effects virtually identical to those of the locus coeruleus on cerebellar Purkinje cells; in the hippocampus, both LC and NE slow pyramidal cell discharge with long latency and long duration actions (Segal & Bloom 1974a, b, 1976a, b, Figure 9), the receptor is also blocked by MJ-1999, prostaglandins of the E series, lithium (Segal 1974), and phenothiazines (Segal & Bloom 1974b, 1976a). Moreover, the action of the pathway is blocked by chronic pretreatment with 6-OHDA or acute pretreatment with reserpine and alpha methyl tyrosine or with inhibitors of dopamine beta hydroxylase (Segal & Bloom 1974b, 1976a, b). Furthermore, in the hippocampus, as in the cerebellum, the NE inhibitory actions appear to be mediated postsynaptically by similar molecular mechanisms (see below; and Bloom 1975a). However, of great interest in regard to generality of release sites and receptive sites in unrestrained rats, the effects of LC stimulation were only observed with cells showing bursting discharges typical of pyramidal cells (Ranck 1973) and were not seen with so-called theta-type cells (Segal & Bloom 1976a, b).

SEPTUM   Segal (1976) has recently reported the effects observed on neurons in the medial septal nuclei and in the diagonal band of Broca when the locus coeruleus is stimulated electrically. LC stimulation was found to produce, as in the areas already described, a long latency (30–100 msec) and long duration (100–300 msec) inhibition of spontaneously active cells. These inhibitory effects were blocked by depletion of NE stores or by pretreatment with 6-OHDA. When septal neurons were examined for changes in basal firing rate 3 or more weeks after transection of the fornix (that is under conditions in which some sprouting of the intact NE innervations has been observed; Moore et al 1971), Segal (1976) observed that firing patterns and rates were significantly slower than in untreated animals, suggesting that the enhanced NE innervation was functionally more effective in producing tonic inhibition. Conversely, when the NE pathway was destroyed by treatment with 6-hydroxydopamine, septal neurons fired significantly faster than normal. The results of this study are thus in general agreement with those reported earlier for other LC-NE target regions.

THALAMUS AND BRAINSTEM   Inhibitions of neurons in two further areas have been reported with LC stimulation. In the cat lateral geniculate nucleus, Nakai & Takaori (1974) observed that interneurons were depressed following locus coeruleus stimulation, while primary optic nerve relay cells gave greater evoked responses during locus stimulation. Effects of LC stimulation were abolished by pretreatment with reserpine, and restored in such cases by parenteral L-DOPA or intraventricular NE. Studies that would confirm the precise termination site of the noradrenergic projection in this and other thalamic nuclei (Lindvall et al 1974) would be very helpful. Characterization of these receptors pharmacologically also remains to be accomplished.

In the spinal trigeminal nucleus, also known to be rich in NE terminals, and a presumed target of LC fibers, Sasa et al (1974, 1975; Table 4), have reported that LC stimulation inhibits orthodromic activation of these cells. This inhibition is also blocked by morphine or reserpine and restored by intravenous L-DOPA or intraventricular NE (Sasa et al 1975).

CYCLIC 3' 5'-ADENOSINE MONOPHOSPHATE AND NE SYNAPTIC EFFECTS   With the demonstration that rat cerebellar Purkinje and hippocampal pyramidal neurons are contacted by LC axons and give uniform inhibitory response to NE and to LC stimulation, subsequent studies investigated further the mechanism of this inhibitory response. In both test systems, cyclic 3'5'-adenosine monophosphate (cyclic AMP) mimicked the ability of NE to depress spontaneous activity (Hoffer et al 1969, 1973, Segal & Bloom 1974a, b, Segal & Oliver 1975, Siggins et al 1971c, d; Figure 8).

In order to pursue the possibility that this action of cyclic AMP in the CNS might indicate another example of a "second messenger" hypothesis (Sutherland et al 1968) for central norepinephrine synapses, many additional experiments have been conducted (see Hoffer et al 1969, 1971b, 1973, Bloom 1975a, b). Biochemical evidence had already suggested that NE

Table 4 Distribution and origin of norepinephrine innervation of brainstem in the rat[a]

Brainstem nucleus or area innervated	Density of innervation[b]	Origin of innervation[c]
Midbrain		
Superior colliculus	2+	LC
Inferior colliculus	2+	LC
Lateral tegmental field	2–3+	LT (LC)
Periaqueductal gray	2–5+	LT + LC
Oculomotor and trochlear nuclei	0–1+	LT
Red nucleus	1+	LC
Substantia nigra, pars compacta	1+	LT (LC)
Substantia nigra, pars reticulata	0	—
Ventral tegmental area	1+	LT + LC
Nucleus linearis raphe	1+	LT
Nucleus dorsalis raphe	2+	LT
Nucleus centralis superior	1+	LT
Tegmental reticular nucleus	2+	LT + LC
Dorsal tegmental nucleus (Gudden)	0	—
Ventral tegmental nucleus (Gudden)	4+	LT
Nuclei of lateral lemniscus	2+	LC
Pons-medulla		
Main sensory trigeminal nucleus	2+	LC
Spinal trigeminal nucleus	2+	LC + LT
Mesencephalic trigeminal nucleus	4+	LT
Nucleus locus coeruleus	2+	LT
Marginal nucleus of brachium conjunctivum	4+	LT
Periventricular gray	1+	LT
Reticular formation, pons	2+	LC + LT
Nucleus raphe magnus	3–4+	LT
Pontine nuclei	2–4+	LT + LC
Superior olivary complex	0–1+	LT
Cochlear nuclei	2+	LC
Vestibular nuclei	0	—
Nucleus tractus solitarius	5+	LT
Cuneate nucleus	0–1+	LT
Gracile nucleus	1–2+	LT
Reticular formation, medulla	2+	LT
Nucleus raphe pallidus	3+	LT
Nucleus raphe obscurus	5+	LT
Motor trigeminal nucleus	4+	LT
Abducens nucleus	0	—
Facial nucleus	3–4+	LT
Hypoglossal nucleus	2–3+	LT
Dorsal motor nucleus of vagus	5+	LT
Area postrema	1+	LT

[a] Observations from P. R. Levitt and R. Y. Moore (unpublished)
[b] Density of innervation, 0–5+, as in Table 3.
[c] Origin of innervation; LC, locus coeruleus; LT, lateral tegmental and dorsal medullary NE neurons. Some of this innervation may be from E neurons. In some areas the innervation overlaps. When one source is in parentheses, it indicates a minor contribution.

could elevate cerebellar cyclic AMP levels through beta-adrenergic receptors (see Rall 1972, also see refs in Bloom 1975b). Recently, Atlas and her colleagues have developed methods for direct microscopic localization of β-receptors (Atlas & Segal 1977, Atlas et al 1977, Melamed et al 1977). With this method, Purkinje neurons in the cerebellum, and pyramidal neurons in the hippocampal and cerebral cortex seem to be the major cell classes showing β-receptors. Furthermore, parenteral administration of phosphodiesterase inhibitors such as aminophylline or theophylline potentiated NE depressions of Purkinje cells, while iontophoretic administration of these methylxanthines, and of papaverine, converted weak excitant actions of iontophoretic cyclic AMP into pronounced depressions (Hoffer et al 1969, 1971a, b, 1973, Siggins et al 1971a; Figure 9). These observations led to the proposal that the action of NE (Hoffer et al 1969), and later that of the NE-mediated locus coeruleus synaptic projection to Purkinje cells (Siggins et al 1971c, d) could be mediated by cyclic AMP. Subsequently, the proposal has been strengthened by observations that the actions of NE (Siggins et al 1971c), the NE pathway (Hoffer et al 1973, Siggins et al 1971d), and the cyclic nucleotide (Siggins et al 1971c) all hyperpolarize Purkinje cells through similar membrane actions in which conductance to passive ion flow is decreased or unchanged. The cyclic AMP mediation of the NE actions also finds support from the observations that prostaglandins and nicotinate (Siggins et al 1971a, b) will selectively block NE effects on Purkinje cells and hippocampal pyramidal cells (Segal & Bloom 1974a,b) as they do on the cyclic AMP mediated adrenergic responses of adipocytes (Sutherland et al 1968). In fact, all the substances shown to block the response of Purkinje neurons to iontophoretic NE, also show very good antagonism towards the activation of cerebellar adenylate cyclase by NE (see Nathanson et al 1976, Nathanson 1977). Even more direct confirmation of the second messenger hypothesis stems from the observation that NE and the NE pathway will increase the number of Purkinje cells reacting positively to a immunocytochemical method detecting bound intracellular cyclic AMP (see Bloom 1975b). The alternative suggestion that actions of cyclic AMP are mediated by conversion to adenosine is disproven by the observations in cerebellum and the cerebral cortex that methylxanthine phosphodiesterase inhibitors potentiate NE and cyclic AMP yet block effects of adenosine or 5'AMP (see Siggins et al 1971a, Hoffer et al 1973, Stone & Taylor 1977). With cultured Purkinje neurons, Gahwiler (1976) has observed potentiation of NE and cyclic AMP depressions with phosphodiesterase inhibitors. He also observed that the thresholds of responses to NE were 100–1000 times lower than for responses to cyclic AMP (applied by superfusion), which is in keeping with predictions from other instances of second messenger mediation (see Sutherland et al 1968, Bloom 1975b, Nathanson 1977).

Finally, cyclic AMP is known to activate a class of enzymes, termed protein kinases (see Greengard 1976), which phosphorylate specific protein substrates; this process may be the mode of expressing altered cyclic AMP levels through functional changes in these protein substrates (see Greengard 1976). In cerebellum, chemical analogs of cyclic AMP mimic the inhibitory effects of NE in direct correlation to their ability to activate brain cyclic AMP dependent protein-kinase (Siggins & Henriksen 1975).

The Lateral Tegmental Norepinephrine System

CELL BODIES   The cell bodies of the lateral tegmental norepinephrine system in the rat were first described by Dahlström & Fuxe (1964). Using their nomenclature, a large group in the vicinity of the lateral reticular nucleus was termed A1, a smaller group within the dorsal accessory inferior olive was termed A3, and a large group beginning about the level of the descending root of the facial nerve and extending rostral to it was termed A5. In addition, they described a group of scattered neurons ventral to the locus coeruleus extending rostral and lateral to it, termed A7. Similar descriptions have been made by Ungerstedt (1971), Palkovits & Jacobowitz (1974), Swanson & Hartman (1975), Dupin et al (1976), and Lindvall & Björklund (1978). In addition, nearly identical cell groups have been identified in several primates (DiCarlo et al 1973, Felten et al 1974, Garver & Sladek 1975), including man (Nobin & Björklund 1973, Olson et al 1973).

The location of the cell bodies in the rat is shown diagrammatically in Figure 10. There are two major groups of NE perikarya within the lateral tegmental field, as that is defined by Berman (1968). One is medullary and located dorsal and lateral to the lateral reticular nucleus. The neurons are medium-sized, multipolar and scattered through the lateral tegmental field. More rostrally, in front of the lateral reticular nucleus they become less numerous but scattered neurons are found in the lateral tegmental field nearly continuously from the caudal medulla to the caudal mesencephalon. The second major group occurs at the pontine level and is first noted as an expansion of the number of cells in the lateral tegmental field at the medial

——➤

Figure 10   Diagrams of coronal sections, A–F, from rostral to caudal through the pons and medulla. The sections were prepared by the GA method. Black dots represent the location and approximate density of NE neuron cell bodies in the brainstem. Abbreviations: A, cerebral aqueduct; AP, area postrema; BP, brachium pontis; CN, cochlear nuclei; IO, inferior olivary complex; IC, inferior colliculus; LC, locus coeruleus; LL, lateral lemniscus; LR, lateral reticular nucleus; LT, lateral tegmental field; NTS, nucleus tractus solitarius; P, pyramid; RB, restiform body; SOC, superior olivary complex; VN, vestibular nuclei; 5M, trigeminal motor nucleus; 5S, sensory trigeminal nucleus; 7, facial nerve; 7N, facial motor nucleus; 10, dorsal motor nucleus of the vagus; 12, hypoglossal nucleus. (From R. Y. Moore, unpublished).

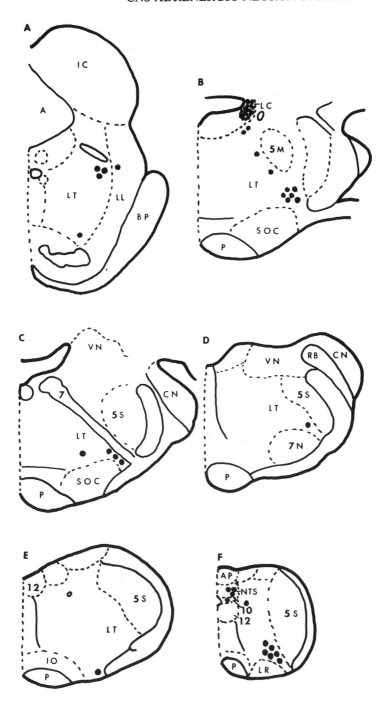

border of the emerging root of the 7th nerve. This group reaches its greatest extent at mid-levels of the superior olivary complex where it occupies the lateral tegmental field lateral to the lateral superior olivary complex. At this level it is continuous with a small group of scattered NE neurons extending from the ventral part of the locus coeruleus through the tegmentum medial to the motor trigeminal nucleus. This scattered cell group, sometimes referred to as the "subcoeruleus" group is continuous rostrally with another group of neurons in the lateral tegmental field, located medial to and partially within the ventral nucleus of the lateral lemniscus and rostral to the locus coeruleus. This group has also been included within the "subcoeruleus" group of some authors (Olson & Fuxe 1972, Ungerstedt 1971). In some animals, notably the cat (Chu & Bloom 1974a, Jones & Moore 1974), there appears to be no distinction between the locus coeruleus and the adjacent cell groups in the dorsolateral isthmic tegmentum.

As reviewed in detail in a later section, the caudal medullary NE cell group and the pontine NE cell groups appear to arise embryologically from two distinct primordia and then undergo a secondary differentiation into subgroups. This further appears to have significance in terms of projections. That is, although both the medullary (including the dorsal medullary) cell groups and the pontine cell groups (including the locus coeruleus) have an ascending and descending axonal collateralization typical of reticular formation neurons in general (cf Scheibel & Scheibel 1957), the caudal groups tend to project predominantly caudally while the rostral groups project predominantly rostrally. There are exceptions to this that we shall note but the general rule appears to hold. No ultrastructural studies of the lateral tegmental cell groups have been carried out.

PATHWAYS—ASCENDING   The pathways arising from the lateral tegmental NE cells are shown diagrammatically in Figure 11. The caudal medullary cell group gives rise to ascending axon collaterals that enter the central tegmental tract and run rostrally in the dorsal, lateral component of that tract. At the level of the superior olivary complex they are joined by a major contribution of axons from the pontine lateral tegmental cell group and, further rostrally, by a contribution from the cells ventral, rostral, and lateral to the locus coeruleus. It should be noted in passing that the ascending CA fibers are mixed in the central tegmental tract with descending fibers. Because of the mixture of ascending and descending components in this system, it has been extremely difficult to elucidate the source of innervation of brainstem nuclei. The pontine and medullary fibers of the central tegmental tract pass through, and ventral to, the decussation of the superior cerebellar peduncles. Rostrally, the fibers disperse among the fibers of the tegmental CA radiations (Lindvall & Björklund 1974b, 1978).

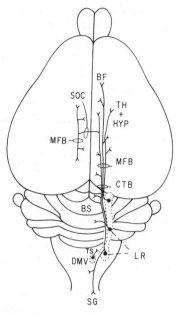

Figure 11   Diagram of the location and projections of the brainstem NE neuron groups, excluding the locus coeruleus, presented in a horizontal view. See text for description. Abbreviations: BF, basal forebrain; BS, brainstem; CTB, central tegmental bundle; DMV, dorsal motor nucleus of the vagus; HYP, hypothalamus; LR, lateral reticular formation; MFB, medial forebrain bundle; SG, spinal cord gray; SOC, supraoptic decussation; TH, thalamus; TS, nucleus tractus solitarius. (From R. Y. Moore, unpublished).

Some of the fibers appear to leave the radiations very shortly after entering them and resume an ascending course in a more ventral and lateral position than previously. Other fibers follow the tegmental radiations ventrally and join the most ventral portions of the central tegmental tract running rostrally into the medial forebrain bundle.

It should be emphasized that both components of ascending NE axons in the central tegmental tract are either contiguous with, or contain, fibers from the ascending locus coeruleus system and the periventricular system (Lindvall & Björklund 1974b, 1978, Jones & Moore 1977). Consequently, the central tegmental tract, in addition to its contributions from other, nonmonoaminergic systems, has mixed components from the CA neuron systems. This again reinforces the view that these systems form a component of the general reticular formation connections and are distinguished only on the basis of transmitter content. The ascending NE fibers in the central tegmental tract that penetrate the tegmental radiations have wide distribution in the mesencephalic tegmentum. Rostrally, at the junction of the mesencephalon and diencephalon, these fibers turn ventrally to pass through and partly medial to the medial lemniscus and then run broadly into the zona incerta and the prerubral field of Forel. From this position some fibers turn laterally in the ventral amygdaloid bundle-ansa peduncularis system to penetrate the internal capsule. This continues along the course of the ascending bundle. The fibers continuing rostrally join those of the remaining ascending lateral tegmental NE axons in the medial forebrain bundle to continue rostrally through the diencephalon. During their course they give rise to one major decussation in the dorsal supraoptic commissure. As the fibers join those of the contralateral medial forebrain bundle they appear both to ascend and descend but their further course and sites of termination cannot be discerned.

PATHWAYS—DESCENDING   The major descending projection from the lateral medullary NE cell groups appears to arise from the caudal medullary cell group and enter the spinal cord as a bulbospinal NE system. As noted above for the ascending systems, however, the descending systems are complex and include components from rostral, lateral tegmental NE cell groups, as well as the locus coeruleus. The bulbospinal system was identified by Carlsson et al (1964) and analyzed in detail by Dahlström & Fuxe (1965) and more recently by Nygren & Olson (1977). The latter authors identified two components to the bulbospinal system. One descends in the ventral funiculus and the ventral part of the lateral funiculus of the spinal cord white matter. These are largely uncrossed. A somewhat smaller system descends in the dorsal part of the lateral funiculus and is partially crossed as Carlsson et al (1964) observed fibers crossing from this system in the dorsal gray commissure of the cord.

TERMINAL AREAS—TELENCEPHALON   In fluorescence histochemical material, axons arising from lateral tegmental NE neuron groups can often be distinguished from DA axons and locus coeruleus NE axons. The varicosities appear larger than those of locus axons, are irregular in size and in their distribution along the axon (Figure 12). There do not appear to be any fibers of the lateral tegmental NE cell groups that innervate the neocortex. Substantial innervation from this system is present, however, in a number of other regions of the telencephalon, particularly the basal forebrain. In the septal area there is a dense plexus of lateral tegmental axons in the ventral portion of the interstitial nucleus of the stria terminalis. This arises in part from the periventricular system and, in part from the medial forebrain bundle. Some fibers continue around the anterior commissure caudal to the main interstitial nucleus and turn dorsally to terminate in the caudal and ventral part of the lateral septal nucleus. The lateral tegmental innervation of the septal area has been demonstrated by several investigators (Olson

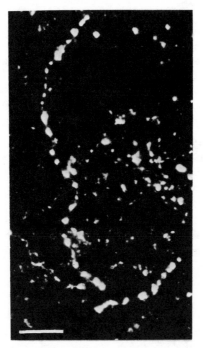

Figure 12 Typical brainstem NE axon in a terminal field, the rostral lateral hypothalamic area, exhibiting large and small, irregularly-spaced varicosities. Marker bar = 35 μm. (From R. Y. Moore, unpublished).

& Fuxe 1972, Maeda & Shimizu 1972, Maeda et al 1972, Lindvall 1975, Moore 1978). In the amygdala there is a rich innervation of lateral tegmental NE terminals in the central amygdaloid nucleus (Fallon et al 1978) whereas other areas, the lateral nucleus, anterior amygdaloid area, and the basolateral nucleus, receive only scattered innervation from this source (Swanson & Hartman 1975, Fallon et al 1978). In addition to this lateral tegmental telencephalic NE innervation there appear to be only scattered terminals in the nucleus of the diagonal band, the deep layers of the olfactory tubercle, and the anterior olfactory nucleus. Consequently, the lateral tegmental NE innervation to telencephalon is much less than that from locus coeruleus, overlaps the locus coeruleus in every area innervated, and is restricted to a few basal forebrain areas.

Diencephalon—thalamus The thalamus receives innervation from the lower brainstem cell groups only via the periventricular system and only to the paraventricular nucleus (Lindvall et al 1974). A component of this is undoubtedly the epinephrine innervation described by Hökfelt et al (1974; see below).

Diencephalon—hypothalamus   The hypothalamus receives a rich NE innervation, nearly all of it from lateral tegmental NE cell groups. The NE content of individual hypothalamic nuclei has been analyzed by Palkovits et al (1974). These data indicate significant NE content in nearly all hypothalamic nuclei, with the highest content occurring in the dorsomedial nucleus, the paraventricular nucleus, and the retrochiasmatic area (each about 50 ng NE per mg protein). No nucleus is devoid of NE, however, and the lowest values for any nucleus, the caudal portion of the arcuate nucleus (12 ng NE per mg protein) is significantly higher than the values recorded for neocortex (4 ng NE per mg protein). Consequently, it would appear from these chemical data that the entire hypothalamus is innervated by NE axons. These data are in general agreement with the fluorescence histochemical literature but there are minor points of difference. Before undertaking a description of the innervation of the hypothalamic nuclei, the source of the innervation should be reviewed. Evidence from a variety of sources indicates that only a very limited part of the innervation arises from neurons of the compact portion of the locus coeruleus. Lesions destroying the locus coeruleus produce little, if any, change in hypothalamic NE content (Kobayashi et al 1975, Jones et al 1977). Studies using the autoradiographic tracing method (Segal et al 1974, Jones & Moore 1977) indicate a projection from the locus coeruleus only to the periventricular nucleus, the paraventricular nucleus, the supraoptic nucleus, and the dorsomedial nucleus. This is in partial agreement with the observations of Olson & Fuxe (1972) who found that severing the projections from the cells of the "subcoeruleus" group denervated essentially the same nuclei. It is likely that the tritiated amino acid injections of Jones & Moore (1977) involved some of these subcoeruleus neurons or alternatively, the very dorsal transections of Olson & Fuxe (1972) may have spared some locus coeruleus projections. What is clear from these studies, and those of Lindvall & Björklund (1974b, 1978), is that the hypothalamic innervation from the locus coeruleus area is limited and the major source of hypothalamic NE innervation is the lower brainstem cell groups of the lateral tegmentum (Olson & Fuxe 1972). The hypothalamic NE innervation in the rat observed by the Falck-Hillarp method has been described by Fuxe (1965), Fuxe & Hökfelt (1969), Hökfelt & Fuxe (1972), and Jacobowitz & Palkovitz (1974). The entire lateral hypothalamic area exhibits varicose, NE fibers from the preoptic region to the mammillary region. The fibers are principally located in the lateral and ventrolateral part of the lateral hypothalamic area and the appearance of the varicosities is such to suggest a terminal innervation. In the medial hypothalmus the entire periventricular complex, from rostral to caudal including the arcuate component, shows a dense innervation, some of which arises from the locus coeruleus. The median eminence receives a lateral tegmental NE innervation (Björklund et al 1970). Rostrally, the medial preoptic area and the anterior hypothalamic area are densely innervated. The anterior hypothalamic nucleus has a sparse innervation and the suprachiasmatic nucleus, while free of innervation within the nucleus itself, is surrounded by a dense shell of NE fibers. This, in effect, may innervate some suprachiasmatic nucleus neurons as their dendrites extend beyond the borders of the nucleus (Szentagothai et al 1968). The supraoptic nucleus also receives a dense innervation largely in the form of a pericellular array of varicose fibers. The region immediately dorsal to the caudal part of the chiasm and the retrochiasmatic area are also densely innervated, due in part of the crossing of NE axons in the dorsal supraoptic decussation. More caudally the ventromedial nucleus appears to remain nearly free of NE innervation throughout its extent. It is surrounded, however, by the densely innervated ventral tuberal and perifornical areas, ventrally and laterally, and by the very densely innervated paraventricular nucleus rostrally and dorsally. Caudally the dorsomedial nucleus forms a very densely innervated zone above the ventromedial nucleus. The dorsal hypothalamic zone is not heavily innervated nor, for the most part, is the mammillary complex. The ventral premammillary nucleus and the tuberomammillary nucleus receive a dense innervation but the remainder of the mammillary region is only sparsely innervated (Table 2). These observations conform quite closely to the distribution of immunohistochemically demonstrable dopamine-β-hydroxylase (Swanson & Hartman 1975). Also, a very similar pattern of innervation has been reported for the cat (Poitras & Parent 1975, Cheung & Sladek 1975) and the macaque monkey (Hoffman et al 1976).

Brainstem   The major NE innervation of brainstem structures arises from nuclei other than the locus coeruleus. The pattern of brainstem NE innervation was initially described by Fuxe (1965) and few studies have been conducted subsequently. Recently, however, a detailed analysis of the organization and origin of the NE neuron innervation of brainstem nuclei and areas has been carried out (P. R. Levitt and R. Y. Moore, unpublished observations) and the observations from this are summarized in Table 4. The nonlocus NE innervation is distributed predominantly to regions of the brainstem other than primary sensory nuclei. Some motor nuclei such as the motor trigeminal nucleus, the facial nucleus, and the dorsal motor nucleus of the vagus receive a dense innervation. Similar density of innervation is present in the pontine and medullary raphe nuclei. Other regions, such as the major portion of the brainstem reticular formation, receive a sparse to moderate innervation from the nonlocus NE neuron groups.

Spinal cord   The innervation of the spinal cord gray is quite homogeneous from segment to segment except in the thoracic and upper lumbar segments. In these there is an extremely dense NE innervation of the intermediolateral cell column. In the remaining cord, the pattern is as follows; there is sparse, scattered NE innervation in the dorsal horn and the commissural gray matter, plexiform in arrangement. In the ventral horn the innervation is similarly plexiform but the density is much greater (Nygren & Olson 1977). There are no ultrastructural studies specifically directed toward the analysis of the terminals of lateral tegmental NE neurons although these have undoubtedly been demonstrated in several studies (cf Bloom 1973, for review).

PHYSIOLOGY   The principal target of the lateral tegmental NE neuron system is the medial hypothalamus, including the median eminence. Evidence is now available to indicate that this innervation has a significant function in the regulation of gonadotropin, growth hormone, and ACTH secretion. This evidence is briefly reviewed.

Gonadotropin secretion   The first studies to suggest a functional role for NE in gonadotropin release were those of Sawyer and co-workers (cf Sawyer et al 1974, Sawyer 1975, for reviews). Their work demonstrates that intraventricular administration of NE results in the release of luteinizing hormone (LH) and ovulation in the rabbit and that this can be blocked by adrenergic blocking drugs. Similarly, in the rat, administration of an adrenergic blocking agent prior to the critical period of the day of proestrus results in inhibition of ovulation. Further evidence along this line was obtained by Ojeda & McCann (1973) who showed that α-methyl-p-tyrosine blocking of CA synthesis blocks both the LH release induced by preoptic stimulation in the female rat, and the post-castration rise in plasma LH and follicle stimulating hormone, in the male rat. Since these effects could be reversed by either L-DOPA or dihydroxyphenylserine administration, and the post-castration rise in serum LH could be blocked by a dopamine-β-hydroxylase inhibitor (Ojeda & McCann 1973), it seems probable that the effects result from excitatory transmission by NE (or possibly E) neurons on appropriate target neurons.

Studies of CA turnover provide further support for the view that the hypothalamic NE innervation has an excitatory function in gonadotropin release. An increased synthesis of NE has been reported during proestrus (Donoso & Gutierrez-Moyano 1970) and Selmanoff et al (1976) find a marked increase in NE content in the suprachiasmatic region but in no other hypothalamic area during the day of proestrus. The median eminence shows an increase from diestrus to proestrus. In addition, in their study DA shows no change in any area during the estrus cycle. In accord with these observations, Löfstrom (1976) has found cyclic changes in NE turnover of terminals in the subependymal layer of the median eminence and medial preoptic region with the greatest turnover occurring during proestrus. Thus, all of the data available are in accord with the view that brainstem NE innervation of the hypothalamus is involved in stimulating pituitary gonadotropin release. The mechanism by which this occurs has not been completely elucidated, but several workers (cf Löfstrom 1976, Fuxe et al 1976, for reviews) have suggested that the NE system is involved in the positive feedback action of the gonadal steroids on gonadotropin secretion. This implies that NE neuron input affects the LH-RH system at one level or another. This could either be in the preoptic-suprachiasmatic region or the median eminence (Sawyer et al 1974). Either or both would be in accord with the available data.

ACTH secretion   Substantial evidence is now available to indicate that the NE neurons that innervate the hypothalamus participate in the regulation of ACTH secretion by functioning as an inhibitory influence (cf Ganong 1974a, b, Van Loon 1973, for reviews). The basis for this is as follows: NE releasing drugs such as amphetamine and L-DOPA inhibit stress-induced ACTH release, an effect mediated by central α-adrenergic receptors; NE depleting drugs such as α-methyl-p-tyrosine or the dopamine-β-hydroxylase inhibitor, FLA63, increase plasma corticosterone levels and that L-DOPA and dihydroxyphenylserine counteract the effect; intraventricular 6-hydroxydopamine causes an elevation in plasma corticosterone (Cuello et al 1974). As with the gonadotropin effects of NE neurons, the ACTH regulation is not understood in terms of the target of the NE neuron input. Recent in vitro work suggests that it is directly upon the corticotropin-releasing factor neuron (Jones et al 1976), but some transsynaptic mechanism cannot be excluded at this time.

Growth hormone secretion   Although the literature is small, there are substantial data to suggest that NE neurons innervating the hypothalamus stimulate growth hormone secretion (cf Ganong 1974a, for review). Administration of L-DOPA in man produces growth hormone release (Perlow et al 1972) and the same occurs in the macaque monkey (Chambers & Brown 1976). In the latter situation, the L-DOPA effect is reproduced by the adrenergic agonist, clonidine, but not by the dopamine agonist, apomorphine. This confirms studies done in the rat (Müller et al 1970) so that the observations are consistent across species. As in the previous neuroendocrine situations, the NE neuron target has not been established but one study has suggested that it is in the region of the ventromedial nucleus (Toivola & Gale 1972). A more recent study has indicated that NE neurons in the posterior periventricular region constitute a facilitatory mechanism for growth hormone secretion (Andersson et al 1977).

ELECTROPHYSIOLOGY   Although the cytology of the NE circuits indicate that there are several sources of NE fibers, all of the probable synaptic target fields evaluated by cellular pharmacology and physiology are those thought to arise from the nucleus locus coeruleus. Among the important problems that must soon come under investigation are questions of the degree to which locus coeruleus synapses are functionally and pharmacologically similar, and whether these LC-derived synapses are the same or different from other cell groups giving rise to NE synapses. For example, cyclic AMP will mimic, and phosphodiesterase inhibitors will potentiate, NE inhibitions in brainstem (Anderson et al 1973) and cerebral cortex (Stone et al 1975, Stone & Taylor 1977).

THE EPINEPHRINE (E) NEURON SYSTEM

Biochemical Studies

At present, E is not distinguishable from other catecholamines using the standard fluorescence histochemical techniques and a large, conflicting biochemical literature on its existence and localization in the mammalian CNS has arisen. This is reviewed briefly since the development of an immunohistochemical method appears to provide a basis for analysis of the E neuron system. E has been localized within specific regions of the mammalian central nervous system by a number of investigators (Vogt 1954, Koslow & Schlumpf 1974, Reid et al 1975, Zivin et al 1975, and Van der Gugten et al 1976) using a variety of methods including bioassay spectrofluorometric assay, gas chromatography-mass spectrometry (GC-MS), and radioisotopic enzymatic assay. In addition, the enzyme that converts NE to E, phenylethanolamine N-methyl-transferase (PNMT), has been demonstrated by enzyme activity assay to be present in the mammalian CNS (Ciaranello et al 1969, Pohorecky et al 1969, Saavedra et al 1974, Reid et al 1975, 1976) in a regional distribution.

With the developments of the sensitive and specific techniques of gas chromatography-mass spectrometry and radiometric enzymatic assay, more accurate determinations of the regional distribution of E have been made. Epinephrine has been identified in the brainstem and spinal cord of the rat by the GC-MS technique (Reid et al 1975). Koslow & Schlumpf (1974) identified E in 8 regions of the rat brain using the GC-MS technique, which can detect less than 1×10^{-14} moles of epinephrine. The highest concentrations of E were found in the periventricular nucleus, the arcuate nucleus, and the paraventricular nucleus. The next higher concentrations of E were found in the ventrolateral medullary reticular formation, the locus coeruleus, the habenular region, and the cerebellum. The percentage of E as a function of NE in the five brain regions (hypothalamic regions were grouped) varied from 6.5% in the cerebellar region to 1.0% in the locus coeruleus. The regional distribution of E in the rat brain has been further studied by Van der Gugten et al (1976) utilizing a combination of the COMT radiometric assay and paper chromatography to separate and identify as little as 10 μg of DA, NE, and E. These investigators studied 92 brain regions and detected E in 29 of them. In these regions, the concentration of E ranged from 2.52 pg/μg protein (nucleus periventricularis) to 0.17 pg/μg protein (inferior olive), while the amount of E ranged from 1.2 to 14.3% of the NE concentration. The telencephalon contained no measurable E with the exception of low concentrations in two septal nuclei, the basal amygdala and the rostral MFB. Within the diencephalon, relatively high concentrations of E were measured in 9 hypothalamic nuclei, in particular the nucleus periventricularis, nucleus dorsomedialis, nucleus paraventricularis, and nucleus supraopticus, with lesser concentrations in the preoptic area, arcuate nucleus-median eminence, and mammillary region. In addition, the midline thalamic nuclei had modest E concentration while low concentrations were found in the interstitial nucleus of the stria medullaris, the posterior thalamic nucleus, and the lateral geniculate nucleus. Measurable amounts of E were also detected in the mesencephalic and pontine CA cell groups. The dorsal medullary region and the nucleus commissuralis had the highest medullary concentration with measurable amounts also present in the facial motor nucleus, the medullary reticular nucleus, the deep cerebellar nuclei, and the inferior olive. These studies show some discrepancies in the areas and amounts of E found, which probably reflect differences in dissection technique and the sensitivity of the method of analysis used.

In recent experiments, Saavedra et al (1974) have analyzed the regional distribution of the E-producing enzyme PNMT with an extremely specific and sensitive method. Their results indicate that PNMT activity is unevenly distributed in the rat brain with the greatest enzyme activity measured in the two areas that contained the medullary cell groups of Dahlström & Fuxe (1964). The region of the locus coeruleus showed intermediate PNMT values, while the pontine cell groups exhibited low enzyme activity. Within the diencephalon, PNMT activity was observed in the preoptic region and in the hypothalamus where the enzyme activity was found to be lower than either the dopamine-β-hydroxylase or the tyrosine hydroxylase activities. PNMT was unevenly distributed within the hypothalamic nuclei studied. The greatest activities were found in the medial basal hypothalamus within the arcuate and paraventricular nuclei and the median eminence. Intermediate activities were observed in the periventricular, dorsomedial, ventromedial, and perifornical nuclei, and the retrochiasmatic area, while low enzyme activities were noted in the remaining hypothalamic nuclei. Low levels of PNMT activity were also found in the substantia nigra, the interstitial nucleus of the stria terminalis, and the septum. No enzyme activity was detected in the cerebral cortex, caudate, amygdala, hippocampus, or olfactory tubercle. These observations do not conform entirely to previous studies but are largely in accord with the work on the distribution of E and the immunohistochemical data reported below.

In addition, Reid et al (1975, 1976) have found PNMT activity in the rat spinal cord that was substrate specific for phenylethanolamines and did not significantly methylate B-phenylethylamine or tryptamine. A transection of the spinal cord at the level of T4–6 resulted in a 70% decrease in the PNMT activity below the lesion, while intercisternal 6-hydroxydopamine injections did not significantly alter the enzyme activity (Reid et al 1976). These results indicated that the PNMT was transported to the spinal cord from neurons located rostral to the lesion and that these CNS PNMT-containing neurons were not particularly sensitive to 6-hydroxydopamine. These findings agree with the immunohistochemical studies of Hökfelt et al (1973, 1974), concerning the location of PNMT positive neurons and their terminal distribution to the spinal cord.

Anatomical Studies

The morphology of the CNS E neuron system has been described by Hökfelt et al (1973, 1974) using an immunohistochemical technique (Table 5). This method provides only indirect evidence for E-containing neurons since, like all immunohistochemical methods, it is dependent upon the specificity of the antibodies to PNMT. It is of considerable interest, however, in that the E neuron system shown by this methodology conforms well to the biochemical data reviewed above.

The PNMT positive neurons identified by Hökfelt et al (1973, 1974) are multipolar reticular neurons with dendritic processes that extend for distances up to 1 mm in the frontal plane. The neuronal perikarya are usually spindle or oval shaped and vary in size from 15–25 μm. These neurons are formed into two groups. The C1 cell group of Hökfelt et al (1973, 1974) is

Table 5   Epinephrine neuron systems in the mammalian brain[a]

Designation of the system	Location of cell bodies	Distribution of terminals
Dorsal tegmental and lateral tegmental	Caudal medulla	Spinal cord, brainstem (locus coeruleus, periaqueductal gray), hypothalamus, thalamus

[a] Observations reported in Hökfelt et al 1974)

similar in morphology and distribution to the medullary lateral tegmental cell group. The perikarya of the small C2 group are also located in the rostral medulla oblongata, close to the midline in the dorsal part of the reticular formation. Rostrally, these perikarya are found ventromedial to the descending vestibular nuclei while, at more caudal levels, some cells are observed immediately ventromedial to the nucleus tractus solitarius. These neurons are located within the dorsal medullary CA cell group described above.

At least one axon bundle arose from these two cell groups. This bundle appears to ascend in the reticular formation along with the NA fibers that originated in the lower brainstem, and comprises the central tegmental tract (Lindvall & Björklund 1974b) and the ventral NA bundle (Olson & Fuxe 1971, Ungerstedt 1971). The organization of the E neuron system is shown diagrammatically in Figure 13.

Hökfelt et al (1973, 1974) have identified PNMT positive varicosities by fluorescence microscopy (average varicosity diameter, $0.5-1.0\mu$m). The varicosities are found only in nuclei known to possess CA-containing terminals (Fuxe 1965). The PNMT positive terminals are primarily observed in certain visceral afferent and efferent nuclei of the brain and spinal cord, in the ventral periventricular gray of the lower brainstem, and in certain nuclei of the hypothalamus and thalamus. The densest network of varicosities is found in the dorsal motor nucleus of the vagus, the nucleus tractus solitarius, the paraventricular thalamic nucleus, and the intermediolateral cell column of the spinal cord. A moderate density of fluorescent varicosities was observed in periventricular gray surrounding the cerebral acqueduct, the ventral floor of the 4th venticle, the dorsolateral 3rd ventricle, the ventral locus coeruleus, the dorsomedial hypothalamic nucleus, and the perifornical area. The lowest density of PNMT positive varicosities was in the rhomboid nucleus of the thalamus, periventricular hypothalamus, ventrolateral arcuate nucleus, posterolateral hypothalamus, medical subthalamus, and periventricular gray surrounding the central canal of the cervical and thoracic spinal cord. However, this anatomical description of PNMT positive neurons and terminals may not represent the entire system since the immunohistochemical technique may not be sufficiently sensitive to reveal all PNMT positive terminals. This conclusion is supported by the work of Saavedra et al (1974) who, in addition to identifying PNMT activity in all the terminal regions described by Hökfelt et al (1974), also detected activity in several other brain areas.

Do the PNMT positive neurons and nerve terminals identified by Hökfelt et al (1973, 1974) represent sufficient evidence for an E neuron system analogous to the NE systems? As noted above, several lines of evidence strongly suggest that they do. Hökfelt et al (1974) argue that PNMT posi-

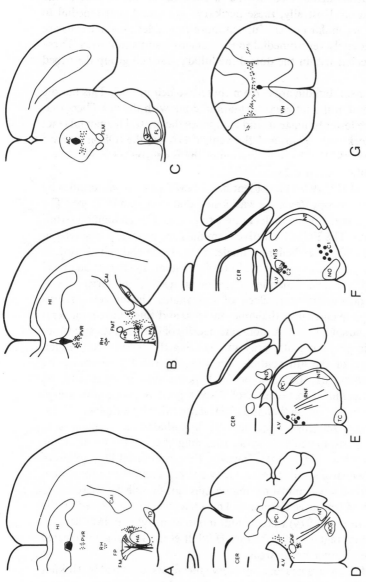

Figure 13 The E neuron system as shown by the immunohistochemical method for PNMT. See text for description. A–G, rostral to caudal coronal sections from rat brain and spinal cord. Abbreviations: AC, cerebral aqueduct; C1, C1 E neuron group; C2, C2 E neuron group; CAI, internal capsule; CER, cerebellum; F, fornix; FL, corticospinal tract; FLM, median longitudinal fasciculus; FMT, mammillothalamic tract; FM and FP, paraventricular hypothalamic nucleus; GNF, genu of facial nerve; HA, anterior hypothalamic nucleus; HD, dorsomedial hypothalamic nucleus; HI, hippocampal formation; HV, ventromedial hypothalamic nucleus; ND, dentate nucleus; NO, inferior olivary complex; NOS, superior olivary complex; NT, spinal trigeminal tract; NTS, nucleus tractus colitarius; PCI, inferior cerebellar peduncle; PVR, paraventricular nucleus of thalamus; RH, rhomboid nucleus of thalamus; RNF, facial nerve; TO, optic tract; TC, corticospinal tract; VH, ventral horn; 4V, fourth ventricle. (From Hökfelt et al 1974).

tive neurons also exhibit a CA-induced fluorescence when examined by the Falck-Hillarp technique, which overcomes the objection that the enzyme detected by immunohistochemistry may be latent and nonfunctional. In particular, group C1 appears identical with the rostral part of the lateral medullary cell group and group C2 is identical to the rostral part of the dorsal medullary CA neuron group. The ascending PNMT positive pathway in the reticular formation is located among the fibers of the ascending CA pathway (Lindvall & Björklund 1974b, Olson & Fuxe 1972, Ungerstedt 1971), and most of the PNMT positive terminals are present in areas rich in CA terminals (Fuxe 1965). In addition, PNMT activity has also been detected by radiometric assay in all of the areas identified by Hökfelt et al (1973, 1974) and Saavedra et al (1974). Consequently, all of the data are in accord with the concept of an E neuron system; it should be emphasized that this E neuron system is small in comparison to the NE and DA systems. Its physiology and function have not been investigated as yet.

CONCLUDING COMMENTS

Contrasts in Systems: DA versus NE and E

The organization of central catecholamine neuron systems differs markedly. DA neuron systems are most numerous both in total neurons and in the number of discrete systems that have been described. DA neuron systems uniformly have a discrete topography and a restricted terminal distribution area. In contrast to this, the NE systems appear to have diffuse and widespread projections with little topography. The E system appears similar but is much smaller in number of neurons and has not yet been studied as exhaustively as the DA or NE systems. Consequently, on morphological grounds, it would appear appropriate to view the DA systems as components of a set of broader systems with distinct and separate functions. For example, the nigro-striatal system can be viewed as one component projecting discretely within the overall neostriatal innervation. The organization of the DA systems is probably best exemplified by the tubero-hypophysial system, which itself may be a series of subsystems serving different functions in neuroendocrine regulation. Similarly, the retinal interplexiform neuron system is a distinct component of an organized system participating in an easily recognized function. In contrast, the NE and E systems appear organized as parts of a very diffuse system, the reticular formation, which has widespread effects upon many other systems. Morphologically two distinct components of the NE systems are evident. First, the locus coeruleus NE system projects widely over vast areas of the neuraxis from olfactory bulb to spinal cord, with an emphasis on cortical structures, particularly on the massive suprasegmental elements of the cerebellar and cerebral cortices. One can only speculate upon the very general functions that such a system must subserve; many such speculations are in the literature but none is supported by conclusive data. Second, the lateral tegmental NE system overlaps the locus coeruleus system in many areas such as spinal cord, brainstem, hypothalamus, and basal forebrain, but has its major projections to "vegetative" areas of the nervous system concerned with homeostatic adaptation of the organism to its environment (see Bates et al 1977, Ward & Gunne 1976a, b, Tribollet et al 1978, Kawamura et al 1978). This is particularly obvious in the NE neuron regulation of hypothalamic-pituitary function. In each case, however, the projections are widespread, the topography of projection apparently diffuse, and the function, thereby, general. Thus, the NE systems can be viewed as containing typical reticular formation neurons characterized in two special ways: first, the neurotransmitter is known and, second, their distribution is more widespread than that of other reticular formation neurons that have been described. The analysis of the function of NE and E neurons systems, particularly at higher levels of integration, remains one major problem yet to be resolved (see Robinson et al 1977b, Mason & Iversen 1977, Ogren & Fuxe 1977).

Cellular Function of Pathways: Transmitters or Modulators

Based largely on the properties revealed from the studies on LC projections to cerebellum, hippocampus, diencephalon, and cerebral cortex, the general properties of the circuits would appear to be: (a) inhibitory; (b) slow onset, with latencies greater than 50 msec; (c) prolonged action, with durations of at least 350 msec; (d) mediated through beta receptors coupled to adenylate cyclase activation. In addition, the LC circuits are known to arise very early in development, and to retain into adulthood a marked propensity to form axon collaterals in response to cellular injury (see Moore et al 1971, Robinson et al 1977a, Pickel et al 1973, 1974b). Although DA systems appear more topographic in projection and less responsive to injury, most of these physiologic properties noted above would seem directly applicable to the DA projections as well (see Björklund & Moore 1978).

One important aspect of the physiology of the central catecholamine circuits, which only now is beginning to be examined, is the question of heterosynaptic interactions; that is, the ability of activity in the NE-containing or DA-containing synaptic systems to affect the responsiveness of a given target cell system to the host of other afferent systems all such target cells receive. Based on studies in the squirrel monkey auditory cortex (Foote et al 1975), the rat cerebellum (Freedman et al 1976, 1977), and the hippocampus of the awake rat (Segal & Bloom 1976a, b), the effects of these systems may not be so easily resolvable into classical terms such as inhibitory or excitatory. For example, in the hippocampus of the awake rat (Segal

& Bloom 1976a, b) and in the cerebellum of the anesthetized rat (Freedman et al 1976), conditioning stimuli in the locus coeruleus or iontophoretic application of NE will potentiate the effects of nonadrenergic inhibitory inputs for considerable periods. However, when the effects of NE and LC are evaluated on excitatory inputs to the same cells, the excitatory effects are also potentiated (Freedman et al 1976, Segal & Bloom 1976b). Such potentiated responses may be similar to the enhanced evoked responses seen following LC conditioning stimuli in lateral geniculate (Nakai & Takaori 1974) and on the increased ratio of evoked responses to spontaneous activity seen during iontophoresis of NE to acoustically reactive units of the squirrel monkey auditory cortex (Foote et al 1975). In short, these latter sets of studies suggest that the combination of electrophysiological properties attributable to the catecholamine circuits may permit them to function in a manner that cannot be compared to simple inhibition, but which rather may be more properly conceived as a "bias" adjusting or "enabling" system that can enhance the effects of target systems. This view provides a valuable adjunct to the vocabulary of cellular communication signals by which neurons communicate, and it may be confidently predicted that this property will come under increasing surveillance in the years ahead.

What Makes Catecholaminergic Neurons Fire?

The overall structure of the catecholaminergic neurons, that is the collection of target cell domains into which catecholaminergic axons project, provides but one dimension of their potential physiological role. A second dimension is provided by the qualitative actions revealed by the experimental activation of these circuits on single unit discharge or the simulations of these synaptic actions by iontophoresis. However, in order to provide a full description of the function of these systems, a third domain must be understood as well: when do these cells fire and within what limits is their activity normally regulated?

As detailed above, locus coeruleus neurons appear to receive an extremely rich array of inputs, some of which have been chemically defined without definition of the cells of origin (Hökfelt et al 1978, Pickel et al 1977), while several others have been anatomically defined (Conrad & Pfaff 1976a, b, Sakai et al 1977, Saper et al 1976, Swanson 1976b, Snider 1975, Gupta et al 1977, Nakamura 1977, Cedarbaum & Aghajanian 1978, Takigawa & Mogenson 1977, Gupta et al 1977) without reference to the transmitter. The locus cells show responsivity to the neuropeptides (Substance P, enkephalin, and others) identified in fibers projecting to LC (Bird & Kuhar 1977, Guyenet & Aghajanian 1977, Scott-Young et al 1978). While many crucial details of the innervation of locus coeruleus neurons remain to be worked out (such as which of these systems go to which LC cells and the extent to which the LC efferent projections may be demonstrated as either parallel overlapping circuits or as selectively topographic circuits), some properties of their discharge patterns have begun to emerge. In anesthetized rats, cells identified as locus coeruleus by post hoc cytology of the recording site (Cedarbaum & Aghajanian 1976, Bird & Kuhar 1977, Graham & Aghajanian 1971, Svensson et al 1975), as well as cells identified as locus coeruleus on the basis of antidromic activation from the lateral hypothalamus (Nakamura & Iwama 1975) or from the cingulate cortex (Faiers & Mogenson 1976) typically show slow (3–5 Hz) tonic rates of discharge, which are, thus far, responsive only to severe functional stimuli like pain (Takigawa & Mogenson 1977, Aghajanian et al 1977, Nakamura 1977) and show variable responses to ethanol (Pohorecky & Brick 1977).

However, in the awake cat, two reports indicate a much less monotonous discharge pattern for locus coeruleus neurons. As mentioned above, the cat is not the optimal species in which to pursue the functional properties of locus coeruleus since the cells are widely scattered and more spatially separated than in rodents or primates. Nevertheless, the results of these two sets of studies do provide a tantalizing glimpse of the forms of physiological that that may some day be defined for neurons. Chu & Bloom (1974a, b) recorded unrestrained cats, correlating the discharge patterns of cells with electroencephalographic activity, and identified recorded cells as "NE-containing" on the basis of combining Prussian Blue staining of the recording site with Falck-Hillarp fluorescence histochemistry. Cells in various components of the complex subdivisions of the feline LC showed generally similar discharge patterns relative to waking and sleeping behavior; slow tonic discharges were common in quiet waking states; somewhat slower discharges were observed during slow wave sleep; and more rapid phasic discharges were seen during the rapid eye movement (REM) stage, with bursts of activity roughly synchronous with the ponto-geniculo-occipital (PGO) spikes. These cells were found to fire quite rapidly (50–70 Hz) in REM sleep bursts, while adjacent cells showed similar patterns relative to the EEG and in the period of waking immediately after REM. There was considerable variance in basal rate and sleep stage rates, with some tendency for the differences to follow a dorso-ventral and medio-lateral gradient.

A small population of the cells described by Chu & Bloom (1974b) followed a different pattern, in which the activity during REM sleep was considerably slower than in waking or slow wave sleep. These latter cells were encountered more frequently by Hobson et al (1975, 1976, McCarley & Hobson 1975) who term them "off" cells. In this study the cells were identified as being in LC on the basis of recording site lesions without fluorescence histochemistry, and appear to be somewhat more posterior than those in the Chu & Bloom (1974b) study. Hobson et al (1976) interpret the existence of presumed LC "off" cells to imply a feedback regulatory function in which the LC tonically inhibits the neurons of the giant pontine reticular fields (thought to be both cholinergic and cholinoceptive); as LC "off" cells slow during prolonged slow wave sleep, eventually stopping as REM emerges, the giant reticular neurons are disinhibited leading to their greater activity during REM (see Hobson et al 1976, Hobson 1976).

As significant as it could be to establish firmly that LC neurons play an active role in the regulation of sleep stages, that phenomenologic relationship would not necessarily clarify the functional role of the LC, as the functional role of sleep per se is an even deeper mystery. Hopefully, in the near future, it may be possible to repeat these observations on species in which LC cells offer a more clustered target for recording activity, and in which additional experimental observations can begin to evaluate correlations in discharge with behavioral variables other than waking and sleeping states.

While the topography of DA neurons suggests that their innervation and efferent distribution may be more specific than that of the LC cells, no direct observations on the activity of these cells in awake unrestrained animals have yet been reported. In anesthetized animals (Bunney & Aghajanian 1977, Groves et al 1975), substantia nigra pars compacta neurons also fire slowly and regularly.

DA, NE, and E Neuron Systems of the CNS: General Remarks

The study of central CA neuron systems now is nearly a quarter of a century old. Progress in the first decade was desultory but, with the development of sensitive biochemical methods for assay of CAs and their synthetic enzymes, pharmacological tools for manipulation of the systems, and, finally, the specific and sensitive Falck-Hillarp florescence histochemical method, the field moved forward rapidly to the state of understanding reflected in this review. In all likelihood, we now know more about the organization and function of central CA neuron systems than about any other central system save for the major primary sensory systems. It is noteworthy, however, that we have little if any information concerning the neurotransmitters utilized by sensory neurons. In contrast, pharmacological manipulation of function is both feasible and understood in the CA neuron systems. This capability provides a basis for the analysis of function at both the cellular level and at higher levels of integration than have been covered here.

Indeed, so much information has now been developed about the organization and function of the central CA neuron systems it is unlikely that this review will ever again be attempted in this series. At the rapid rate with which new data are being added, soon it will no longer be possible to review the anatomy and physiology of all central CA neuron systems in any depth in so limited a space. Rather, future reviews will undoubtedly focus on one or another of the systems or, perhaps, on only one facet of a system. The purpose of this review and its predecessor (Moore & Bloom 1978) has been to record briefly our current understanding of the anatomy and physiology of CA neuron systems in the mammalian brain. Hopefully, this effort will prove helpful to others in advancing our knowledge of the functional interaction of the CA neuron systems with other central systems in the elaboration of both homeostatic and behavioral adaptive mechanisms.

ACKNOWLEDGMENTS

The preparation of this review and some of the work presented in it has been supported by USPHS Grant NS-12080 and NSF Grant BNS76-09318.

Literature Cited

Aghajanian, G. K., Cedarbaum, J. M., Wang, R. Y. 1977. Evidence for norepinephrine-mediated collateral inhibition of locus coeruleus neurons. *Brain Res.* 136:570–77

Andén, N.-E., Dahlstrom, A., Fuxe, K., Larsson, K., Olson, L., Ungerstedt, U. 1966. Ascending monoamine neurons to the telencephalon and diencephalon. *Acta Physiol. Scand.* 67:313–26

Anderson, E. G., Haas, H. L., Hosli, L. 1973. Comparison of effects of noradrenaline and histamine with cyclic AMP on brain stem neurones. *Brain Res.* 49: 471–75

Andersson, K., Fuxe, K., Eneroth, P., Gustafsson, J. A., Skett, P. 1977. On the catecholamine control of growth hormone regulation. Evidence for discrete changes in dopamine and noradrenaline turnover following growth hormone administration. *Neurosci. Lett.* 5:83–89

Atlas, D., Segal, M. 1977. Simultaneous visualization of noradrenergic fibers and adrenoreceptors in pre and postsynaptic regions. *Brain Res.* 135:347–50

Atlas, D., Teichberg, V. I., Changeux, J. P. 1977. Direct evidence for beta-adrenoreceptors on the Purkinje cells of mouse cerebellum. *Brain Res.* 128:532–36

Bates, D., Weinshilbaum, Campbell, R. J., Sundt, T. M., Jr. 1977. The effect of lesions in the locus coeruleus on the physiological responses of the cerebral blood vessels in cats. *Brain Res.* 136: 431–43

Berman, A. 1968. *The Brain Stem of the Cat,* Madison: Univ. Wisconsin Press

Bevan, J. A. 1977. Some functional consequences of variation in adrenergic synaptic cleft width and nerve density and distribution. *Fed. Proc.* 36:2439–43

Bird, S. J., Kuhar, M. J. 1977. Iontophoretic application of opiates to the locus coeruleus. *Brain Res.* 122:523–33

Björklund, A., Falck, B., Hromek, F., Owman, C., West, K. A. 1970. Identification and terminal distribution of the tubero-hypophyseal monoamine systems in the rat by means of stereotaxic and microspectro-fluorimetric techniques. *Brain Res.* 17:1–23

Björklund, A., Moore, R. Y. 1978. *The Central Adrenergic Neuron,* New York: Raven Press. In press

Blackstad, T., Fuxe, K., Hökfelt, T. 1967. Noradrenaline nerve terminals in the hippocampal region of the rat and guinea pig. *Z. Zellforsch. Mikrosk. Anat.* 78:463–73

Bloom, F. E. 1968. Electrophysiological pharmacology of single nerve cells. In *Psychopharmacology—A Ten Year Progress Report,* ed. D. H. Efron, pp. 355–74. Washington, D.C.: Govt. Printing Office.

Bloom, F. E. 1973. Ultrastructural identification of catecholamine-containing central synaptic terminals. *J. Histochem. Cytochem.* 21:333–48

Bloom, F. E. 1974. To spritz or not to spritz: the doubtful value of aimless iontophoresis. *Life Sci.* 14:1819–34

Bloom, F. E. 1975a. In *Handbook of Psychopharmacology,* ed. L. L. Iversen, S. D.

Iversen, S. H. Snyder, 6:1–22. New York: Raven Press

Bloom, F. E. 1975b. The role of cyclic nucleotides in central synaptic function. *Rev. Physiol. Biochem. Pharmacol.* 74: 1–103

Bloom, F. E. 1978. Central noradrenergic systems: Physiology and pharmacology. In *Psychopharmacology—A 20 Year Progress Report,* ed. M. E. Lipton, K. C. Killam, A. DiMascio, pp. 131–142. New York: Raven Press

Bloom, F. E., Battenberg, E. L. F. 1976. A rapid, simple and more sensitive method for the demonstration of central catecholamine-containing neurons and axons by glyoxylic acid induced fluorescence: II. A detailed description of methodology. *J. Histochem. Cytochem.* 24:561–71

Bloom, F. E., Hoffer, B. J. 1973. In *Frontiers in Catecholamine Research,* ed. E. Usdin, S. Snyder, pp. 637–42. New York: Pergamon

Bloom, F. E., Hoffer, B. J., Siggins, G. R. 1971. Studies on norepinephrine-containing afferents to Purkinje cells of rat cerebellum: I. Localization of fibers and their synapses. *Brain Res.* 25:501–21

Braak, H. 1975. On the pars cerebellaris loci coerulei within the cerebellum of man. *Cell Tissue Res.* 160:279–83

Bunney, B. S., Aghajanian, G. K. 1977. Studies on cerebral cortex neurons. In *Pharmacology of Non-striatal Dopaminergic Neurons,* ed. E. Costa, M. Trabucchi, G. L. Gessa, pp. 65–70. New York: Raven Press

Cajal, S. R. 1911. *Histologie du Systeme Nerveux,* Vol. II. pp. 1–993. Paris: Maloine

Carlsson, A., Falck, B., Fuxe, K., Hillarp, N.-A. 1964. Cellular localization of monoamines in the spinal cord. *Acta Physiol. Scand.* 60:112–19

Cedarbaum, J. M., Aghajanian, G. K. 1976. Noradrenergic neurons of the locus coeruleus: inhibition by epinephrine and activation by the alpha antagonist piperoxane. *Brain Res.* 112:413–19

Cedarbaum, J. M., Aghajanian, G. K. 1978. Afferent projections to the rat locus coeruleus as determined by a retrograde tracine technique. *J. Comp. Neurol.* 178:1–16

Chambers, J. W., Brown, G. M. 1976. Neurotransmitter regulation of growth hormone and ACTH in the rhesus monkey: effects of biogenic amines. *Endocrinology* 98:420–28

Cheung, Y., Sladek, J. R. Jr. 1975. Catecholamine distribution in the feline hypothalamus. *J. Comp. Neurol.* 164: 339–60

Chu, N.-S., Bloom, F. E. 1974a. The catecholamine-containing neurons in the cat dorsolateral pontine tegmentum: distribution of the cell bodies some axonal projections. *Brain Res.* 66:1–21

Chu, N.-S., Bloom, F. E. 1974b. Activity patterns of catecholamine-containing pontine neurons in the dorsolateral tegmentum of unrestrained cats. *J. Neurobiol.* 5:527–44

Ciaranello, R. D., Barchas, R. E., Byers, G. S., Stemmle, D. W., Barchas, J. D. 1969. Enzymatic synthesis of adrenaline in mammalian brain. *Nature* 221: 368–69

Conrad, L. C. A., Pfaff, D. W. 1976a. Efferents from medial basal forebrain and hypothalamus in the rat. I. An autoradiographic study of the medial preoptic area. *J. Comp. Neurol.* 167: 185–220

Conrad, L. C. A., Pfaff, D. W. 1976b. Efferents from medial basal forebrain and hypothalamus in the rat. II. An autoradiography study of the anterior hypothalamus. *J. Comp. Neurol.* 167: 221–62

Coyle, J. T., Molliver, M. E. 1977. Major innervation of newborn rat cortex by monoaminergic neurons. *Science* 196: 444–46

Cuello, A. C., Shoemaker, W. J., Ganong, W. F. 1974. Effect of 6-hydroxydopamine on hypothalamic norepinephrine and dopamine content, ultrastructure of the median eminence and plasma corticosterone. *Brain Res.* 78:57–69

Dahlström, A., Fuxe, K. 1964. Evidence for the existence of monoamine-containing neurons in the central nervous system. I. Demonstration of monoamines in the cell bodies of brain stem neurons. *Acta Physiol. Scand.* Suppl. 232, 62:1–55

Dahlström, A., Fuxe, K. 1965. Evidence for the existence of monoamine neurons in the central nervous system. II. Experimentally induced changes in the intraneuronal amine levels of the bulbospinal neuron systems. *Acta Physiol. Scand.* Suppl. 247, 64:1–36

Descarries, L., Droz, B. 1970. Intraneural distribution of exogenous norepinephrine in the central nervous system of the rat. *J. Cell Biol.* 49:385–99

Descarries, L., Lapierre, Y. 1973. Noredrenergic axon terminals in the cerebral cortex of rat. I. Radioautographic visualization after topical application of DL-[H] norepinephrine. *Brain Res.* 51: 141–60

Descarries, L., Watkins, K. C., Lapierre, Y. 1977. Noradrenergic axon terminals in the cerebral cortex of the rat. III. Topometric ultrastructural analysis. *Brain Res.* 133:197–222

DiCarlo, V., Hubbard, J. E., Pate, P. 1973. Fluorescence histochemistry of monoamine-containing cell bodies in the brain stem of the squirrel monkey (*Saimuiri sciureus*). IV. An atlas. *J. Comp. Neurol.* 152:347–72

Dismukes, K. 1977. New look at the aminergic nervous system. *Nature* 269:557–58

Domesick, V. B. 1970. The fasciculus cinguli in the rat. *Brain Res.* 20:19–32

Donoso, A. O., Guiterrez-Moyana, M. B. 1970. Adrenergic activity in hypothalamus and ovulation. *Proc. Soc. Exp. Biol. Med.* 135:633–35

Dupin, J. C., Descarries, L., deChamplain, J. 1976. Radioautographic visualization of central catecholamine neurons in newborn rat after intravenous administration of tritiated norepinephrine. *Brain Res.* 103:588–96

Engberg, I., Marshall, K. C. 1971. Mechanism of noradrenaline hyperpolarization in spinal cord motoneurons of the cat. *Acta Physiol. Scand.* 83:142–44

Faiers, A. A., Mogenson, G. J. 1976. Electrophysiological identification of neurons in locus coeruleus. *Exp. Neurol.* 53:254–66

Fallon, J. H., Koziell, D. A., Moore, R. Y. 1978. Catecholamine innervation of the basal forebrain. II. Amygdala, suprarhinal cortex and autorhinal cortex. *J. Comp. Neurol.* 180:509–32

Fallon, J. H., Moore, R. Y. 1978. Catecholamine innervation of the basal forebrain. III. Olfactory bulb, anterior olfactory nuclei, olfactory tubercle and piriform cortex. *J. Comp. Neurol.* 170:533–44

Felten, D., Laties, A., Carpenter, M. 1974. Localization of monoamine containing cell bodies in the squirrel monkey brain. *Am. J. Anat.* 138:153–66

Foote, S. L., Freedman, R., Oliver, A. P. 1975. Effects of putative neurotransmitters on neuronal activity in monkey auditory cortex. *Brain Res.* 86:229–42

Freedman, R., Hoffer, B. J. 1975. Phenothiazine antagonism of the noradrenergic inhibition of cerebellar Purkinje neurons. *J. Neurobiol.* 6:277–88

Freedman, R., Hoffer, B. J., Puro, D., Woodward, D. J. 1976. Noradrenaline modulation of the responses of the cerebellar Purkinje cell to afferent synaptic activity. *Br. J. Pharmacol* 57:603–5

Freedman, R., Hoffer, B. J., Woodward, D. J. 1975. A quantitative microiontopho-

retic analysis of the responses of central neurons to noradrenaline: interactions with cobalt, manganese, verapamil and dichloroisoprenaline. *Br. J. Pharmacol.* 54:529–539

Freeman, R., Hoffer, B. J., Woodward, D. J., Puro, D. 1977. Interaction of norepinephrine with cerebellar activity evoked by mossy and climbing fibers. *Exp. Neurol.* 55:269–88

Fuxe, K. 1965. Evidence for the existence of monoamine neurons in the central nervous system. II. The distribution of monoamine terminals in the central nervous system. *Acta Physiol. Scand.* (Suppl. 247) 64:37–84

Fuxe, K., Hamburger, B., Hökfelt, T. 1968. Distribution of noradrenaline nerve terminals in cortical areas of the rat. *Brain Res.* 8:125–31

Fuxe, K., Hökfelt, T., Ungerstedt, U. 1969. Distribution of monoamines in the mammalian central nervous system by histochemical studies. In *Metabolism of Amines in Brain*, ed. G. Hooper, pp. 10–23. London: Macmillan.

Fuxe, K., Hökfelt, T. 1969. Catecholamines in the hypothalamus and pituitary gland. In *Frontiers in Neuroendocrinology,* ed. L. Martini, W. F. Ganong, pp. 47–96. New York: Oxford Press.

Fuxe, K., Hökfelt, T., Agnati, L., Löfstrom, A., Everitt, B. J., Johansson, O., Jonsson, G., Wuttke, W., Goldstein, M. 1976. Role of monoamines in the control of ganodotropin secretion. In *Neuroendocrine Regulations of Fertility,* ed. T. C. Anand-Kumar, pp. 124–140. Basel: Karger.

Fuxe, K., Hökfelt, T., Agnati, L. F., Johansson, O., Goldstein, M., Jonsson, G., Wattke, K. 1978. Mapping out central catecholamine neurons. Immunohistochemical studies on catecholamine synthesizing enzymes. In: *Psychopharmacology—A Generation of Progress,* ed. M. A. Lipton, A. DiMascio, K. F. Killian, pp. 67–94. New York: Raven Press.

Gahwiler, B. H. 1976. Inhibitory action of noradrenaline and cyclic AMP in explants of rat cerebellum. *Nature* 259: 483–84

Ganong, W. F. 1974a. Brain mechanisms regulating the secretion of the pituitary gland. In *The Neurosciences—Third Study Program,* ed. F. O. Schmitt, F. G. Worden, pp. 549–64. Cambridge: MIT Press.

Ganong, W. F. 1974b. The role of catecholamines and acetylcholine in the regulation of endocrine function. *Life Sci.* 15:1401–14

Garver, D. L., Sladek, J. R. Jr. 1975. Momoamine distribution in primate brain. I. catecholamine-containing perikarya in the brain stem of Macaca speciosa. *J. Comp. Neurol.* 159:289–304

Geller, H., Hoffer, B. J. 1977. Effect of calcium removal on monoamine-elicited depressions of cultured tuberal neurons. *J. Neurobiol.* 8:61–67

Graham, A. W., Aghajanian, G. K. 1971. Effects of amphetamine on single cell activity in a catecholamine nucleus, the locus coeruleus. *Nature* 234:100–2

Greengard, P. 1976. Possible role for cyclic nucleotides and phosphorylated membrane proteins in postsynaptic actions of neurotransmitters. *Nature* 260:101–8

Grillo, M. 1966. Electron microscopy of sympathetic tissues. *Pharmacol. Rev.* 18: 387–400

Groves, P. M., Wilson, C. J., Young, S. J., Rebec, G. V. 1975. Self-stimulation by dopaminergic neurons. *Science* 190: 522–29

Gupta, K. C., Moolenaar, G.-M., Holloway, J. A. 1977. The nucleus locus coeruleus: afferent connections. *Fed. Proc.* 36:385

Guyenet, P. G., Aghajanian, G. K. 1977. Excitation of neurons in the nucleus locus coeruleus by Substance P and related peptides. *Brain Res.* 136:178–84

Hobson, J. A. 1976. The Sleep-Dream Cycle: A Neurobiological Cycle In *Pathobiology Annual,* pp. 369–403. New York: Appleton-Century Crofts.

Hobson, J. A., McCarley, R. W., Pivik, R. T., Freedman, R. 1975. Selective firing by cat pontine brain stem neurons in desynchronized sleep. *J. Neurophysiol.* 37:497–511

Hobson, J. A., McCarley, R. W., Wyzinski, P. W. 1976. Sleep cyclic oscillation: reciprocal discharge by two brain stem neuronal groups. *Science* 189:55–58

Hoffer, B. J., Bloom, F. E. 1976. In *Chemical Transmission in the Mammlian Central Nervous System,* ed. C. H. Hockman, D. Bieger, pp. 327–48 Baltimore: Univ. Park Press.

Hoffer, B. J., Neff, N. H., Siggins, G. R. 1971a. Microiontophoretic release of norepinephrine from micropipettes. *Neuropharmacology* 10:175–80

Hoffer, B. J., Siggins, G. R., Bloom, F. E. 1969. Prostaglandins E1 and E2 antagonize norepinephrine effects on cerebellar Purkinje cells: microelectrophoretic study. *Science* 166:1418–20

Hoffer, B. J., Siggins, G. R., Bloom, F. E. 1971b. Studies on norepinephrine containing afferents to Purkinje cells of rat cerebellum: II. Sensitivity of Purkinje cells to norepinephrine and related substances administered by microiontophoresis. *Brain Res.* 25:523–34

Hoffer, B. J., Siggins, G. R., Woodward, D. J., Bloom, F. E. 1971c. Spontaneous discharge of Purkinje neurons after destruction of catecholamine-containing afferents by 6-hydroxydopamine. *Brain Res.* 30:425–30

Hoffer, B. J., Siggins, G. R., Oliver, A. P., Bloom, F. E. 1973. Activation of the pathway from locus coeruleus to rat cerebellar Purkinje neurons: pharmacological evidence of noradrenergic central inhibition. *J. Pharmacol. Exp. Ther.* 184:553–69

Hoffman, G. E., Sladek, J. R. Jr., Felten, D. L. 1976. Monoamine distribution in primate brain. III. Catecholamine-containing varicosities in the hypothalamus of Macaca Mulatta. *Am. J. Anat.* 147: 501–14

Hökfelt, T. 1967. On the ultrastructural localization of noradrenaline in the central nervous system of the rat. *Z. Zellforsch. Mikrosk. Anat.* 79:110–17

Hökfelt, T. 1968. In vitro studies on central and peripheral monoamine neurons at the ultrastructural level. *Z. Zellforsch. Mikrosk. Anat.* 91:1–74

Hökfelt, T., Elde, R., Johansson, O., Ljungdahl, A., Schuktzberg, M., Fuxe, K. 1978. The distribution of peptide-containing neurons in the nervous system. In *Psychopharmacology—A generation of Progress,* ed. K. Killam, A. DiMascio, M. Lipton, pp. 39–68. New York: Raven Press

Hökfelt, T., Fuxe, K. 1972. Brain endocrine interactions on the morphology and the neuro-endocrine role of hypothalamus catecholamine neurons. In *Median Eminence. Structure and Function,* ed. K. M. Knigge, E. E. Scott, A. Weindl, pp. 181–223. Basel: Karger.

Hökfelt, T., Fuxe, K., Goldstein, M., Johansson, O. 1973. Evidence for adrenaline neurons in the rat brain. *Acta Physiol. Scand.* 89:286–88

Hökfelt, T., Fuxe, K., Goldstein, M., Johansson, O. 1974. Immunohistochemical evidence for the existence of adrenaline neurons in the rat brain. *Brain Res.* 66:235–51

Hökfelt, T., Johnsson, O., Fuxe, K., Goldstein, M., Park, D. 1977. Immunohistochemical studies on the localization and distribution of monoamine neuron systems in the rat brain II. Tyrosine hy-

164   MOORE & BLOOM

droxylase in the telencephalon. *Med. Biol.* 55:21–40

Hubbard, J. E., DiCarlo, V. 1973. Fluorescence histochemistry of monoamine-containing cell bodies in the brainstem of the squirrel monkey (*Saimirisciureus*). I. The locus coeruleus. *J. Comp. Neurol.* 147:553–66

Jacobowitz, D. M. 1978. In *Psychopharmacology—A 20 Year Progress Report*, ed. M. E. Lipton, K. C. Killiam, A. DiMascio, pp. 119–130. New York: Raven Press.

Jacobowitz, D. M., Palkovits, M. 1974. Topographic atlas of catecholamine and acetylcholinesterase-containing neurons in the rat brain. I. Forebrain (telencephalon, diencephalon). *J. Comp. Neurol.* 157:13–28

Jones, B. E., Halaris, A. E., McIlhany, M., Moore, R. Y. 1977. Ascending projections of the locus coeruleus in the rat. I. Axonal transport in central noradrenaline neurons. *Brain Res.* 127:1–22

Jones, B. E., Moore, R. Y. 1974. Catecholamine-containing neurons of the nucleus locus coeruleus in the cat. *J. Comp. Neurol.* 157:43–52

Jones, B. E., Moore, R. Y. 1977. Ascending projections of the locus coeruleus in the rat. II. Autoradiographic study. *Brain Res.* 127:23–53

Jones, M. T., Hillhouse, E. W., Burden, J. 1976. Effect of various putative neurotransmitters on the secretion of corticotrophin-releasing hormone from the rat hypothalamus in vitro—a model of the neurotransmitters invoked. *J. Endocrinol.* 69:1–10

Kawamura, H., Gunn, C. G., Frohlich, E. D. 1978. Cardiovascular alteration by nucleus locus coeruleus in spontaneously hypertensive rat. *Brain Res.* 140:137–48

Kobayashi, R. M., Palkovits, M., Jacobowitz, D. M., Kopin, I. J. 1975. Biochemical mapping of the noradrenergic projection from the locus coeruleus. *Neurology* 25:223–33

Koda, L. Y., Bloom, F. E. 1977. A light and electron microscopic study of noradrenergic terminals in the rat dentate gyrus. *Brain Res.* 120:327–35

Koda, L. Y., Schulman, J. A., Bloom F. E. 1978b. Ultrastructural identification of noradrenergic terminals in rat hippocampus: unilateral destruction of the locus coeruleus with 6-hydroxydopamine. *Brain Res.* In press

Koda, L. Y., Wise, R. A., Bloom, R. E. 1978a. Light and electron microscopic changes in the rat dentate gyrus after

lesions or stimulation of the ascending locus coeruleus pathway. *Brain Res.* 144:363–68

Koslow, S. H., Schlumpf, M. 1974. Quantitation of adrenaline in rat brain nuclei and areas by mass fragmentography. *Nature* 251:530–31

Kromer, L. F., Moore, R. Y. 1977. Cochlear nucleus innervation by central norepinephrine neurons in the rat. *Brain Res.* 118:531–37

Kuypers, H. G. J. M., Maisky, V. A. 1975. Retrograde axonal transport of horseradish peroxidase from spinal cord to brain stem cell groups in the cat. *Neurosci. Lett.* 1:9–14

Landis, S., Bloom, F. E. 1975. Ultrastructural identification of noradrenergic boutons in mutant and normal mouse cerebellar cortex. *Brain Res.* 96:299–305

Landis, S., Shoemaker, W. J., Bloom, F. E., Schlumpf, M. 1975. Catecholamines in mutant mouse cerebellum: Flourescence microscopic and chemical studies. *Brain Res.* 93:253–66

Lapierre, Y., Beaudet, A., Demianczuk, N., Descarries, L. 1973. Noradrenergic axon terminals in the cerebral cortex of the rat. II. Quantitative data revealed by light and electron microscopic radioautography of the frontal cortex. *Brain Res.* 63:175–82

Lauder, J. M., Bloom, F. E. 1975. Ontogeny of monoamine neurons in the locus coeruleus, raphe nuclei and substantia nigra of the rat. II. Synaptogenesis. *J. Comp. Neurol.* 163:251–65

Lenn, N. J. 1965. Electron microscopic observations on monoamine-containing brain stem neurons in normal and drug-tested rats. *Anat. Rec.* 153:399–406

Levitt, P., Moore, R. Y. 1978. Noradrenaline neuron innervation of the neocortex in the rat. *Brain Res.* 139:219–32

Lindvall, O. 1975. Mesencephalic dopaminergic afferents to the lateral septal nucleus of the rat. *Brain Res.* 87:89–95

Lindvall, O., Björklund, A. 1974a. The glyoxylic acid fluorescence histochemical method: a detailed account of the methodology for the visualization of central catecholamine neurons. *Histochemistry* 39:97–127

Lindvall, O., Björklund, A. 1974b. The organization of the ascending catecholamine neuron systems in the rat brain. As revealed by the glyoxylic acid fluorescence method. *Acta Physiol. Scand.* 412:1–48

Lindvall, O., Björklund, A. 1978. Organization of catecholamine neurons in the rat central nervous system. In *Handbook of*

Psychopharmacology, ed. L. Iversen, S. Iversen, S. H. Snyder. New York: Plenum Press. In press

Lindvall, O., Björklund, A., Nobin, A., Stenevi, U. 1974. The adrenergic innervation of the rat thalamus as revealed by the glyoxylic acid fluorescence method. *J. Comp. Neurol.* 154:317–48

Löfstrom, A. 1976. Catecholamine turnover alterations in discrete areas of the median eminence at the 4- and 5-day cyclic rat. *Brain Res.* 120:113–31

Maeda, T., Pin. C., Salvert, D., Ligier, M., Jouvet, M. 1972. Les neurones cortenant des catecholamines du tegmentum portique et leurs voies de projection chez le chat. *Brain Res.* 57:119–52

Maeda, T., Shimizu, N. 1972. Projections ascendentes du locus coeruleus et e'autres neurones aminergiques pontiques au niveau de prosencephale du rat. *Brain Res.* 36:19–35

Mason, T., Iversen, S. D. 1977. An investigation of the role of cortical and cerebellar noradrenaline in associative motor learning in the rat. *Brain Res.* 134:513–27

McCarley, R. W., Hobson, J. A. 1975. Neuronal excitability modulation over the sleep cycle: a structural and mathematical model. *Science* 189:58–60

Melamed, E., Lahav, M., Atlas, D. 1977. β-adrenergic receptors in rat cerebral cortex: histochemical localization by a fluorescent blocker. *Brain Res.* 128:379–84

Moore, R. Y. 1975. Monoamine neurons innervating the hippocampal formation and septum: organization and response to injury. In *The Hippocampus,* ed. R. L. Isaacson, K. H. Pribam, 5:215–237. New York: Plenum.

Moore, R. Y. 1978. Catecholamine innervation of the basal forebrain. I. The septal area. *J. Comp. Neurol.* 177:665–84

Moore, R. Y., Björklund, A., Stenevi, U. 1971. Plastic changes in the adrenergic innervation of the rat septal area in response to denervation. *Brain Res.* 33:13–35

Moore, R. Y., Bloom, R. E. 1978. Central catecholamine neuron systems: Anatomy and physiology of the dopamine systems. *Ann. Rev. Neurosci.* 1:129–69

Mugnaini, E., Dahl, A. L. 1975. Mode of distribution of aminergic fibers in the cerebellar cortex of the chicken. *J. Comp. Neurol.* 162:417–32

Müller, E. E., Pecile, A., Felici, M., Cocchi, D. 1970. Norepinephrine and dopamine injection into lateral brain ventricle of the rat and growth hormone releasing

activity in the hypothalamus and plasma. *Endocrinology* 86:1376–82

Nakai, Y., Takaori, S. 1974. Influence of NE-neurons from LC on lateral geniculate activities of cats. *Brain Res.* 71:47–60

Nakamura, S., Iwama, K. 1975. Antidromic activation of the rat locus coeruleus neurons from hippocampus, cerebral and cerebellar cortices. *Brain Res.* 99:372–76

Nakamura, S. 1977. Some electrophysiological properties of neurones of rat locus coeruleus. *J. Physiol.* 267:641–58

Nathanson, J., Freedman, R., Hoffer, B. J. 1976. Lithium inhibits brain adenylate cyclase and blocks noradrenergic depression of Purkinje cell discharge independent of calcium. *Nature* 261:330–31

Nathanson, J. 1977. Cyclic nucleotides and nervous system function. *Physiol. Rev.* 57:158–256

Nicholson, J. L., Bloom, F. E. 1974. Ontogeny of monoamine neurons in the locus coeruleus, raphe nuclei and substantia nigra of the rat. I. Cell differentiation. *J. Comp. Neurol.* 155:469–82

Nobin, A., Björklund, A. 1973. Topography of the monoamine neuron systems in the human brain as revealed in fetuses. *Acta Physiol. Scand.* 388:1–40

Nygren, L. G., Olson, L. 1977. A new major projection from locus coeruleus: the main source of noradrenergic nerve terminals in the ventral and dorsal columns of the cord. *Brain Res.* 132:85–94

Orgren, S.-O., Fuxe, K. 1977. On the role of brain noradrenaline and the pituitary adrenal axis in avoidance learning. I. Studies with corticosterone. *Neurosci. Lett.* 5:291–96

Ojeda, S. R., McCann, S. M. 1973. Evidence for the participation of a catecholaminergic mechanism in the post-castration rise in plasma gonadotropins. *Neuroendocrinology* 12:295–355

Olson, L., Fuxe, K. 1971. On the projections from the locus coeruleus noradrenaline neurons: the cerebellar innervation. *Brain Res.* 28:165–71

Olson, L., Fuxe, K. 1972. Further mapping out of central noradrenaline systems. Projections of the "subcoeruleus" area. *Brain Res.* 43:289–95

Olson, L., Boreus, L. O., Seiger, A. 1973. Histochemical demonstration and mapping of 5-hydroxytryptamine- and catecholamine-containing neuron systems in the human fetal brain. *Z. Anat. Entwicklungs gesch.* 139:259–82

Olszewski, J., Baxter, D. 1954. *Cytoarchitecture of the Human Brain Stem.* pp. 170–173. Philadelphia: Lippincott

Palkovits, M., Brownstein, M., Saavedra, J. M. 1974. Norepinephrine and dopamine content of hypothalamic nuclei of the rat. *Brain Res.* 77:137–49

Palkovits, M., Jacobowitz, D. M. 1974. Topographic atlas of catecholamine and acetylcholinesterase-containing neurons in rat brain. II. Hindbrain (mesencephalon, rhombencephalon). *J. Comp. Neurol.* 157:29–42

Perlow, M., Sassin, I. F., Boyar, D., Hellman, L., Weitzman, E. D. 1972. Release of human growth hormone, follicle stimulating hormone, and luteinizing hormone in response to L-dihydrophenylalanine (L-DOPA) in normal man. *Dis. Nerv. Syst.* 33:804–10

Phillis, J. W., Lake, N., Yarborough, G. G. 1973. Calcium mediation of the inhibitory effects of biogenic amines on cerebral cortical neurons. *Brain Res.* 53: 465–69

Pickel, V. M., Joh, T. H., Reis, D. J. 1977. A serotonergic innervation of noradrenergic neurons in nucleus locus coeruleus: demonstration by immunocytochemical localization of the transmitter specific enzymes tyrosine and tryptophan hydroxylase. *Brain Res.* 131:197–214

Pickel, V. M., Krebs, W. H., Bloom, F. E. 1973. Proliferation of norepinephrine-containing axons in rat cerebellar cortex after peduncle lesions. *Brain Res.* 59: 169–79

Pickel, V. M., Segal, M., Bloom, F. E. 1974a. A radioautographic study of the efferent pathways of the nucleus locus coeruleus. *J. Comp. Neurol.* 155:15–42

Pickel, V. M., Segal, M., Bloom, F. E. 1974b. Axonal proliferation following lesions of cerebellar peduncles. A combined fluorescence microscopic and radioautographic study. *J. Comp. Neurol.* 155:43–60

Pohorecky, L. A., Brick, J. 1977. Activity of neurons in the locus coeruleus of the rat: inhibition by ethanol. *Brain Res.* 131:174–79

Pohorecky, L. A., Zigmond, M., Karten, H., Wurtman, R. J. 1969. Enzymatic conversion of norepinephrine to epinephrine by the brain. *J. Pharmacol. Exp. Ther.* 165:190–95

Poitras, D., Parent, A. A. 1975. A fluorescence microscopic study of the distribution of monoamines in the hypothalamus of the cat. *J. Morphol.* 145: 387–407

Rall, T. W. 1972. Role of adenosine 3',5'-monophosphate (cyclic AMP) in actions of catecholamines. *Pharmacol. Rev.* 24:399–409

Ramon-Moliner, E. 1977. Non-synaptic chemical neurotransmission. *Experientia* 33:1342–44

Ranck, J. B. Jr. 1973. Studies on single neurons in dorsal hippocampal formation and septum in unrestrained rats. Part I. Behavioral correlates and firing repertoires. *Exp. Neurol.* 41:461–531

Reid, J. L., Zivin, J. A., Foppen, F. H., Kopin, I. J. 1975. Catecholamine neurotransmitters and synthetic enzymes in the spinal cord of the rat. *Life Sci.* 16:975–84

Reid, J. L., Zivin, J. A., Kopin, I. J. 1976. The effects of spinal cord transsection and intracisternal 6-hydroxydopamine on phenylethanolamine-N-methyl transferase (PNMT) activity in rat brain stem and spinal cord. *J. Neurochem.* 26:629–31

Robinson, R. G., Bloom, R. E. Battenberg, E. L. F. 1977a. A fluorescent histochemical study of changes in noradrenergic neurons following experimental cerebral infarction in the rat. *Brain Res.* 132:259–72

Robinson, R. E., Vanderwolf, C. H., Pappas, B. A. 1977b. Are the dorsal noradrenergic bundle projections from the locus coeruleus important for neocortical or hippocampal activation. *Brain Res.* 138:75–98

Russell, G. V. 1955. The nucleus locus coeruleus (dorsalis tegmenti). *Tex. Rep. Biol. Med.* 13:939–88

Saavedra, J. M., Palkovits, M., Brownstein, M. J., Axelrod, J. 1974. Localization of phenylethanolamine N-methyl transferase in the rat brain nuclei. *Nature* 248:695–96

Sakai, K., Touret, M., Salvert, D., Leger, L., Jouvet, M. 1977. Afferent projections to cat locus coeruleus as visualized by the horse radish peroxidase technique. *Brain Res.* 119:21–41

Salmoiraghi, G. C., Stefanis, C. 1971. Central synapses and suspected transmitters. *Int. Rev. Neurobiol.* 10:1–30

Saper, C. B., Swanson, L. W., Cowan, W. M. 1976. The efferent connections of the ventromedial nucleus of the hypothalamus in the rat. *J. Comp. Neurol.* 167:409–42

Sasa, M., Muneyiko, K., Ikeda, H., Takaori, S. 1974. Noradrenaline-mediated inhibition by locus coeruleus of spinal trigeminal neurons. *Brain Res.* 80: 443–60

Sasa, M., Muneyiko, K., Takaori, S. 1975. Morphine interference with noradrenaline-mediated inhibition from the locus coeruleus. *Life Sci.* 17:1373–80

Sawyer, C. H. 1975. Some recent developments in brain-pituitary-ovarian physiology. *Neuroendocrinology* 17:97–124

Sawyer, C. H., Hilliard, J., Kanematsu, S., Scaramuzzi, R., Blake, C. A. 1974. Effects of intraventricular infusions of norepinephrine and dopamine on LH release and ovulation in the rabbit. *Neuroendocrinology* 15:328–37

Scheibel, M. E., Scheibel, A. B. 1957. In *Reticular Formation of the Brain,* ed. H. H. Jasper, L. D. Proctor, R. S. Knighton, W. C. Moseley, R. T. Costello, pp. 31–55. Boston: Little, Brown.

Scott-Young, W. III, Bird, S. J., Kuhar, M. J. 1978. Iontophoresis of methionine enkephaline in the locus coeruleus area. *Brain Res.* In Press

Segal, M. 1974. Lithium and the monoamine transmitters in the rat hippocampus. *Nature* 250:71–73

Segal, M. 1976. Brain stem afferents to the rat medial septum. *J. Physiol.* 261:617–31

Segal, M., Bloom, R. E. 1974a. The action of norepinephrine in the rat hippocampus: I. Iontophoretic studies. *Brain Res.* 72:79–97

Segal, M., Bloom, F. E. 1974b. The action of norepinephrine in the rat hippocampus: II. Activation of the input pathway. *Brain Res.* 72:99–114

Segal, M., Bloom, F. E. 1976a. The action of norepinephrine in the rat hippocampus: III. Hippocampal cellular responses to locus coeruleus stimulation in the awake rat. *Brain Res.* 107:499–511

Segal, M., Bloom, F. E. 1976b. The action of norepinephrine in the rat hippocampus: IV. The effects of locus coeruleus stimulation on evoked hippocampal unit activity. *Brain Res.* 107:513–25

Segal, M., Oliver, A. P. 1975. Transmembrane changes in hippocampal neurons: hyperpolarizing actions of norepinephrine, cyclic AMP and locus coeruleus. *Proc. Soc. Neurosci.* 361

Segal, M., Pickel, V., Bloom, F. 1974. The projections of the nucleus locus coeruleus: an autoradiographic study. *Life Sci.* 13:817–21

Selmanoff, M. R., Pranik-Holdaway, M. J., Weiner, R. I. 1976. Concentrations of dopamine and norepinephrine in discrete hypothalamic nuclei during the rat estrus cycle. *Endocrinology* 99:326–29

Siggins, G. R., Henriksen, S. J. 1975. Inhibition of rat Purkinje neurons by analogues of cyclic adenosine monophosphate: correlation with protein kinase activation. *Science* 179:585–88

Siggins, G. R., Hoffer, B. J., Bloom, F. E. 1971a. Studies on norepinephrine-containing afferents to Purkinje cells of rat cerebellum: III. Evidence for mediation of norepinephrine effects by cyclic 3',5'-adenosine monophosphate. *Brain Res.* 25:535–53

Siggins, G. R. Hoffer, B. J., Bloom, F. E. 1971b. Prostaglandin-norepinephrine interactions in brain: Microelectrophoretic and histochemical correlates. *Ann. NY Acad. Sci.* 180:302–23

Siggins, G. R., Oliver, A. P., Hoffer, B. J., Bloom, F. E. 1971c. Cyclic adenosine monophosphate and norepinephrine: Effects on transmembrane properties of cerebellar Purkinje cells. *Science* 171:192

Siggins, G. R., Hoffer, B. J., Oliver, A. P., Bloom, F. E. 1971d. Activation of a central noradrenergic projection to cerebellum. *Nature* 233:481–83

Sladek, J. R. Jr., Walker, P. 1977. Serotonin-containing neuronal perikarya in the primate locus coeruleus and subcoeruleus. *Brain Res.* 134:359–66

Snider, R. S. 1975. A cerebellar-ceruleus pathway. *Brain Res.* 88:59–67

Sotelo, C., Palay, S. 1971. Altered axons and axon terminals in the lateral vestibular nucleus of the rat: possible example of axonal remodeling. *Lab. Invest.* 25:653–72

Stone, T. W., Taylor, D. A., Bloom. R. E. 1975. Cyclic AMP and cyclic GMP may mediate opposite neuronal responses in the rat cerebral cortex. *Science* 187:845–47

Stone, T. W., Taylor, D. A. 1977. Microiontophoretic studies of the effects of cyclic nucleotides on excitability of neurones in rat cerebral cortex. *J. Physiol.* 266:523–43

Sutherland, E. W., Robinson, G. A., Butcher, R. 1968. Some aspects of the biological role of adenosine 3',5'-monophosphate (cyclic AMP). *Circulation* 3:279–306

Svensson, T. H., Bunney, B. S., Aghajanian, G. K. 1975. Inhibition of both NA and 5-HT neurons in brain by the alpha agonist clonidine. *Brain Res.* 92:291–306

Swanson, L. W. 1976a. An autoradiographic study of the efferent projections of the preoptic region in the rat. *J. Comp. Neurol.* 167:227–56

Swanson, L. W. 1976b. The locus coeruleus: a cytoarchitectonic, golgi and immunohistochemical study in the albino rat. *Brain Res.* 110:39–56

Swanson, L. W., Connelly, M. A., Hartman, B. K. 1977. Ultrastructural evidence for central monoaminergic innervation of blood vessels in the paraventricular nucleus of the hypothalamus. *Brain Res.* 136:166–73

Swanson, L. W., Hartman, B. K. 1975. The central adrenergic system. An immunofluorescence study of the location of cell bodies and their efferent connections in the rat utilizing dopamine-beta-hydroxylase as a marker. *J. Comp. Neurol.* 163:467–506

Szentagothai, J., Flerko, B., Mess, B., Halasz, B. 1968. *Hypothalamic Control of the Anterior Pituitary.* Budapest: Akademiai Kiado

Takigawa, M., Mogenson, G. J. 1977. A study of inputs to antidromically identified neurons of the locus coeruleus. *Brain Res.* 135:217–30

Tohyama, M. 1976. Comparative anatomy of cerebellar catecholamine innervations from teleosts to mammals. *J. Hirnforsch.* 17:43–60

Toivola, P. T. K., Gale, C. C. 1972. Central adrenergic regulation of growth hormone secretion in baboons. *Int. J. Neurosci.* 4:53–63

Tranzer, J. P., Thoenen, H. 1967. Electron-microscopic localization of 5-hydroxy-dopamine (3,4,5-trihydroxy-phenylethylamine), a new "false" sympathetic transmitter. *Experientia* 23:743–45

Tribollet, E., Clarke, G., Dreifuss, J. J., Lincoln, D. W. 1978. The role of central adrenergic receptors in the reflex release of oxytocin. *Brain Res.* 142:69–84

Ungerstedt, U. 1971. Stereotaxic mapping of the monoamine pathways in the rat brain. *Acta Physiol. Scand.* 367:1–48

Van der Gugten, J., Palkovits, M., Wijen, H. L. J. M., Versteeg, D. H. G. 1976. Regional distribution of adrenaline in rat brain. *Brain Res.* 107:171–75

Van Loon, F. G. 1973. In *Frontiers in Neuroendocrinology,* ed. L. Martin, W. F. Ganong, pp. 209–49. New York: Oxford Press

Vogt, M. 1954. The concentration of sympathin in different parts of the central nervous system under normal conditions and after the administration of drugs. *J. Physiol.* 123:451–81

Ward, D. G., Gunne, C. G. 1976a. Locus coeruleus complex elicitation of a pressor response and a brain stem region necessary for its occurrence. *Brain Res.* 107:401–6

Ward, D. G., Gunne, C. G. 1976b. Locus coeruleus complex: differential modulation of depressor mechanisms. *Brain Res.* 107:407–11

Woodward, D. J., Hoffer, B. J., Siggins, G. R., Bloom, F. E. 1971. The ontogenetic development of synaptic junctions, synaptic activation and responsiveness to neurotransmitter substances in rat cerebellar Purkinje cells. *Brain Res.* 34:73–79

Zivin, J. A., Reid, J. L., Saavedra, J. M., Kopin, I. J. 1975. Quantitative localization of biogenic amines in the spinal cord. *Brain Res.* 99:293–301

Ann. Rev. Neurosci. 1979. 2:169–91
Copyright © 1979 by Annual Reviews Inc. All rights reserved

PHYSIOLOGY OF THE RETINA ♦11521

Akimichi Kaneko

Department of Physiology, Keio University, School of Medicine, Shinanomachi, Shinjuku-ku, Tokyo 160, Japan

INTRODUCTION

Because of its orderly structure, the vertebrate retina is one of the few parts of the nervous system in which function is well correlated with structural organization. One major reason for this has been the introduction of intracellular staining techniques that can be used in conjunction with physiological studies. Thus a major aspect of the progress made since a previous review article (Witkovsky 1971) on the physiology of the vertebrate retina, has been at the level of single cells. Since space is limited, this article stresses single cell studies on the lower vertebrates, with particular attention to the correlation between physiology and morphology. This limited view is supplemented by recent excellent reviews on various aspects of the retina. *The Handbook of Sensory Physiology* Vol. VII/2 by Fuortes (1972) and *The Vertebrate Retina* by Rodieck (1973) in particular cover almost the entire field. Color vision is discussed in Daw's review (1973) and Werblin (1974) summarizes questions on adaptation.

Improved Techniques

First, I wish to list several technical improvements that have made significant contributions to research on the retina in recent years.

The introduction of Procion yellow for intracellular staining made it possible to reliably identify cells recorded by microelectrodes (Stretton & Kravitz 1968). Symposium proceedings have been published (Kater & Nicholson 1973) on the technique and application of this method. Intracellular staining is particularly useful for identifying retinal cells, since they cannot be penetrated under visual control using an ordinary light microscope, first because the tissue is not transparent and second because inspec-

0147-006X/79/0315-0169$01.00

tion light causes bleaching of the visual pigments. In special instances when visual control of microelectrode manipulation is imperative, infrared converters are used (cf Hagins et al 1970). Once a correlation between cellular type and physiological response is established by intracellular staining, one can subsequently identify recorded cells by their response characteristics. Since Procion yellow easily diffuses inside an injected cell, it stains the entire cell. Recently not only cell type, but also subclasses, have been identified and correlated with function (Naka & Ohtsuka 1975, Famiglietti et al 1977).

Superfusion of the isolated retina enables one to modify the ionic environment of retinal neurons to study the mechanisms of response generation and synaptic transmission. This technique has been particularly useful in the study of photoreceptors (Hagins et al 1970, Cervetto 1973, Brown & Pinto 1974). Superfusion is also advantageous in maintaining the retina in good condition for a long time. Also in pharmacological studies, superfusion enables quantitative application of chemicals such as transmitter candidates or blocker substances (Cervetto & Piccolino 1974, Kaneko & Shimazaki 1975, Miller & Dacheux 1976a, b, c).

The construction and application of microelectrodes have also been improved. In particular, glass microelectrodes with beveled tips have made penetration much easier and have improved recording stability (Brown & Flaming 1975). Beveling one barrel of a double-barreled microelectrode reduces the coupling resistence between the current and recording electrodes, making measurement of membrane properties more accurate (Werblin 1975a). Also for faster and easier filling it is now common to insert a fine glass fiber into the capillary tubing before making microelectrodes (Tasaki et al 1968).

Introduction and wide application of computers have made complex and time-consuming calculations into a handy process, as in the calculation of kernels with the white noise method (Marmarelis & Naka 1973a, b. c), or in noise analysis of receptors and bipolar cells (Simon et al 1975, Schwartz 1977). The study of ganglion cells by extracellular recording is largely aided by the use of computers. Such application is so widespread that no further comments are necessary.

IDENTIFICATION OF RETINAL NEURONS COMPOSING THE STRAIGHT SIGNAL PATHWAYS

Cajal (1893) suggested purely on a morphological basis that the straightest, and therefore the most direct, pathway of signal transmission in the retina is from photoreceptors to bipolar cells, to ganglion cells. Electron microscopy has confirmed this hypothesis (Dowling & Boycott 1966, Dowling 1968) and physiological studies have also supported this view.

Photoreceptors

Bortoff (1964), using *Necturus* retina, was the first to show that photoreceptors hyperpolarize in response to illumination. Following this report, many investigators have confirmed this finding both in rods and in cones of various animals (carp cones, Tomita 1965; frog rods, Toyoda et al 1970, Brown & Pinto 1974; turtle cones, Baylor & Fuortes 1970; turtle rods, Schwartz 1973; salamander cones and rods, Lasansky & Marchiafava 1974; axolotl rods, Grabowski & Pak 1975, Grabowski et al 1972; *Necturus* cones and rods, Werblin & Dowling 1969, Fain & Dowling 1973). In the case of several lower vertebrates, the site of recording has been morphologically identified as a receptor. Because the current distribution in the extracellular space around all vertebrate receptors is similar (Hagins et al 1970), the photoreceptors of higher vertebrates are also believed to have hyperpolarizing receptor potentials. The mechanism of generation of receptor potentials, is summarized in another chapter in this volume (Hubbell & Bownds, this volume).

The photoreceptor is composed of three different parts: the outer segment, the inner segment, and the synaptic terminals. Incident light energy is transduced into neuronal activity at the outer segment. The inner segment is believed to maintain the ionic composition of photoreceptor cells by an active ionic pump. Its energy is supplied by the mitochondria packed in this region. For signal transmission, the inner segment functions as a pathway from the outer segment to the photoreceptor terminal. In a voltage clamp experiment, Werblin (1975a) has demonstrated a regenerative hyperpolarization in the mudpuppy photoreceptor, suggesting that the photo-signal is amplified in the inner segment.

The terminal of photoreceptors has a unique structure. The thin process that comes down from the inner segment expands at the terminal and is called, in the case of cones, the pedicle, or for rods, the spherule. Here, synapses are formed with bipolar and horizontal cells. In the receptor terminal, synaptic vesicles are clustered around dense presynaptic ribbons. Opposite the ribbon, three postsynaptic elements are arranged in such a way that the central process is flanked by the two lateral processes. The central process is usually the dendritic tip of a bipolar cell (Stell 1967, Boycott & Kolb 1973), but, at least in the goldfish, horizontal cell dendrites can also be the central element (Stell 1976). The lateral processes come from horizontal cells (Dowling 1968, Stell 1967). Both the central and lateral processes invaginate the receptor terminal to make an invaginating synapse.

Lasansky (1971) first suggested that other contacts at the receptor cell base also function as synapses. In turtle cones, invaginating synapses are not seen with flat bipolar cells; instead, these bipolar cells make superficial contacts at the base of the cone pedicle. Although the membrane specialization at these contacts is atypical and clusters of vesicles are few, this is thought to be the only site of signal transmission from cones to the flat bipolar cell.

As mentioned above, the receptor potential is a graded hyperpolarization. Because regenerative spikes do not occur, this synapse must show a unique operation, different from that of well-known synapses in other parts of the nervous system. The mechanism of synaptic transmission is discussed in a later section.

RECEPTIVE FIELD ORGANIZATION   Extensive study of cone receptive fields was first made in the turtle by Baylor et al (1971). They demonstrated that the receptor potential shows spatial summation up to a radius of about 40 μm. They also demonstrated that the spatial summation results from electrical coupling between cones. The distance of current spread was close to that found for spatial summation of the photic responses. Gap junctions are found between the contiguous cones (Raviola & Gilula 1973) and electrical coupling occurs only between cones of the same type (Baylor & Hodgkin 1973).

Spatial summation in rods covers a much wider area than in cones (200 μm in *Chelydra serpentina*) and the electrical coupling is stronger (Schwartz 1973, 1975, 1976, Copenhagen & Owen 1976a, b). In *Bufo marinus,* contiguous rods have gap junctions on the radial fins (Fain et al 1976), but in *Chelydra* the interreceptor contacts are probably made at the tip of the thin process originating from the receptor terminal region (Copenhagen & Owen 1976a).

Illumination of more remote areas results in a decrease of the receptor potential during the steady phase. Thus the receptive field of turtle cones has an antagonistic center-surround organization (Baylor & Fuortes 1970, Baylor et al 1971). More recently, similar interactions have been found in gecko (Pinto & Pak 1974a, b) and perch retinas (Burkhardt 1977), so it seems probable that this receptive field organization is universal among vertebrates. The antagonistic lateral effect is thought to be mediated by horizontal cells through a reciprocal synapse. The cone response to peripheral illumination has a time course that is similar to that recorded in horizontal cells (O'Bryan 1973). Baylor et al (1971) clearly demonstrated that hyperpolarization of a horizontal cell by extrinsic current injection produced a depolarizing potential change in turtle cones.

Horizontal cell feedback is color specific (Fuortes et al 1973), and plays an important role in constructing the chromatic response properties of horizontal cells. In the turtle, each horizontal cell receives input from a corresponding type of cone: L-type from red-sensitive cones, R/G type from green-sensitive cones, and B/G type from blue-sensitive cones. (For further description of the types, see the section on horizontal cells.) All cones have a recurrent connection with L-type horizontal cells. Thus the depolarizing response in R/G type and G/B type horizontal cells is mediated by feedback from the L-type horizontal cells to the corresponding cones.

It is difficult to detect the feedback effects of horizontal cells in other animals. It has been suggested that the lateral interaction is very susceptible to damage (Pinto & Pak 1974a), but in addition, the ease of detection may depend on the morphology of the photoreceptors. For example, the carp, in which the lateral effect is hardly detected, has a long, thin, tortuous process connecting the inner segment and pedicle, while the turtle has a stout, short connection. Since one does not expect to penetrate the small receptor terminals, the site of potential recording is probably the inner segment in most examples. The electrical signal of the lateral interaction is supposed to be evoked at the receptor terminal and conducted passively to the inner segment. Therefore, the degree of decrement will be highly determined by the dimensions of the connecting channel and by its electrical properties.

Bipolar Cells

As the name signifies, the bipolar cell has two processes, running in opposite directions from the cell body, which is located in the inner nuclear layer. The dendrite branches into fine terminals and connects with photoreceptors. The dendritic field varies from species to species and from one cell to another, but ranges from somewhere between several microns and 100 μm in diameter. The bipolar cell axon runs to the inner plexiform layer and makes contacts with ganglion and amacrine cells. The axon does not branch extensively, but usually has just a few short appendages that come out from the axonal stem (cf. Cajal 1893).

Bipolar cells show graded responses without regenerative spikes (Werblin & Dowling 1969, Kaneko 1970, Schwartz 1974, Richter & Simon 1975). They are classified into two groups according to response polarity. The first type of bipolar cells is depolarized by a spot illumination: it is referred to as depolarizing or on-type. The second type is hyperpolarized by a spot illumination and is called hyperpolarizing or off-type.

RECEPTIVE FIELDS   Bipolar cells have a concentrically organized, antagonistic receptive field. Illumination of the surround produces hyperpolarization in the on-type cells and depolarization in the off-type cells. Since the center is more sensitive than the surround, diffuse illumination evokes the center response. To evoke a surround response, it is necessary to illuminate a wide area of the surround (Kaneko 1970), and in some animals to inactivate the center with a bright adapting spot (Werblin & Dowling 1969).

In a comparison of the limit of spatial summation and the extent of the dendritic field, the receptive field center was found to be almost identical to the dendritic field, and it was concluded from this observation that the center response is generated by direct input from receptors (Werblin & Dowling 1969, Werblin 1970, Kaneko 1973). The receptive field surround far exceeds the dendritic expansion and is thought to be mediated by horizontal cells. Simulating the horizontal cell response by injection of extrinsic current, J. Toyoda (personal communication) has recorded a response in bipolar cells with the same polarity as that caused by surround illumination.

The mechanism of the horizontal cell action on bipolar cells is still not well understood. Since the horizontal cells give inhibitory feedback to receptors (Baylor et al 1971), receptors themselves have a center-surround organization. The bipolar cell may simply reflect this activity. But there is evidence suggesting that feedback action on the cone is insufficient to account for the peripheral antagonism seen in bipolar cells. In the turtle, the decrease in sensitivity caused by surround illumination is not equal in cones and in bipolar cells (Richter & Simon 1975). Also in mudpuppy (Dowling & Werblin 1969), cat, and rabbit (Brown et al 1966), direct synapses are reported from horizontal cells to bipolar cells.

MORPHOLOGICAL DIFFERENCE OF ON- AND OFF-TYPE BIPOLAR CELLS   On- and off-type bipolar cells have different morphologies. In the carp, intracellular staining has demonstrated that the primary morphological distinction between off and on bipolar cells is the level at which their axon terminals are found in the inner plexiform layer: off-type axons terminate in sublamina a (close to the inner nuclear layer) and on-type in sublamina b (near the ganglion cell layer). This distinction cuts across the classification between large mixed bipolar cells and small cone bipolar cells (Famiglietti et al 1977). A similar classification is also applicable to ganglion and amacrine cells (see following sections).

CONVERGENCE OF ROD AND CONE SIGNALS   Bipolar cells are also classified according to size, and this classification is highly correlated with the classification by input. In the cyprinid retina, Golgi-EM studies (Stell 1967) have shown the large bipolar cells, which was originally a "rod bipolar cell," to have connections with both rods and cones. The small bipolar cells, on the other hand, receive cone input only, as originally suggested (Stell 1967). When conditions were changed from scotopic to photopic, it was found that the spectral sensitivity peak of mixed bipolar cells was shifted from 523 nm (the absorption maximum of porphyrhopsin, Munz & Schwanzara 1967) to 620 nm (the absorption maximum of the red-sensitive cone pigment, Marks 1965), suggesting that they receive inputs from both rods and cones. The response polarity remained the same under the scotopic and photopic conditions. These results indicate that it is the postsynaptic cell (bipolar cell) that determines the response polarity (Kaneko et al 1978). A similar organization has been reported for the ganglion cell receptive fields of the goldfish (Beauchamp & Daw 1972, Raynauld 1972).

Ganglion Cells

Ganglion cells are the only cells in the retina with a typical axon. The cell bodies are located at the innermost part of the retina, usually in a monolayer. The axons compose the optic nerve, which conveys the spike discharges, encoded from the graded signals at the ganglion cells, to the higher visual centers. Morphologically, ganglion cells are classified into several types according to their dendritic pattern (Cajal 1893, Polyak 1941).

Ganglion cell responses are described as "on," "off," and "on-off" in reference to the timing of illumination, with the idea that a cell receives excitatory and inhibitory inputs at some particular period of illumination (Hartline 1938, Kuffler 1953, Barlow 1953).

RECEPTIVE FIELDS   Since its demonstration by Kuffler (1953), the antagonistic center-surround organization has been thought to be the most fundamental structure of the ganglion cell receptive field. This structure is widely found in lower vertebrates such as goldfish (Wagner, MacNichol & Wolbarsht 1960, Daw 1968) and in higher mammals (cat, Kuffler 1953; ground squirrel, Michael 1968; monkey, Hubel & Wiesel 1960; human, Weinstein et al 1971). These ganglion cells are either on-center or off-center cells. No strict correlation of the receptive field size and the dendritic field has been made on identified ganglion cells, but from statistical observations it is believed that the size of the center is the same as that of the dendritic field (Brown & Major 1966, Famiglietti & Kolb 1976).

An entirely different idea has been proposed by Lettvin's group (Lettvin et al 1959, Maturana et al 1960) to describe ganglion cell responses. They took the functional significance of spike generation as the most important criterion, and classified frog ganglion cells into "convex edge detectors," "dimming detectors," etc. Similar cells have been reported in birds (Maturana & Frenk 1963) and rabbits (Barlow et al 1964). To a flashing spot, many of these cells show only on-off responses. They respond better to moving stimuli, and often to movement in a particular direction (Barlow et al 1964, Wyatt & Daw 1975).

In these animals, the visual signal is highly processed within the retina, even to the extent seen in the mammalian visual cortex. Complexity of the receptive field organization is thought to be correlated with the number and variety of amacrine cells. In amphibia, birds, and rabbits, direct connection of bipolar cells to ganglion cells is rare; instead, the activity of bipolar cells is relayed by a series of amacrine cells to the ganglion cells (Dowling 1968, Dubin 1970).

MORPHOLOGICAL DIFFERENCE OF ON- AND OFF-TYPE GANGLION CELLS   In cat and fish retinas, in which ganglion cells with concentric receptive fields dominate, on-center and off-center ganglion cells have been shown to have different morphologies (Nelson et al 1978, Famiglietti et al 1977). This is the result of functional separation of sublaminae *a* and *b* of the inner plexiform layer, which carry off and on signals respectively, as described above. Thus the on-center ganglion cells extend their dendrites in sublamina *b* where on-type bipolar axons terminate, while the dendrites of the off-center ganglion cells are confined to sublamina *a*.

These findings suggest that the synaptic connections between bipolar and ganglion cells are all excitatory, or that the polarity of the response is maintained through this transmission. Physiological support for this hypothesis comes from the observation that depolarization of an on-type bipolar cell by extrinsic current evoked spikes in on-center ganglion cells, while depolarization of an off-type bipolar cell evoked spikes in off-center ganglion cells (Naka 1977, Baylor & Fettiplace 1977). No cross-interaction was detected between pairs of different response polarity. Furthermore, the hyperpolarization produced in off-center ganglion cells by illumination was found to be the result of "disfacilitation," or the removal of sustained excitatory input imposed during darkness (Miller & Dacheux 1976b, c).

The destination of dendrites of the on-off type ganglion cells is expected to be in both sublaminae, if this type of cell receives synaptic inputs directly from bipolar cells. But as mentioned above, many of these ganglion cells are thought to have closer connections with amacrine cells (Werblin & Dowling 1969). Since the dendritic arborization of on-off amacrine cells is found in both sublaminae (Famiglietti et al 1977), the on-off type ganglion cells can receive input from on-off amacrine cells regardless of the position of the dendrites.

IDENTIFICATION OF RETINAL NEURONS COMPOSING THE LATERAL PATHWAYS

Considered alone, the direct signal pathways from receptors to bipolar cells to ganglion cells are fundamentally isolated from each other. Lateral interaction between these straight pathways are made through two lateral elements, the horizontal cells in the outer plexiform layer, and the amacrine cells in the inner plexiform layer.

Horizontal Cells

Horizontal cells are located in the most distal part of the inner nuclear layer. Their dendritic processes invaginate photoreceptor terminals.

Horizontal cells respond to light with graded sustained responses that have a high degree of spatial summation. They all show hyperpolarization to white light, but their response polarity to chromatic illumination depends on the wavelength of the incident light. According to their spectral response properties, horizontal cells are classified into three groups: the L-type that show hyperpolarizing responses to all spectral light; the biphasic C-type, which are hyperpolarized by short wavelengths and depolarized by long wavelengths; and triphasic C-type, which are hyperpolarized by the monochromatic light of both spectrum ends and depolarized by the intermediate spectral region (MacNichol & Svaetichin 1958, Tomita 1965). Before their origin had been determined, these responses were called S-potentials after Svaetichin (1953), who first described them in the fish retina. By intracellular staining these responses have now been identified as coming from horizontal cells (Werblin & Dowling 1969, Kaneko 1970, 1971, Kaneko & Yamada 1972, Steinberg & Schmidt 1970, Matsumoto & Naka 1972, Miller et al 1973, Simon 1973, Mitarai et al 1974, Hashimoto et al 1976).

Horizontal cells are also morphologically classified into subtypes, and particularly in the teleost and elasmobranch, several subtypes are described. Here morphological and physiological classification of carp and goldfish horizontal cells are summarized, as examples, since these animals are often used in the study of horizontal cells. In the fish retina, horizontal cells of different subtypes are arranged in layers. Cajal (1893) classified them into two large groups, external and intermediate horizontal cells. (In fact, Cajal described a third type of horizontal cell, the internal, but the structure he referred to is now known to be part of the external horizontal cell, as discussed below.)

EXTERNAL HORIZONTAL CELLS   The external horizontal cells of carp and goldfish are connected to cones (Stell 1967). Within the layer of external horizontal cells, cells showing different spectral sensitivity are arranged in layers whose order, from the receptor side, is L-, biphasic C-, and triphasic C-types (Mitarai et al 1974, Hashimoto et al 1976). Not only are the cells layered, but they differ in shape when seen face-on. Such layering is most clearly seen in the *Eugerres* retina, in which each of the three layers of horizontal cells is connected to red-, green-, and blue-sensitive cones separately (Laufer & Millán 1970).

Axons of external horizontal cells.   In external horizontal cells, a thin axon comes out and runs laterally from the cell body. After passing the underlying layer of intermediate horizontal cells, the axon increases its diameter to about 10 to 15 μm in the carp and goldfish. The axon then continues to run parallel to the retinal layers and tapers off. No branching is seen and the final ending is not well defined. Cajal (1893) called these thick tubular structures internal horizontal cells, but recently their continuity to the external horizontal cell has been confirmed both in the Golgi preparation (Stell 1975) and by Procion yellow staining (Hashimoto et al 1976). Therefore, these structures should be called horizontal cell axons, not internal horizontal cells.

From the axons of external horizontal cells one can record responses similar to those from the cell body. Three spectral response types are seen. The only differences in the response properties of the axon compared to those of the cell body are that the amplitude of the axon response is smaller, and that axons show larger spatial summation (Kaneko 1970).

The functional role of the horizontal cell axons is still puzzling. Naka (1976) has recently reported *en passant* synapses of horizontal cell axons to amacrine cell bodies in the catfish, but such synapses are not found in other animals. Since the horizontal cell axons occupy a large space in the fish retina, one is curious as to their functional role.

INTERMEDIATE HORIZONTAL CELLS   Intermediate horizontal cells of the teleost retina are located just vitread to the external horizontal cells. They extend longer dendrites to rod terminals (Stell 1967). Responses from the intermediate horizontal cells are recorded only in the scotopic retina and their spectral sensitivity agrees well with the absorption of porphyrhopsin, peaking at 523 nm (Munz & Schwanzara 1967). The sensitivity peak does not shift with chromatic adaptation. With a strong background they are saturated. All of these characteristics suggest the rod input to the intermediate horizontal cell (Kaneko & Yamada 1972, Mitarai et al 1974, Hashimoto et al 1976). In contrast to external horizontal cells, the intermediate horizontal cells do not have axons. This agrees with the finding that scotopic responses are not obtained from the layer of the horizontal cell axons.

ELECTRICAL COUPLING   One of the characteristics of fish horizontal cells is a high degree of spatial summation (Tomita 1965, Naka & Rushton 1967, Norton et al 1968, Kaneko 1970, 1971, cf. Lamb 1976 for the turtle). The spatial summation is almost linear and is caused by electrical coupling between neighboring horizontal cells. (But for some nonlinear fraction of spatial summation, see Marchiafava & Pasino 1973.) Direct demonstration of coupling was made in the dogfish retina by simultaneous recording from a pair of horizontal cells. The current injected into a cell spread to its neighbors with some decrement. Furthermore, injected Procion yellow was found to diffuse into neighboring cells. In the electron microscope, gap junctions are widely found between contiguous horizontal cells (Yamada & Ishikawa 1965, Witkovsky & Dowling 1969, Stell 1972, Raviola 1976). Electrical coupling is limited to horizontal cells of the same type: external with external, and internal with internal (Kaneko 1971), and probably limited to horizontal cells of the same spectral response properties.

INTERACTION WITH PHOTORECEPTORS   As discussed above, horizontal cells are thought to provide a pathway for lateral interaction in receptors. The interaction of horizontal cells with receptors is color specific (Fuortes et al 1973, Fuortes & Simon 1974). Horizontal cells are also specific for cones or rods. In the carp, the external horizontal cell is connected only to cones and the intermediate horizontal cell to rods (Stell 1967). This indicates that, at least in the cyprinids, rod and cone mosaics are independent in their lateral interactions. Convergence of the two signals occurs first at the level of mixed bipolar cells (Kaneko et al 1978).

In the cat and rabbit, the horizontal cell axon enlarges in diameter after running a certain distance from the cell body in the outer plexiform layer. The dendrites connect to cone terminals, while branched axon terminals connect to rods (Fisher & Boycott 1974, Kolb 1974). Though the two parts are connected with a thin axon (Steinberg 1969), local interactions between receptors of the same type are probably more effective.

INTERACTION WITH BIPOLAR CELLS   It has been reported for the rabbit, mudpuppy, and catfish that horizontal cell dendrites synapse directly on bipolar cells (Brown et al 1966, Dowling & Werblin 1969, Naka 1976). These synapses were found on bipolar cell dendrites in their course from the cell body to the dendritic tip in the receptor terminals or only near the end of the process within the invagination. It is thus hard to differentiate direct versus indirect effects of horizontal cells on bipolar cells. One possible approach might be a pharmacological test, if one could identify the transmitter substances of photoreceptors and horizontal cells.

Amacrine Cells

Amacrine cell bodies are located in the vitread (proximal) part of the inner nuclear layer and they extend dendrites into the inner plexiform layer. The amacrine cell response was first described as a transient depolarization for both on and off illumination (Werblin & Dowling 1969, Kaneko 1970), but later, depolarizing and hyperpolarizing sustained responses were also found (Kaneko 1973, Naka & Ohtsuka 1975). Most amacrine cells lack center-surround organization.

One feature of amacrine cells that is different from other retinal periph-eral neurons is that amacrine cells show spike discharges (Werblin & Dowl-ing 1969, Kaneko 1970, Werblin 1977). These spikes are, however, atypical and are often abortive. They sometimes appear as noise-like fluctuations in membrane potential. It would be interesting to know if this noisy compo-nent reflects the noise recorded from receptors and from off-type bipolar cells (Simon et al 1975, Lamb & Simon 1976). Miller & Dacheux (1976d) claim that large action potentials are produced at the cell soma and small spikes at the dendrites.

MORPHOLOGICAL DIFFERENCE OF ON-, OFF-, AND ON-OFF TYPE AMACRINE CELLS   The morphological correlation of amacrine cells of different response types was made first in the catfish (Naka & Ohtsuka 1975). In the flat-mount preparation, the dendrites of transient (on-off) type amacrine cells were seen crossing each other, but not those of the sustained (on- or off-) type cells. This face-on appearance of amacrine cells reflects the fact that the arborization of transient type cells is multistratified. A similar finding was made in the carp (Murakami & Shimoda 1977). The functional significance of the morphological characteristics of different amacrine cells became clear with the finding that the inner plexiform layer is divided into sublaminae a and b, each dealing with off and on signals separately (Fami-glietti et al 1977). The dendrites of the on-type cell are confined to the sublamina b, those of the off-type cell to the sublamina a, and the on-off type cells were found sending their dentrites to both sublaminae. This agrees with the different appearance of various types of amacrine cells in the face-on view. Because the axon terminals of on- and off-type bipolar cells are confined to the corresponding sublamina of the inner plexiform layer, it is tempting to think that the bipolar to amacrine synapses are all excita-tory. This results in the simplest wiring network between the bipolar and amacrine cells.

It is still unclear how the transient responses in the on-off type amacrine cells are made. Simple linear summation of the sustained signals from bipolar cells would not result in a transient depolarization. In fact, a large distortion from linear summation is found in the dynamic characteristics of on-off amacrine cells (Marmarelis & Naka 1973a, b, c, Schellart & Spekreijse 1972, Toyoda 1974).

Interplexiform Cells

In examining the distribution of amine-containing cells in the fish retina, Ehinger et al (1969) found a new type of cell, located in the inner nuclear layer and extending its processes to both the inner plexiform layer and to the outer plexiform layer, where they end on the cell bodies of horizontal cells (interplexiform cell, for other animals, cf. Ehinger & Falck 1969, Boycott et al 1975, Dowling & Ehinger 1975). These cells contain dopamine, but their electrical responses have not been determined. Application of dopamine to goldfish external horizontal cells (L-type) depolarized the membrane potential and reduced the amplitude of the response to illumination (Dowling et al 1976). It is supposed that the interplexiform cell controls the effectiveness of lateral interaction mediated by horizontal cells.

FUNCTIONAL ORGANIZATION OF THE SYNAPSE AT THE RECEPTOR TERMINALS

Transmitter Release from Receptor Terminals

It was at first puzzling to understand how the hyperpolarizing receptor potential is transmitted to the next stage. From several pieces of evidence, it is now believed that transmitter is liberated in the dark when the receptors are maintained in a depolarized state (Trifonov 1968). The transmitter substance, which is still unidentified, depolarizes horizontal cells.

The first piece of evidence was the observation that radial (transretinal) current pulses across the retina, from the receptor side to the vitreous side, can evoke a depolarization in horizontal cells (Trifonov 1968, Byzov & Trifonov 1968). This effect was produced by increased transmitter release from receptor terminals that were depolarized by the applied electric field. The current was more effective during illumination than during darkness. Current in the opposite direction hyperpolarized the receptor terminals, and by reducing the amount of transmitter released, also hyperpolarized horizontal cells. In this instance, however, the current was more effective when applied in the dark. This difference in effectiveness may be interpreted as follows: in the dark the terminal membrane is depolarized to such a level that the transmitter release is nearly maximal, so additional depolarization is not very effective. During bright illumination transmitter release is almost stopped, and an additional hyperpolarization is ineffective in producing a further decrease in release (Kaneko & Shimazaki 1976, Byzov & Cervetto 1977).

It is interesting that radial current pulses of short duration are effective only when applied in the direction to depolarize the terminal membrane, but are ineffective when applied in the opposite direction. It is tempting to speculate that the receptor terminal membrane is able to amplify depolarizing pulses. It will be interesting to see if this mechanism involves calcium ion entry.

The hypothesis that transmitter release from photoreceptors occurs in the dark is further supported by the finding that horizontal cells are hyperpolarized when chemical transmission is blocked by calcium depletion from the extracellular medium or by the addition of calcium antagonists (Dowling & Ripps 1973, Cervetto & Piccolino 1974, Kaneko & Shimazaki 1975, Trifonov et al 1974). On the other hand, enhanced release caused by lanthanum ions (cf. Heuser & Miledi 1971) depolarized horizontal cells to a level similar to that observed in the dark (Kaneko & Shimazaki 1975).

The third piece of supporting evidence comes from morphological observations. Two kinds of observations have been made that can be interpreted in terms of the hypothesis of recycling vesicle membrane proposed by Heuser and Reese (1973). According to this hypothesis, faster recycling is expected in the dark. Schaeffer & Raviola (1976) found that the membrane area of the turtle cone pedicle increases in the dark and forms tortuous infoldings; with maintained illumination the invaginations are shallow. A second series of experiments examined the kinetics of uptake of the extracellular marker, horseradish peroxidase (HRP). Incorporation of HRP into coated vesicles was faster in the dark than during illumination (Schacher et al 1974, Ripps et al 1976).

Transmitter Action on Horizontal Cells

The transmitter released from photoreceptors depolarizes horizontal cells by increasing sodium permeability (Kaneko & Shimazaki 1975, Waloga & Pak 1978). Removal of extracellular sodium strongly hyperpolarizes horizontal cells even in the dark. The membrane conductance of the horizontal cell is high in the dark and is decreased by illumination (Toyoda et al 1969, Trifonov et al 1974, Werblin 1975b). The reversal potential is found to be nearly zero in the fish (Trifonov et al 1974) and between +15 and +50 mV in the mudpuppy (Werblin 1975b). After removal of synaptic input, horizontal cells are hyperpolarized to the equilibrium potential of potassium ions. (Kaneko & Shimazaki 1975).

Transmitter Action on On-type Bipolar Cells

When chemical transmission is blocked, on-type bipolar cells are depolarized, suggesting that the transmitter acts to hyperpolarize bipolar cells (Kaneko & Shimazaki 1976, Dacheux & Miller 1976). Toyoda et al (1977) suggest that the action of the cone transmitter is entirely different from that of rods. They propose that the action of the cone transmitter is to increase potassium permeability and that of the rods is to decrease sodium permeability. Both effects are synergetic in hyperpolarizing the on-type bipolar cell. Their conclusion is based on membrane resistance measurements and the level of reversal potentials found by polarizing the bipolar cells with extrinsic currents. The effects of varying external ionic composition must be examined before drawing a final conclusion, but such experiments are very difficult, since a change in the ionic environment directly affects the receptor responses. The mechanism of response generation in off-type bipolar cells has not been systematically studied.

Receptor-Bipolar Synapses: Morphological Differences between On- and Off-types

Relevant to synaptic mechanisms, the differences in morphology of on- and off-type bipolar cells are important to note. In freeze-fracture electron microscopy of the turtle retina, Raviola & Gilula (1975) found that the postsynaptic membrane of the invaginating bipolar cell has an intramembranous structure usually found in the inhibitory postsynaptic membrane (cf. Landis et al 1974). On the other hand, the flat bipolar cell, which makes basal contacts, is equipped with a structure typically found in the excitatory postsynaptic membrane. From these observations, they suggested that the invaginating bipolars are on types and the flat bipolar cells show off-type responses.

Following the on- and off-type bipolar cells identified from the location of axon terminals in the goldfish Golgi preparation, Stell et al (1977) found that the dendritic tip of the on-center bipolar cells was either a central element of the triad of the invaginating synapse or made narrow-cleft junctions with the receptor terminals. The dendritic tip of the off-type bipolar cell made wide-cleft junctions, but it was never found to be the central element of the invaginating synapse.

Transmitter Substance

The transmitter substance of receptors has not been identified. Aspartate and glutamate effect horizontal and bipolar cells with the polarity expected for the endogenous transmitter substance (Cervetto & MacNichol 1972, Murakami et al 1972, 1975), but it is still doubtful if one of these is *the* transmitter because of their abundance in the nervous system and because of the surprisingly high concentrations necessary to produce effects. There is even a report that aspartate depolarizes gecko photoreceptors (Kleinschmidt 1973).

Turtle cones have been shown to synthesize acetylcholine (ACh) (Lam 1972a), but the specific activity of choline acetyltransferase in isolated cone preparations was later found to be lower than that of the entire retina (Lam 1976). Recently, the possibility that ACh is the cone transmitter in the turtle has been raised again by finding that atropine blocks horizontal cell responses (Gerschenfeld & Piccolino 1977, Piccolino & Gershenfeld 1977). Strangely, bipolar cells are not affected by atropine. This hypothesis is difficult to understand, unless one accepts the notion that more than one transmitter is released from cone terminals.

Transmitter of Horizontal Cells

The feedback action of horizontal cells on to photoreceptors works through a chemical synapse, but this transmitter is also unidentified. At the moment, investigators are paying attention to gamma-aminobutyric acid (GABA), since horizontal cells both actively take up GABA from the bathing medium (Lam & Steinman 1971, Ehinger 1976) and also synthesize GABA (Graham 1972, Lam 1972b, 1975).

FUNCTIONAL ORGANIZATION OF THE SYNAPSE IN THE INNER PLEXIFORM LAYER

The inner plexiform layer forms the second neuropil of the retina and is composed of three neuronal elements: the axon terminal of bipolar cells, dendrites of amacrine cells and dendrites of ganglion cells. At each synaptic complex, the bipolar cell terminal is the presynaptic element, and contains a synaptic ribbon surrounded by a cluster of vesicles. Facing the ribbon, two postsynaptic elements compose the diad: dendrites of amacrine cells and of ganglion cells. This is the most typical combination in mammals and in fish, but in the frog both postsynaptic elements are often amacrine cells (Dowling 1968, Dowling & Boycott 1966).

Conventional synapses are also found in the inner plexiform layer: either between amacrine and ganglion cells or between two amacrine cells. Sometimes series of amacrine cells are found connected by conventional synapses (Dubin 1970).

Transmitter Substance and Function of the Inner Plexiform Layer

DIRECTIONAL SELECTIVITY   Directional selectivity must be made at this synaptic layer, since it is not found in bipolar cells. Wyatt & Daw (1976) found that inhibitory (GABAergic) neurons contribute to the formation of directional selectivity. After administration of picrotoxin, directionality in rabbit ganglion cells was lost. On the other hand, strychnine, an antagonist of glycine, did not affect directional units, but did decrease the surround response in off-center ganglion cells (cf. Miller et al 1977 for the mudpuppy).

The excitatory transmitter to directionally selective and on-center rabbit ganglion cells is probably ACh (Masland & Ames 1976, Masland & Livingstone, 1976). One to 10 μM ACh strongly increases spike discharges in those ganglion cells, even after endogenous chemical transmission is blocked. It is interesting that ACh is ineffective for most of the off-type ganglion cells. It seems, therefore, probable that there are separate classes of amacrine cells, each with a different transmitter, associated with different functional classes of ganglion cells.

Reciprocal Synapse

When the bipolar cell terminal is examined in the electron microscope, one finds many presynaptic endings on the bipolar cell end bulb. These synapses have a conventional structure, and are identified as coming from amacrine cells. Thus at the terminal, bipolar cells receive feedback innervation from amacrine cells. Although morphologically defined, the functional role of these synapses is still puzzling. In the cyprinid fish, the bipolar cell terminal is often as large as the cell body, and sometimes one can penetrate this structure by microelectrodes (Murakami & Shimoda 1978). No detectable difference was found between the responses recorded from the cell body and those from the terminal bulb, although no direct comparison has been made in records obtained from the same cell. It is not even known whether the reciprocal synapse is excitatory or inhibitory.

CLOSING REMARKS

In the closing remarks of a review article of 1972, Stell (1972) raised about 20 questions that awaited solutions with then-current techniques. At the moment four of these have been answered with conclusive evidence: interreceptor contacts; the mechanism of horizontal cell action in the receptor synaptic complexes; horizontal cell coupling; and identification of horizontal cell axons in teleostean retina. Four other questions have been partly answered: the function of dopaminergic cells on the teleostean horizontal cells; the role of stratification of the inner plexiform layer; circuitry subserving specific functions such as color, movement etc; and chemical transmitters in some neurons.

The main neuronal circuitry of the retina has now been made clear, but most of our understanding still remains superficial. The retina provides us with unique questions related to various aspects of neuroscience. To answer these questions we must widen our views and in the future try to employ more combined techniques.

Literature Cited

Barlow, H. B. 1953. Summation and inhibition in the frog's retina. *J. Physiol. London* 119:69–88

Barlow, H. B., Hill, R. M., Levick, W. R. 1964. Retinal ganglion cells responding selectively to direction and speed of image motion in the rabbit. *J. Physiol. London* 173:377–407

Baylor, D. A., Fettiplace, R. 1977. Transmission from photoreceptors to ganglion cells in turtle retina. *J. Physiol. London* 271:391–424

Baylor, D. A., Fuortes, M. G. F. 1970. Electrical responses of single cones in the retina of the turtle. *J. Physiol. London* 207:77–92

Baylor, D. A., Fuortes, M. G. F., O'Bryan, P. M. 1971. Receptive fields of cones in the retina of the turtle. *J. Physiol. London* 214:265–94

Baylor, D. A., Hodgkin, A. L. 1973. Detection and resolution of visual stimuli by turtle photoreceptors. *J. Physiol. London* 234:163–98

Beauchamp, R. D., Daw, N. W. 1972. Rod and cone input to single goldfish optic nerve fibers. *Vision Res.* 12:1201–12

Bortoff, A. 1964. Localization of slow potential responses in the *Necturus* retina. *Vision Res.* 4:627–35

Boycott, B. B., Dowling, J. E., Fisher, S. K., Kolb, H., Laties, A. M. 1975. Interplexiform cells of the mammalian retina and their comparison with catecholamine-containing retinal cells. *Proc. R. Soc. London Ser. B.* 191:353–68

Boycott, B. B., Kolb, H. 1973. The connections between bipolar cells and photoreceptors in the retina of the domestic cat. *J. Comp. Neurol.* 148:91–114

Brown, J. E., Major, D. 1966. Cat retinal ganglion cell dendritic fields. *Exp. Neurol.* 15:70–78

Brown, J. E., Major, D., Dowling, J. E. 1966. Synapses of horizontal cells in rabbit and cat retinas. *Science* 153:1639–41

Brown, J. E., Pinto, L. H. 1974. Ionic mechanism for the photoreceptor potential of the retina of *Bufo marinus*. *J. Physiol. London* 236:575–91

Brown, K. T., Flaming, D. G. 1975. Instrumentation and technique for beveling fine micropipette electrodes. *Brain Res.* 86:172–80

Burkhardt, D. A. 1977. Responses and receptive-field organization of cones in perch retinas. *J. Neurophysiol.* 40:53–62

Byzov, A. L., Cervetto, L. 1977. Effects of applied currents on turtle cones in darkness and during the photoresponse. *J. Physiol. London* 265:85–102

Byzov, A. L., Trifonov, Y. A. 1968. The response to electric stimulation of horizontal cells in the carp retina. *Vision Res.* 8:817–22

Cajal, S. R. y. 1893. La rétine de vertébrés. *La Cellule* 9:119–246

Cervetto, L. 1973. Influence of sodium, potassium and chloride ions on intracellular responses of turtle photoreceptors. *Nature* 241:401–3

Cervetto, L., MacNichol, E. F. Jr. 1972. Inactivation of horizontal cells in turtle retina by glutamate and aspartate. *Science* 178:767–68

Cervetto, L., Piccolino, M. 1974. Synaptic transmission between photoreceptors and horizontal cells in the turtle retina. *Science* 183:417–19

Copenhagen, D. R., Owen, W. G. 1976a. Coupling between rod photoreceptors in a vertebrate retina. *Nature* 260:57–59

Copenhagen, D. R., Owen, W. G. 1976b. Functional characteristics of lateral interactions between rods in the retina of the snapping turtle. *J. Physiol. London* 259:251–82

Dacheux, R. F., Miller, R. F. 1976. Photoreceptor-bipolar cell transmission in the perfused retina eyecup of the mudpuppy. *Science* 191:963–64

Daw, N. W. 1968. Colour-coded ganglion cells in the goldfish retina: extension of their receptive fields by means of new stimuli. *J. Physiol. London* 197:567–92

Daw, N. W. 1973. Neurophysiology of color vision. *Physiol. Rev.* 53:571–611

Dowling, J. E. 1968. Synaptic organization of the frog retina: an electron microscopic analysis comparing the retinas of frogs and primates. *Proc. R. Soc. London Ser. B* 170:205–28

Dowling, J. E., Boycott, B. B. 1966. Organization of the primate retina: electron microscopy. *Proc. R. Soc. London Ser. B* 166:80–111

Dowling, J. E., Ehinger, B. 1975. Synaptic organization of the amine-containing interplexiform cells of the goldfish and cebus monkey retinas. *Science* 188:270–73

Dowling, J. E., Ehinger, B., Hedden, W. L. 1976. The interplexiform cell: a new type of retinal neuron. *Invest. Ophthalmol.* 15:916–26

Dowling, J. E., Ripps, H. 1973. Effect of magnesium on horizontal cell activity in the skate retina. *Nature* 242:101–3

Dowling, J. E., Werblin, F. S. 1969. Organization of retina of the mudpuppy, *Necturus maculosus*. I. Synaptic structure. *J. Neurophysiol.* 32:315–38

Dubin, M. W. 1970. The inner plexiform layer of the vertebrate retina: a quantitative and comparative electron microscopic analysis. *J. Comp. Neurol.* 140:479–506

Ehinger, B. 1976. Selective neuronal accumulation of Ω-amino acids in the rabbit retina. *Brain Res.* 107:541–54

Ehinger, B., Falck, B. 1969. Adrenergic retinal neurons of some new world monkeys. *Z. Zellforsch. Mikrosk. Anat.* 100:364–75

Ehinger, B., Falck, B., Laties, A. M. 1969. Adrenergic neurons in teleost retina. *Z. Zellforsch. Mikrosk. Anat.* 97: 285–97

Fain, G., Dowling, J. E. 1973. Intracellular recordings from single rods and cones in the mudpuppy retina. *Science* 180: 1178–81

Fain, G. L., Gold, G. H., Dowling, J. E. 1976. Receptor coupling in the toad retina. *Cold Spring Harbor Symp. Quant. Biol.* 40:547–61

Famiglietti, E. V. Jr., Kaneko, A., Tachibana, M. 1977. Neuronal architecture of ON and OFF pathways to ganglion cells in carp retina. *Science* 198:1267–69

Famiglietti, E. V. Jr., Kolb, H. 1976. Structural basis for ON- and OFF-center responses in retinal ganglion cells. *Science* 194:193–95

Fisher, S. K., Boycott, B. B. 1974. Synaptic connexions made by horizontal cells within the outer plexiform layer of the retina of the cat and the rabbit. *Proc. R. Soc. London Ser. B* 186:317–31

Fuortes, M. G. F. 1972. *Handbook of Sensory Physiology. VII/2. Physiology of photoreceptor organs.* Berlin, Heidelberg, New York: Springer. 765 pp.

Fuortes, M. G. F., Schwartz, E. A., Simon, E. J. 1973. Colour-dependence of cone responses in the turtle retina. *J. Physiol. London* 234:199–216

Fuortes, M. G. F., Simon, E. J. 1974. Interactions leading to horizontal cell responses in the turtle retina. *J. Physiol. London* 240:177–98

Gerschenfeld, H. M., Piccolino, M. 1977. Muscarinic antagonists block cone to horizontal cell transmission in turtle retina. *Nature* 268:257–59

Grabowski, S. R., Pak, W. L. 1975. Intracellular recordings of rod responses during dark-adaptation. *J. Physiol. London* 247:363–91

Grabowski, S. R., Pinto, L. H., Pak, W. L. 1972. Adaptation in retinal rods of axolotl: intracellular recordings. *Science* 176:1240–43

Graham, L. T. Jr. 1972. Intraretinal distribution of GABA content and GAD activity. *Brain Res.* 36:476–79

Hagins, W. A., Penn, R. D., Yoshikami, S. 1970. Dark current and photocurrent in retinal rods. *Biophys. J.* 10:380–412

Hartline, H. K. 1938. The response of single optic nerve fibres of the vertebrate eye to illumination of the retina. *Am. J. Physiol.* 121:400–15

Hashimoto, Y., Kato, A., Inokuchi, M., Watanabe, K. 1976. Re-examination of horizontal cells in the carp retina with procion yellow electrode. *Vision Res.* 16:25–29

Heuser, J. E., Miledi, R. 1971. Effect of lanthanum ions on function and structure of frog neuromuscular junctions. *Proc. R. Soc. London Ser. B* 179:247–60

Heuser, J. E., Reese, T. S. 1973. Evidence for recycling of synaptic vesicle membrane during transmitter release at the frog neuromuscular junction. *J. Cell Biol.* 57:315–44

Hubbell, W. L., Bownds, M.D. 1979. *Ann. Rev. Neurosci.* 2:17–34

Hubel, D. H., Wiesel, T. N. 1960. Receptive fields of optic nerve fibers in the spider monkey. *J. Physiol. London* 154:572–80

Kaneko, A. 1970. Physiological and morphological identification of horizontal, bipolar and amacrine cells in goldfish retina. *J. Physiol. London* 207:623–33

Kaneko, A. 1971. Electrical connexions between horizonal cells in the dogfish retina. *J. Physiol. London* 213:95–105

Kaneko, A. 1973. Receptive field organization of bipolar and amacrine cells in the goldfish retina. *J. Physiol. London* 235:133–53

Kaneko, A., Famiglietti, E. V. Jr., Tachibana, M. 1978. Physiological and morphological identification of signal pathways in the carp retina. In *Neurobiology of Chemical Transmission,* ed. M. Otsuka, Z. W. Hall. New York: Wiley In press

Kaneko, A., Shimazaki, H. 1975. Effects of external ions on the synaptic transmission from photoreceptors to horizontal cells in the carp retina. *J. Physiol. London* 252:509–22

Kaneko, A., Shimazaki, H. 1976. Synaptic transmission from photoreceptors to bipolar and horizontal cells in the carp retina. *Cold Spring Harbor Symp. Quant. Biol.* 40:537–46

Kaneko, A., Yamada, M. 1972. S-potentials in the dark-adapted retina of the carp. *J. Physiol. London* 227:261–73

Kater, S. B., Nicholson, C. 1973. *Intracellular Staining in Neurobiology.* New

York, Heidelberg, Berlin: Springer. 332 pp.

Kleinschmidt, J. 1973. Adaptation properties of intracellularly recorded *Gekko* photoreceptor potentials. In *Biochemistry and Physiology of Visual Pigments*. ed. H. Langer, pp. 219–24. Berlin, Heidelberg, New York: Springer. 366 pp.

Kolb, H. 1974. The connections between horizontal cells and photoreceptors in the retina of the cat: electron microscopy of Golgi preparations. *J. Comp. Neurol.* 155:1–14

Kuffler, S. W. 1953. Discharge patterns and functional organization of mammalian retina. *J. Neurophysiol.* 16:37–68

Lam, D. M. K. 1972a. Biosynthesis of acetylcholine in turtle photoreceptors. *Proc. Natl. Acad. Sci. USA* 69:1987–91

Lam, D. M. K. 1972b. The biosynthesis and content of gamma-aminobutyric acid in the goldfish retina. *J. Cell Biol.* 54:225–31

Lam, D. M. K. 1975. Biosynthesis of γ-aminobutyric acid by isolated axons of cone horizontal cells in the goldfish retina. *Nature* 254:345–47

Lam, D. M. K. 1976. Synaptic chemistry of identified cells in the vertebrate retina. *Cold Spring Harbor Symp. Quant. Biol.* 40:571–79

Lam, D. M. K., Steinman, L. 1971. The uptake of [γ-^{3}H] aminobutyric acid in the goldfish retina. *Proc. Natl. Acad. Sci. USA* 68:2777–81

Lamb, T. D. 1976. Spatial properties of horizontal cell responses in the turtle retina. *J. Physiol. London* 263:239–55

Lamb, T. D., Simon, E. J. 1976. The relation between intercellular coupling and electrical noise in turtle photoreceptors. *J. Physiol. London* 263:257–86

Landis, D. M. D., Reese, T. S., Raviola, E. 1974. Differences in membrane structure between excitatory and inhibitory components of the reciprocal synapse in the olfactory bulb. *J. Comp. Neurol.* 155:67–92

Lasansky, A. 1971. Synaptic organization of cone cells in the turtle retina. *Phil. Trans. R. Soc. London Ser. B* 262: 365–81

Lasansky, A., Marchiafava, P. L. 1974. Light-induced resistance changes in retinal rods and cones of the tiger salamander. *J. Physiol. London* 236:171–91

Laufer, M., Millán, E. 1970. Spectral analysis of L-type S-potentials and their relation to photopigment absorption in a fish (*Eugerres plumieri*) retina. *Vision Res.* 10:237–51

Lettvin, J. Y., Maturana, H. R., McCulloch, W. S., Pitts, W. H. 1959. What the frog's eye tells the frog's brain. *Proc. IRE* 47:1940–51

MacNichol, E. F. Jr., Svaetichin, G. 1958. Electric responses from the isolated retina of fishes. *Am. J. Ophthalmol.* 46:26–46

Marchiafava, P. L., Pasino, E. 1973. The spatial dependent characteristics of the fish S-potentials evoked by brief flashes. *Vision Res.* 13:1355–65

Marks, W. B. 1965. Visual pigments of single goldfish cones. *J. Physiol. London* 178:14–32

Marmarelis, P. Z., Naka, K. I. 1973a. Nonlinear analysis and synthesis of receptive-field responses in the catfish retina. I. Horizontal cell → ganglion cell chain. *J. Neurophysiol.* 36:605–18

Marmarelis, P. Z., Naka, K. I. 1973b. Nonlinear analysis and synthesis of receptive-field responses in the catfish retina. II. One-input white-noise analysis. *J. Neurophysiol.* 36:619–33

Marmarelis, P. Z., Naka, K. I. 1973c. Nonlinear analysis and synthesis of receptive-field responses in the catfish retina. III. Two-input white-noise analysis. *J. Neurophysiol.* 36:634–48

Masland, R. H., Ames, A. III. 1976. Responses to acetylcholine of ganglion cells in an isolated mammalian retina. *J. Neurophysiol.* 39:1220–35

Masland, R. H., Livingstone, C. J. 1976. Effect of stimulation with light on synthesis and release of acetylcholine by an isolated mammalina retina. *J. Neurophysiol.* 39:1210–19

Matsumoto, N., Naka, K. I. 1972. Identification of intracellular responses in the frog retina. *Brain Res.* 42:59–71

Maturana, H. R., Frenk, S. 1963. Directional movement and horizontal edge detectors in the pigeon retina. *Science* 142:977–79

Maturana, H. R., Lettvin, J. Y., McCulloch, W. S., Pitts, W. H. 1960. Anatomy and physiology of vision in the frog (*Rana pipiens*). *J. Gen. Physiol.* 43: Suppl. 2, pp. 129–75

Michael, C. R. 1968. Receptive fields of single optic nerve fibers in a mammal with an all-cone retina. I. Contrast-sensitive units. *J. Neurophysiol.* 31:249–56

Miller, R. F., Dacheux, R. F. 1976a. Synaptic organization and ionic basis of on and off channels in mudpuppy retina. I. Intracellular analysis of chloride-sensitive electrogenic properties of receptors, horizontal cells, bipolar cells and amacrine cells. *J. Gen. Physiol.* 67:639–59

Miller, R. F., Dacheux, R. F. 1976b. Synaptic organization and ionic basis of on and off channels in mudpuppy retina. II. Chloride-dependent ganglion cell mechanisms. *J. Gen. Physiol.* 67:661–78

Miller, R. F., Dacheux, R. F. 1976c. Synaptic organization and ionic basis of on and off channels in mudpuppy retina. III. A model of ganglion cell receptive field organization based on chloride-free experiments. *J. Gen. Physiol.* 67:679–90

Miller, R. F., Dacheux, R. F. 1976d. Dendritic and somatic spikes in mudpuppy amacrine cells: identification and TTX sensitivity. *Brain Res.* 104:157–62

Miller, R. F., Dacheux, R. F., Frumkes, T. E. 1977. Amacrine cells in *Necturus* retina: evidence for independent γ-aminobutyric acid- and glycine-releasing neurons. *Science* 198:748–50

Miller, W. H., Hashimoto, Y., Saito, T., Tomita, T. 1973. Physiological and morphological identification of L- and C-type S-potentials in the turtle retina. *Vision Res.* 13:443–47

Mitarai, G., Asano, T., Miyake, Y. 1974. Identification of five types of S-potential and their corresponding generating sites in the horizontal cells of the carp retina. *Jpn J. Ophthalmol.* 18:161–76

Munz, F. W., Schwanzara, S. A. 1967. A nomogram for retinene₂-based visual pigments. *Vision Res.* 7:111–20

Murakami, M., Ohtsu, K., Ohtsuka, T. 1972. Effects of chemicals on receptors and horizontal cells in the retina. *J. Physiol. London* 227:899–913

Murakami, M., Ohtsuka, T., Shimazaki, H. 1975. Effects of aspartate and glutamate on the bipolor cells in the carp retina. *Vision Res.* 15:456–58

Murakami, M., Shimoda, Y. 1977. Identification of amacrine and ganglion cells in the carp retina. *J. Physiol. London* 264:801–18

Murakami, M., Shimoda, Y. 1978. Intracellular double staining: localization of recording site in single retinal neurons. *Brain Res.* 144:164–68

Naka, K. I. 1976. Neuronal circuitry in the catfish retina. *Invest. Ophthalmol.* 15:926–35

Naka, K. I. 1977. Functional organization of catfish retina. *J. Neurophysiol.* 40:26–43

Naka, K. I., Ohtsuka, T. 1975. Morphological and functional identifications of catfish retinal neurons. II. Morphological identification. *J. Neurophysiol.* 38:72–91

Naka, K. I., Rushton, W. A. H. 1967. The generation and spread of S-potentials in fish (Cyprinidae). *J. Physiol. London* 192:437–61

Nelson, R., Famiglietti, E. V. Jr., Kolb, H. 1978. Intracellular staining reveals different levels of stratification for on- and off-center ganglion cells in cat retina. *J. Neurophysiol.* 41:472–83

Norton, A. L., Spekreijse, H., Wolbarsht, M. L., Wagner, H. G. 1968. Receptive field organization of the S-potential. *Science* 160:1021–22

O'Bryan, P. M. 1973. Properties of the depolarizing synaptic potential evoked by peripheral illumination in cones of the turtle retina. *J. Physiol. London* 235:207–23

Piccolino, M., Gerschenfeld, H. M. 1977. Lateral interactions in the outer plexiform layer of turtle retinas after atropine block of horizontal cells. *Nature* 268:259–61

Pinto, L. H., Pak, W. L. 1974a. Light-induced changes in photoreceptor membrane resistance and potential in gecko retinas. I. Preparations treated to reduce lateral interactions. *J. Gen. Physiol.* 64:26–48

Pinto, L. H., Pak, W. L. 1974b. Light-induced changes in photoreceptor membrane resistance and potential in gecko retinas. II. Preparations with active lateral interactions. *J. Gen. Physiol.* 64:49–69

Polyak, S. 1941. *The Retina.* Chicago: Univ. Chicago Press. 1390 pp.

Raviola, E. 1976. Intercellular junctions in the outer plexiform layer of the retina. *Invest. Ophthalmol.* 15:881–95

Raviola, E., Gilula, N. B. 1973. Gap junctions between photoreceptor cells in the vertebrate retina. *Proc. Natl. Acad. Sci. USA* 70:1677–81

Raviola, E., Gilula, N. B. 1975. Intramembrane organization of specialized contacts in the outer plexiform layer of the retina. A freeze-fracture study in monkeys and rabbits. *J. Cell Biol.* 65:192–222

Raynauld, J.-P. 1972. Goldfish retina: Sign of the rod input in opponent color ganglion cells. *Science* 177:84–85

Richter, A., Simon, E. J. 1975. Properties of centre-hyperpolarizing, red-sensitive bipolar cells in the turtle retina. *J. Physiol. London* 248:317–34

Ripps, H., Shakib, M., MacDonald, E. D. 1976. Peroxidase uptake by photoreceptor terminals of the skate retina. *J. Cell Biol.* 70:86–96

Rodieck, R. W. 1973. *The Vertebrate Retina. Principles of Structure and Function.* San Francisco: Freeman. 1044 pp.

Schacher, S. M., Holtzman, E., Hood, D. C. 1974. Uptake of horseradish peroxidase by frog photoreceptor synapses in the dark and the light. *Nature* 249:261–63

Schaeffer, S. F., Raviola, E. 1976. Ultrastructural analysis of functional changes in the synaptic endings of turtle cone cells. *Cold Spring Harbor Symp. Quant. Biol.* 40:521–28

Schellart, N. A. M., Spekreijse, H. 1972. Dynamic characteristics of retinal ganglion cell responses in goldfish. *J. Gen. Physiol.* 59:1–21

Schwartz, E. A. 1973. Responses of single rods in the retina of the turtle. *J. Physiol. London* 232:503–14

Schwartz, E. A. 1974. Responses of bipolar cells in the retina of the turtle. *J. Physiol. London* 236:211–24

Schwartz, E. A. 1975. Rod-rod interaction in the retina of the turtle. *J. Physiol. London* 246:617–38

Schwartz, E. A. 1976. Electrical properties of the rod syncytium in the retina of the turtle. *J. Physiol. London* 257:379–406

Schwartz, E. A. 1977. Voltage noise observed in rods of the turtle retina. *J. Physiol. London* 272:217–46

Simon, E. J. 1973. Two types of luminosity horizontal cells in the retina of the turtle. *J. Physiol. London* 230:199–211

Simon, E. J., Lamb, T. D., Hodgkin, A. L. 1975. Spontaneous voltage fluctuations in retinal cones and bipolar cells. *Nature* 256:661–62

Steinberg, R. H. 1969. Rod and cone contributions to S-potentials from the cat retina. *Vision Res.* 9:1319–29

Steinberg, R. H., Schmidt, R. 1970. Identification of horizontal cells as S-potential generators in the cat retina by intracellular dye injection. *Vision Res.* 10:817–20

Stell, W. K. 1967. The structure and relationships of horizontal cells and photoreceptor-bipolar synaptic complexes in goldfish retina. *Am. J. Anat.* 121:401–24

Stell, W. K. 1972. The morphological organization of the vertebrate retina. In *Handbook of Sensory Physiology. Vol. VII/2. Physiology of Photoreceptor Organs*, ed. M. G. F. Fuortes, pp. 111–213. Berlin, Heidelberg, New York: Springer. 765 pp.

Stell, W. K. 1975. Horizontal cell axons and axon terminals in goldfish retina. *J. Comp. Neurol.* 159:503–20

Stell, W. K. 1976. Functional polarization of horizontal cell dendrites in goldfish retina. *Invest. Ophthalmol.* 15:895–908

Stell, W. K., Ishida, A. T., Lightfoot, D. O. 1977. Structural basis for on- and off-center responses in retinal bipolar cells. *Science* 198:1269–71

Stretton, A. O. W., Kravitz, E. A. 1968. Neuronal geometry: determination with a technique of intracellular dye injection. *Science* 162:132–34

Svaetichin, G. 1953. The cone action potential. *Acta Physiol. Scand.* 29:Suppl. 106, pp. 565–600

Tasaki, K., Tsukahara, Y., Ito, S., Wayner, M. J., Yu, W. Y. 1968. A simple, direct and rapid method for filling microelectrodes. *Physiol. Behav.* 3:1009–10

Tomita, T. 1965. Electrophysiological study of the mechanisms subserving color coding in the fish retina. *Cold Spring Harbor Symp. Quant. Biol.* 30:559–66

Toyoda, J. 1974. Frequency characteristics of retinal neurons in the carp. *J. Gen. Physiol.* 63:214–34

Toyoda, J., Fujimoto, M., Saito, T. 1977. Responses of second-order neurons to photic and electric stimulation of the retina. In *Vertebrate Photoreception,* ed. H. B. Barlow, P. Fatt, pp. 231–50. London, New York, & San Francisco: Academic. 379 pp.

Toyoda, J., Hashimoto, H., Anno, H., Tomita, T. 1970. The rod responses in the frog as studied by intracellular recording. *Vision Res.* 10:1093–1100

Toyoda, J., Nosaki, H., Tomita, T. 1969. Light-induced resistance changes in single photoreceptors of *Necturus* and *Gekko. Vision Res.* 9:453–60

Trifonov, Y. A. 1968. Study of synaptic transmission between the photoreceptor and the horizontal cell using electrical stimulation of the retina. *Biofizika* 13:809–17

Trifonov, Y. A., Byzov, A. L., Chailahian, L. M. 1974. Electrical properties of subsynaptic and nonsynaptic membranes of horizontal cells in fish retina. *Vision Res.* 14:229–41

Wagner, H. G., MacNichol, E. F. Jr., Wolbarsht, M. L. 1960. The response properties of single ganglion cells in goldfish retina. *J. Gen. Physiol.* 43: Pt. II, pp. 45–62

Waloga, G., Pak, W. L. 1978. Ionic mechanism for the generation of horizontal cell potentials in isolated axolotl retina. *J. Gen. Physiol.* 71:69–92

Weinstein, G. W., Hobson, R. R., Baker, F. H. 1971. Extracellular recordings from human retinal ganglion cells. *Science* 171:1021–22

Werblin, F. S. 1970. Response of retinal cells to moving spots: intracellular recording in *Necturus maculosus. J. Neurophysiol.* 33:342–50

Werblin, F. S. 1974. Organization of the vertebrate retina: receptive fields and sensitivity control. In *The Eye. Vol. 6. Comparative Physiology* eds. H. Davson, L. T. Graham Jr. pp. 257–81. New York, London: Academic. 411 pp.

Werblin, F. S. 1975a. Regenerative hyperpolarization in rods. *J. Physiol. London* 244:53–81

Werblin, F. S. 1975b. Anomalous rectification in horizontal cells. *J. Physiol. London* 244:639–57

Werblin, F. S. 1977. Regenerative amacrine cell depolarization and formation of on-off ganglion cell response. *J. Physiol. London* 264:767–85

Werblin, F. S., Dowling, J. E. 1969. Organization of the retina of the mudpuppy, *Necturus maculosus.* II. Intracellular

recording. *J. Neurophysiol.* 32:339–55

Witkovsky, P. 1971. Peripheral mechanisms of vision. *Ann. Rev. Physiol.* 33:257–80

Witkovsky, P., Dowling, J. E. 1969. Synaptic relationships in the plexiform layers of carp retina. *Z. Zellforsch. Mikrosk. Anat.* 100:60–82

Wyatt, H. J., Daw, N. W. 1975. Directionally sensitive ganglion cells in the rabbit retina: specificity for stimulus direction, size, and speed. *J. Neurophysiol.* 38:613–26

Wyatt, H. J., Daw, N. W. 1976. Specific effects of neurotransmitter antagonists on ganglion cells in rabbit retina. *Science* 191:204–5

Yamada, E., Ishikawa, T. 1965. The fine structure of the horizontal cells in some vertebrate retinae. *Cold Spring Harbor Symp. Quant. Biol.* 30:383–92

Ann. Rev. Neurosci. 1979. 2:193–225
Copyright © 1979 by Annual Reviews Inc. All rights reserved

VISUAL PATHWAYS ♦11522

R. W. Rodieck

Department of Ophthalmology RJ-10, School of Medicine,
University of Washington, Seattle, WA 98195

INTRODUCTION

This review discusses the functional pathways traced by the fibers of the different types of retinal ganglion cells. Knowledge of these pathways has rapidly increased in the last few years, primarily because of two developments. The first is the recognition that there exist functionally distinct classes of mammalian ganglion cells, and the realization that they project to different regions of the brain. It is an interesting historical point that this development has come so late. Since the work of Ramón y Cajal in the last century, it has been known that, in all vertebrates there are a variety of morphologically distinct ganglion cells with dendritic patterns disposed in distinct ways within the inner plexiform layer. As long ago as 1933, G. H. Bishop (1933) described two conduction velocity groups in the fibers of the rabbit optic nerve. Noting that peripheral sensory nerve fibers within different conduction velocity groups subserve different functions, he speculated that the same might hold for the optic nerve as well. Neurophysiological techniques have basically changed little in the last quarter century, since Kuffler (1953) first described the center-surround organization of cat retinal ganglion cells. This first development thus appears to be primarily the result of a shift in view rather than a technical advance or refinement.

The second development is the use of new, or greatly refined techniques for tracing anatomical pathways in the nervous system. Of particular importance are the use of anterograde axonal transport of radioactively labeled amino acids for tracing the pathway from the cell body to the axonal terminals, the use of retrograde axonal transport of horseradish peroxidase for tracing the pathway from axon terminals to cell body, and the use of radioactively labeled deoxyglucose for determining relative levels of neuronal activity in conscious animals, under different conditions of sensory stimulation.

193

0147-006X/79/0315-0193$01.00

Three main topics are covered in this review: (*a*) the different morphological and functional types of ganglion cells in the retina; (*b*) the primary projections of these types onto different brain centers; (*c*) the pathways from some of these centers to the cerebral cortex. No attempt is made here to review the wide range of subcortical visual pathways, except to the degree that they can be traced as functionally distinct pathways from the retina. Van Essen (1979), in this volume, discusses the different visual areas of the cerebral cortex and some of their projections. Columnar organization of the visual cortex has been recently reviewed by Hubel & Wiesel (1977) and is not covered here. The influence of visual experience has been reviewed by Barlow (1975) and is not discussed here, nor are the aberrant pathways found in albino animals.

One of the most important current questions on cortical function concerns the way that the different functional pathways, and the different cortical areas to which they project, relate to the notion of hierarchical organization of the visual cortex, as developed by Hubel & Wiesel (1962, 1965, 1968). There are presently a great number of findings bearing on this topic, worthy of a full and detailed review. While the findings discussed here relate to this topic, it will not be covered per se.

The focal point of this review is the cat's visual system, since our understanding of this system is the most complete. Findings in other species are considered where they pertain to generalizations from the cat, or limitations thereof, or where knowledge for the cat is significantly limited relative to some other animal.

CAT RETINAL GANGLION CELLS

Morphological Types

Using the Golgi technique, Boycott & Wässle (1974) have shown that the cat retina contains at least three morphologically distinct types of ganglion cells. The α type, which comprises about 4–7% of the population, have the largest somata and are distributed more or less uniformly across the retina, with a mild concentration toward the visual streak, and a stronger concentration toward the area centralis (Wässle, Levick & Cleland 1975). The β type, which comprises about half the population, have only a slightly different morphological form, but within any given retinal area have smaller somata than α cells and have the smallest dendritic tree of any class. γ cells, which comprise about half the population, are distinguished by a small soma and a large dendritic tree, comparable in extent to that of α cells. The high concentration of cells in the area centralis is primarily due to β cells and, to a slightly lesser extent, to γ cells (Fukuda & Stone 1974), whereas the visual streak is formed primarily by γ cells (Rowe & Stone 1976b).

Famiglietti & Kolb (1976) found that the α and β classes (their Classes I & II) can be further subdivided in terms of the depth of the branching patterns of the dendrites within the inner plexiform layer (IPL). The dendrites of some α and β cells terminate in the outer third, where the flat cone bipolars terminate, the others in the inner two thirds of the IPL, where the invaginating cone bipolars terminate (Boycott & Kolb 1973, Kolb & Famiglietti 1974). There is evidence in both cat (Famiglietti & Kolb 1976, Nelson et al 1978) and carp (Famiglietti et al 1977) that off-center ganglion cells have dendrites terminating in the outer third of the IPL, whereas on-center cells receive their signals from the inner two thirds of this layer.

Functional Types

In terms of receptive-field properties, there are at least 12 distinct types of cat ganglion cells, 9 with antagonistic center-surround organization and 3 with other forms of organization. The distinguishing receptive-field properties of the various types are summarized in Table 1.

CENTER-SURROUND   These 9 types may be grouped into four on-center/off-center pairs and one color-coding type.

X (brisk-sustained) & Y (brisk-transient)   These two pairs were the first to be distinguished (Enroth-Cugell & Robson 1966, Cleland et al 1971, Saito et al 1971). On- and off-center X cells appear identical to each other in every other regard, so too for Y cells—except that there appears to be a group of on-center Y cells with smaller receptive fields near the area centralis (Cleland & Levick 1974a).

Tonic (sluggish-sustained) & phasic (sluggish-transient)   These two pairs were subsequently recognized (Stone & Fukuda 1974a, Cleland & Levick 1974b), although they constitute about half the cells of the cat retina. On-center and off-center tonic cells are not a well-matched pair in that they differ in their coding of light intensity and in their central projections (see below).

Color-coding   Whereas all the concentric cells described above receive their signals only from rods and from green (556 nm) cones, a small minority ($< 1\%$) also receive from blue (450 nm) cones (Daw & Pearlman 1970, Pearlman & Daw 1970, Cleland & Levick 1974b, Cleland et al 1976, Rowe & Stone 1976a, Wilson et al 1976). All show an on-center for blue cones and most show an off-center for green cones. Conduction latency to optic tract stimulation is slightly greater than for X cells (Kirk et al 1975).

Table 1 Properties of cat retinal ganglion cells

	Y	X	phasic	tonic	color coding	suppressed by contrast	direction selective	local edge detector
Percentage encountered when recording from optic tract	80%	20%	not encountered					
Percentage encountered when recording from retina	25%	55%	4.6%	7.4%	<1%	<1%	~1%	~5%
Best estimate of percentage of population	~4–7%	~45%	~20%	~30%	observations insufficient to draw conclusions			
Concentration in retinal distribution	area centralis (weak)	area centralis (strong)	visual streak	visual streak	?	?	?	visual streak
Latency to optic tract stimulation	1.0–2.4 ms	2.5–5.9 ms	6.1–18.7 ms	4.6–24 ms	3.8–14.2 ms	3.9–13.9 ms	6.1–12.4 ms	6.6–15.9 ms
Receptive field center diameter	0.6–2.5°	0.1–1.3°	0.5–2°		1.5–2.3°	0.5–2°	0.7–1.7°	0.5–2.3°
Soma diameter (20° horizontal eccentricity)	~30 μm	14–24 μm	>8 μm, <24 μm		unknown, probably small			
Morphological form	α	β	γ		?	?	?	?
Shift (McIlwain) effect	strong	weak	absent					

Table 1 *(Continued)*

	Contrast reversal test		not tested			
	no null position	null position	no distinctive pattern	not tested	complete suppression except for jerky movements	low frequency as for X / high frequency as for Y
Response to drifting grating	twice that of grating frequency	same as that of grating frequency	no distinctive pattern	not tested	complete suppression except for jerky movements	low frequency as for X / high frequency as for Y
Response to standing contrast	mainly transient	sustained	transient	sustained	sustained suppression	transient / transient
Percentage in temporal retina passing contralaterally	5%	0%	~70%	~20%	?	>90% / ~80%

Notes: *Percentage recorded from optic tract* based on Enroth-Cugell & Robson's (1966) sample of 128 cells. *Percentage recorded from retina* based on Cleland & Levick's (1974a, b) sample of 960 cells. *Percentage of population* Y (Wässle et al 1975, Stone 1978), others (Fukuda & Stone 1974). *Retinal distribution* (Cleland & Levick 1974b, Cleland et al 1973, Fukuda & Stone 1974, Rowe & Stone 1976b, Wässle et al 1975). *Latency to optic tract stimulation* (Cleland et al 1971, Cleland & Levick 1974a, b, Kirk et al 1975, Rowe & Stone 1976a, b, Stone & Fukuda 1974a, Stone & Hoffmann 1972). Latency increases with eccentricity from the optic disk, so that the range of latencies for a given retinal region is somewhat smaller than indicated by these values, which are those of Cleland & Levick (1974b). *Receptive-field center diameter* Center diameter increases with eccentricity from the area centralis, so that the range of diameters for a given retinal region is somewhat smaller than indicated by these values based on the studies of Stone & Fukuda (1974a) and of Cleland & Levick (1974a, b). *Soma diameter (20° horizontal eccentricity)* based on Kelly & Gilbert (1975), but see other studies (Boycott & Wässle 1974, Cleland et al 1975a, Fukuda & Stone 1974, Rowe & Stone 1976a, b, Stone & Fukuda 1974a, b, Wässle et al 1975) for additional information, including variation in somal and dendritic sizes vs. eccentricity from area centralis. *Morphological form* see text. *Shift (McIlwain) effect* (Barlow et al 1977, Cleland et al 1971, Fischer et al 1975, Ikeda & Wright 1972, Winters & Hamasaki 1976). *Contrast reversal test* (Enroth-Cugell & Robson 1966). *Response to standing contrast* (e.g. Cleland et al 1971, Cleland & Levick 1974a, b, Hochstein & Shapley 1976). *Response to drifting grating* (Enroth-Cugell & Robson 1966) This distinction can be applied to X and Y cells only under mesopic conditions (for references see Rowe & Stone 1977). *Percentage in temporal retina passing contralaterally* see text.

The retinas of rabbits (Caldwell & Daw 1978a) and squirrels (Michael 1968) also contain only two cone types, blue and green, with the blue type more common than in the cat. In the rabbit these always show a blue on-center, as for the cat, but are X-like by all other criteria. In the monkey retina or dorsal lateral geniculate nucleus, almost all cells that show a blue-cone input are blue on-center (Wiesel & Hubel 1966, DeMonasterio & Gouras 1975, Dreher et al 1976, Malpeli & Schiller 1978); these are also X-like by other criteria (Dreher et al 1976).

OTHER TYPES   At least three types (or more if certain varieties are included) show receptive-field organization not based on the antagonistic center-surround plan first described by Kuffler.

Suppressed-by-contrast   Rarely encountered ($<$ 1%), this type has a continuously maintained activity that is suppressed by the presence of any form of contrast in its visual field (Rodieck 1967, Stone & Hoffmann 1972, Cleland & Levick 1974b, Stone & Fukuda 1974a, Kirk et al 1975, Cleland et al 1976, Rowe & Stone 1976a, Wilson et al 1976). Cleland & Levick (1974b) report two varieties (uniformity-detector and edge-inhibitory-off-center), but the validity of this distinction is questioned by Rowe & Stone (1976a). This cell type is basically identical to the rabbit uniformity detector (Levick 1967).

Direction-selective   Cells of this type ($<$ 1%) respond to a certain direction of movement, independent of contrast, and are completely unresponsive to movement in the opposite direction (Stone & Hoffmann 1972, Cleland & Levick 1974b, Stone & Fukuda 1974a, Rowe & Stone 1976a). Stone & Fukuda (1974a) report two varieties (on-off and on), based on their response to flashing spots; Cleland & Levick (1974b) report only one (on-off). Unlike rabbit direction-selective units, most of which have fast-conducting fibers (Caldwell & Daw 1978a), those of the cat have slow-conducting axons and do not have preferred axes of direction selectivity (Oyster 1968).

Local-edge-detector (excited-by-contrast, on-off-center-phasic-W)   Cells of this type ($<$ 5%) (Stone & Hoffmann 1972, Cleland & Levick 1974b, Stone & Fukuda 1974a, Kirk et al 1975, Cleland et al 1976, Rowe & Stone 1976a, Wilson et al 1976) are essentially the same of those of the rabbit (Levick 1967, Caldwell & Daw 1978a).

At present we have only vague hints as to what functional roles the different types of cat ganglion cells subserve, though phrases such as "X is for pattern, Y for movement" are often repeated. About as much as can be said at present is that:

1. The highest density of Y cells ($\sim$ 200/mm^2, ref. Wässle et al 1975), which is found at the area centralis, is too low by a factor of about 15 to account for the cat's behaviorally measured spatial acuity of about 6 cycles/deg.
2. The same reasoning holds for the rarely-encountered cells, presuming that their sparsity is genuine and not the result of electrode sampling.
3. The relatively higher sensitivity of Y cells to movement suggests that they may set the cat's threshold to movement.
4. On-center tonic cells appear to be the only type capable of signaling information about the ambient level of light intensity (Barlow & Levick 1969, Cleland & Levick 1974b, Stone & Fukuda 1974a) and are thus likely candidates for the control of the pupilloconstrictor reflex and the light-mediated component of various circadian rhythms (see below).

Identification and Classification

Although there is close agreement among different laboratories as to the properties of the various ganglion-cell types, there is no corresponding agreement as to how they should be named or classified. In itself, naming is not substantive; taxonomists use essentially arbitrary names whose choice is resolved in terms of historical precedence.

As discussed by Tyner (1975) and Rowe & Stone (1977) classification involves all the possible distinguishing features of each of the recognized types within the scheme. Identification, on the other hand, is the procedure by which a given cell is typed—here one applies a few selected tests, ideally only one. Neither classification nor identification implies the functional roles these cell types subserve in vision. These roles are presently unknown, and a great deal more will have to be learned about the visual process before they are known with a degree of certainty.

Although the phylogenetic and ontogenetic origins of the different cell types are unknown, they will probably provide the ultimate basis for the classification scheme. It remains possible that new cell types may be found, or that cells now classed as one type may have to be subdivided. Thus, the existing classification schemes may have to be modified or revised as new evidence comes to hand.

When Enroth-Cugell & Robson (1966) discovered that cat center-surround cells could be further distinguished from the single pair known till then, they happened to do so by means of a contrast-reversal test; some cells (X) were capable of giving a null response to this test, other cells (Y) were not (Table 1). Some workers, apparently unaware of the distinction between identification and classification, came to treat this distinguishing feature, or some other one, as definitional (i.e. if null then X, if not then Y). The outcome of this notion is that every center-surround cell, in the cat or elsewhere, must be either X or Y. Identification and classification become merged in this approach, and exactitude in classifying is achieved at the expense of biological relevancy.

As noted in Table 1, the axons of Y cells have the fastest conduction velocities; those of X cells are slower. Still slower are those of the other two pairs of center-surround types, as well as all the other types. These slowest cells also have certain similarities in their retinal distribution, naso-temporal division, and central projections; they came to be collectively termed W-cells (e.g. Stone & Fukuda 1974a). But it is now recognized that there are differences between the conduction velocities, retinal distributions, naso-temporal divisions, and central projections of each of the types grouped in this category, as shown in Table 1 and discussed below. In my view the time has come to dispense with the W grouping and instead to treat each type, or pair of types, separately.

Correspondence Between Functional and Morphological Classes

By exhaustively mapping a small patch of retina electrophysiologically for all the Y cells, and later locating all the α cells in a stained whole mount of the patch, Cleland et al (1975a) demonstrated a one-to-one correspondence between the functional class of Y cells and the morphological class of α cells. There is, at any eccentricity, a good correspondence between the size of the dendritic field of the α cells and the receptive-field center size of the Y cells (Cleland & Levick 1974a).

X cells have the smallest receptive-field centers of any functional class (Cleland & Levick 1974a, Stone & Fukuda 1974a) and β cells have the smallest dendritic-field diameters of any morphological class (Boycott & Wässle 1974). Furthermore, by injecting horseradish peroxidase (HRP) into laminae A and A1 of the dorsal lateral geniculate nucleus (LGNd), where the fibers of X and Y cells terminate, Kelly & Gilbert (1975) found retrograde labeling of the middle- and large-sized ganglion cell somata, but not the smallest. There is thus a strong correlation between the functional X class and the morphological β class.

Tonic and phasic cells show a wide range of receptive-field center diameters, as do other types (Cleland & Levick 1974a, b, Stone & Fukuda 1974a) and this range overlaps reasonably well with the range of γ cell dendritic-field diameters (Boycott & Wässle 1974), each plotted against eccentricity from the area centralis. Injection of HRP into the superior colliculus causes the small- and large-sized ganglion cells to be labeled, but not the middle-sized somas (Kelly & Gilbert 1975). All cell types except X cells project to the superior colliculus, and this absence of X cells correlates with the absence of labeling in the middle-sized somata. The large-cell labeling is presumably entirely accounted for by the Y projection to the superior colliculus. The great majority of the remaining functional types are the tonic and phasic cells, so these must have small somata, and can presumably be identified with γ cells. However, since X and Y cells are morphologically distinct, one wonders whether there are not morphological distinctions between these two types as well. Boycott & Wässle (1974), in addition to describing α, β, and γ cells, also noted a few (δ cells) with small somata that did not exhibit the morphology of γ cells, and further commented that their classification scheme may not be complete. Thus the identification of both tonic and phasic types with γ cells remains speculative. The morphological identification of the rarer types is unclear, but it would be surprising if the different functional types did not arise from correspondingly distinct morphological types.

Crossed and Uncrossed Projections

Primate ganglion cells that lie temporal to the fovea have axons that pass uncrossed to the ipsilateral optic tract, whereas those nasal to the fovea decussate at the optic chiasm to project into the contralateral optic tract. There is a thin vertical strip, of about 1°, where these projections overlap, that is where cells projecting to either tract intermingle, as well as a few aberrant cells scattered on the rim of the fovea (Stone et al 1973, Bunt et al 1977).

In the cat, however, each major type of ganglion cell has a different pattern of naso-temporal division (Stone & Fukuda 1974b, Kirk et al 1976a, b). X cells have the same form of division as found for primate ganglion cells, with a 1° median strip of overlap passing vertically through the center of the area centralis. For all the remaining types, all the cells in the nasal retina send fibers to the contralateral optic tract, just as for X cells. However, in the temporal retina there are some that send fibers to the contralateral optic tract, and their line of naso-temporal division is located 1–2° temporal to that of the X cells. About 5% of the Y cells, 20% of the tonic ones, and 70% of the phasic cells do so. Most of the local-edge detectors (~80%) and almost all of the direction-selective cells also send their axons into the contralateral optic tract.

Phylogenetically, the contralateral projection of optic nerve fibers is older than the ipsilateral projection, and it is interesting to note the various cells that have to a great degree retained their crossed projection in an animal with forward directed eyes and considerable binocular overlap.

Similar Cell Types in Other Species

There is growing evidence that some of the differences observed between different types of cat ganglion cells also occur in other animals. Dreher et al (1976) proposed the terms *X-like* and *Y-like* to characterize cells in the monkey lateral geniculate nucleus that could be distinguished on the basis of a number of tests developed to identify cat X and Y cells.

Two ganglion cell types have been reported in the retina of the eel *Anguilla rostrata* (Shapley & Gordon 1978). One type resembles cat X cells in its response to contrast reversal and to drifting gratings. The other type is distinct from any seen in the cat, though it corresponds most closely to local edge detectors (on-off W cells).

Ganglion cells in *Necturus* (Tuttle & Scott 1978) and in the frog (Gordon & Shapley 1978) can be distinguished on the basis of their responses to contrast reversal. Ganglion cells with X- and Y-like properties have been briefly reported in the owl (Pettigrew 1978).

The optic nerve of the Australian brush-tailed possum contains at least three conduction velocity groupings of fibers (Rowe et al 1976). Those of the fastest group have Y-like receptive-field properties, those of the remaining two groups show sustained responses to stationary visual stimuli but have not been further distinguished.

In the rat, three groups of ganglion cells, based on somal diameter, have been distinguished: large (L), medium (M), and small (S) (Bunt, Lund & Lund 1974, Fukuda 1977). The M and S cells concentrate to form an area centralis, the L cells are uniformly distributed across the retina. Three corresponding conduction-velocity groups are seen in the optic tract and all three groups project to the dorsal lateral geniculate nucleus and probably to the superior colliculus.

The mink retina (Dubin & Turner 1977) and LGNd (Sanderson 1974) are so similar to those of the cat as to strongly suggest similar ganglion cell types in this species (Guillery & Oberdorfer 1977), and probably in carnivores generally.

As noted above, there are many correspondences between the ganglion cells of the rabbit and cat (Caldwell & Daw 1978a, b, Caldwell, Daw & Wyatt 1978). Cells with X-like and Y-like properties have been reported in the dorsal lateral geniculate nucleus of the tree shrew (Sherman et al 1975), the owl monkey (Sherman et al 1976), and the macaque (Dreher et al 1976). A variety of ganglion cell types occur in the monkey retina (DeMonasterio & Gouras 1975, DeMonasterio, Gouras & Tolhurst 1975); most of these exhibit either X- or Y-like properties (Gouras 1968, 1969, Schiller & Malpeli 1977, DeMonasterio 1978). A few have been reported to exhibit W-like properties and these have a high probability of being antidromically activated from the superior colliculus. Conversely, those that exhibit color-coding properties, which are all X-like, do not project to the superior colliculus (Schiller & Malpeli 1977).

CENTRAL PROJECTIONS OF GANGLION CELLS

In all vertebrate classes there appears to be a common set of projections to six regions: (*a*) hypothalamus (suprachiasmatic nuclei—SN), (*b*) accessory optic nuclei (except some elasmobranchs and teleosts), (*c*) pretectum, (*d*) optic tectum (superior colliculus), (*e*) ventral thalamus (ventral lateral geniculate nucleus—LGNv), (*f*) dorsal thalamus (dorsal lateral geniculate nucleus—LGNd) (Riss & Jakway 1970, Ebbesson 1970, 1972).

The retinal projection to the dorsal thalamus and then on to the cerebral cortex is highly developed in mammals having good vision, and most of the central work on vision has concentrated on this pathway. But there has been a growing interest in other retinal projections and a recognition of the importance of the roles they play in vision.

Suprachiasmatic Nucleus (SN)

The suprachiasmatic nuclei lie just above the optic chiasm and just lateral to the third ventricle. Each nucleus receives a bilateral projection of retinal fibers, which form both Gray type I and type II synapses (Güldner & Wolff 1978b) onto dendrites and dendritic spines of SN neurons (Hendrickson et al 1972, Moore & Lenn 1972, Moore 1973, 1978). Studies on the rat have shown that many, if not all of these fibers are collateral branches of fibers going on to the optic tract (Mason & Lincoln 1976, Mason et al 1977, Millhouse 1977). A range of fiber diameters has been reported, suggesting that they may arise from more than a single type of ganglion cell. However, R. Y. Moore (personal communication) finds that retrograde transport of HRP, diffused onto the rat SN, labels a group of small cells occurring throughout the retina. The SN also receives a visual input from the LGNv, which overlaps the retinal projection (Swanson et al 1974, Ribak & Peters 1975), as well as receiving a number of nonvisual inputs (Moore 1978). SN cells project to a number of areas, which are otherwise nonvisual (Swanson & Cowan 1975).

Neurons of the SN are reported to show small dendritic side branches that appear to form Gray type II synapses onto their cell of origin (Güldner & Wolff 1978a). The response of SN cells to retinal illumination has been briefly reported (Lincoln et al 1975, Sawaki 1978).

A number of studies have demonstrated a pivotal role for the SN in the neural control of a variety of circadian rhythms: all but the latest of these studies have been reviewed by Moore (1978). Ablation of the SN abolishes the circadian rhythms of drinking (Stephan & Zucker 1972), eating (Nagai et al 1978), locomotion (Stephan & Zucker 1972, Stetson & Watson-Whitmyre 1976), sleep/wakefulness (Ibuka & Kawamura 1975), levels of pineal serotonin N-acetyltransferase (Moore & Klein 1974), and adrenal corticosterone (Moore & Eichler 1972) and disrupts reproductive function (Brown-Grant & Raisman 1977, Raisman & Brown-Grant 1977). Surprisingly, ablation of the SN is reported to leave the daily temperature rhythm of rats unaltered (Dunn et al 1977).

Using the deoxyglucose technique, Schwartz & Gainer (1977) have shown that the SN has its highest glucose consumption in the daytime, even in the absence of environmental light cues. Thus the rhythmic activity of the SN itself does not depend on the visual input. Rather the diurnal cycle of light intensity, conveyed by the visual input, is used to synchronize this rhythm.

As noted earlier, the only cells in the cat retina that appear capable of signaling light intensity are the on-center tonic cells, which happen to have small somata. Visual signals can reach the SN either directly from the retina or indirectly via the LGNv. In the cat the LGNv receives its retinal input primarily from on-center tonic cells (see below). As mentioned above, all rat ganglion cells projecting directly to the SN are small. Thus current evidence suggests that cat on-center tonic cells and a similar type of ganglion cell in the rat, provide the essential visual signal to synchronize the activity of the SN to the diurnal cycle of light and dark.

Accessory Optic System

In the cat a fasciculus of thin fibers from the contralateral eye branches from the optic tract at the level of the brachium, and passes over the surface of the cerebral peduncle to terminate in a small, well-defined group of cells, the *medial terminal nucleus* (MTN) (Hayhow 1959). Lying interspersed along this tract are numerous cells that receive an input from the fibers of the tract, an aggregation of these, near the lateral border being termed the *lateral terminal nucleus* (LTN) and a sparser group near its proximal end, termed the *dorsal terminal nucleus* (DTN) (Hayhow, Webb & Jervie 1960).

All vertebrates appear to possess a similar system, although there is some variation in the number of terminal nuclei and in the fasciculi through which they receive their retinal input. The fasciculi, collectively termed the *accessory optic tract,* together with the interposed cells and terminal nuclei, comprise the *accessory optic system* (this definition is slightly modified from that of Hayhow (1966) in order to avoid including the SN as part of the accessory optic system). In the rat, at least, there does not appear to be any retinotopic organization of the MTN (Kostović 1971).

In the pigeon the MTN (nucleus of the basal optic root) receives large diameter retinal axons arising from displaced ganglion cells (Karten et al 1977) and projects directly to the vestibulocerebellum (Brauth & Karten

1977). Directionally selective cells have been briefly reported in the pigeon MTN (cited by Brauth & Karten 1977).

In the rabbit MTN, Walley (1967) found very large receptive fields, some as large as 90°, most of which were directionally selective. Purkinje cells of the rabbit flocculus are known to receive a visual input indirectly by way of mossy fibers (Maekawa & Takeda 1975, 1976; see also Lisberger & Fuchs 1974, Ghelarducci et al 1975, Miles & Fuller 1975). In chinchillas, Winfield et al (1978) have demonstrated a projection from the MTN to the flocculus, and suggest that it may provide a visual mossy-fiber input to the granule cells of the cerebellar cortex. The retinal fibers projecting to the chinchilla MTN arise from large ganglion cells, both displaced and within the ganglion cell layer (J. A. Winfield & A. Hendrickson, personal communication).

Cells of the rabbit LTN and DTN, together with some from the pretectal nucleus of the optic tract, project to cells in the ipsilateral dorsal cap and (probably) beta nucleus of the inferior olive (Mizuno et al 1973, 1974, Takeda & Maekawa 1976). The olivary cells project in turn as climbing fibers onto Purkinje cells in the contralateral flocculus and nodulus of the vestibulocerebellum (Maekawa & Simpson 1972, 1973, Alley et al 1975). Electrophysiological recording demonstrates that these climbing fibers convey direction-specific information about movements in large parts of the visual world, are optimally responsive to slow movements of less than 1° (Simpson & Alley 1974), and show other features similar to rabbit on-center directional ganglion cells. These ganglion cells show three preferred directions (Oyster 1968), which appear to correspond to the directions of movement of the visual field associated with rotation about the principal axes of each of the three semicircular canals (Simpson & Hess 1977).

Floccular Purkinje cells are known to inhibit cells of the vestibular nuclei. Primary vestibular fibers are likewise known to branch to excite both the cells of the vestibular nuclei and Purkinje cells, indirectly via mossy fibers. The cells of the vestibular nuclei then project to the oculomotor nuclei.

Recordings from floccular Purkinje cells indicate that they play an essential role in coordinating head and eye movements (Lisberger & Fuchs 1974, Miles & Fuller 1975). When the head turns to the left the eyes turn to the right by the same amount so as to maintain a stable gaze. The pathway for this vestibulo-ocular reflex is a three-neuron reflex from primary vestibular fibers to vestibular nuclei to oculomotor nuclei. Such eye movements can occur in the dark, as well as after complete removal of the cerebellum, but can be suppressed when the eye tracks a moving target.

The role of the Purkinje cells of the flocculus appears to be to set the gain of this reflex so that the eyes move by just the right amount to maintain a steady gaze (e.g. Ito 1977). Gonshor & Melvill Jones (1976a, b) have demonstrated that visual experience plays a powerful role in setting the magnitude of this gain. If subjects wear reversing (Dove) spectacles then the magnitude of gain decreases within a few days and actually reverses within a few weeks. When the subject then turns his head to the left his eyes now also turn to the left, and do so even in the dark. The same effect has been demonstrated in cats (Melvill Jones & Davies 1976) and a similar effect has been found in monkeys (Miles & Fuller 1974). Removal of the flocculus abolishes this plasticity (Ito et al 1974, Hassul et al 1976, Robinson 1976). The plasticity and adaptability of this goal-directed reflex of adults has recently been reviewed by Melvill Jones (1977).

Pretectum

The pretectal region is a complex collection of cells of varying size and morphology, showing blurred patterns of aggregation, and receiving localized fiber projections, not only from the retina, but from the superior colliculus (Graham 1977), the LGNv (Edwards et al 1974, Graybiel 1974b, Swanson et al 1974), the visual cortex (Kawamura et al 1974, Updyke 1977), the frontal eye fields of monkeys (Astruc 1971, Künzle & Akert 1977), and the dorsal column nuclei and spinal gray of rats (Lund & Webster 1967a, b). Pretectal cells project to a great variety of areas, recently summarized by Berman (1977).

Based on degeneration methods and cytoarchitectonics, there appear to be about seven or so nuclear subdivisions in mammals (Scalia 1972) including the cat (Kanaseki & Sprague 1974). Three of these are reported to receive bilateral retinal inputs: the nucleus of the optic tract (NOT), the olivary pretectal nucleus (PO), and the posterior pretectal nucleus (PP). Berman (1977) could not discern a retinal projection to PP but did find that the terminal field projection to NOT consisted of two or three finger-like strips. Since the boundary between PP and NOT is not sharply marked in the literature, it remains possible that the posterior strip of her NOT corresponds to the retinal projection to the PP of other authors.

By placing stimulating electrodes in the cat pretectal region and antidromically activating the retinal cells they studied, Cleland & Levick (1974a, b) showed that the pretectal region receives inputs from all known ganglion cell types. Some, if not all, of the X fibers projecting to the pretectum are known to branch and go to the LGNd as well (Fukuda & Stone 1974).

It has been held that the retino-pretectal projections in the cat are not retinotopically organized (Laties & Sprague 1966); however, it has been demonstrated that they are in the rat (Scalia & Arango 1977). Furthermore Updyke (1977) has shown that the projections from visual cortical areas 17,

18, and 19 are topographically organized. Thus it would be a rather startling finding if the absence of a correspondingly organized retinal projection were confirmed.

Hoffmann & Schoppmann (1975) and Hoffmann et al (1977) have recently recorded from cells in the cat NOT and shown that the conduction velocity of their retinal input was slower than that of X or Y cells. All the cells they described had receptive fields in the visual streak area of the retina and exhibited functional properties markedly different from those found in the superior colliculus. These cells had a high firing rate independent of background illumination, responded well to very slow stimulus motion, and were direction selective for very slow movements. Large textured patterns seemed optimally suited to evoke strong tonic responses to movement. In their latest report Hoffmann et al (1977) describe these cells as excited by movements from the periphery to the center of the visual field and inhibited by movements in the opposite direction, although the strongest activation was from temporal to center in the contralateral visual hemifield. These properties suggest that this group of cells subserves roles in both fixation and in optokinetic nystagmus. Most were binocular, favoring the contralateral eye. However, in monocularly deprived cats, or in normal cats after a lesion to the visual cortex, all of these cells lost their ipsilateral input and responded only to movement from temporal to center in the contralateral visual hemifield, indicating that they acquired their ipsilateral input via the visual cortex. These observations fit well with those of Wood et al (1973), who found that after removal of the visual cortex, cats showed a normal monocular optokinetic nystagmus to temporo-nasal movement, but a strongly impaired response to movement in the opposite direction.

In the rabbit, Collewijn (1974, 1975a, b) has likewise demonstrated a close relation between optokinetic nystagmus and the properties of cells in the NOT. The rabbit has little binocular overlap, and these cells all showed properties strikingly similar to those of monocularly deprived cats. In particular, they responded only to movements from temporal to nasal. This direction of movement is much more effective in eliciting optokinetic nystagmus in the rabbit than the opposite direction. Total ablation of the superior colliculus, sparing the pretectum, did not interfere with optokinetic nystagmus. Furthermore, electrical stimulation of the NOT elicited vigorous horizontal eye movements, even in animals in which the optic nerves had been previously sectioned and allowed to degenerate.

Since the work of Magoun and Ranson in the 1930s, it has been known that the pupilloconstrictor reflex survives total ablation of the superior colliculus, but not of the pretectum. As noted earlier it is likely that the on-center tonic cells play a critical role in this regard. There have been a number of reports of pretectal cells in the cat responsive to light intensity, which have been recently reviewed by Sprague et al (1973, see also Clarke & Ikeda 1978). Cells subserving this function are reported to lie in the NOT. Considering their distinctive response properties and retinal requirements, it would be surprising if the regions subserving optokinetic nystagmus and pupilloconstriction were not anatomically segregated in the pretectum, but this remains to be established.

Superior Colliculus

The superior colliculus (SC) is a layered structure (e.g. Kanaseki & Sprague 1974), which can be usefully viewed as being composed of two main subdivisions: *superficial* and *deep* (Harting et al 1973). The deep layers, not further considered here, are those lying below the stratum opticum; they do not receive a retinal projection, but do receive afferents from, and project to, a great variety of areas, visual and otherwise (e.g. Graham 1977, Edwards et al 1978). The superficial layers receive afferents primarily from the retina (e.g. Kanaseki & Sprague 1974, Graybiel 1975, 1976), visual cortex (Garey et al 1968, Kawamura et al 1974, Gilbert & Kelly 1975, McIlwain 1977, Updyke 1977), and from a cell group in the lateral tegmentum, termed the *parabigeminal nucleus* (Graybiel 1978c, Edwards et al 1978). This nucleus also receives a massive collicular projection, which is topographically organized (Graybiel 1972, Harting et al 1973, Graham 1977), as is the parabigeminal nucleus itself (Sherk 1978). Graybiel has suggested that this nucleus may represent a sort of satellite system of the superior colliculus:

"According to this plan the major strata in tectum and tegmentum could be considered as connected in parallel, as though the parabigemino-tegmental complex were a 'miniature colliculus' linked in register with its more prominent counterpart." (Graybiel 1978c).

The cortical projections arise from areas 17, 18, 19, and the lateral suprasylvian area, and are in topographic register with the retinal projection (Garey et al 1968, McIlwain 1973a, b, 1977, Updyke 1977). These cortico-tectal projections arise from a specific group of pyramidal cells in layer V (e.g. Palmer & Rosenquist 1974, Gilbert & Kelly 1975, Magalhães-Castro et al 1975), which, in area 17 at least, have complex receptive-field properties (Palmer & Rosenquist 1974) of a particular type, and have been termed *special complex* (Gilbert 1977). Removal of area 17 markedly decreases the number of binocularly driven, and directionally selective, collicular cells (Wickelgren & Sterling 1969).

Fibers from the contralateral eye terminate in a dense continuous band in the upper part of the superficial gray layer, those from the ipsilateral eye terminate in register in the lower part of this layer, in bands or slabs (Graybiel 1975, 1976) similar in width and form to those of the visual cortex (Hubel & Wiesel 1977, Shatz et al 1977). The projections of the substantia nigra pars reticulata to the deeper layers of the SC also shows similar bands or slabs (Graybiel 1978a), suggesting that periodic vertical segmentation may be a general feature of collicular organization (see also Graybiel 1978b).

It is a striking, but well-confirmed finding that the axons of all types of ganglion cell, except X cells, terminate in the SC (Hoffmann 1972, 1973, Cleland & Levick 1974a, Fukuda & Stone 1974, Kelly & Gilbert 1975, Magalhães-Castro et al 1976). All the cells in the superficial layers of the colliculus respond only to visual stimuli, acquiring their receptive-field properties partly from their retina input and partly from the visual cortex. These cells are typical of those found along sensory pathways, having small dendritic fields with little overlap. In addition to the pathway to the deeper collicular layers, the cells of the superficial layers (stratum griseum superficiale) project to the nucleus of the optic tract and olivary pretectal nucleus of the pretectum, to the pulvinar-lateral posterior complex, and to the dorsal and ventral lateral geniculate nuclei (e.g. Graham 1977). Of particular importance is the topographically ordered collicular projection to the pulvinar/lateral-posterior complex (Graybiel 1972, Kawamura 1974, Kawamura & Kobayashi 1975, Graham 1977) to a region termed the *interadjacent zone* by Updyke (1977). Since the pulvinar/lateral-posterior complex projects to the visual cortex, this pathway provides a route to the visual cortex that is in parallel with the geniculate pathway (see below).

Ventral Lateral Geniculate Nucleus (LGNv)

The LGNv (pregeniculate nucleus of primates) is part of the ventral thalamus and is a general feature of the mammalian visual system (Niimi et al 1963, Jordan & Holländer 1972). The retinal input is bilateral, consisting of fine fibers (Hayhow 1958) and is retinotopically ordered (Montero et al 1968, Mathers & Mascetti 1975, Spear et al 1977). In cats the retinal projection consists of two nonadjacent terminal fields; the larger dorsal field receives nonoverlapping projections from each eye; the smaller ventral terminal field receives a weaker projection, primarily from the contralateral eye (Holländer & Sanides 1976). The LGNv also receives inputs from the visual cortex (Heath & Jones 1971, Kawamura et al 1974, Updyke 1977), superior colliculus (Graybiel 1972, 1974b, Graham 1977), pretectum (Graybiel 1974a, Berman 1977), and deep cerebellar nuclei (Graybiel 1974b).

Although LGNv cells do not project to the cerebral cortex, they do project to a variety of areas, including the pretectum, superior colliculus, pontine nuclei, suprachiasmatic nuclei, lateral terminal nuclei of the accessory optic tract, zona incerta, and the contralateral LGNv (Hendrickson 1973, Graybiel 1974b, Edwards et al 1974, Swanson et al 1974, Ribak & Peters 1975).

In cats only slowly conducting retinal fibers project to the LGNv, which thus excludes both X and Y cells (Spear et al 1977). A variety of receptive field types are present (Hughes & Ater 1977, Spear et al 1977); these have larger receptive fields than retinal cells, suggesting some degree of excitatory convergence. Almost all of the concentric type have on-center regions and many respond to light intensity in a manner suggesting an input from tonic cells. Cells with similar receptive field properties have been described in the LGNv of the rabbit (Mathers & Mascetti 1975) and rat (Hale & Sefton 1978), although for the rat all three conduction-velocity fiber groups project from the retina to the LGNv.

The functional roles that the LGNv subserve are unknown. Polyak (1957) suggested that the pathway of the pupilloconstrictor reflex passed from retina through the LGNv to the pretectum, instead of directly. There is some evidence for this in rats, where the destruction of the LGNv abolishes the pupilloconstrictor reflex (Legg 1975) and impairs intensity discrimination (Horel 1968, Legg & Cowey 1977), while leaving horizontal/vertical discrimination unimpaired.

In monkeys and cats, LGNv cells discharge with a burst of activity at the onset of saccadic eye movements in the dark (Buttner & Fuchs 1973, Putkonen et al 1973, Magnin & Fuchs 1977), indicating that the LGNv is somehow involved with eye movements.

Dorsal Lateral Geniculate Nucleus (LGNd)

A coronal view of the cat LGNd is represented schematically in Figure 1, together with part of its cortical projection. There are two separate retinotopic projections to the LGNd, one to the medial interlaminar nucleus (MIN), the other to the main laminae (lamLGNd). The MIN receives its retinal input almost entirely from Y cells, whereas the lamLGNd receives from X and Y cells in its upper part, and from tonic and phasic, as well as other cell types, in its lower.

The great majority of geniculate cells projecting to the visual cortex (relay cells) show receptive-field properties very similar to those of ganglion cells, indicating that there is essentially no mixing of the excitatory signals of the various cell types at this level. Depending on the type of relay cell, and its location in the LGNd, there are a variety of geniculocortical projections, both to different cortical areas and to different layers within them. Only what is known of the projection to area 17 is shown in detail in Figure 1, since it is the best studied, but there is no reason to believe that the projections to other visual areas are any less complex.

lamLGNd   Two lamination schemes are shown in Figure 1, the AC system is that of Guillery (1970) as revised by Hickey & Guillery (1974); the AMB system is that of Rioch (1929) and others as revised by Famiglietti (1975).

There is growing evidence, both in terms of input types and output projections, that the upper part of lamina C (Lamina M) is closely associated with the overlying A laminae (Wilson et al 1976, Guillery & Oberdorfer 1977, Holländer & Vanegas 1977, Mitzdorf & Singer 1977), whereas cells of the lower part of C (B_0) show properties similar to those within the underlying laminae (Mason 1976, Wilson et al 1976). The AMB system better reflects these recent findings and is used here.

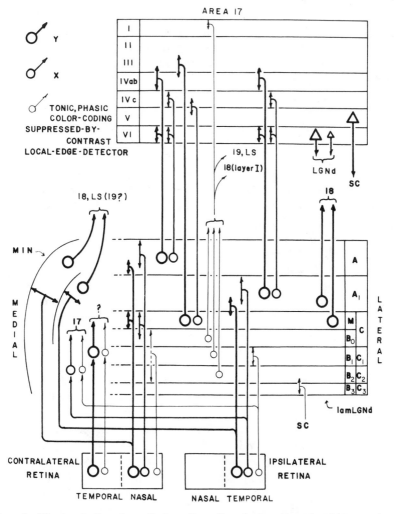

Figure 1   Direct projection of cat retinal ganglion cells to the dorsal lateral geniculate nucleus and from there to area 17 (striate area) of the visual cortex. *lamLGNd* main laminae of the dorsal lateral geniculate nucleus, *LS* lateral suprasylvian cortical area, MIN medial interlaminar nucleus, *SC* superior colliculus.

A, A₁, & M laminae   There is no zone of binocular overlap between the terminal fields of the contralateral and ipsilateral projections to A and A_1 respectively (Hickey & Guillery 1974, Famiglietti 1975). Evidence conflicts as to whether there is such a zone between A_1 and M (Hickey & Guillery 1974, Famiglietti 1975). Binocularly driven cells are reported there (Sanderson et al 1971, Sanderson 1971) although these may be nearby lamellar cells having dendrites that cross lamellar borders (Guillery 1966). Each of these three laminae receive retinal projections from both X and Y cells but not from other types (Cleland et al 1971, Fukada & Saito 1972, Hoffmann et al 1972, Fukuda & Stone 1974, Kelly & Gilbert 1975, Mason 1976, Wilson et al 1976, but see also Cleland et al 1976). All retinal X and Y cells are reported to terminate there (Gilbert & Kelly 1975). In both A and A_1 the X fibers terminate in the dorsal and central parts of the lamina, whereas the Y fibers terminate close to its ventral border (Mitzdorf & Singer 1977). The cells in these three laminae receive their excitatory input from one or a few retinal fibers (Bishop et al 1958, Creutzfeldt 1968, Cleland et al 1971) with only a small minority of cells receiving an excitatory mixture of X and Y signals (Cleland et al 1971). Consequently, the great majority of cells in these laminae show either X or Y receptive-field properties. The Y cells correspond to Guillery's (1966) morphological class 1, whereas the X cells correspond to his class 2 (LeVay & Ferster 1977, Ferster & LeVay 1978). Class 2 cells are further distinguished by the presence of a large cytoplasmic inclusion, termed a laminar body (LeVay & Ferster 1977, but see also Kalil & Worden 1978).

Cells in these laminae fire only in response to an incoming retinal action potential, and in this sense "relay" the retinal action potential to the visual cortex (Hubel & Wiesel 1961, Cleland et al 1971). Though excited by only one retinal fiber, or those of a few nearby cells, these relay cells can be inhibited by nearby activity from the same eye (suppressive field) (Hubel & Wiesel 1961, Levick et al 1972) or from the homonyous region of the other eye (binocular inhibition) (Singer 1970, Sanderson et al 1971, Rodieck & Dreher 1978). Both of these forms of inhibition arise from a single binocularly driven mechanism (G. Lambert, R. W. Rodieck, B. Dreher, unpublished observations). Rodieck (1973) reports that this inhibition is mediated entirely by Y retinal fibers, but Hoffmann et al (1972) and Singer & Bedworth (1973) believe there is an X contribution as well. Possible neural pathways mediating this inhibition is a complex topic not reviewed here. The magnitude of the inhibition is mediated by nonretinal inputs to the LGNd from the midbrain reticular formation or from the visual cortex. These influences, and the possible inhibitory pathways, have recently been reviewed in detail (Singer 1977, Burke & Cole 1978).

The conduction velocity distinction between tract fibers from different retinal cell types is repeated in the LGNd relay fibers projecting to the visual cortex (Cleland et al 1971, 1975b, Hoffmann & Stone 1971, Hoffmann et al 1972, Fukada & Saito 1972, Wilson & Stone 1975, Wilson et al 1976) so that the geniculocortical Y fibers are the fastest.

Relay cells of laminae A and A_1 project to cortical area 17 or to area 18, but not elsewhere (e.g. Rosenquist et al 1974, Gilbert & Kelly 1975, LeVay & Gilbert 1976, Holländer & Vanegas 1977) and it is possible that this is true for the cells of the M lamina as well. The terminals of these fibers form ocular dominance columns in layers IV and VI (LeVay & Gilbert 1976, Shatz et al 1977, LeVay et al 1978, Ferster & LeVay 1978).

X cells from A or A_1 project only to area 17 (Stone & Dreher 1973, LeVay & Ferster 1977) where they terminate in layers IVc and VI (Ferster & LeVay 1978). Layer IVc primarily contains small stellate cells with spiny dendrites (O'Leary 1941, Ferster & LeVay 1978, Lund et al 1978) that have receptive fields that are either concentric or simple (Hubel & Wiesel 1962, Kelly & Van Essen 1974, Gilbert 1977). Y cells projecting to area 17 from laminae A or A_1 terminate in the lower part of layer III, in IVab and in VI (Ferster & LeVay 1978). In IVab they arborize among a population of both simple and complex cells.

Area 17 receives a projection from the M lamina (LeVay & Ferster 1977) arising from the smaller (X) cells and from a fraction of the larger (Y) cells. It is not known in which cortical layers these fibers terminate. The disposition shown in Figure 1 is a supposition on my part, based on the literature (LeVay & Gilbert 1976, Ferster & LeVay 1978).

Area 18 receives its AM laminar projection only from Y relay cells (Stone & Dreher 1973). Current evidence indicates that 17 and 18 receive from essentially separate groups of Y cells (LeVay & Ferster 1977). Area 18 receives its ipsilateral input from A_1 and its contralateral input primarily from M, with a minor projection from A (LeVay & Gilbert 1976, LeVay & Ferster 1977, Holländer & Vanegas 1977).

Thus for the contralateral pathway, lamina A projects almost entirely to 17, lamina M mainly to 18. For the ipsilateral projection from A_1, it is as if the groups projecting to 17 or to 18 had either merged or failed to separate to the degree seen for the contralateral projection.

B laminae   Laminae B_0–B_2 receive a direct retinal projection, but lamina B_3 probably does not (Hickey & Guillery 1974, Famiglietti 1975). The retinal fibers that end in B_0–B_2 are fine (Hayhow 1958, Guillery 1966, 1970) and are slowly conducting (Wilson & Stone 1975, Cleland et al 1975b, Wilson et al 1976, Mitzdorf & Singer 1977). The B laminae contain not only a full retinotopic projection of the contralateral visual hemifield, as for the A laminae, but also include a representation of the ipsilateral hemifield arising from ganglion cells in the temporal retina of the contralateral eye (Sanderson & Sherman 1971). The extent of overlap of the terminal fields of the projections from the two eyes remains unsettled. Guillery (1970, Hickey & Guillery 1974) reports that there is probably no overlap zone in his C laminae, whereas Famiglietti (1975) reports the presence of crossed fibers terminating throughout B_0-B_2. Fine fibers from the stratum griseum superficiale of the superior colliculus project to B_3 and possibly B_2 as well (Niimi et al 1970, Graybiel & Nauta 1971, Graham 1977).

The retinal fibers projecting to the B laminae arise mainly from phasic and tonic ganglion cells, but a few color-coding, suppressed-by-contrast, and local-edge-detector types are also reported (Wilson & Stone 1975, Cleland et al 1975b, 1976, Wilson et al 1976). There appears to be no segregation of different types within the B laminae. Y cells do not occur in these laminae, nor apparently do direction-selective cells. The situation regarding X cells is uncertain, if they do project there, then they do so only sparsely, and probably only to B_0, although possibly to B_1 as well.

The principal cells in the B laminae are small and constitute Guillery's (1966) class 4. They have thin geniculocortical axons (Ferster & LeVay 1978) that conduct action potentials slowly (Wilson & Stone 1975, Cleland et al 1975b, 1976, Wilson et al 1976).

LeVay & Gilbert (1976) labeled the M and B laminae with radioactive proline and found projections to areas 17, 18, and 19 and to several areas of the lateral suprasylvian complex. Cells of the B laminae project to 17 and 18, for they can be antidromically activated by electrical stimulation of 17 and 18 and are labeled after HRP injections of either area (LeVay & Ferster 1977).

In area 17 the projection from the M and B laminae splits into three main bands, the lower two of which may receive the X and Y inputs from the M lamina, as discussed above. The fibers projecting to the upper part of layer I are very fine (Ferster & LeVay 1978) and it thus seems plausible that this is where fibers of the B laminae terminate. In area 18 and the suprasylvian cortex the label from the B laminae was confined to layers I and IV.

Area 19 receives its geniculocortical input entirely from very fine fibers terminating in layers IV, V and VI (Rossignol & Colonnier 1971). It therefore seems probable that area 19 receives its geniculate input entirely from the small class 4 cells of the B laminae and possibly also from some of the small cells of the MIN (but see Hollander & Vanegas 1977).

MEDIAL INTERLAMINAR NUCLEUS (MIN)   In addition to the main retinotopic laminar projection, the LGNd of carnivores contains a cytoarchitectonically distinct zone, which receives a second retinotopic projection (cf Hayhow 1958, Kinston et al 1969, Guillery 1970, Sanderson 1971). Although an MIN is a general feature among carnivores (Sanderson 1971), a homologous region has not been described in other mammals. Retinal fibers from the two eyes terminate in separate laminae of the MIN, the ipsilateral projection has a retinotopic map of the same extent as for the A laminae (i.e. vertical meridian to periphery), but the contralateral map is like that of the B laminae in that it also includes a representation of the 'wrong' hemifield, arising from decussating fibers of the temporal retina (Sanderson 1971, Sanderson & Sherman 1971). The great majority of MIN neurons belong to Guillery's class 1, though a group of smaller cells has also been reported (LeVay & Ferster 1977).

The MIN receives almost exclusively Y retinal fibers (Mason 1975, Palmer et al 1975, Dreher & Sefton 1975, 1978), though a few X, tonic, phasic, and local-edge-detectors project there as well (Dreher & Sefton 1978). Receptive fields of MIN Y cells are larger than those of retinal Y cells, indicating some degree of excitatory summation of retinal inputs. The few other cell types present are found primarily in that part of the MIN that borders the B laminae (Dreher & Sefton 1978), a region with small cells, some of which contain laminar bodies, that project only to area 17 (LeVay & Ferster 1977).

As noted above, all retinal Y fibers terminate in the A lamLGNd (Gilbert & Kelly 1975). Thus the Y retinal fibers that project to the superior colliculus must branch to do so, and electrophysiological studies confirm this (Hoffmann 1973, Singer & Bedworth 1973, Fukuda & Stone 1974). Most of the MIN Y cells can also be activated orthodromically from the superior colliculus, indicating that many retinal Y fibers must terminate in at least three distinct regions.

The large class 1 cells of the MIN project to area 18 (Burrows & Hayhow 1971, Rosenquist et al 1974, Maciewicz 1975, LeVay & Ferster 1977) and to the lateral suprasylvian (Clare Bishop) area (Maciewicz 1974, Rosenquist et al 1974). The question of an MIN projection to area 19 is in dispute. As noted above, area 19 is reported to receive only very fine fibers from the LGNd; consistent with this, Gilbert & Kelly (1975) did not observe a projection to 19 from MIN, but others report one (e.g. Höllander & Vanegas 1977).

COMPARISON WITH MONKEY LGNd   The primate LGNd has X-like cells in the parvocellular layers and Y-like cells in the magnocellular layers (Dreher et al 1976, Sherman et al 1976). X-like and Y-like LGNd cells also project along slow and fast pathways (Marrocco & Brown 1975) to distinct and nonoverlapping layers of visual cortex (Hubel & Wiesel 1972). There are thus a number of similarities between the cat A laminae and the main layers of the monkey LGNd, with the monkey showing a greater degree of anatomical segregation.

The monkey LGNd also has a sparse group of cells ventral to the main laminae, that receives a retinal input (e.g. Kaas et al 1972). There is evidence that these cells project to the striate cortex (Ogren & Hendrickson 1976).

Pulvinar/Lateral-Posterior Complex

In the cat this is an extensive and important visual area lying medial to the LGNd, which receives fibers from, and projects to, a variety of visual areas, cortical and otherwise. The terminal fields of its various inputs are topographically organized and, as Updyke (1977) has stressed, this provides better and clearer grounds for subdividing this region than the earlier divisions and nomenclatures, which were based on cytoarchitectonics.

Berman & Jones (1977) have recently used the autoradiographic technique to demonstrate that this region receives a direct bilateral retinal projection in the form of a thin interrupted sheet, which is continuous at its caudal end with the MIN. In macaque monkeys a similar retino-pulvinar projection to the lateral edge of the inferior pulvinar has been reported (Campos-Ortega et al 1972), but not confirmed (Trojanowski & Jacobson 1975).

CONCLUSION

There are a variety of ganglion cell types that project via different pathways to a number of sites in the brain. As far as these pathways have been traced there is no evidence for the mixing of excitatory signals from cells of different types, rather the pathways are in parallel. Cells with receptive-field organization other than the antagonistic center-surround form are sparse in the cat and must thus serve roles other than that of coding visual detail, but what these roles are remains a mystery.

Why there should be four on/off pairs of center-surround types is also a mystery. X and Y cells are somewhat similar in a number of ways, including their projection to the striate area; there are corresponding similarities for tonic and phasic cells. However, in terms of receptive-field properties and retinal distribution, Y and phasic cells have resemblances as do X and tonic cells. One wonders whether X and Y cells may have developed phylogenetically from the older tonic and phasic types, but there is no evidence that has much bearing on this notion.

Perhaps the simplest visual pathway is that of the X cells; except for the minor input to the pretectum, they appear to project straight through the LGNd to area 17. Removal of all of areas 17 and 18 essentially removes the X pathway, yet such a lesion causes only modest decrements in the visual discrimination of cats (Sprague et al 1977). Apparently the X pathway and its cortical region are not essential for the cat to have reasonable vision. It is possible that area 17 and the X pathway are specialized mainly for fine acuity (rather than an essential stage in the processing of visual form), requiring a large cortical representation for that reason alone. There are many tonic and phasic ganglion cells, which project to other visual areas, either via the B laminae of the LGNd or via the pathway through the SC to the pulvinar/lateral-posterior complex and then to the visual cortex. It is possible that the behavioral decrement observed on removal of areas 17 and 18 can be accounted for mainly by the difference in peak spatial density of X cells compared to that of tonic cells.

Literature Cited

Alley, K., Baker, R., Simpson, J. I. 1975. Afferents to the vestibulo-cerebellum and the origin of the visual climbing fibers in the rabbit. *Brain Res.* 98: 582–89

Astruc, J. 1971. Corticofugal connections of area 8 (frontal eye field) in *Macaca mulatta. Brain Res.* 33:241–56

Barlow, H. B. 1975. Visual experience and cortical development. *Nature* 258:199–204

Barlow, H. B., Derrington, A. M., Harris, L. R., Lennie, P. 1977. The effects of remote retinal stimulation on the responses of cat retinal ganglion cells. *J. Physiol. London* 269:177–94

Barlow, H. B., Levick, W. R. 1969. Changes in the maintained discharge with adaptation level in the cat retina. *J. Physiol. London* 202:699–718

Berman, N. 1977. Connections of the pretectum in the cat. *J. Comp. Neurol.* 174:227–54

Berman, N., Jones, E. G. 1977. A retino-pulvinar projection in the cat. *Brain Res.* 134:237–48

Bishop, G. H. 1933. Fiber groups in the optic nerve. *Am. J. Physiol.* 106:460–74

Bishop, P. O., Burke, W., Davis, R. 1958. Synaptic discharge by single fiber in mammalian visual system. *Nature* 182:728–30

Boycott, B. B., Kolb, H. 1973. The connection between bipolar cells and photoreceptors in the retina of the domestic cat. *J. Comp. Neurol.* 148:91–114

Boycott, B. B., Wässle, H. 1974. The morphological types of ganglion cells of the domestic cat's retina. *J. Physiol. London* 240:397–419

Brauth, S. E., Karten, H. J. 1977. Direct accessory optic projections to the vestibulo-cerebellum: A possible channel

for oculomotor control systems. *Exp. Brain Res.* 28:73–84

Brown-Grant, K., Raisman, G. 1977. Abnormalities in reproductive function associated with the destruction of the suprachiasmatic nuclei in female rats. *Proc. R. Soc. London Ser. B* 198:279–96

Bunt, A. H., Lund, R. D., Lund, J. S. 1974. Retrograde axonal transport of horseradish peroxidase by ganglion cells of the albino rat retina. *Brain Res.* 73: 215–28

Bunt, A. H., Minckler, D. S., Johanson, G. W. 1977. Demonstration of bilateral projection of the central retina of the monkey with horseradish peroxidase neuronography. *J. Comp. Neurol.* 171:619–30

Burke, W., Cole, A. M. 1978. Extraretinal influences on the lateral geniculate nucleus. *Rev. Physiol. Biochem. Pharmacol.* 80:106–66

Burrows, G. R., Hayhow, W. R. 1971. The organization of the thalamo-cortical visual pathways in the cat. *Brain Behav. Evol.* 4:220–72

Buttner, V., Fuchs, A. F. 1973. Influence of saccadic eye movements on unit activity in simian lateral geniculate and pregeniculate nuclei. *J. Neurophysiol.* 36:127–41

Caldwell, J. H., Daw, N. W. 1978a. New properties of rabbit retinal ganglion cells. *J. Physiol. London* 276:257–76

Caldwell, J. H., Daw, N. W. 1978b. Effects of picrotoxin and strychnine on rabbit retinal ganglion cells: Changes in centre surround receptive fields. *J. Physiol. London* 276:299–310

Caldwell, J. H., Daw, N. W., Wyatt, H. J. 1978. Effects of picrotoxin and strychnine on rabbit retinal ganglion cells: Lateral interactions for cells with more complex receptive fields. *J. Physiol. London* 276:288–98

Campos-Ortega, J. A., Hayhow, W. R., Clüver, R. F. de V. 1972. A note on the problem of retinal projections to the inferior pulvinar of primates. *Brain Res.* 22:126–30

Clarke, R. J., Ikeda, H. 1978. Properties of neurones in the nucleus of the optic tract of the pretectum in the cat. *J. Physiol. London* 275:46–47P

Cleland, B. G., Dubin, M. W., Levick, W. R. 1971. Sustained and transient neurones in the cat's retina and lateral geniculate nucleus. *J. Physiol. London* 217:473–96

Cleland, B. G., Levick, W. R. 1974a. Brisk and sluggish concentrically organized ganglion cells in the cat's retina. *J. Physiol. London* 240:421–56

Cleland, B. G., Levick, W. R. 1974b. Properties of rarely encountered types of ganglion cells in the cat's retina and an overall classification. *J. Physiol. London* 240:457–92

Cleland, B. G., Levick, W. R., Morstyn, R., Wagner, H. G. 1976. Lateral geniculate relay of slowly conducting retinal afferents to cat visual cortex. *J. Physiol. London* 255:299–320

Cleland, B. G., Levick, W. R., Sanderson, K. J. 1973. Properties of sustained and transient ganglion cells in the cat retina. *J. Physiol. London* 228:649–80

Cleland, B. G., Levick, W. R., Wässle, H. 1975a. Physiological identification of a morphological class of cat retinal ganglion cells. *J. Physiol. London* 248:151–71

Cleland, B. G., Morstyn, R., Wagner, H. G., Levick, W. R. 1975b. Long-latency retinal input to lateral geniculate neurones of the cat. *Brain Res.* 91:306–10

Collewijn, H. 1974. Oculomotor areas in the rabbit's brain stem. *Brain Res.* 66:362–63

Collewijn, H. 1975a. Direction-selective units in the rabbit's nucleus of the optic tract. *Brain Res.* 100:489–508

Collewijn, H. 1975b. Oculomotor areas in the rabbit's midbrain and pretectum. *J. Neurobiol.* 6:3–22

Creutzfeldt, O. D. 1968. Functional synaptic organization in the lateral geniculate body and its implication for information transmission. In *Structure and Function of Inhibitory Neuronal Mechanisms,* ed. C. von Euler, S. Skoglund, V. Söderberg, pp. 117–22. Oxford: Pergamon. 563 pp.

Daw, N. W., Pearlman, A. L. 1970. Cat color vision: Evidence for more than one cone process. *J. Physiol. London* 211:125–37

DeMonasterio, F. M. 1978. Receptive-field properties of opponent-color X and Y ganglion cells of macaque retina. *Invest. Ophthalmol. Vis. Sci.* 17: Supplement ARVO Abstracts p130

DeMonasterio, F. M., Gouras, P. 1975. Functional properties of ganglion cells of the rhesus monkey retina. *J. Physiol. London* 251:167–95

DeMonasterio, F. M., Gouras, P., Tolhurst, D. J. 1975. Trichromatic colour opponency in ganglion cells of the rhesus monkey retina. *J. Physiol. London* 251:197–216

Dreher, B., Fukada, Y., Rodieck, R. W. 1976. Identification, classification and anatomical segregation of cells with X-like and Y-like properties in the lateral geniculate nucleus of old-world primates. *J. Physiol. London* 258:433–52

Dreher, B., Sefton, A. J. 1975. Receptive field properties of cells in cat's medial interlaminar nucleus (MIN). *Proc. Aust. Physiol. Pharmacol. Soc.* 6:209

Dreher, B., Sefton, A. J. 1978. Properties of neurones in cat's dorsal lateral geniculate nucleus: a comparison between medial interlaminar and laminated parts of the nucleus. *J. Comp. Neurol.* In press

Dubin, M. W., Turner, L. 1977. Anatomy of the retina of the mink (*Mustela vison*). *J. Comp. Neurol.* 173:275–88

Dunn, J. D., Castro, A. J., McNulty, J. A. 1977. Effect of suprachiasmatic ablation on the daily temperature rhythm. *Neurosci. Lett.* 6:345–48

Ebbesson, S. O. E. 1970. On the organization of central visual pathways in vertebrates. *Brain Behav. Evol.* 3:178–94

Ebbesson, S. O. E. 1972. A proposal for a common nomenclature for some optic nuclei in vertebrates and the evidence for a common origin of two such cell groups. *Brain Behav. Evol.* 6:75–91

Edwards, S. B., Ginsburgh, C. L., Henkel, C. K. 1978. Sources of subcortical projections to the cat superior colliculus. *Anat. Rec.* 190:388

Edwards, S. B., Rosenquist, A. C., Palmer, L. A. 1974. An autoradiographic study of ventral lateral geniculate projections in the cat. *Brain Res.* 72:282–87

Enroth-Cugell, C., Robson, J. G. 1966. The contrast sensitivity of retinal ganglion cells of the cat. *J. Physiol. London* 187:517–52

Famiglietti, E. V. Jr. 1975. Another look at lateral geniculate lamination in the cat. *Neurosci. Abstr.* 1:41

Famiglietti, E. V. Jr., Kaneko, A., Tachibana, M. 1977. Neuronal architecture of on and off pathways to ganglion cells in carp retina. *Science* 198: 1267–69

Famiglietti, E. V. Jr., Kolb, H. 1976. Structural basis for on- and off-center responses in retinal ganglion cells. *Science* 194:193–95

Ferster, D., LeVay, S. 1978. The axonal arborizations of lateral geniculate neurons in the striate cortex of the cat. *J. Comp. Neurol.* In press

Fischer, B., Kruger, J., Droll, W. 1975. Quantitative aspects of the shift-effect in cat retinal ganglion cells. *Brain Res.* 83:391–403

Fukada, Y., Saito, H. 1972. Phasic and tonic cells in the cat's lateral geniculate nucleus. *Tohoku J. Exp. Med.* 106:209–10

Fukuda, Y. 1977. A three-group classification of rat retinal ganglion cells; histological and physiological studies. *Brain Res.* 119:327–44

Fukuda, Y., Stone, J. 1974. Retinal distribution and central projections of Y-, X-, and W-cells of the cat's retina. *J. Neurophysiol.* 37:749–72

Garey, L. J., Jones, E. G., Powell, T. P. S. 1968. Interrelationship of striate and extrastriate cortex with the primary relay sites of the visual pathway. *J. Neurol. Neurosurg. Psychiatry* 31:135–57

Ghelarducci, B., Ito, M., Yagi, N. 1975. Impulse discharges from flocculus Purkinje cells of alert rabbits during visual stimulation combined with horizontal head rotation. *Brain Res.* 87:66–72

Gilbert, C. D. 1977. Laminar differences in receptive field properties of cells in cat primary visual cortex. *J. Physiol. London* 268:391–421

Gilbert, C. D., Kelly, J. P. 1975. The projections of cells in different layers of the cat's visual cortex. *J. Comp. Neurol.* 163:81–106

Gonshor, A., Melvill Jones, G. 1976a. Short term adaptive changes in the human vestibulo-ocular reflex arc. *J. Physiol. London* 256:361–79

Gonshor, A., Melvill Jones, G. 1976b. Extreme vestibulo-ocular adaptation induced by prolonged optical reversal of vision. *J. Physiol. London* 256:381–414

Gordon, J., Shapley, R. 1978. Quantitative analysis of spatial summation in the frog retina. *Invest. Ophthalmol. Vis. Sci.* 17: Supplement ARVO Abstracts p128

Gouras, P. 1968. Identification of cone mechanisms in monkey ganglion cells. *J. Physiol. London* 199:533–47

Gouras, P. 1969. Antidromic responses of orthodromically identified ganglion cells in monkey retina. *J. Physiol. London* 204:407–19

Graham, J. 1977. An autoradiographic study of the efferent connections of the superior colliculus in the cat. *J. Comp. Neurol.* 173:629–54

Graybiel, A. M. 1972. Some extrageniculate pathways in the cat. *Invest. Ophthalmol.* 11:322–32

Graybiel, A. M. 1974a. Some efferents of the pretectal region of the cat. *Anat. Rec.* 178:365

Graybiel, A. M. 1974b. Visuo-cerebellar and cerebello-visual connections involving the ventral lateral geniculate nucleus. *Exp. Brain Res.* 20:303–6

Graybiel, A. M. 1975. Anatomical organization of retinotectal afferents in the cat: An autoradiographic study. *Brain Res.* 96:1–23

Graybiel, A. M. 1976. Evidence for banding of the cat's ipsilateral retinotectal connection. *Brain Res.* 114:318–27

Graybiel, A. M. 1978a. Organization of the nigrotectal connection: An experimental tracer study in the cat. *Brain Res.* 143:339–48

Graybiel, A. M. 1978b. A stereometric pattern of distribution of acetylthiocholinesterase in the deep layers of the superior colliculus. *Nature* 272:539–41

Graybiel, A. M. 1978c. A satellite system of the superior colliculus: The parabigeminal nucleus and its projections to the superficial collicular layers. *Brain Res.* 145:365–74

Graybiel, A. M., Nauta, W. J. H. 1971. Some projections of superior colliculus and visual cortex upon the posterior thalamus in the cat. *Anat. Rec.* 169:328

Guillery, R. W. 1966. A study of golgi preparations from the dorsal lateral geniculate nucleus of the adult cat. *J. Comp. Neurol.* 128:21–50

Guillery, R. W. 1970. The laminar distribution of retinal fibers in the dorsal lateral geniculate nucleus of the cat: A new interpretation. *J. Comp. Neurol.* 138:339–68

Guillery, R. W., Oberdorfer, M. D. 1977. A study of fine and coarse retino-fugal axons terminating in the geniculate C laminae and in the medial interlaminar nucleus of the mink. *J. Comp. Neurol.* 176:515–26

Güldner, F.-H., Wolff, J. R. 1978a. Self-innervation of dendrites in the rat suprachiasmatic nucleus. *Exp. Brain Res.* 32:77–82

Güldner, F.-H., Wolff, J. R. 1978b. Retinal afferents form Gray-type-I and type-II synapses in the suprachiasmatic nucleus (rat). *Exp. Brain Res.* 32:83–89

Hale, P. T., Sefton, A. J. 1978. A comparison of the visual and electrical response properties of cells in the dorsal and ventral lateral geniculate nuclei. *Brain Res.* 153:591–95

Harting, J. K., Hall, W. C., Diamond, I. T., Martin, G. F. 1973. Anterograde degeneration study of the superior colliculus in *Tupaia glis:* Evidence for a subdivision between superficial and deep layers. *J. Comp. Neurol.* 148:361–86

Hassul, M., Daniels, P. D., Kimm, J. 1976. Effects of bilateral flocculectomy on the vestibulo-ocular reflex in the chinchilla. *Brain Res.* 118:339–43

Hayhow, W. R. 1958. The cytoarchitecture of the lateral geniculate body in the cat in relation to the distribution of crossed and uncrossed optic fibers. *J. Comp. Neurol.* 110:1–63

Hayhow, W. R. 1959. An experimental study of the accessory optic fiber system in the cat. *J. Comp. Neurol.* 113:281–313

Hayhow, W. R. 1966. The accessory optic system in the marsupial Phalanger, *Trichosurus vulpecula. J. Comp. Neurol.* 126:653–72

Hayhow, W. R., Webb, C., Jervie, A. 1960. The accessory optic fiber system in the rat. *J. Comp. Neurol.* 115:187–215

Heath, C. J., Jones, E. G. 1971. The anatomical organization of the suprasylvian gyrus of the cat. *Ergeb. Anat. Entwicklungsgesch.* 45–3:1–64

Hendrickson, A. 1973. The pregeniculate nucleus of the monkey. *Anat. Rec.* 175:341

Hendrickson, A. E., Wagoner, N., Cowan, W. M. 1972. An autoradiographic and electron microscopic study of retino-hypothalamic connections. *Z. Zellforsch. Mikrosk. Anat.* 135:1–26

Hickey, T. L., Guillery, R. W. 1974. An autoradiographic study of retinogeniculate pathways in the cat and the fox. *J. Comp. Neurol.* 156:239–54

Hochstein, S., Shapley, R. M. 1976. Quantitative analysis of retinal ganglion cell classifications. *J. Physiol. London* 262: 237–64

Hoffmann, K.-P. 1972. The retinal input to the superior colliculus in the cat. *Invest. Ophthalmol.* 11:467–73

Hoffmann, K.-P. 1973. Conduction velocity in pathways from the retina to superior colliculus in the cat: A correlation with receptive-field properties. *J. Neurophysiol.* 36:409–24

Hoffmann, K.-P., Behrend, L., Schoppmann, A. 1977. Visual responses of neurons in the nucleus of the optic tract of visually deprived cats. *Neurosci. Abstr.* 3:563

Hoffmann, K.-P., Schoppmann, A. 1975. Retinal input to direction selective cells in the nucleus tractus opticus of the cat. *Brain Res.* 99:359–66

Hoffmann, K.-P., Stone, J. 1971. Conduction velocity of afferents to cat visual cortex: A correlation with cortical receptive field properties. *Brain Res.* 32:460–66

Hoffmann, K.-P., Stone, J., Sherman, S. M. 1972. Relay of receptive-field properties in dorsal lateral geniculate nucleus of the cat. *J. Neurophysiol.* 35:518–31

Holländer, H., Sanides, D. 1976. The retinal projection to the ventral part of the lateral geniculate nucleus. An experimental study with silver-impregnation methods and axoplasmic protein tracing. *Exp. Brain Res.* 26:32–42

Holländer, H., Vanegas, H. 1977. The projection from the lateral geniculate nucleus onto the visual cortex in the cat. A quantitative study with horseradish-peroxidase. *J. Comp. Neurol.* 173: 519–36

Horel, A. J. 1968. Effects of subcortical lesions on brightness discrimination acquired by rats without visual cortex. *J. Comp. Physiol. Psychol.* 65:103–9

Hubel, D. H., Wiesel, T. N. 1961. Integrative action in the cat's lateral geniculate body. *J. Physiol. London* 155:385–98

Hubel, D. H., Wiesel, T. N. 1962. Receptive fields, binocular interaction and functional architecture in the cat's visual cortex. *J. Physiol. London* 160:106–54

Hubel, D. H., Wiesel, T. N. 1965. Receptive fields and functional architecture in two non-striate visual areas (18 and 19) of the cat. *J. Neurophysiol.* 28:229–89

Hubel, D. H., Wiesel, T. N. 1968. Receptive fields and functional architecture of monkey striate cortex. *J. Physiol. London* 195:215–43

Hubel, D. H., Wiesel, T. N. 1972. Laminar and columnar distribution of geniculo-cortical fibers in the macaque monkey. *J. Comp. Neurol.* 146:421–50

Hubel, D. H., Wiesel, T. N. 1977. Functional architecture of macaque monkey visual cortex. *Proc. R. Soc. London Ser. B.* 198:1–59

Hughes, C. P., Ater, S. B. 1977. Receptive field properties in the ventral lateral geniculate nucleus of the cat. *Brain Res.* 132:163–66

Ibuka, N., Kawamura, H. 1975. Loss of circadian rhythm in sleep-wakefulness cycle in the rat by suprachiasmatic nucleus lesions. *Brain Res.* 96:76–81

Ikeda, H., Wright, M. J. 1972. Functional organization of the periphery effect in retinal ganglion cells. *Vision Res.* 12: 1859–79

Ito, M. 1977. Neuronal events in the cerebellar flocculus associated with an adaptive modification of the vestibulo-ocular reflex of the rabbit. In *Control of Gaze by Brain Stem Neurons, Developments in Neuroscience,* ed. R. Baker, A. Berthoz, 1:391–98. Amsterdam: Elsevier. 514 pp.

Ito, M., Shiida, T., Yagi, N., Yamamoto, M. 1974. Visual influence on rabbit horizontal vestibulo-ocular reflex presumably effected via the cerebellar flocculus. *Brain Res.* 65:170–74

Jordan, H., Holländer, H. 1972. The structure of the ventral part of the lateral geniculate nucleus. A cyto- and myeloarchitectonic study in the cat. *J. Comp. Neurol.* 145:259–72

Kaas, J. H., Guillery, R. W., Allman, J. M. 1972. Some principles of organization in the dorsal lateral geniculate nucleus. *Brain Behav. Evol.* 6:253–99

Kalil, R., Worden, I. 1978. Cytoplasmic laminated bodies in the lateral geniculate nucleus of normal and dark reared cats. *J. Comp. Neurol.* 178:469–86

Kanaseki, T., Sprague, J. M. 1974. Anatomical organization of pretectal nuclei and tectal laminae in the cat. *J. Comp. Neurol.* 158:319–38

Karten, H. J., Fite, K. V., Brecha, N. 1977. Specific projection of displaced retinal ganglion cells upon the accessory optic system in the pigeon (*Columbia livia*). *Proc. Natl. Acad. Sci. USA* 74:1753–56

Kawamura, S. 1974. Topical organization of the extrageniculate visual system in the cat. *Exp. Neurol.* 45:451–61

Kawamura, S., Kobayashi, E. 1975. Identification of laminar origin of some tectothalamic fibers in the cat. *Brain Res.* 91:281–85

Kawamura, S., Sprague, J. M., Niimi, K. 1974. Corticofugal projections from the visual cortices to the thalamus, pretectum and superior colliculus in the cat. *J. Comp. Neurol.* 158:339–62

Kelly, J. P., Gilbert, C. D. 1975. The projections of different morphological types of ganglion cells in the cat retina. *J. Comp. Neurol.* 163:65–80

Kelly, J. P., Van Essen, D. C. 1974. Cell structure and function in the visual cortex of the cat. *J. Physiol. London* 238:515–47

Kinston, W. J., Vadas, M. A., Bishop, P. O. 1969. Multiple projection of the visual field to the medial portion of the dorsal lateral geniculate nucleus and the adja-

cent nuclei of the thalamus of the cat. *J. Comp. Neurol.* 136:295–316

Kirk, D. L., Cleland, B. G., Levick, W. R. 1975. Axonal conduction latencies of cat retinal ganglion cells. *J. Neurophysiol.* 38:1395–1402

Kirk, D. L., Levick, W. R., Cleland, B. G., Wässle, H. 1976a. Crossed and uncrossed representation of the visual field by brisk-sustained and brisk-transient cat retinal ganglion cells. *Vision Res.* 16:225–31

Kirk, D. L., Levick, W. R., Cleland, B. G. 1976b. The crossed or uncrossed destination of axons of sluggish-concentric and non-concentric cat retinal ganglion cells, with an overall synthesis of the visual field representation. *Vision Res.* 16:233–36

Kolb, H., Famiglietti, E. V. Jr. 1974. Rod and cone pathways in the inner plexiform layer of cat retina. *Science* 186:47–49

Kostović, I. 1971. The terminal distribution of accessory optic fibers in the rat. *Brain Res.* 31:202–6

Kuffler, S. 1953. Discharge patterns and functional organization of mammalian retina. *J. Neurophysiol.* 16:37–68

Künzle, H., Akert, K. 1977. Efferent connections of cortical, area 8 (frontal eye field) in *Macaca fascicularis*. A reinvestigation using the autoradiographic technique. *J. Comp. Neurol.* 173:147–64

Laties, A. M., Sprague, J. M. 1966. The projection of optic fibers to the visual centers in the cat. *J. Comp. Neurol.* 12:35–70

Legg, C. R. 1975. Effects of subcortical lesions on the pupillary light reflex in the rat. *Neuropsychologia* 13:373–76

Legg, C. R., Cowey, A. 1977. The role of the ventral lateral geniculate nucleus and posterior thalamus in intensity discrimination in rats. *Brain Res.* 123:261–73

LeVay, S., Ferster, D. 1977. Relay cell classes in the lateral geniculate nucleus of the cat and the effects of visual deprivation. *J. Comp. Neurol.* 172:563–84

LeVay, S., Gilbert, C. D. 1976. Laminar patterns of geniculocortical projection in the cat. *Brain Res.* 113:1–19

LeVay, S., Stryker, M. P., Shatz, C. J. 1978. Ocular dominance columns and their development in layer IV of the cat's visual cortex: A quantitative study. *J. Comp. Neurol.* 179:223–44

Levick, W. R. 1967. Receptive fields and trigger features of ganglion cells in the visual streak of the rabbit's retina. *J. Physiol. London* 188:285–307

Levick, W. R., Cleland, B. G., Dubin, M. W.

1972. Lateral geniculate neurons of cat: Retinal inputs and physiology. *Invest Ophthalmol.* 11:302–11

Lincoln, D. W., Church, J., Mason, C. A. 1975. Electrophysiological activation of suprachiasmatic neurones by changes in retinal illumination. *Acta Endocrinol. (Kbh.) Suppl.* 199:184

Lisberger, S. G., Fuchs, A. F. 1974. Response of flocculus Purkinje cells to adequate vestibular stimulation in the alert monkey: Fixation *vs.* compensatory eye movements. *Brain Res.* 69:347–53

Lund, J. S., Henry, G. H., MacQueen, C. L., Harvey, A. R. 1978. Anatomical organization of the primary visual cortex (area 17) of the cat. A comparison with area 17 of the macaque monkey. *J. Comp. Neurol.* (Submitted for publication)

Lund, R. D., Webster, K. E. 1967a. Thalamic afferents from the dorsal column nuclei. An experimental anatomical study in the rat. *J. Comp. Neurol.* 130:301–12

Lund, R. D., Webster, K. E. 1967b. Thalamic afferents from the spinal cord and trigeminal nuclei. *J. Comp. Neurol.* 130:313–28

Maciewicz, R. J. 1974. Afferents to the lateral suprasylvian gyrus of the cat traced with horseradish peroxidase. *Brain Res.* 78:139–43

Maciewicz, R. J. 1975. Thalamic afferents to areas 17, 18 and 19 of cat cortex traced with horseradish peroxidase. *Brain Res.* 84:308–12

Maekawa, K., Simpson, J. I. 1972. Climbing fiber activation of Purkinje cells in the flocculus by impulses transferred through the visual pathway. *Brain Res.* 39:245–51

Maekawa, K., Simpson, J. I. 1973. Climbing fiber responses evoked in vestibulocerebellum of rabbit from visual system. *J. Neurophysiol.* 36:649–66

Maekawa, K., Takeda, T. 1975. Mossy fiber responses evoked in the cerebellar flocculus of rabbits by stimulation of the optic pathway. *Brain Res.* 98:590–95

Maekawa, K., Takeda, T. 1976. Electrophysiological identification of the climbing and mossy fiber pathways from the rabbit's retina to the contralateral cerebellar flocculus. *Brain Res.* 109:169–74

Magalhães-Castro, H. H., Murata, L. A., Magalhães-Castro, B. 1976. Cat retinal ganglion cells projecting to the superior colliculus as shown by the horseradish peroxidase method. *Exp. Brain Res.* 25:541–49

Magalhães-Castro, H. H., Saraiva, P. E. S., Magalhães-Castro, B. 1975. Identification of corticotectal cells of the visual cortex of cats by means of horseradish peroxidase. *Brain Res.* 83:474–79

Magnin, M., Fuchs, A. F. 1977. Discharge properties of neurons in the monkey thalamus tested with angular acceleration, eye movement and visual stimuli. *Exp. Brain Res.* 28:293–99

Malpeli, J. G., Schiller, P. H. 1978. Lack of blue off-center cells in the visual system of the monkey. *Brain Res.* 141:385–89

Marrocco, R. T., Brown, J. B. 1975. Correlation of receptive field properties of monkey LGN cells with the conduction velocity of retinal afferent input. *Brain Res.* 92:137–44

Mason, C. A., Lincoln, D. W. 1976. Visualization of the retino-hypothalamic projection in the rat by cobalt precipitation. *Cell Tissue Res.* 168:117–31

Mason, C. A., Sparrow, N., Lincoln, D. W. 1977. Structural features of the retinohypothalamic projection in the rat during normal development. *Brain Res.* 132:141–48

Mason, R. 1975. Cell properties in the medial interlaminar nucleus of the cat's lateral geniculate nucleus in relation to the transient/sustained classification. *Exp. Brain Res.* 22:327–29

Mason, R. 1976. Functional organization in the cat's dorsal lateral geniculate complex. *J. Physiol. London* 258:66P–67P

Mathers, L. H., Mascetti, G. G. 1975. Electrophysiological and morphological properties of neurons in the ventral lateral geniculate nucleus of the rabbit. *Exp. Neurol.* 46:506–20

McIlwain, J. T. 1973a. Topographic relationships in projection from striate cortex to superior colliculus of the cat. *J. Neurophysiol.* 36:690–701

McIlwain, J. T. 1973b. Retinotopic fidelity of striate cortex-superior colliculus interactions in the cat. *J. Neurophysiol.* 36:702–710

McIlwain, J. T. 1977. Topographic organization and convergence in corticotectal projections from areas 17, 18 and 19 in the cat. *J. Neurophysiol.* 40:189–98

Melvill Jones, G. 1977. Plasticity in the adult vestibulo-ocular reflex arc. *Philos. Trans. R. Soc. London Ser. B* 278: 319–34

Melvill Jones, G., Davies, P. 1976. Adaptation of cat vestibulo-ocular reflex to 200 days of optically reversed vision. *Brain Res.* 103:551–54

Michael, C. R. 1968. Receptive fields of single optic nerve fibers in a mammal with an all-cone retina. III. Opponent color units. *J. Neurophysiol.* 31:268–82

Miles, F. A., Fuller, J. H. 1974. Adaptive plasticity in the vestibulo-ocular responses of the rhesus monkey. *Brain Res.* 80:512–16

Miles, F. A., Fuller, J. H. 1975. Visual tracking and the primate flocculus. *Science* 189:1000–2

Millhouse, O. E. 1977. Optic chiasm collaterals afferent to the suprachiasmatic nucleus. *Brain Res.* 137:351–55

Mitzdorf, U., Singer., W. 1977. Laminar segregation of afferents to lateral geniculate nucleus of the cat: An analysis of current source destiny. *J. Neurophysiol.* 40:1227–44

Mizuno, N., Mochizuki, K., Akimoto, C., Matsushima, R. 1973. Pretectal projections to the inferior olive in the rabbit. *Exp. Neurol.* 39:498–506

Mizuno, N., Nakamura, Y., Iwahori, N. 1974. An electron microscope study of the dorsal cap of the inferior olive in the rabbit, with special reference to the pretecto-olivary fibers. *Brain Res.* 77:385–95

Montero, V. M., Brugge, J. F., Beitel, R. E. 1968. Relation of the visual field to the lateral geniculate body of the albino rat. *J. Neurophysiol.* 31:221–36

Moore, R. Y. 1973. Retinohypothalamic projection in mammals: A comparative study. *Brain Res.* 49:403–9

Moore, R. Y. 1978. Central neural control of circadian rhythms. *Front. Neuroendocrinol.* 5:185–206

Moore, R. Y., Eichler, V. B. 1972. Loss of a circadian adrenal corticosterone rhythm following suprachiasmatic lesions in the rat. *Brain Res.* 42:201–6

Moore, R. Y., Klein, D. C. 1974. Visual pathways and the central neural control of a circadian rhythm in pineal serotonin N-acetyltransferase activity. *Brain Res.* 71:17–33

Moore, R. Y., Lenn, N. J. 1972. A retinohypothalamic projection in the rat. *J. Comp. Neurol.* 146:1–14

Nagai, K., Nishio, T., Nakagawa, H., Nakamura, S., Fukuda, Y. 1978. Effect of bilateral lesions of the suprachiasmatic nuclei on the circadian rhythm of food-intake. *Brain Res.* 142:384–89

Nelson, R., Famiglietti, E. V. Jr., Kolb, H. 1978. Intracellular staining reveals different levels of stratification for on- and off-center ganglion cells in cat retina. *J. Neurophysiol.* 41:472–83

Niimi, K., Kanaseki, T., Takimoto, T. 1963. The comparative anatomy of the ventral nucleus of the lateral geniculate body

in mammals. *J. Comp. Neurol.* 121:313–23

Niimi, K., Miki, M., Kawamura, S. 1970. Ascending projections of the superior colliculus in the cat. *Okajimas Folia Anat. Jpn.* 47:269–87

Ogren, M., Hendrickson, A. 1976. Pathways between striate cortex and subcortical regions in *Macaca mulatta* and *Saimiri sciureus:* Evidence for a reciprocal pulvinar connection. *Exp. Neurol.* 53:780–800

O'Leary, J. L. 1941. Structure of the area striata of the cat. *J. Comp. Neurol.* 75:131–64

Oyster, C. W. 1968. The analysis of image motion by the rabbit retina. *J. Physiol. London* 199:613–35

Palmer, L. A., Rosenquist, A. C. 1974. Visual receptive fields of single striate units projecting to the superior colliculus in the cat. *Brain Res.* 67:27–42

Palmer, L. A., Rosenquist, A. C., Tusa, R. 1975. Visual receptive fields in the lam LGNd, MIN and PN of the cat. *Neurosci. Abstr.* 1:54

Pearlman, A. L., Daw, N. W. 1970. Opponent color cells in the cat lateral geniculate nucleus. *Science* 167:84–86

Pettigrew, J. D. 1978. Binocular visual processing in the owl's telencephalon. *Proc. R. Soc. London Ser. B* In press

Polyak, S. 1957. *The Vertebrate Visual System.* Chicago: Univ. Chicago Press. 1390 pp.

Putkonen, P. T. S., Magnin, M., Jeannerod, M. 1973. Directional responses to head rotation in neurons from the ventral nucleus of the lateral geniculate body. *Brain Res.* 61:407–11

Raisman, G., Brown-Grant, K. 1977. The 'suprachiasmatic syndrome': endocrine and behavioural abnormalities following lesions of the suprachiasmatic nuclei in the female rat. *Proc. R. Soc. London Ser. B* 198:297–314

Ribak, C. E., Peters, A. 1975. An autoradiographic study of the projections from the lateral geniculate body of the rat. *Brain Res.* 92:341–68

Rioch, D. M. 1929. Studies on the diencephalon of carnivora. I. The nuclear configuration of the thalamus, epithalamus and hypothalamus of the dog and cat. *J. Comp. Neurol.* 49:1–119

Riss, W., Jakway, J. S. 1970. A perspective on the fundamental retinal projections of vertebrates. *Brain Behav. Evol.* 3:30–55

Robinson, D. A. 1976. Adaptive gain control of vestibuloocular reflex by the cerebellum. *J. Neurophysiol.* 39:954–69

Rodieck, R. W. 1967. Receptive fields in the cat retina: A new type. *Science* 157: 90–92

Rodieck, R. W. 1973. Inhibition in cat lateral geniculate nucleus, role of Y system. *Proc. Aust. Physiol. Pharmacol. Soc.* 4:167

Rodieck, R. W., Dreher, B. 1978. Visual suppression from nondominant eye in the lateral geniculate nucleus: A comparison of cat and monkey. *Exp. Brain Res.* In press

Rosenquist, A. C., Edwards, S. B., Palmer, L. A. 1974. An autoradiographic study of the projections of the dorsal lateral geniculate nucleus and the posterior nucleus in the cat. *Brain Res.* 80:71–93

Rossignol, S., Colonnier, M. 1971. A light microscope study of degeneration patterns in cat cortex after lesions of the lateral geniculate nucleus. *Vision Res.* Suppl. 3:329–38

Rowe, M. H., Stone, J. 1976a. Properties of ganglion cells in the visual streak of the cat's retina. *J. Comp. Neurol.* 169:99–126

Rowe, M. H., Stone, J. 1976b. Conduction velocity groupings among axons of cat retinal ganglion cells, and their relationship to retinal topography. *Exp. Brain Res.* 25:339–57

Rowe, M. H., Stone, J. 1977. Naming of neurones. Classification and naming of cat retinal ganglion cells. *Brain Behav. Evol.* 14:185–216

Rowe, M. H., Tancred, E., Freeman, B., Stone, J. 1976. Properties of ganglion cells in the retina of the brush-tailed possum, *Trichosurus vulpecula*. *Neurosci. Abstr.* 2:1089

Saito, H., Shimahara, T., Fukuda, Y. 1971. Phasic and tonic responses in the cat optic nerve fibers—stimulus-response relation. *Tohoku J. Exp. Med.* 104: 313–23

Sanderson, K. J. 1971. The projection of the visual field to the lateral geniculate and medial interlaminar nuclei in the cat. *J. Comp. Neurol.* 143:101–18

Sanderson, K. J. 1974. Lamination of the dorsal lateral geniculate nucleus in carnivores of the weasel (*Mustelidae*), raccoon (*Procyonidae*) and fox (*Canidae*) families. *J. Comp. Neurol.* 153:239–66

Sanderson, K. J., Bishop, P. O., Darian-Smith, I. 1971. The properties of the binocular receptive fields of lateral geniculate neurons. *Exp. Brain Res.* 13:178–207

Sanderson, K. J., Sherman, S. M. 1971. Nasotemporal overlap in visual field projected to lateral geniculate nucleus in the cat. *J. Neurophysiol.* 34:453–66

Sawaki, Y. 1978. Retinohypothalamic projection: Electrophysiological evidence for the existence in female rats. *Brain Res.* 120:336–41

Scalia, F. 1972. The termination of retinal axons in the pretectal region of mammals. *J. Comp. Neurol.* 145:223–58

Scalia, F., Arango, V. 1977. The map of the retina in the pretectal complex. *Anat. Rec.* 187:778

Schiller, P. H., Malpeli, J. G. 1977. Properties and tectal projections of monkey retinal ganglion cells. *J. Neurophysiol.* 40:428–45

Schwartz, W. J., Gainer, H. 1977. Suprachiasmatic nucleus: Use of ^{14}C-labeled deoxyglucose uptake as a functional marker. *Science* 197:1089–91

Shapley, R. M., Gordon, J. 1978. The eel retina: Ganglion cell classes and spatial mechanisms. *J. Gen. Physiol.* 71:139–55

Shatz, C. J., Lindström, S., Wiesel, T. N. 1977. The distribution of afferents representing the right and left eyes in the cat's visual cortex. *Brain Res.* 131:103–6

Sherk, H. 1978. Visual response properties and visual field topography in the cat's parabigeminal nucleus. *Brain Res.* 145:375–79

Sherman, S. M., Norton, T. T., Casagrande, V. A. 1975. X- and Y-cells in the dorsal lateral geniculate nucleus of the tree shrew (*Tupaia glis*). *Brain Res.* 93:152–57

Sherman, S. M., Wilson, J. R., Kass, J. H., Webb, S. V. 1976. X- and Y-cells in the dorsal lateral geniculate nucleus of the owl monkey (*Aotus trivirgatus*). *Science* 192:475–77

Simpson, J. I., Alley, K. 1974. Visual climbing fiber input to the rabbit vestibulocerebellum: a source of direction specific information. *Brain Res.* 82: 302–8

Simpson, J. I., Hess, R. 1977. Complex and simple visual messages in the flocculus. In *Control of Gaze by Brain Stem Neurons, Developments in Neuroscience* ed. R. Baker, A. Berthoz, 1:351–60. Amsterdam: Elsevier. 514 pp.

Singer, W. 1970. Inhibitory binocular interaction in the lateral geniculate body of the cat. *Brain Res.* 18:165–70

Singer, W. 1977. Control of thalamic transmission by corticofugal and ascending reticular pathways in the visual system. *Physiol. Rev.* 57:386–420

Singer, W., Bedworth, N. 1973. Inhibitory interaction between X and Y units in the cat lateral geniculate nucleus. *Brain Res.* 49:291–307

Spear, P. D., Smith, D. C., Williams, L. L. 1977. Visual receptive-field properties of single neurons in cat's ventral lateral geniculate nucleus. *J. Neurophysiol.* 40: 390–409

Sprague, J. M., Berlucchi, G., Rizzolatti, G. 1973. The role of the superior colliculus and pretectum in vision and visually guided behavior. In *Handbook of Sensory Physiology, Volume VII/3 Central Processing of Visual Information Part B: Visual Centers in the Brain,* ed. R. Jung, pp. 27–101. Berlin: Springer. 738 pp.

Sprague, J. M., Levy, J., DiBerardino, A., Berlucchi, G. 1977. Visual cortical areas mediating form discrimination in the cat. *J. Comp. Neurol.* 172:441–88

Stephan, F. K., Zucker, I. 1972. Circadian rhythms in drinking behavior and locomotor activity of rats are eliminated by hypothalamic lesions. *Proc. Natl. Acad. Sci. USA* 69:1583–86

Stetson, M. H., Watson-Whitmyre, M. 1976. Nucleus suprachiasmaticus: The biological clock in the hamster? *Science* 191:197–99

Stone, J. 1978. The number and distribution of ganglion cells in the cat's retina. *J. Comp. Neurol.* 180:753–72

Stone, J., Dreher, B. 1973. Projection of X- and Y-cells of the cat's lateral geniculate nucleus to areas 17 and 18 of visual cortex. *J. Neurophysiol.* 36:551–67

Stone, J., Fukuda, Y. 1974a. Properties of cat retinal ganglion cells: A comparison of W-cells with X- and Y-cells. *J. Neurophysiol.* 37:722–48

Stone, J., Fukuda, Y. 1974b. The naso-temporal division of the cat's retina re-examined in terms of Y- , X- and W-cells. *J. Comp. Neurol.* 155:377–94

Stone, J., Hoffmann, K.-P. 1972. Very slow-conducting ganglion cells in the cat's retina: A major, new functional type? *Brain Res.* 43:610–16

Stone, J., Leicester, J., Sherman, S. M. 1973. The naso-temporal division of the monkey retina. *J. Comp. Neurol.* 150: 333–48

Swanson, L. W., Cowan, W. M. 1975. The efferent connections of the suprachiasmatic nucleus of the hypothalamus. *J. Comp. Neurol.* 160:1–12

Swanson, L. W., Cowan, W. M., Jones, E. G. 1974. An autoradiographic study of the efferent connections of the ventral lateral geniculate nucleus in the albino rat and the cat. *J. Comp. Neurol.* 156: 143–64

Takeda, T., Maekawa, K. 1976. The origin of the pretecto-olivary tract. A study using the horseradish peroxidase method. *Brain Res.* 117:319–25

Trojanowski, J. Q., Jacobson, S. 1975. Peroxidase labeled subcortical afferents to pulvinar in rhesus monkey. *Brain Res.* 97:144–50

Tuttle, J. R., Scott, L. C. 1978. X-like and Y-like ganglion cells in the *Necturus* retina. *Invest. Ophthalmol. Vis. Sci.* 17: Supplement ARVO Abstracts p128

Tyner, C. F. 1975. The naming of neurons. Applications of taxonomic theory to the study of cellular populations. *Brain Behav. Evol.* 12:75–96

Updyke, B. V. 1977. Topographic organization of the projections from cortical areas 17, 18, and 19 onto the thalamus, pretectum and superior colliculus in the cat. *J. Comp. Neurol.* 173:81–122

Van Essen, D. C. 1979. Visual areas of the mammalian cerebral cortex. *Ann. Rev. Neurosci.* 2:227–63

Walley, R. E. 1967. Receptive fields in the accessory optic system of the rabbit. *Exp. Neurol.* 17:27–43

Wässle, H., Levick, W. R., Cleland, B. G. 1975. The distribution of the alpha type of ganglion cells in the cat's retina. *J. Comp. Neurol.* 159:419–38

Wickelgren, B. G., Sterling, P. 1969. Influence of visual cortex on receptive fields in the superior colliculus of the cat. *J. Neurophysiol.* 32:16–23

Wiesel, T. N., Hubel, D. H. 1966. Spatial and chromatic interactions in the lateral geniculate body of the rhesus monkey. *J. Neurophysiol.* 29:1115–56

Wilson, P. D., Rowe, M. H., Stone, J. 1976. Properties of relay cells in cat's lateral geniculate nucleus: a comparison of W-cells with X- and Y-cells. *J. Neurophysiol.* 39:1193–1209

Wilson, P. D., Stone, J. 1975. Evidence of W-cell input to the cat's visual cortex via the C laminae of the lateral geniculate nucleus. *Brain Res.* 92:472–78

Winfield, J. A., Hendrickson, A., Kimm, J. 1978. Anatomical evidence that the medial terminal nucleus of the accessory optic tract in mammals provides a visual mossy fiber input to the flocculus. *Brain Res.* 151:175–82

Winters, R. W., Hamasaki, D. I. 1976. Temporal characteristics of peripheral inhibition of sustained and transient ganglion cells in cat retina. *Vision Res.* 16:37–45

Wood, C. C., Spear, P. D., Braun, J. J. 1973. Direction-specific deficits in horizontal optokinetic nystagmus following removal of visual cortex in the cat. *Brain Res.* 60:231–37

Ann. Rev. Neurosci. 1979. 2:227–63

VISUAL AREAS OF THE ♦11523
MAMMALIAN CEREBRAL CORTEX

David C. Van Essen

Division of Biology, California Institute of Technology,
Pasadena, California 91125

INTRODUCTION

The cerebral neocortex is an especially intriguing part of the nervous system, partly because of its involvement in complex perceptual and cognitive functions, and partly because its relatively uniform histological structure belies the great diversity of roles it performs. The most thoroughly studied cortical regions have been the various sensory areas; of these the best understood are the visual areas, particularly the striate area, or primary visual cortex. One advantage associated with studying the visual system is the ease with which functional properties of neurons can be analyzed using a wide variety of patterned visual stimuli. Another is the fact that the main pathways leading from the retina to the cortex are arranged in a relatively straighforward fashion and have been intensively studied both anatomically and physiologically (see review by Rodieck, this volume). Consequently, in recent years it has been possible to obtain a remarkably detailed portrait of the organization and function of the striate cortex, owing to the contributions of many workers, most notably Hubel and Wiesel.

Outside the striate area lies a broad cortical zone, the extrastriate cortex of the occipital lobe. This region receives inputs from the striate cortex, is primarily visual in function, and is believed to be involved in a higher or more abstract level of analysis than that carried out by the striate cortex. Since the classical cytoarchitectural studies of Brodman (1905, 1909) it has been widely accepted that the extrastriate visual cortex of higher mammals is organized as two concentric rings, areas 18 and 19, which surround the striate cortex (area 17) and which, together with area 17, occupy most or all of the occipital lobe. Over the past decade, however, studies on a number of species have shown that higher mammals have many more than three

227

visual cortical areas; there are as many as a dozen or so identified in some species. These areas form a mosaic occupying all of the occipital lobe plus parts of temporal cortex. Moreover, it appears that the internal topographic organization of some of these visual areas is distinctly more complex than had previously been suspected.

The purpose of this review is to discuss the organization and function of the mammalian visual cortical areas including the striate cortex, but with particular emphasis on extrastriate visual areas. Part I concerns the overall plan of organization of visually responsive regions of occipital and temporal cortex. Specifically, it deals with the number of distinct visual areas found in various representative species, the manner of representation of the visual field within individual areas, and the relationship between topographically defined subdivisions and cortical architecture. Part II concerns the anatomical pathways connecting individual visual areas with various subcortical visual centers and with other cortical areas. Such information provides a basis for discussion of the hierarchical levels of visual information processing and the degree of parallel processing within the cortex. In addition, possible homologies among visual areas in different species will be considered. Part III concerns the functional properties of neurons in different visual areas and possible roles of individual areas in visual analysis and perception. Studies of cortical development and plasticity, relating especially to the innately determined organization of the striate cortex and the effects of visual experience, have progressed enormously in recent years (see Barlow 1975, Hubel & Wiesel 1977), but are not covered in this review.

PART I. THE ORGANIZATION OF VISUAL CORTICAL AREAS

Over the past century numerous methods have been devised for distinguishing discrete subdivisions of the cerebral cortex. For the visual cortex the three most useful approaches have been based, respectively, on (a) cortical architecture, which is concerned with regional variations in the histological structure of the cortex, (b) visual topography, which deals with the way the visual field is represented on the cortical surface, and (c) anatomical pathways, which reveal the major inputs and outputs of each region. Some cortical areas, most notably the striate cortex, are easy to recognize using any of these approaches, and there is precise agreement concerning the areal boundaries revealed by different techniques. In other visual cortical regions there have been varying degrees of uncertainty, ambiguity, and conflicting evidence concerning the number and location of distinct areas. If a particular region of cortex can be partitioned into one or more well-defined representations of the visual field, then it is reasonable to regard such topographically defined areas as fundamental units of cortical organization. Even detailed topographic information is not always sufficient to permit the unambiguous demarcation of visual areas, however, and in regions where visual topography is particularly complex or degraded, the determination of fundamental cortical subdivisions can best be made using a combination of criteria from several independent approaches.

The stage at which visual information is integrated with inputs from other sensory modalities varies considerably among different mammals. In primates, for example, the mixing of visual with auditory and/or somatosensory inputs apparently occurs only at an advanced level of processing outside the occipital lobe, whereas in other species, such as the cat, multimodal integration takes place gradually, starting at the primary visual area (see below). This review is restricted to cortical areas believed to be predominantly or exclusively visual in function, and not with "polysensory" regions in which the mixing of sensory modalities is more extensive.

Visual Areas I and II

The existence of multiple cortical representations of sensory surfaces is a general feature of mammalian visual, somatosensory, and auditory systems (see Woolsey 1971, Kaas 1977). In the visual system this principle is manifested by the presence in the occipital lobe of a pair of adjoining visual areas whose locations and patterns of internal organization are similar in several important respects in all species studied to date. The larger area is situated at or near the posterior pole of the hemisphere and is variously called: (a) the striate cortex, because of the existence in many species of a prominent stripe of myelinated fibers in layer IV; (b) area 17 or, less commonly, OC, from the architectonic nomenclature for the striate area (Brodmann 1905, 1909, von Economo & Koskinas 1925); (c) visual I, or V1, based on approaches that map visual topography; and (d) primary visual cortex, because it receives direct and powerful inputs from the dorsal lateral geniculate nucleus of the thalamus and is the principal cortical recipient of such inputs (except in the cat). The smaller of the two areas is situated lateral to the striate cortex in nonprimates and surrounds the striate area along its lateral and anterior margins in primates. This area is generally called visual II or V2. The use of architectonic terms such as area 18 or OB is acceptable for those species in which the second topographic representation has been shown specifically to coincide with the histologically defined zone. In some species, however, visual II and area 18 are not coextensive (see below), and the terms should not be used interchangeably. For the entire set of visually responsive regions outside the striate cortex the general term "extrastriate visual cortex" is preferable to other alternatives such as "prestriate cortex" (since these areas are not anterior to striate cortex in all species) or "circumstriate cortex" (since in general they do not completely encircle the striate cortex).

The striate cortex invariably contains a single point-to-point representation of the entire contralateral visual hemifield plus, in some species, a narrow (1–10°) strip of the ipsilateral hemifield next to the vertical meridian. The perimeter of the visual hemifield (i.e. a closed loop along the vertical meridian and around the far periphery) is represented along the border of striate cortex in a similar fashion in all mammals studied. In particular, the intersection of the horizonal and vertical meridians, which represents the center of gaze for species with frontally directed eyes, is situated laterally, and the representation of the far periphery occurs along the medial border of striate cortex. The proportion of striate cortex devoted to central parts of the visual field varies greatly among species in a manner that is closely related to the density of retinal ganglion cells at different eccentricities, but the relationship between cortical magnification factor and retinal ganglion cell density is highly nonlinear in some species (e.g. Malpeli & Baker 1975, Myerson et al 1977, Tusa et al 1978). Visual II does not encircle the striate cortex completely, but rather adjoins it only along the portion representing the vertical meridian. Along this common border the two visual maps are in register. Receptive fields in either area are farther from the vertical meridian at recording sites more distant from the visual I/II boundary, but the representation in visual II is not always a simple mirror image of that in the striate cortex. Most or all features of this basic pattern of organization for the two areas have been demonstrated for a number of species including, among nonprimates, the cat (Talbot & Marshall 1941, Talbot 1942, Hubel & Wiesel 1965, Tusa et al 1978), rabbit (Thompson et al 1950, Woolsey et al 1973), rat (Adams & Forrester 1968, Montero et al 1973b), hedgehog (Kaas et al 1970), squirrel (Hall et al 1971), tree shrew (Kaas et al 1972), sheep (Clarke & Whitteridge 1976), mouse (Dräger 1975, Wagor et al 1977), hamster (Tiao & Blakemore 1976), opossum (Sousa et al 1978), and guinea pig (Choudhury 1978). Among primates this arrangement has been seen in the macaque (Talbot & Marshall 1941, Daniel & Whitteridge 1961, Zeki 1969, Cragg 1969), squirrel monkey (Cowey 1964), owl monkey (Allman & Kaas 1971b, 1974a), and bushbaby (Allman et al 1973, see Allman 1977).

The topographic organization of visual II differs from that of visual I in several respects and, interestingly, is not identical in all mammals. In general, visual II is a smaller area containing a coarser representation of a smaller portion of the visual field than is found for visual I. The portion of the visual field whose representation is reduced in visual II is the temporal periphery, i.e. the region farthest from the vertical midline. In the hedgehog, squirrel, and tree shrew, visual I and II appear to be organized in a simple mirror image fashion, but more commonly the two representations are not topologically equivalent. In many species the representation of the horizontal meridian forms the border of visual II along the side away from visual I. Thus, the upper and lower visual quadrants in visual II are represented in separate regions that are contiguous only along a short segment near the representation of the center of gaze. This type of representation, in which neighboring loci in the visual field that are on opposite sides of the horizontal meridian are represented in quite separate loci in the cortex, has been termed a second-order transformation of the visual hemifield (Allman & Kaas 1974a). It is contrasted to a first-order transformation, in which neighboring loci in the visual hemifield invariably map onto adjacent loci in the cortex, as is the case for visual I in all species studied. A second-order transformation, with a dual, or split, representation along most of the horizontal meridian, has been found in visual II of the squirrel monkey (Cowey 1964, Spatz & Tigges 1972a), macaque (Zeki 1969, Cragg, 1969), owl monkey (Allman & Kaas 1974a), bushbaby (see Allman 1977), and mouse (Wagor et al 1977).

A different type of complexity in the topographic organization of visual II relates to the occurrence of irregularities in the normal point-to-point mapping of the visual field aside from those found along the horizontal meridian. Islands of irregular topography have been found in visual II of the cat (Donaldson & Whitteridge 1977, Tusa, Rosenquist & Palmer, in preparation; see below), and macaque (D. C. Van Essen & J. L. Bixby, unpublished observations), and occasional aberrant receptive fields have been reported in visual II of the hedgehog (Kaas et al 1970). The functional significance of this type of organization is unclear, but it may be related to the more striking complexities of visual topography found in certain other extrastriate areas to be described below.

Evolution of Multiple Visual Representations

An expansion in the amount of neocortex devoted to vision in occipital and temporal parts of the hemisphere has been a prominent development during mammalian evolution. The acquisition of high visual acuity by many species as a result of a greatly increased retinal ganglion cell density near the center of gaze is invariably associated with an increased surface area of striate cortex, mainly in the region representing central vision. For example, the striate cortex occupies a surface area of only about 20 mm^2 (25% of the neocortex) in the hedgehog, a phylogenetically primitive species with poor vision, and a comparable percentage but even less absolute surface area in small rodents with low visual acuity, such as the rat and mouse (Kaas et al 1970, Krieg 1946, Dräger 1975). In contrast, the striate cortex has an estimated mean surface area of 380 mm^2 in the cat, 1300 mm^2 (20% of the neocortex) in the macaque, and 2100 mm² (3% of the neocortex) in the human (Tusa et al 1978, Daniel & Whitteridge 1961, Stensaas et al 1974). The expansion of extrastriate visual cortex has generally outstripped that of the striate cortex: the total extent of extrastriate visual areas is much less than that of striate cortex in primitive mammals, but is much larger in more advanced species (e.g. Kaas et al 1970, Blinkov & Glezer 1968). Part of the enlargement of extrastriate visual cortex is linked to an increase in size of individual visual areas, but of greater significance is the appearance during mammalian evolution of additional representations of the visual field in occipital and temporal cortex.

In general, it appears that species with poorly developed visual systems (including certain insectivores, marsupials, and rodents) have few extrastriate visual areas, of which one is medial to, and one or more lateral to the striate cortex. More highly visual species (including certain carnivores and primates) have many more visual representations, most of which are situated lateral and/or anterior to striate cortex. There are a dozen or more identified visual areas in the cat, eight in the owl monkey, and six in the macaque. It is likely that additional areas will be discovered in cortical regions not yet thoroughly explored in these species. Because of the considerable diversity in arrangement of extrastriate visual areas, primate and nonprimate species are discussed separately, and within each group the simpler patterns of organization are considered first. Particular attention is devoted to the three highly evolved species that have been studied in greatest detail, namely the cat, owl monkey, and macaque.

PRIMITIVE NONPRIMATES   The simplest known arrangement of visual cortical areas is that of the hedgehog, an insectivore whose forebrain is regarded as one of the most primitive and unspecialized among extant mammals (see Diamond & Hall 1969). The occipital cortex of the hedgehog is dominated by visual areas I and II, which occupy most of the posterior neocortex outside known auditory and somatosensory regions (Kaas et al 1970). There is, in addition, a narrow strip of cortex that is medial to, and receives inputs from, the striate area (Gould & Ebner 1978) and a small patch of posterior temporal cortex whose function has not been determined. Thus there are two known visual cortical areas in the hedgehog, plus a probable third area, and only a restricted region in which additional specifically visual areas might be found. In the opossum, another primitive mammal, the striate cortex projects to three nearby cortical regions, one situated medially and two laterally, which suggests the presence of at least three extrastriate visual areas (Benevento & Ebner 1971). The possibility of considerable complexity in the topographic organization of the areas lateral to striate cortex is raised by the pattern of interhemispheric connections in the opossum. Complete lesions of the visual cortex in one hemisphere produce several irregular bands of degenerating terminals lateral to the striate cortex in the other hemisphere (Benevento & Ebner 1971). If interhemispheric fibers terminate preferentially along representations of the vertical meridian in the opossum's visual cortex, as is the case in other species (see below), then this pattern of connections would indicate either that there are more than two visual representations lateral to the striate cortex or that the individual visual areas must themselves have a complex internal organization. Physiological studies that might distinguish among the various possibilities have not been reported.

Evidence for multiple visual cortical representations has recently been obtained for a number of nocturnal rodents: the mouse, rat, hamster, and guinea pig. All four species apparently have a single visual representation in a region medial to the striate cortex that is approximately coextensive with Krieg's (1946) cytoarchitectonic area 18 (Montero et al 1973a, Dräger 1975, Tiao & Blakemore 1976, Choudhury 1978). This area is unlikely to be homologous to area 18 in other species, however, so that the use of another term such as Vm (Tiao & Blakemore 1976) is preferable. Visual topography in the region lateral to the striate cortex is more complex (Montero et al 1973a,b). In the mouse there are at least two lateral extrastriate visual areas, one of which (visual II) has a common border with the striate cortex and the other of which occupies a small region anterolateral to visual II (Wagor et al 1977).

ADVANCED NONPRIMATES   In mammals with more highly developed visual capacities than the species discussed above, the expansion of extrastriate visual cortex has taken place largely in the region lateral to visual I and II. In the grey squirrel, for example, visual II is adjoined along most or all of its lateral border by a single visual representation, visual III. Visual III is coextensive with area 19, as defined architectonically, and is topographically organized in an approximate mirror image fashion to that of visual II (Hall et al 1971). Additional visually responsive cortex is present in temporal cortex ventrolateral to visual III, which suggests the existence of at least a third lateral extrastriate visual area. Physiological recordings reported for the tree shrew have been less extensive than for the squirrel, but they indicate that at least one visual area exists lateral to visual II (Kaas et al 1972). In the rabbit, anatomical and physiological studies suggest that at least three, and possibly as many as six, visual areas exist in occipital and temporal cortex lateral to the striate area (Woolsey et al 1973, Montero & Murphy 1976, Towns et al 1977, Mathers et al 1977, Chow et al 1977). In none of these species has a visually responsive region medial to the striate cortex been described.

Visual areas in the cat   Studies of visual topography in the cat were the first to reveal the existence of a second visual representation in the cerebral cortex (Talbot 1942), and later of a third representation (Hubel & Wiesel 1965) plus additional visually responsive regions (Marshall et al 1943, Clare & Bishop 1954). More recent studies have demonstrated that in fact there are many additional, topographically organized visual areas lateral to visual I-III in occipital and temporal cortex (Tusa et al 1975, Palmer et al 1978). These areas differ considerably with respect to their absolute size, topographical organization, and relative emphasis on representations of particular portions of the visual field (e.g. central vs peripheral; horizontal vs vertical meridians).

Visual I in the cat, as in other mammals, contains a simple point-to-point representation of the contralateral visual hemifield. In the region lateral to the striate cortex, visual II and visual III are in register with Otsuka & Hassler's (1962) architectonically defined areas 18 and 19, respectively (Hubel & Wiesel 1965, Donaldson & Whitteridge 1977, Tusa et al, in preparation). The question of how far these areas extend around to the medial side of striate cortex has been difficult to resolve on anatomical criteria (Otsuka & Hassler 1962, Sanides & Hoffman 1969), but physiological mappings indicate that most of visual II (area 18) and all of visual III (area 19) are situated lateral to striate cortex (Tusa et al, in preparation). Visual II and visual III are topographically organized in an unusually complex, but mirror symmetric, pattern (Donaldson & Whitteridge 1977, Leventhal et al 1978, Tusa et al, in preparation). The border between visual II and III, which in most other species represents either the horizontal meridian or the temporal periphery, has features of both patterns in the cat, in that the horizontal meridian is represented posteriorly and an irregular portion of the periphery is represented anteriorly. Associated with this is the presence of "islands" of irregular topography whose number and position vary among individual cats. Although visual I and II can each be recognized on the basis of certain distinctive histological features, subtle variations in cortical architecture are visible within both areas. Sanides & Hoffman (1969) have described four architectonic subdivisions of their "parastriate" cortical zone, which is approximately coextensive with visual II. They also report seven subdivisions of their "peristriate" zone, but several of these subdivisions lie outside topographically defined visual III. This structural heterogeneity is of considerable interest in view of the topographic complexities of visual II and III, but it is difficult to establish clear correlations between topographically and architectonically defined subdivisions.

Numerous additional visual areas besides visual I-III have been described in the cat. On the medial wall of the hemisphere, in the splenial sulcus ventral to the striate cortex, there is a distinctive architectonic region, called the area prostriata by Sanides & Hoffman (1969), which has been reported to be visually responsive (Kalia & Whitteridge 1973, but see Tusa et al 1978). The position and reported topographic organization of this area suggest that it may be equivalent to the medial visual area, Vm, of more primitive mammals. Two other sets of visual areas have been described in the region lateral to visual I-III (Tusa et al 1975, Palmer et al 1978). One group, containing four partial representations of the visual hemifield, is situated within areas 20 and 21 of Heath & Jones (1971). A second group, the lateral suprasylvian visual areas, occupies the medial and lateral banks of the middle suprasylvian sulcus and the ventral and dorsal banks of the posterior suprasylvian sulcus (see Figure 1A). This region appears to contain at least six separate visual areas, although its topographic organization is too complex to permit its partitioning in an unambiguous fashion. The six areas are arranged in two mirror symmetric rows (Figure 1B). The positions of the areas in the medial-ventral row agree with architectonic

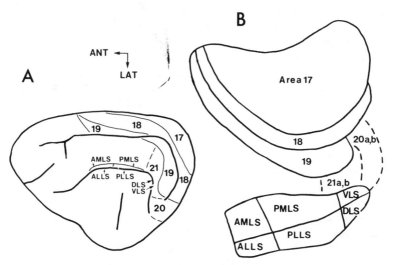

Figure 1   Visual cortical areas of the cat. *A* Positions of visual areas on the exposed dorsal and lateral surfaces of the left hemisphere. The borders of areas 20 and 21 (dotted lines) have not been established precisely, and each of these areas may contain two topographically organized subdivisions. The lateral suprasylvian region consists of at least three pairs of areas; the anteromedial and anterolateral syprasylvian areas (AMLS, ALLS), the posteromedial and posterolateral areas (PMLS, PLLS), and the ventral and dorsal areas (VLS, DLS). *B* A schematic "unfolded" map of cat visual cortical areas, showing the topological relationships among areas and also their approximate relative surface areas. Adapted, with permission, from Tusa et al 1978; Palmer et al 1978; and R. J. Tusa, A. C. Rosenquist, L. A. Palmer, in preparation.

subdivisions described by Sanides & Hoffmann (1969), but the areas in the lateral-dorsal row are not in register with known architectonic zones.

Each pair of lateral suprasylvian areas has a distinctive topographic organization. The most anterior areas, AMLS and ALLS, are organized in the simplest fashion, with each area containing a single point-to-point representation of much of the contralateral hemifield (i.e. a first-order transformation). Compared to other cortical visual areas, relatively little cortex in these areas is devoted to central vision. In contrast, the most posterior pair of areas, VLS and DLS, have an exaggerated emphasis on central parts of the visual field and also of regions near the horizontal meridian. The remaining two lateral suprasylvian areas, PMLS and PLLS, have a more complex organization than the other areas, in that individual points within the visual field are represented in two separate regions within each area. An alternative arrangement is to partition both PMLS and PLLS into separate anterior and posterior subdivisions, thereby increasing the number of identified lateral suprasylvian areas to eight. Although there is no compelling evidence for either formulation of lateral suprasylvian organization, the latter one seems preferable, as one could then distinguish between four anterior areas on the one hand, all of which share certain features such as a reduced emphasis on central vision, and, on the other hand, four posterior areas, each of which has an extreme emphasis on the center of gaze and the horizontal meridian. The lateral suprasylvian region is bounded laterally by auditory cortex, posteriorly by visual areas 20 and 21, anteriorly by polysensory cortex, and medially by a narrow strip of cortex of unknown function lying adjacent to visual III (Landgren & Silfvenius 1968, Dow & Dubner 1971, Palmer et al 1978). Thus, the limitations of available space suggest that the total number of predominantly visual areas in the cat cortex may not be much more that the fourteen or so that have already been identified.

The topographic organization of the various extrastriate visual areas in the cat is reflected to a considerable degree in the pattern of interhemispheric connections. Although detailed studies of topography and interhemispheric inputs have not been made in the same individual animals, comparisons can be made between published maps of visual topography (Tusa et al 1978 and in preparation, Palmer et al 1978) and of callosal fiber terminations (Ebner & Myers 1965, Garey et al 1968). In the region of visual areas I-III callosal fibers terminate in a strip along the visual I/II border and again along the vertical meridian representation of visual III. The two bands are separate except in a region of more diffuse connections near the center of gaze representation. In the splenial sulcus, where visual II does not adjoin the far peripheral representation in the striate cortex, the transcallosal inputs are sparse and situated slightly away from the boundary of the striate cortex. In the lateral suprasylvian region transcallosal inputs are more widespread, but they tend to be densest in regions where the vertical meridian and center of gaze are represented. Thus, interhemispheric connections provide an important, albeit imperfect, measure of visual topography in the cat.

PRIMATE VISUAL AREAS   Recent studies of visual topography in a variety of primates have revealed the existence of more than one visual representation within Brodmann's area 19 and, in some species, also within area 18. The precise arrangement of visual areas has been examined most completely for the owl monkey, a New-World primate (Allman & Kaas 1971a, b, 1974a, b, 1975, 1976); a description of this pattern provides a suitable framework for comparison with other species.

The owl monkey   As in other primates, the striate cortex in the owl monkey is situated at the posterior pole of the occipital lobe, with central visual fields represented laterally on the exposed hemispheric surface and the visual periphery represented within the calcarine sulcus (Allman & Kaas 1971b). Visual II is coextensive with Brodmann's area 18 and almost completely encircles the striate cortex, except for a short gap in visual II next to the striate cortical representation of the far periphery (Allman & Kaas 1974a). This intercalated segment has a distinctive cytoarchitecture and is presumably equivalent to the area prostriata described by Sanides (1970) in Old-World primates.

Immediately adjacent to visual II in the owl monkey is a belt of cortex containing at least five visual representations (Figure 2A, B). Three of these, the medial (M), dorsomedial (DM) and dorsolateral (DL) visual areas,

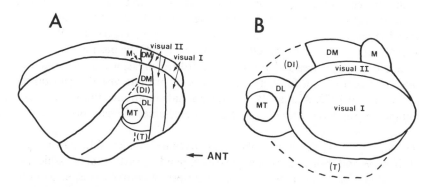

Figure 2   Visual cortical areas of the owl monkey. *A* A view of visual areas in the two hemispheres as seen from above and to the left of the midline. Well-defined visual areas, whose borders are indicated by solid lines, include visual I and II, and the middle temporal, dorsolateral, dorsomedial, and medial areas (MT, DL, DM, and M). Regions whose topographic organization is incompletely worked out include dorsointermediate (DI) and tentorial (T) cortex. *B* A schematic unfolded map of visual areas in the owl monkey, showing relative sizes and topological relationships among areas. Adapted, with permission, from Allman (1977).

contain first- or second-order transformations of the entire visual hemifield (Allman & Kaas 1974b, 1975, 1976). The other two regions, the dorsointermediate and tentorial zones, are topographically organized, but in a manner that has not been fully elucidated. On the basis of its proximity to visual II this entire region has been termed the "third tier" of visual cortical areas. Some third tier areas, such as the medial and dorsolateral areas, have visual representations in register with that in visual II along their common borders. For other areas, particularly the dorsomedial area, there is a mismatch of the representations along the border with visual II. An additional visual representation, the middle temporal area (MT) has been found in a region anterior to the third tier areas (Allman & Kaas 1971a). Thus, there are at least eight visual areas in the owl monkey, and additional topographically defined areas may yet be found in regions of parietal and inferotemporal cortex known to be visually responsive. The various extrastriate areas differ in their topographic organization, particularly with regard to whether there is a split representation of the horizontal meridian, and in the relative amounts of cortex devoted to central vs peripheral vision. Many of the topographically defined visual areas, including area MT and some of the third-tier areas, can be recognized anatomically by their distinctive architectural features (Allman & Kaas 1971a, 1974b, 1975). This finding emphasizes both the value of cortical architecture as a guide to visual organization and the dangers of overreliance on classical anatomical maps of the cortex.

The arrangement of visual areas in other New-World monkeys, such as the marmoset and the squirrel monkey is similar in several important respects to that found in the owl monkey. Common to all of these species is the presence of a middle temporal visual area, MT, and, based on the projections from known visual areas, at least two third-tier areas adjoining visual II (Spatz et al 1970, Spatz & Tigges 1972a, b, Tigges et al 1973a, b, 1974, Spatz 1975, 1977). An Old-World prosimian, the bushbaby, also has a middle temporal area and at least two third-tier areas, which suggests that this arrangement is a fundamental feature of primate visual organization (Allman et al 1973, Allman 1977 and personal communication). The exact number of third-tier areas has not been determined for any primate.

The macaque   The occipital lobe of the macaque monkey differs from that of New-World primates in its greatly expanded surface area, which is associated with a much more elaborate pattern of convolutions, and in the fact that multiple visual representations have been found within Brodmann's area 18 as well as in area 19. The geometrical complexities introduced by the extensive convolutions of the cortex have made it difficult to map visual topography in the macaque by physiological techniques alone. Much of our present understanding of the organization of this region has come from studies of anatomical connections, applied either alone or in conjunction with physiological recordings (Cragg 1969, Zeki 1969, 1970, 1971, Van Essen & Zeki 1978). Within Brodmann's area 18 there are at least four visual representations, V2, V3, V4, and V3A, the last area so named because it was only recently discovered and lies between two previously identified areas. Within area 19 only one well-defined visual area has been identified. This area, in the posterior bank of the superior temporal sulcus, is probably equivalent to the middle temporal area (MT) of New-World primates (see below) and will be referred to by that term. MT occupies only a limited portion of area 19, and it is likely that additional visual areas will be found in relatively unexplored parts of both area 18 and area 19. Visually responsive cortex also occupies large portions of the temporal and parietal lobes (areas 20, 21, and 7), but the visual field is represented very coarsely in these regions (Gross et al 1972, Lynch et al 1977, Robinson & Goldberg 1977), and it is not known whether distinct topographically defined subdivisions are present.

The positions and internal organization of extrastriate areas in the macaque have been studied most extensively in the dorsal half of the occipital lobe (Figure 3A, B). The most revealing approach to determining the borders of visual areas in this region has been to combine extensive physiological mapping of visual topography with a detailed analysis of interhemispheric connections in the same animal. Callosal fibers in the macaque terminate preferentially, although not always exclusively, in regions of vertical meridian representation, so that the overall pattern of interhemispheric connections provides a means of assessing visual topography over a wide area of cortex (Van Essen & Zeki 1978). The combined physiological and anatomical information from such experiments can best be displayed on two-dimensional representations of the cortical surface reconstructed from a series of histological sections (Figure 3C, D). The resulting cortical map preserves all topological relationships among visual areas and even provides reasonable approximations of the actual physical dimensions of different areas.

Of particular interest is the question of whether area V3 is the only visual area adjoining V2 on the side opposite the striate cortex, as is the case in the cat, or whether it is one of several third-tier visual areas, as in New-World primates. It is significant in this regard that the callosal fibers terminating along the superior and inferior vertical meridian representations of V3 form two entirely separate bands situated in ventral and dorsal parts, respectively, of the occipital lobe (Van Essen & Zeki 1978). The simplest interpretation of this result is that V3 does not occupy a continuous strip of cortex alongside V2 and that other areas, in particular V3A and V4, directly abut against V2 in the region between the dorsal and ventral subdivisions of V3. Thus, it appears that a third tier of visual areas may exist in the macaque, although the question of possible homologies with third-tier areas in New-World monkeys has not been resolved. A distinctive feature of V3 in the macaque is an apparently reduced emphasis on central vision relative to the visual periphery, as has also been described for the medial visual area in the owl monkey and some of the lateral suprasylvian areas in the cat.

An important aspect of visual organization in the macaque is the complexity of visual topography in V4 and, to a lesser degree, in other extrastri-

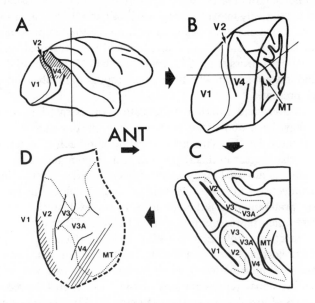

Figure 3   Visual cortical areas in the macaque, as seen in the intact hemisphere (*A, B*), in histological sections (*C*), and in an "unfolded" reconstruction of the dorsal half of extrastriate occipital cortex (*D*). *A* Lateral view of the right hemisphere, showing the exposed portions of visual areas VI, V2, and V4. The exposed portion of dorsal extrastriate occipital cortex has been shaded in *A* and *D* in order to facilitate understanding of the relationship between regions as seen on the two-dimensional map relative to their positions in the intact hemisphere. Vertical line indicates the location of a cut in the coronal plane made to provide the view in *B*. *B* An expanded view, from an anterolateral angle, of the occipital lobe, showing areas V1, V2, V4, and the buried middle temporal area, MT, in the superior temporal sulcus. Thin line shows the level of horizontal section illustrated in *C*. *C* A horizontal section through the occipital lobe, showing the approximate locations of known visual areas at this level. Dotted lines indicate the portion of extrastriate cortex from this particular section included in the reconstructed map in *D*. *D* A flattened, two-dimensional reconstruction of dorsal extrastriate occipital cortex. Dotted lines show the altered contour of the section illustrated in *C*. Thin solid lines indicate the known borders of visual areas. Shaded regions are portions of the cortex visible from the lateral side of the hemisphere. Adapted from Van Essen & Zeki (1978).

ate areas. Although the visual representation is coarser in V4 than in other areas such as V1 and V2, a clear topographic organization is present insofar as neighboring cortical points represent nearby parts of the visual field. The organization is complex, however, in the sense that individual loci in the visual field are represented more than once in V4, but not in such a way that the region can be split easily into well-defined subdivisions (Zeki 1971, Van Essen & Zeki 1978). As in the cat, then, visual topography in the macaque provides valuable information, but not unambiguous criteria, for subdividing the cortex. Other lines of evidence suggesting a complex organization of individual visual cortical areas in the macaque are the presence of local heterogeneity in cell function (see Part III) and the existence of multiple architectonic subdivisions within single topographically defined areas (Beck 1934, D. C. Van Essen & J. L. Bixby, unpublished observations).

Human visual areas   Our present knowledge of the organization of visual areas in the human occipital lobe is rather limited, due largely to the inaccessibility of the human brain to the techniques which have been most productive in elucidating visual cortical organization of experimental mammals. Although classical anatomists recognized only two or three architectonic subdivisions of the human occipital lobe (e.g. Campbell 1905, Brodmann 1909), more recent anatomical and physiological observations hint at levels of complexity that may be comparable to those found in other primates. The most suggestive results have been the observation of numerous architectonic subdivisions of occipital cortex in sections stained for lipofuchsin granules (Braak 1977) and the indications of complex topographic organization revealed by stimulation of extrastriate occipital cortex (Brindley et al 1972). Whether such approaches can ever be extended far enough to yield a clear understanding of the organization of the human visual cortex is difficult to predict.

PART II. VISUAL PATHWAYS

A knowledge of the major pathways connecting different visual areas is valuable in many respects, especially as a basis for distinguishing individual areas, for suggesting homologies between visual areas in different species, and for determining the various routes along which visual information can be transmitted, thereby permitting an assessment of different hierarchical levels of analysis within the cortex. A comprehensive tabulation of known visual cortical connections in different species is beyond the scope of this review. Emphasis is placed, instead, on pathways that illustrate fundamental aspects of visual cortical organization.

Connections With Subcortical Centers

In all mammals visual information is transmitted from the retina to the cortex along two somewhat independent pathways. The more direct route involves the projection from the dorsal lateral geniculate nucleus (dLGN) to the striate cortex and, in some species, to certain extrastriate visual areas. This pathway appears to mediate fine-grained pattern analysis. The longer pathway is from the retina to the superior colliculus, from the colliculus to the pulvinar or the lateral posterior nucleus of the thalamus, and from there mainly to extrastriate visual areas. The function of this pathway, at least at the collicular level, appears related mainly to visual attention and, in primates, to the control of eye movements (Trevarthen 1968, Schneider 1969, Schiller & Stryker 1972, Mohler & Wurtz 1977). Although in most species there is relatively little overlap in the thalamocortical termination of the two pathways, the existence of major projections from the striate cortex to the colliculus, pulvinar, and extrastriate cortex (see below) insure that there can be significant interactions between these systems at several levels.

In primates, the geniculocortical projection is confined exclusively to the striate cortex (Wilson & Cragg 1967, Garey & Powell 1971, Hubel et al 1975, Kaas et al 1976, Wilson et al 1977, Rowe et al 1977). The finding of an apparent geniculate projection to visual II in experiments on the squirrel monkey involving retrograde transport (Wong-Riley 1976) is inconsistent with other published results on the same species and may have resulted from accidental labeling of the striate cortex in the calcarine sulcus (see Tigges et al 1977). The primate pulvinar, which contains several anatomically distinct subdivisions (see Allman 1977) projects extensively, although not uniformly, to occipital and temporal extrastriate visual cortex; it also has a relatively sparse projection to the striate cortex, mainly to layers I and II (Lin et al 1974, Benevento & Rezak 1976, Ogren & Hendrickson 1976, 1977, Trojanowski & Jacobson 1976, Curcio & Harting 1978). In most nonprimates the striate cortex is the principal target of the dLGN, but there may be a limited input to visual II in the mouse, tree shrew, squirrel, and hedgehog (Dräger 1974, Hubel 1975, Weber et al 1977, Gould & Ebner 1978). The lateral posterior nucleus, which receives tectal inputs and is believed to be homologous to part of the pulvinar in primates, projects mainly to visual II and other extrastriate areas (Harting et al 1972, Kaas et al 1972, Coleman et al 1977). In the cat, the dLGN (including the medial interlaminar nucleus, which also receives retinal afferents) projects to areas 17, 18, 19, and the lateral suprasylvian areas. In addition, all of these areas receive inputs of varying strength from extrageniculate thalamic visual nuclei, including the pulvinar, posterior, and lateral posterior nuclei, parts of which are themselves targets of superior collicular projections (Gary & Powell 1967, Wilson & Cragg 1967, Graybiel 1972, Gilbert & Kelly 1975, Le Vay & Gilbert 1976, Holländer & Vanegas 1977, Berson & Graybiel 1978). Thus, the overlap of these two sets of projections is much greater in the cat than in other mammals that have been examined.

The detailed anatomical organization of the major visual input to the cortex, the geniculostriate pathway, has been examined intensively in a number of species, especially the macaque and the cat. These studies have provided valuable insights into the arrangement of ocular dominance columns and the functional significance of cortical laminations, but they have been reviewed elsewhere (Hubel & Wiesel 1977) and are not discussed in detail here.

The existence of powerful, highly organized projections from visual cortical areas to numerous subcortical targets has been revealed with striking clarity by the application of the recently developed retrograde and anterograde tracing techniques. The most prominent of these descending pathways are the projections to the dorsal lateral geniculate nucleus (exclusively from cells in layer VI of both striate and extrastriate areas) and the projections from cells in layer V, again in both striate and certain extrastriate areas, to the superior colliculus and the pulvinar (Spatz et al 1970, Spatz & Tigges 1973, Holländer 1974, Spatz & Erdmann 1974, Gilbert & Kelly 1975, Lund et al 1976, Ogren & Hendrickson 1977, Lin & Kaas 1977, Updyke 1977).

Intracortical Visual Connections

In all mammals the striate cortex sends direct projections to surrounding extrastriate regions, invariably including visual II and generally other areas as well. Among lower mammals the additional known projections of the striate cortex are to the medial visual area (Vm) in the opossum, hedgehog, and rat, and to one or more areas lateral to visual II in the opossum, rat, and rabbit (Benevento & Ebner 1971, Gould & Ebner 1978, Montero et al 1973a, Towns et al 1977). However, relatively little is known about the projections from any of the extrastriate visual areas in these species. Studies of visual cortical connections have generally concentrated on the cat and certain primates, the latter of which are discussed first because of their less extensive interconnections.

PRIMATE VISUAL CORTICAL PROJECTIONS Connections among the numerous visual cortical areas in primates have been studied mainly in the macaque and in several New-World monkeys. In general, each visual area receives inputs from more than one cortical area and in turn has multiple targets within the cortex, so that the various areas cannot be arranged in a strict serial order simply on the basis of a sequential set of projections. Nevertheless, some areas are clearly closer than others to the primary visual inputs, and in primates it is convenient to distinguish four hierarchical levels based on the existence and relative strengths of different anatomical pathways (Figure 4). At the first level is the striate cortex, which is the only area to receive a direct geniculate input. At the second level is visual II, which receives its principal cortical input from the striate area (Cowey 1964, Zeki 1969, Spatz et al 1970, Schiller & Malpeli 1977). In addition to projecting to other extrastriate areas visual II sends a moderate projection back to the striate cortex that terminates mainly in layer I, but also in all other laminae except layer IV (Tigges et al 1974, 1977, see Ogren & Hendrickson 1977). Third-level areas are considered to be those which receive a direct input from striate cortex and, in addition, receive major inputs from other cortical areas. Areas in this category are MT and possibly the medial area (M) in New-World monkeys, and both V3 and MT in the macaque (Spatz et al 1970, Tigges et al 1974, Martinez-Millán & Holländer 1975, Wagor et al 1975, Zeki 1969, 1971, 1978b). Fourth-level areas are those receiving strong inputs from second- and/or third-level areas but little or

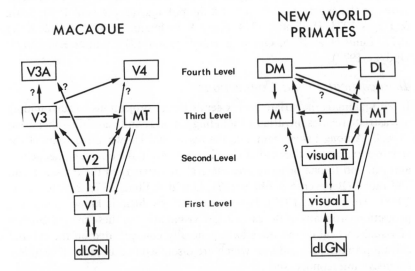

Figure 4 Major pathways linking visual cortical areas in the macaque (*left*) and in New World monkeys (*right*). The connections shown among third tier areas (DL, DM and M) in new-world monkeys are based mainly on evidence from the owl monkey, plus additional observations on areas presumed to be homologous in the squirrel monkey and marmoset. Question marks denote projections that have not been demonstrated with certainty. Relatively sparse projections are indicated with thin arrows. Connections with visual subdivisions of the pulvinar have been omitted from the diagrams, as they are complex and not well understood.

nothing directly from the striate cortex. At this level are various third-tier visual areas, including areas DL and DM in New-World monkeys, plus V4 and V3A in the macaque (Zeki 1969, 1971, 1978a, b, Spatz & Tigges 1972b, Tigges et al 1974, Wagor et al 1975, Kaas & Lin 1977). Although it might seem unduly confusing to assign some of the third-tier visual areas to a fourth hierarchical level, there is no reason to expect an absolute correlation between the physical proximity of areas and their relative status in the processing of visual information. In general, there are numerous routes along which signals can be transmitted to reach any particular visual area, and it is a challenge to the physiologist to determine the relative functional importance of the various pathways demonstrated by neuroanatomists.

The higher levels of visual processing carried out in cortical regions outside the occipital lobe are mediated by pathways to several parts of the temporal, parietal, and frontal lobes from extrastriate occipital cortex. In the macaque there are visual projections to area 7 in the parietal lobe, areas 20 and 21 in the temporal lobe, and the frontal eye fields (area 8) in the frontal lobe (Kuypers et al 1965, Pandya & Kuypers 1969, Jones & Powell 1970, Zeki 1978b). Relatively little is known about the contributions of individual visual areas to these pathways, however. In New-World monkeys there are projections to the parietal and temporal lobes from areas DM and MT, and to the frontal lobe from visual II and MT (Spatz & Tigges 1972a, Tigges et al 1974, Wagor et al 1975). Projections back to extrastriate occipital cortex have been described from presumed visual regions of parietal and temporal cortex in the macaque and from posterior parietal cortex in the owl monkey (Kuypers et al 1965, Pandya & Kuypers 1969, Kaas et al 1977). There is no indication, however, of a significant input from nonvisual cortical sensory areas to the primate occipital lobe (Jones & Powell 1970).

VISUAL PROJECTIONS IN THE CAT   Visual cortical areas in the cat are, in general, much more richly interconnected than in primates. In addition to the aforementioned direct geniculate projections to extrastriate areas as well as to the striate cortex, reciprocal connections have been demonstrated between nearly all combinations of visual cortical areas that have been studied in the cat (Hubel & Wiesel 1965, Wilson 1968, Garey et al 1968, Heath & Jones 1970, 1971, Kawamura 1973, Gilbert & Kelly 1975). The complexity of visual pathways makes it difficult to distinguish well-defined hierarchical levels of visual processing, and lesion studies (see Part III) indicate that extrastriate areas can function to a substantial degree without their normal inputs from the striate cortex. Nevertheless, there are strong reasons for not regarding all geniculate-recipient areas (visual I, II, III, and some of the lateral suprasylvian areas) as being at the same hierarchical level, i.e. as a set of primary visual cortical areas (see, however, Tretter et al 1975). In particular, the striate cortex occupies a special position at a lower level than the others, since it is the largest area, has the finest-grained visual representation, receives the strongest input from the predominant geniculate layers (the A laminae), and, most importantly, projects strongly to the other cortical areas while receiving relatively weak projections from each of them in return (Heath & Jones 1971, Rosenquist et al 1974, Gilbert & Kelly 1975, Le Vay & Gilbert 1976). The other geniculate-recipient areas differ in the laminar origin of their geniculate inputs, in the organization of their connections with extrageniculate thalamic nuclei, and in the relative strengths of their own intracortical interconnections (Heath & Jones 1971, Rosenquist et al 1974, Gilbert & Kelly 1975, Maciewicz 1975, Holländer & Vanegas 1977, Updyke 1977). The functional significance of these differences is not obvious, so it is unclear whether one can reasonably group all of these areas into a single, second hierarchical level. In any event, they probably represent a lower level of processing than regions such as areas 20 and 21 and some of the lateral suprasylvian areas, which have no geniculate input and relatively little input from the striate cortex (Wilson 1968, Heath & Jones 1971, Kawamura 1973). Thus, it is possible to distinguish at least three hierarchical levels of visual cortical areas in the cat, although the specific criteria employed differ from those most useful for primates.

Homologies Among Visual Areas

The existence of numerous visual cortical areas in higher mammals leads naturally to the question of which areas can be regarded as homologous, in the strict sense of having a common embryological origin, or as equivalent, in the sense of having similar functions, connections, and patterns of internal organization. Such information is valuable for understanding the evolution of visual cortex and also for assimilating the increasingly detailed information available from studies on different species. Relatively little is known about the embryonic development of different visual cortical areas, but indirect physiological and anatomical evidence suggesting the homology, or at least equivalence, of striate cortex and also of visual II in all mammals is discussed in preceding sections. It is important to recognize, however, that for both visual I and II there exist significant species differences in their architecture, internal organization, and principal inputs and outputs. A notable illustration of such differences concerns the complete mixing of ocular inputs in layer IV or the striate cortex in some species (mouse, squirrel, owl monkey, and squirrel monkey), their segregation into ocular dominance columns in other species (macaque, spider monkey, and, to some extent, cat and bushbaby), and their partial segregation into separate laminae, rather than columns, in the tree shrew (Dräger 1974, Hubel et al 1975, Casagrande & Harting 1975, Hubel 1975, Kaas et al 1976, Weber et al 1977, Shatz et al 1977, Florence & Casagrande 1978).

Aside from visual I and II the most interesting comparison of visual areas concerns MT, the striate-receptive area in temporal cortex of primates. The similarities in position, size, visual topography, myeloarchitecture, and connections of area MT in several New-World primates (owl monkey, squirrel monkey, and marmoset) and in a prosimian, the bushbaby, are so strong as to leave little doubt that these areas are equivalent (Allman & Kaas 1971a, Allman et al 1973, Spatz & Tigges 1972b, Spatz et al 1970, Tigges et al 1973b). Their relationship to the striate-receptive temporal area in the macaque is less obvious, however, because of the difficulty in comparing positions in a convoluted hemisphere to those in a smooth one, and because the architecture and connections of this area in the macaque are less well understood. It is very striking, however, that in both the macaque and the marmoset the projections to these areas from striate cortex arise only from layers IVb and VI (terminology of Brodmann 1905) and not from layers II and III, the normal source of connections to other cortical areas (Spatz 1975, 1977, Lund et al 1976). On the basis of this unique pattern of connections, the similarities in topographic organization (Zeki 1969, 1971, Ungerleider & Mishkin 1978, Allman & Kaas 1971a), and the functional similarities mentioned in Part III, use of the term MT for this region in the macaque as well as New-World primates seems warranted.

PART III. FUNCTIONS OF VISUAL AREAS

The remarkable specificity of anatomical connections throughout the visual system has its physiological counterpart in the highly selective functional properties of neurons in each visual area. The analysis of single cell function in different visual areas thus yields important clues concerning the overall strategies employed in the process of form perception. Not surprisingly, the best understood cortical area is the striate cortex, but enough information on receptive field properties in extrastriate visual areas is available to provide intriguing suggestions of the varieties of visual functions carried out in different parts of the occipital and temporal lobes.

Striate Cortex

Single unit studies on the cat and the macaque have suggested that two of the principal functions of the striate cortex in these species are to combine the inputs from the two eyes in a way that yields a fused binocular image of the visual world and to analyze the visual world with respect to the orientation of local contours (Hubel & Wiesel 1962, 1968). Associated with these respective functions are two independent, overlapping columnar systems within the cortex, one for ocular dominance and one for stimulus orientation. Each system consists of a set of radially oriented columns or slabs of cortex, with cells in each column sharing a particular eye or orientation preference (Hubel et al 1975, 1978). Within each column several physiological cell types are found which differ in their receptive field organization. Hubel & Wiesel (1962) originally distinguished two broad classes of cells in the cat's striate cortex. "Simple" cells are orientation selective, but otherwise similar in many respects to the geniculate neurons that project to the cortex. "Complex" cells are also orientation selective and have properties suggestive of a major input from simple cells. Both simple and complex cells have been reported in all other mammals studied, including the macaque, rabbit, opossum, mouse, and hamster (Hubel & Wiesel 1968, Wurtz 1969, Chow et al 1971, Rocha-Miranda et al 1973, Dräger 1975, Tiao & Blakemore 1976). In its most elementary form, though, this hierarchical scheme is clearly an oversimplification, as was suggested from the outset (Hubel & Wiesel 1962). Not only are there various subtypes of simple and complex cells, but in many species there are cells that do not fit into the simple/complex classification. Numerous recent studies (see below) have enriched our understanding of these and other intricate details of cortical organization; nevertheless the fundamental notion of hierarchical processing and the basic distinction between simple and complex cell classes retain their usefulness as a valid framework for studying the striate cortex.

Layer IV of the cat receives a direct geniculate input and contains mostly simple cells, whereas complex cells are concentrated in deeper and more superficial laminae. The functional properties of these cells are correlated with their structure, since many simple cells are stellate in form, and most complex cells are pyramidal (Kelly & Van Essen 1974). Whether the simple cells found outside layer IV, especially in Layer VI (Gilbert 1977) are predominantly stellate cells is not known. In the macaque both nonoriented and simple cells occur in layer IV, so presumably the relationship between cell structure and function is not the same as in the cat. In both species certain functional subclasses of complex cell, such as the specialized types projecting to the superior colliculus, are found in only one or two layers (Palmer & Rosenquist 1974, Dow 1974, Gilbert 1977). The restricted laminar distribution of particular cell types is especially significant in view of the striking differences in the connections of individual cortical layers discussed in Part II.

The occurrence in some cells of end-inhibition, i.e. a reduction or abolition of responses to optimally oriented stimuli of increasing length, was originally regarded as indicative of a distinct functional class of "hypercomplex" cells (Hubel & Wiesel 1965, 1968). More recent studies suggest, however, that in the striate cortex there is a continuous spectrum in the degree of end-inhibition occurring in different simple and complex cells, except for those in layer VI, where end-inhibition is minimal (Schiller et al 1976a, Gilbert 1977, Rose 1977). Thus, end-inhibition apparently is not a property of a distinct cell class in the striate cortex. However, this conclusion is not directly applicable to hypercomplex cells in extrastriate visual areas, where they were first described.

Another important aspect of striate cortical function concerns the handling of information that is relayed to the cortex along different functional channels, namely the X- , Y- , and W-cell pathways in the cat and a similar system in primates. Each channel originates as a set of physiological and morphologically distinct retinal ganglion cells and is represented in the dLGN by a set of relay cells having a characteristic morphology and laminar distribution (see Rodieck, this volume). In the macaque there is anatomical evidence (Hubel & Wiesel 1972, Lund & Boothe 1975, Lund et al 1976) that different laminae of the striate cortex are involved in the processing of information from the parvocellular and magnocellular geniculate layers, which contain X-like and Y-like cells, respectively.

In the cat the laminar segregation of X, Y, and W cell types is less distinct at both the geniculate and cortical levels, but nonetheless there is physiological and anatomical evidence for at least a limited separation of these channels within the striate cortex (Hoffmann & Stone 1971, Le Vay & Gilbert 1976). One suggestion has been that simple and complex cells receive their dominant input directly from geniculate X- and Y-cells, respectively, and that the mode of operation of the striate cortex is therefore one of strictly parallel processing rather than hierarchical analysis (Hoffman & Stone 1971, Maffei & Fiorentini 1973, Movshon 1975). However, a hierarchical scheme of processing is entirely consistent with the notion of separate cortical X and Y channels, especially since other groups have reported X-like properties for some complex cells as well as simple cells and Y-like properties for other simple and complex cells (Kelly & Van Essen 1974, Ikeda & Wright 1975, Singer et al 1975, Goodwin & Henry 1978). Although it is clear that many complex cells, at least in deeper layers, receive a direct geniculate input (Hoffmann & Stone 1971, Singer et al 1975), this observation rules out only a strictly serial type of analysis and not a more general hierarchical scheme. The key issue is whether complex cells receive major inputs from simple cells and are at a genuinely higher level of processing. Strong reasons for suspecting this to be the case stem from the facts that simple, but not complex, cells are concentrated in layers receiving direct geniculate inputs and that the receptive field properties of complex cells, particularly their sensitivity to stimulus width, are difficult to explain without assuming a major input from simple cells. Thus, it seems likely that the striate area operates by the hierarchical (albeit not strictly serial) analysis of information transmitted to the cortex along separate X, Y, and W channels. It is not known how extensive the interactions are between these channels at the cortical level, nor is it known how the W-cell inputs to the cortex are processed.

An additional function that has been suggested for the striate cortex is to analyze the disparity of retinal images in the two eyes to provide for stereoscopic depth discrimination. In primates the range over which stereopsis is operative extends from a few seconds to several degrees of disparity near the center of gaze and from a few minutes to many degrees in the periphery (Blakemore 1970, Sarmiento 1975). Presumably, then, neurons somewhere in the visual pathway must be selective for disparities within this range. Studies of binocular interactions in the striate cortex of the cat, sheep, and macaque have focused on four distinct issues relating to the neurophysiological basis of stereopsis. First is the question of retinal correspondence, that is, whether binocularly driven cortical cells have receptive fields in precisely corresponding parts of the two retinae. It seems generally agreed that some striate cortical neurons have receptive fields with demonstrably nonzero binocular disparity, but, because of the difficulty of monitoring eye positions accurately, it remains controversial whether their incidence is very low (Hubel & Wiesel 1970, 1973, Clarke et al 1976, Clarke & Whitteridge 1978) or rather high (Barlow et al 1967, Nikara et al 1968, Joshua & Bishop 1970, see also Poggio & Fischer 1977). It should be noted, though, that many cells whose receptive field disparities are classified as experimentally indistinguishable from zero might in fact have functionally significant nonzero disparities. A second issue concerns disparity tuning, that is, the sensitivity of neurons to the disparity of stimuli presented simultaneously to the two eyes. Many striate cortical neurons in the cat are sharply tuned for disparity in that they respond well during binocular stimulation only within a narrow range of disparities much smaller than the overall receptive field dimensions (Pettigrew et al 1968). Sharp disparity tuning has also been reported for many cells in the striate cortex of the macaque when the animal is unanesthetized (Poggio & Fischer 1977), but not during anesthesia (Hubel & Wiesel 1970). Once again it has been difficult to determine what percentage of disparity-tuned cells have optimal disparities different from zero. Since disparity-tuned cells that prefer stimuli in exact retinal correspondence might be involved in the process of binocular fusion rather than actual depth perception, the functional significance of disparity tuning in the striate cortex remains uncertain. A third issue concerns cells that have an absolute requirement for simultaneous binocular stimulation, that is, which cannot be activated through either eye alone. Such "binocular-only" cells are present, but rare, in the striate cortex

(Poggio & Fischer 1977, Clarke & Whitteridge 1978). These cells generally are sharply tuned for disparity and often have nonzero preferred disparities, so it is likely that they are specifically involved in stereopsis. A fourth issue concerns differences in preferred orientations for stimuli to the two eyes, a property which could be used to detect stimuli lying at an angle to the frontoparallel plane. Although cells with disparities in binocular orientation preferences are present in the cat's striate cortex (Blakemore et al 1972), they are not sharply tuned for particular orientation disparities and are unlikely to play a major role in stereopsis (Nelson et al 1977).

It is likely, in view of these various considerations, that the striate cortex has at least a limited role in depth discrimination involving small retinal disparities, but that coarse stereopsis, involving disparities of a large fraction of a degree or more, must be handled mainly in extrastriate areas (see below). The overall importance of the striate cortex for stereopsis, other than as a conduit for the transmission of monocular information to other areas, remains unclear, however.

The existence within the visual system, and particularly within the striate cortex, of functionally distinct neural channels tuned to different spatial frequencies of visual stimulation, has been suggested by recent psychophysical and physiological studies (e.g. Campbell & Robson 1968, Blakemore & Campbell 1969, Maffei & Fiorentini 1973, Schiller et al 1976b). Such channels presumably reflect the occurrence of cells having different preferred stimulus widths and may be linked to X, Y, and W channels in the visual pathway, given the different receptive field sizes associated with these cell classes. The notion that spatial frequency channels reflect a mode of cortical processing akin to Fourier analysis (e.g. Pollen & Ronner 1975, Glezer et al 1976) seems premature at best, however, since no cortical neurons yet encountered have the requisite narrowness of spatial frequency tuning nor the generalization over a large enough number of spatial cycles to be particularly suggestive of a genuine Fourier-type analysis (cf Julesz 1975, Tyler 1978).

Extrastriate Visual Areas

A basic issue confronting neurophysiologists interested in extrastriate visual areas is to determine the specific visual functions mediated by each cortical area. A knowledge of the major anatomical pathways discussed in Part II permits the assignment of each area to a particular level in a hierarchical scheme, but the determination of their actual functions requires a knowledge of neuronal receptive field properties in different areas and in different layers of the same area.

The initial functional study of visual areas II and III in the cat by Hubel & Wiesel (1965) showed a virtual absence of simple cells in both areas, with visual II containing mostly complex cells and visual III containing many complex, hypercomplex and "higher-order hypercomplex" cells. These results, combined with observations on the projections of the striate cortex, supported the notion that visual I, II, and III carry out progressively more advanced levels of form analysis. It is evident, however, that processing in these areas is not strictly serial, given (a) the existence of direct geniculate inputs to visual II and III in the cat, (b) the reciprocal nature of cortical and subcortical connections of these visual areas (see Part II), and (c) the persistence of visual responsiveness of many cells in visual II following bilateral lesions of the striate cortex (Donaldson & Nash 1973, Dreher & Cottee 1975). It is likely, moreover, that visual II is involved in other functions besides form analysis, since in the cat and sheep it has a higher incidence of cells with large binocular disparities than does visual I (Clarke et al 1976, Clarke & Whitteridge 1978), and since it also contains cells specialized for movement analysis (Pettigrew 1973, Cynader & Regan 1978). Visual II and III in the cat differ from visual I in their sources of geniculate inputs: visual II receives inputs mainly from Y-cells and visual III from the C laminae and the MIN, which contain mainly W- and Y-cells, whereas visual I is a target for all three cell types (see Rodieck, this volume). It is not known, however, whether the intracortical projections to visual II and III from the striate cortex arise selectively from specific populations of X-, Y-, or W-like cells. Nor is it clear what the relative importance of direct inputs from the dLGN may be, compared to those from other cortical areas, and from extrageniculate thalamic nuclei, in the generation of highly specialized receptive field properties, especially in visual III.

Other interesting aspects of visual function in the cat's extrastriate cortex include the preponderance of direction-selective cells in the lateral suprasylvian region (Spear & Baumann 1975, Camarda & Rizzolatti 1976) and the suggestion of advanced levels of form analysis in areas 20 and 21 (Markuszka 1978). It is also noteworthy that there are specific auditory inputs to many visual cortical neurons in the cat, even in visual I and II (Morrell 1972, Fishman & Michael 1973). It is not known whether these inputs act to direct visual attention to relevant external stimuli or whether they serve some other purpose in the integration of visual and auditory information.

In the macaque, studies of cellular function in several extrastriate visual areas have shown striking regional variations in the distribution of certain receptive field types. Perhaps the clearest illustration of this is the relative abundance of direction-selective cells in area MT compared to other visual areas. Most cells in MT of the macaque show a strong preference for stimuli moving in a particular direction, either within the plane of fixation or along the axis towards or away from the animal (Zeki 1974a, b). A similar emphasis on direction selectivity has been found in area MT of the owl monkey (Newsome et al 1978). Thus it is interesting that area MT has reciprocal connections with layer IVb of the striate cortex (Spatz 1975, 1977, Lund et al 1976), since this sublayer contains many directionally selective cells (Dow 1974) and receives a preferential input from Y-like cells in the magnocellular subdivision of the dLGN by way of a relay in cortical layer IVcα (Lund & Boothe 1975, see Rodieck, this volume).

A second example of regional specialization in the macaque concerns the nonuniform distribution of color-specific cells in different visual areas. Color-specific cells form a substantial fraction of the cells in the striate cortex, ranging from about half of the cells near the foveal representation to a considerably lower level in the periphery (Hubel & Wiesel 1968, Dow & Gouras 1973, Gouras 1974, Poggio et al 1975, Bertulis et al 1977). Outside the striate cortex such cells are abundant in area V4 and also in a neighboring part of the superior temporal sulcus that may be a subdivision of V4 (Zeki 1973, 1977, 1978c, Van Essen & Zeki 1978). They are much less common in V2 and are rare in V3, V3A, and MT (Zeki 1974a, 1977, 1978c, Zeki & Sandeman 1976, Baizer et al 1977, Van Essen & Zeki 1978). An accurate determination of the overall incidence of color-specific cells in V4 is hampered by the nonrandom (clustered) distribution of these cells and by uncertainties concerning the boundaries of V4.

Another aspect of extrastriate cortical function in the macaque is the processing of information for stereoscopic depth discrimination. Disparity-tuned, binocular-only cells ("binocular depth" cells) are present in the posterior bank of the lunate sulcus, including area V2 and probably also V3 (Hubel & Wiesel 1970 and personal communication; Poggio & Fischer 1977). The importance of these areas for stereopsis, relative to the striate cortex and to other extrastriate areas, cannot be resolved without more precise information about the incidence of the various types of disparity-sensitive cells in each area.

The notion that each visual cortical area might be the principal mediator of one or more well-defined functions has an obvious intuitive appeal and is supported by the evidence just described for regional specializations in cellular receptive field properties. Nevertheless, at present it is only a working hypothesis, not an established principle, and should serve primarily to stimulate further physiological and behavioral studies of the visual cortex.

Lesion Studies

An important means for assessing the overall role of particular cortical areas in visual perception, particularly in humans, has been to observe the behavioral deficits produced by restricted cortical lesions. For example, the central role of the striate cortex in visual perception is evident from the severe deficits caused by lesions of this area, especially in primates (see Polyak 1957, Weiskrantz 1972). Although total removal of the striate cortex abolishes conscious visual perception in humans and leads to serious behavioral defects in macaques, it is significant that in both species considerable visual function persists in the form of abilities for spatial localization and even crude pattern discrimination (Humphrey & Weiskrantz 1967, Schilder et al 1971, Pöppel et al 1973, Weiskrantz et al 1974). The residual visual capacities are presumed to be dependent on functions mediated by the superior colliculus and its targets, but this point has not been tested explicitly. In nonprimates removal of the striate cortex causes behavioral deficits that are less severe than in primates. This difference is probably related to the more prominent role of the superior colliculus in normal visually guided behavior of nonprimates and, in species such as the cat, to the presence of direct geniculate inputs to several extrastriate cortical areas (Schneider 1969, Sprague et al 1977).

Restricted lesions of extrastriate cortex in the occipital lobe of primates generally produce much less dramatic visual defects than striate cortical lesions, and it has sometimes been questioned whether the extrastriate occipital areas play any significant role in visual analysis and perception (e.g. Pribram et al 1969). Fairly complete lesions of this region in the macaque prevent pattern discrimination (Mishkin 1972, Keating 1975), but the absence of noticeable deficits following less extensive lesions should be interpreted with caution in view of such factors as (a) the difficulty of removing any particular visual area in its entirety, especially if it is buried within deep cortical sulci; (b) the difficulty of choosing sensitive behavioral tests for assessing what may be subtle changes in highly specialized functions; and (c) the possibility of considerable redundancy in the cortical pathways available for carrying out particular types of visual processing. In any event, restricted lesions in the visual regions of the temporal and parietal lobes can disrupt specific visual functions, such as certain aspects of pattern discrimination (Iwai & Mishkin 1969, Gross 1973), stereopsis (Cowey & Porter 1978), or visual attention (see Mountcastle et al 1975). Another interesting example concerns rare cases of loss of color perception following inferior occipital lobe lesions in humans (Meadows 1974, A. L. Pearlman, J. Birch-Cox, J. C. Meadows, submitted for publication). Although the specific region whose loss is responsible for cerebral color blindness is not well defined, these clinical cases are of particular interest in relation to physiological studies of color processing in V4 and other areas of the macaque.

CONCLUDING REMARKS

As recently as two decades ago the cerebral cortex could be regarded as a magnificent but mysterious structure whose mode of operation was only poorly understood. Our knowledge of the organization and function of cortical sensory and motor areas, particularly the visual areas, has expanded enormously since then, and many aspects of cerebral function now seem explicable, in a general sense, in terms of the known connections and physiological properties of specific ensembles of cortical neurons. This progress can be attributed in large part to the development of physiological methods for recording single-unit activity in the cortex and, more recently, of refined anatomical techniques for the anterograde and retrograde tracing of neural pathways.

The most thoroughly understood cortical area is the striate cortex, whose major known functions are to provide for fused binocular images of the visual world by combining the inputs from the two eyes, to carry out early stages of form analysis by analyzing contours in the visual world and to initiate the process of stereoscopic depth discrimination. Information is processed to a large extent in a hierarchical fashion within the striate cortex, but there appear to be several functionally distinct channels that receive separate inputs, process information relatively independently of one another, and distribute their outputs to different cortical and subcortical targets. Analysis of the exquisitely organized laminar and columnar systems within the striate area has provided an invaluable key to elucidating the function of this region of cortex.

A major development over the past decade in understanding the visual cortex outside the striate area has been the realization that in higher mammals there exists a surprisingly large number of distinct, topographically organized visual cortical areas, some of which have unexpectedly complex patterns of internal organization. The number and arrangement of visual areas vary considerably among species, and progress can be expected in the near future in the study of comparative aspects of extrastriate visual cortical organization. Another major frontier is related to the functions associated with different visual areas. It is clear that there are striking regional variations in the receptive field properties of extrastriate visual neurons, but the assignment of specific functions to particular extrastriate visual areas cannot yet be made with great confidence. A satisfactory understanding of how each visual cortical area operates will in general depend on detailed information from several approaches, including single-unit functional studies in conscious as well as anesthetized animals, determination of major input and output pathways, and assessment of specific visual defects associated with well-defined cortical lesions. Whether such approaches will eventually suffice to explain the highest levels of phenomena relating to visual perception and recognition remains an open question.

ACKNOWLEDGMENTS

I am grateful for the critical comments on the manuscript provided by J. M. Allman, J. L. Bixby, D. P. Corey, K. Fryxell, A. J. Hudspeth, J. Maunsell, W. T. Newsome, J. D. Pettigrew, A. C. Rosenquist, S. L. Shotwell, and R. J. Tusa, and for the assistance provided by E. Hanson, P. Knudsen, and A. I. Van Essen. Work from the author's laboratory is supported by grant R01 EY02091 from the National Eye Institute, National Institutes of Health.

Literature Cited

Adams, A. D., Forrester, J. M. 1968. The projection of the rat's visual field on the cerebral cortex. *Q. J. Exp. Physiol.* 53:327–36

Allman, J. M. 1977. Evolution of the visual system in the early primates. *Prog. Psychobiol. Physiol. Psychol.* 7:1–53

Allman, J. M., Kaas, J. H. 1971a. A representation of the visual field in the caudal third of the middle temporal gyrus of the owl monkey (*Aotus trivirgatus*). *Brain Res.* 31:85–105

Allman, J. M., Kaas, J. H. 1971b. Representation of the visual field in striate and adjoining cortex of the owl monkey (*Aotus trivirgatus*). *Brain Res.* 35:89–106

Allman, J. M., Kaas, J. H. 1974a. The organization of the second visual area (V II) in the owl monkey: A second order transformation of the visual hemifield. *Brain Res.* 76:247–65

Allman, J. M., Kaas, J. H. 1974b. A crescent-shaped cortical visual area surrounding the middle temporal area (MT) in the owl monkey (*Aotus trivirgatus*). *Brain Res.* 81:199–213

Allman, J. M., Kaas, J. H. 1975. The dorsomedial cortical visual area: A third tier area in the occipital lobe of the owl monkey (*Aotus trivirgatus*). *Brain Res.* 100:473–87

Allman, J. M., Kaas, J. H. 1976. Representation of the visual field on the medial wall of occipital-parietal cortex in the owl monkey. *Science* 191:572–75

Allman, J. M., Kaas, J. H., Lane, R. H. 1973. The middle temporal visual area (MT) in the bushbaby, *Galago senegalensis*. *Brain Res.* 57:197–202

Baizer, J. S., Robinson, D. L., Dow, B. M. 1977. Visual responses of area 18 neurons in awake, behaving monkey. *J. Neurophysiol.* 40:1024–37

Barlow, H. B. 1975. Visual experience and cortical development. *Nature* 258:199–204

Barlow, H. B., Blakemore, C., Pettigrew, J. D. 1967. The neural mechanism of binocular depth discrimination. *J. Physiol. London* 193:327–42

Beck, E. 1934. Der Occipitallappen des Affen (Macacus rhesus) und des Menschen in seiner cytoarchitektonischen Struktur. *J. Psychol. Neurol.* 46:193–323

Benevento, L. A., Ebner, F. F. 1971. The areas and layers of cortico-cortical terminations in the visual cortex of the Virginia opossum. *J. Comp. Neurol.* 141:157–90

Benevento, L. A., Rezak, M. 1976. The cortical projections of the inferior pulvinar and adjacent lateral pulvinar in the rhesus monkey (*Macaca mulatta*): An autoradiographic study. *Brain Res.* 108:1–24

Berson, D. M., Graybiel, A. M. 1978. Parallel thalamic zones in the LP-pulvinar complex of the cat identified by their afferent and efferent connections. *Brain Res.* 147:139–48

Bertulis, A., Guld, C., Lennox-Buchthal, M. A. 1977. Spectral and orientation specificity of single cells in foveal striate cortex of the vervet monkey, *Cercopithecus aethiops*. *J. Physiol. London* 268:1–20

Blakemore, C. 1970. The range and scope of binocular depth discrimination in man. *J. Physiol. London* 211:599–622

Blakemore, C., Campbell, F. W. 1969. On the existence of neurones in the human visual system selectively sensitive to the orientation and size of retinal images. *J. Physiol. London* 226:725–40

Blakemore, C., Fiorentini, A., Maffei, L. 1972. A second neural mechanism of binocular depth discrimination. *J. Physiol. London* 226:725–40

Blinkov, S. M., Glezer, I. I. 1968. *The Human Brain in Figures and Tables.* New York: Plenum. 482 pp.

Braak, H. 1977. The pigment architecture of the human occipital lobe. *Anat. Embryol.* 150:229–50

Brindley, G. S., Donaldson, P. E. K., Falconer, M. A., Rushton, D. N. 1972. The extent of the region of occipital cortex that when stimulated gives phosphenes fixed in the visual field. *J. Physiol. London* 225:P57–58

Brodmann, K. 1905. Beitrage zur histologischen Localisation der Grosshirnrinde. Dritte Mitteilung. Die Rindenfelder der niederen Affen. *J. Psychol. Neurol.* 4:177–226

Brodmann, K. 1909. *Vergleichende Lokalisationslehre der Grosshirnrinde.* Leipzig: Barth. 324 pp.

Camarda, R., Rizzolatti, G. 1976. Visual receptive fields in the lateral suprasylvian area (Clare-Bishop area) of the cat. *Brain Res.* 101:427–43

Campbell, A. W. 1905. *Histological Studies on the Localisation of Cerebral Function.* Cambridge: Cambridge Univ. Press. 360 pp.

Campbell, F. W., Robson, J. G. 1968. Application of Fourier analysis to the visibility of gratings. *J. Physiol. London* 197:551–66

Casagrande, V. A., Harting, J. K. 1975. Transneuronal transport of tritiated fucose and proline in the visual pathways of tree shrew *Tupaia glis. Brain Res.* 96:367–72

Choudhury, B. P. 1978. Retinotopic organization of the guinea pig's visual cortex. *Brain Res.* 144:19–29

Chow, K. L., Douville, A., Mascetti, G., Grobstein, P. 1977. Receptive field characteristics of neurons in a visual area of the rabbit temporal cortex. *J. Comp. Neurol.* 171:135–46

Chow, K. L., Masland, R. H., Stewart, D. L. 1971. Receptive field characteristics of striate cortical neurons in the rabbit. *Brain Res.* 33:337–52

Clare, M. H., Bishop, G. H. 1954. Responses from an association area secondarily activated from optic cortex. *J. Neurophysiol.* 17:271–77

Clarke, P. G. H., Donaldson, I. M. L., Whitteridge, D. 1976. Binocular visual

mechanisms in cortical areas I and II of the sheep. *J. Physiol. London* 256: 509–26

Clarke, P. G. H., Whitteridge, D. 1976. The cortical visual areas of the sheep. *J. Physiol. London* 256:497–508

Clarke, P. G. H., Whitteridge, D. 1978. A comparison of stereoscopic mechanisms in cortical visual areas V1 and V2 of the cat. *J. Physiol. London* 275:P92–93

Coleman, J., Diamond, I. T., Winer, J. A. 1977. The visual cortex of the opossum: The retrograde transport of horseradish peroxidase to the lateral geniculate and lateral posterior nuclei. *Brain Res.* 137:233–52

Cowey, A. 1964. Projection of the retina on to striate and prestriate cortex in the squirrel monkey (*Saimiri sciureus*). *J. Neurophysiol.* 27:366–96

Cowey, A., Porter, J. 1978. Brain damage and global stereopsis. *Proc. R. Soc. London Ser. B* In press

Cragg, B. G. 1969. The topography of the afferent projections in circumstriate visual cortex of the monkey studied by the Nauta method. *Vision Res.* 9:733–47

Curcio, C. A., Harting, J. K. 1978. Organization of pulvinar afferents to area 18 in the squirrel monkey: evidence for stripes. *Brain Res.* 143:155–61

Cynader, M., Regan, D. 1978. Neurones in cat parastriate cortex sensitive to the direction of motion in three-dimensional space. *J. Physiol. London* 274:549–66

Daniel, P. M., Whitteridge, D. 1961. The representation of the visual field on the cerebral cortex in monkeys. *J. Physiol. London* 159:203–21

Diamond, I. T., Hall, W. C. 1969. Evolution of neocortex. *Science* 164:251–62

Donaldson, I. M. L., Nash, J. R. G. 1973. Interaction between visual cortical areas: the effect of a chronic lesion in area 17 on the properties of area 18 units in the cat. *J. Physiol. London* 234: P77–78

Donaldson, I. M. L., Whitteridge, D. 1977. The nature of the boundary between cortical visual areas II and III in the cat. *Proc. R. Soc. London Ser. B* 199:445–62

Dow, B. M. 1974. Functional classes of cells and their laminar distribution in monkey visual cortex. *J. Neurophysiol.* 37:927–46

Dow, B. M., Dubner, R. 1971. Single-unit responses to moving visual stimuli in middle suprasylvian gyrus of the cat. *J. Neurophysiol.* 34:47–55

Dow, B. M., Gouras, P. 1973. Color and spatial specificity of single units in rhesus

258   VAN ESSEN

monkey foveal striate cortex. *J. Neurophysiol.* 36:79–99

Dräger, U. C. 1974. Autoradiography of tritiated proline and fucose transported transneuronally from the eye to the visual cortex in pigmented and albino mice. *Brain Res.* 82:284–92

Dräger, U. 1975. Receptive fields of single cells and topography in mouse visual cortex. *J. Comp. Neurol.* 160:269–90

Dreher, B., Cottee, L. J. 1975. Visual receptive-field properties of cells in area 18 of cat's cerebral cortex before and after acute lesions in area 17. *J. Neurophysiol.* 38:735–50

Ebner, F. F., Myers, R. E. 1965. Distribution of corpus callosum and anterior commisure in cat and raccoon. *J. Comp. Neurol.* 124:353–66

Fishman, M. C., Michael, C. R. 1973. Integration of auditory information in the cat's visual cortex. *Vision Res.* 13: 1415–19

Florence, S. L., Casagrande, V. A. 1978. A note on the evolution of ocular dominance columns in primates. *Assoc. Res. Vision Ophthalmol.* p. 291 (Abstr.)

Garey, L. J., Jones, E. G., Powell, T. P. S. 1968. Interrelationships of striate and extrastriate cortex with the primary relay sites of the visual pathway. *J. Neurol. Neurosurg. Psychiatry* 31:135–57

Garey, L. J., Powell, T. P. S. 1967. The projection of the lateral geniculate nucleus upon the cortex in the cat. *Proc. R. Soc. London Ser. B* 169:107–26

Garey, L. J., Powell, T. P. S. 1971. An experimental study of the termination of the lateral geniculo-cortical pathway in the cat and monkey. *Proc. R. Soc. London Ser. B* 179:41–63

Gilbert, C. D. 1977. Laminar differences in receptive field properties of cells in cat primary visual cortex. *J. Physiol. London* 268:391–421

Gilbert, C. D., Kelly, J. P. 1975. The projections of cells in different layers of the cat's visual cortex. *J. Comp. Neurol.* 163:81–106

Glezer, V. D., Cooperman, A. M., Ivanov, V. A., Tscherbach, T. A. 1976. An investigation of spatial frequency characteristics of the complex receptive fields in the visual cortex of the cat. *Vision Res.* 16:789–97

Goodwin, A. W., Henry, G. H. 1978. The influence of stimulus velocity on the responses of single neurones in the striate cortex. *J. Physiol. London* 277:467–82

Gould, H. J., Ebner, F. F. 1978. Connections of the visual cortex in the hedgehog (Paraechinus hypomelas). II. Cor-

ticocortical projections. *J. Comp. Neurol.* 177:473–502

Gouras, P. 1974. Opponent-colour cells in different layers of foveal striate cortex. *J. Physiol. London* 238:583–602

Graybiel, A. M. 1972. Some ascending connections of the pulvinar and nucleus lateralis posterior of the thalamus in the cat. *Brain Res.* 44:99–125

Gross, C. G. 1973. Inferotemporal cortex and vision. *Prog. Physiol. Psychol.* 5:77–123

Gross, C. G., Rocha-Miranda, C. E., Bender, D. B. 1972. Visual properties of neurons in inferotemporal cortex of the macaque. *J. Neurophysiol.* 35:96–111

Hall, W. C., Kaas, J. H., Killackey, H., Diamond, I. T. 1971. Cortical visual areas in the grey squirrel (*Sciurus carolinensis*): a correlation between cortical evoked potential maps and architectonic subdivisions. *J. Neurophysiol.* 34:437–52

Harting, J. K., Hall, W. C., Diamond, I. T. 1972. Evolution of the pulvinar. *Brain Behav. Evol.* 6:424–52

Heath, C. J., Jones, E. G. 1970. Connexions of area 19 and the lateral suprasylvian area of the visual cortex of the cat. *Brain Res.* 19:302–5

Heath, C. J., Jones, E. G. 1971. The anatomical organization of the suprasylvian gyrus of the cat. *Ergeb. Anat. Entwicklungs gesch.* 43:1–64

Hoffmann, K.-P., Stone, J. 1971. Conduction velocity of afferents to cat visual cortex: a correlation with cortical receptive field properties. *Brain Res.* 32:460–66

Holländer, H. 1974. Projections from the striate cortex to the diencephalon in the squirrel monkey (*Saimiri sciureus*). A light microscopic radioautographic study following intracortical injection of [³H] leucine. *J. Comp. Neurol.* 155:425–40

Holländer, H., Vanegas, H. 1977. The projection from the lateral geniculate nucleus onto the visual cortex in the cat. A quantitative study with horseradish peroxidase. *J. Comp. Neurol.* 173: 519–36

Hubel, D. H. 1975. An autoradiographic study of the retino-cortical projections in the tree shrew (*Tupaia glis*). *Brain Res.* 96:41–50

Hubel, D. H., Wiesel, T. N. 1962. Receptive fields, binocular interaction and functional architecture in the cat's visual cortex. *J. Physiol. London* 160:106–54

Hubel, D. H., Wiesel, T. N. 1965. Receptive fields and functional architecture in two non-striate visual areas (18 and 19) of the cat. *J. Neurophysiol.* 28:229–89

Hubel, D. H., Wiesel, T. N. 1968. Receptive fields and functional architecture of monkey striate cortex. *J. Physiol. London* 195:215–43

Hubel, D. H., Wiesel, T. N. 1970. Cells sensitive to binocular depth in area 18 of the macaque monkey cortex. *Nature* 225:41–42

Hubel, D. H., Wiesel, T. N. 1972. Laminar and columnar distribution of geniculocortical fibers in the macaque monkey. *J. Comp. Neurol.* 146:421–50

Hubel, D. H., Wiesel, T. N. 1973. A reexamination of stereoscopic mechanisms in area 17 of the cat. *J. Physiol. London* 232:P29–30

Hubel, D. H., Wiesel, T. N. 1977. Functional architecture of macaque monkey visual cortex. *Proc. R. Soc. London Ser. B* 198:1–59

Hubel, D. H., Wiesel, T. N., Le Vay, S. 1975. Functional architecture of area 17 in normal and monocularly deprived macaque monkeys. *Cold Spring Harbor Symp. Quant. Biol.* 40:581–89

Hubel, D. H., Wiesel, T. N., Stryker, M. P. 1978. Anatomical demonstration of orientation columns in macaque monkey. *J. Comp. Neurol.* 177:361–80

Humphrey, N. K., Weiskrantz, L. 1967. Vision in monkeys after removal of the striate cortex. *Nature* 215:595–97

Ikeda, H., Wright, M. J. 1975. The relationship between the 'sustained-transient' and the 'simple-complex' classifications of neurones in area 17 of the cat. *J. Physiol. London* 244:P59–60

Iwai, E., Mishkin, M. 1969. Further evidence on the locus of the visual area in the temporal lobe of the monkey. *Exp. Neurol.* 25:585–94

Jones, E. G., Powell, T. P. S. 1970. An anatomical study of converging sensory pathways within the cerebral cortex of the monkey. *Brain* 93:793–820

Joshua, D. E., Bishop, P. O. 1970. Binocular single vision and depth discrimination. Receptive field disparities for central and peripheral vision and binocular interaction on peripheral single units in cat striate cortex. *Exp. Brain Res.* 10:389–416

Julesz, B. 1975. Two-dimensional spatial-frequency-tuned channels in visual perception. In *Signal Analysis and Pattern Recognition in Biomedical Engineering,* ed. G. F. Inbar, pp. 177–79. New York: Wiley. 324 pp.

Kaas, J. H. 1977. Sensory representations in mammals. In *Function and Formation of Neural Systems,* ed. G. S. Stent, pp. 65–80. Berlin: Dahlem Konferenzen. 365 pp.

Kaas, J. H., Hall, W. C., Diamond, I. T. 1970. Cortical visual areas I and II in the hedgehog: Relation between evoked potential maps and architectonic subdivisions. *J. Neurophysiol.* 33:595–615

Kaas, J. H., Hall, W. C., Killackey, H., Diamond, I. T. 1972. Visual cortex of the tree shrew (*Tupaia glis*): architectonic subdivisions and representations of the visual field. *Brain Res.* 42:491–96

Kaas, J. H., Lin, C.-S. 1977. Cortical projections of area 18 in owl monkeys. *Vision Res.* 17:739–41

Kaas, J. H., Lin, C.-S., Casagrande, V. A. 1976. The relay of ipsilateral and contralateral retinal input from the lateral geniculate nucleus to striate cortex in the owl monkey: a transneuronal transport study. *Brain Res.* 106:371–78

Kaas, J. H., Lin, C.-S., Wagor, E. 1977. Cortical projections of posterior parietal cortex in owl monkeys. *J. Comp. Neurol.* 171:387–408

Kalia, M., Whitteridge, D. 1973. The visual areas in the splenial sulcus of the cat. *J. Physiol. London* 232:275–83

Kawamura, K. 1973. Corticocortical fiber connections of the cat cerebrum. III. The occipital region. *Brain Res.* 51:41–60

Keating, E. G. 1975. Effects of prestriate and striate lesions on the monkey's ability to locate and discriminate visual forms. *Exp. Neurol.* 47:16–25

Kelly, J. P., Van Essen, D. C. 1974. Cell structure and function in the visual cortex of the cat. *J. Physiol. London* 238:515–47

Krieg, W. J. S. 1946. Connections of the cerebral cortex. I. The albino rat. A: Topography of the cortical areas; and B: Structure of the cortical areas. *J. Comp. Neurol.* 84:221–323

Kuypers, H. G. J. M., Szwarcbart, M. K., Mishkin, M., Rosvold, H. E. 1965. Occipitotemporal corticocortical connections in the rhesus monkey. *Exp. Neurol.* 11:246–62

Landgren, S., Silfvenius, H. 1968. Projections of the eye and the neck region on the anterior suprasylvian cerebral cortex of the cat. *Acta Physiol. Scand.* 74:340–47

Le Vay, S., Gilbert, C. D. 1976. Laminar patterns of geniculocortical projection in the cat. *Brain Res.* 113:1–19

Leventhal, A. G., Hale, P. T., Dreher, B. 1978. The representation of the visual field in areas 18 and 19 of the cat. *Proc. Aust. Physiol. Pharmacol. Soc.* 9:P59 (Abstr.)

Lin, C. S., Kaas, J. H. 1977. Projections from cortical visual areas 17, 18, and MT onto the dorsal lateral geniculate nucleus in owl monkeys. *J. Comp. Neurol.* 173:457–74

Lin, C. S., Wagor, E., Kaas, J. H. 1974. Projections from the pulvinar to the middle temporal visual area (MT) in the owl monkey, *Aotus trivirgatus. Brain Res.* 76:145–49

Lund, J. S., Boothe, R. G. 1975. Interlaminar connections and pyramidal neuron organization in the visual cortex, area 17, of the macaque monkey. *J. Comp. Neurol.* 159:305–34

Lund, J. S., Lund, R. D., Hendrickson, A. E., Bunt, A. H., Fuchs, A. F. 1976. The origin of efferent pathways from the primary visual cortex, area 17, of the macaque monkey as shown by retrograde transport of horseradish peroxidase. *J. Comp. Neurol.* 164:287–304

Lynch, J. C., Mountcastle, V. B., Talbot, W. H., Yin, T. C. T. 1977. Parietal lobe mechanisms for directed visual attention. *J. Neurophysiol.* 40:362–89

Maciewicz, R. J. 1975. Thalamic afferents to areas 17, 18 and 19 of cat cortex traced with horseradish peroxidase. *Brain Res.* 84:308–12

Maffei, L., Fiorentini, A. 1973. The visual cortex as a spatial frequency analyzer. *Vision Res.* 13:1255–67

Malpeli, J. G., Baker, F. H. 1975. The representation of the visual field in the lateral geniculate nucleus of *Macaca mulatta. J. Comp. Neurol.* 161:569–94

Markuszka, J. 1978. Visual properties of neurons in the posterior suprasylvian gyrus of the cat. *Exp. Neurol.* 59:146–61

Marshall, W. H., Talbot, S. A., Ades, H. W. 1943. Cortical response of the anesthetized cat to gross photic and electrical afferent stimulation. *J. Neurophysiol.* 6:1–16

Martinez-Millán, L., Holländer, H. 1975. Cortico-cortical projections from striate cortex of the squirrel monkey (*Saimiri sciureus*). A radioautographic study. *Brain Res.* 83:405–17

Mathers, L. H., Douville, A., Chow, K. L. 1977. Anatomical studies of a temporal visual area in the rabbit. *J. Comp. Neurol.* 171:147–56

Meadows, J. C. 1974. Disturbed perception of colours associated with localized cerebral lesions. *Brain* 97:615–32

Mishkin, M. 1972. Cortical visual areas and their interactions. In *Brain and Human Behavior*, ed. A. G. Karczmar, J. C. Eccles, pp. 187–208. Berlin: Springer. 475 pp.

Mohler, C. W., Wurtz, R. H. 1977. Role of striate cortex and superior colliculus in visual guidance of saccadic eye movements in monkeys. *J. Neurophysiol.* 40:74–94

Montero, V. M., Bravo, H., Fernández, V. 1973a. Striate-peristriate cortico-cortical connections in the albino and gray rat. *Brain Res.* 53:202–7

Montero, V. M., Murphy, E. H. 1976. Cortico-cortical connections from the striate cortex in the rabbit. *Anat. Rec.* 184:483

Montero, V. M., Rojas, A., Torrealba, F. 1973b. Retinotopic organization of striate and peristriate visual cortex in the albino rat. *Brain Res.* 53:197–201

Morrell, F. 1972. Visual system's view of acoustic space. *Nature* 238:44–46

Mountcastle, V. B., Lynch, J. C., Georgopoulos, A., Sakata, H., Acuna, C. 1975. Posterior parietal associations cortex of the monkey: command functions for operations within extrapersonal space. *J. Neurophysiol.* 38:871–908

Movshon, J. A. 1975. The velocity tuning of single units in cat striate cortex. *J. Physiol. London* 249:445–68

Myerson, J., Manis, P. B., Miezin, F. M., Allman, J. M. 1977. Magnification in striate cortex and retinal ganglion cell layer of owl monkey: a quantitative comparison. *Science* 198:855–57

Nelson, J. I., Kato, H., Bishop, P. O. 1977. Discrimination of orientation and position disparities by binocularly activated neurons in cat striate cortex. *J. Neurophysiol.* 40:260–83

Newsome, W. T., Baker, J. F., Miezin, F. M., Myerson, J., Petersen, S. E., Allman, J. M. 1978. Functional localization of neuronal response properties in extrastriate visual cortex of the owl monkey. *Assoc. Res. Vision Ophthalmol.* p. 174 (Abstr.)

Nikara, T., Bishop, P. O., Pettigrew, J. D. 1968. Analysis of retinal correspondence by studying receptive fields of binocular single units in cat striate cortex. *Exp. Brain Res.* 6:353–72

Ogren, M. P., Hendrickson, A. E. 1976. Pathways between striate cortex and subcortical regions in *Macaca mulatta* and *Saimiri sciureus:* Evidence for a reciprocal pulvinar connection. *Exp. Neurol.* 53:780–800

Ogren, M. P., Hendrickson, A. E. 1977. The distribution of pulvinar terminals in visual areas 17 and 18 of the monkey. *Brain Res.* 137:343–50

Otsuka, R., Hassler, R. 1962. Über Aufbau und Gliederung der corticalen Sehsphäre bei der Katze. *Arch. Psychiat. Nervenkr.* 203:212–34

Palmer, L. A., Rosenquist, A. C. 1974. Visual receptive fields of single striate cortical units projecting to the superior colliculus in the cat. *Brain Res.* 67:27–42

Palmer, L. A., Rosenquist, A. C., Tusa, R. J. 1978. The retinotopic organization of lateral suprasylvian visual areas in the cat. *J. Comp. Neurol.* 177:237–56

Pandya, D. N., Kuypers, H. G. J. M. 1969. Cortico-cortical connections in the rhesus monkey. *Brain Res.* 13:13–36

Pettigrew, J. D. 1973. Binocular neurones which signal change of disparity in area 18 of cat visual cortex. *Nature* 241:123–24

Pettigrew, J. D., Nikara, T., Bishop, P. O. 1968. Binocular interaction on single units in cat striate cortex: simultaneous stimulation by single moving slit with receptive fields in correspondence. *Exp. Brain Res.* 6:391–410

Poggio, G. F., Baker, F. H., Mansfield, R. J. W., Sillito, A., Grigg, P. 1975. Spatial and chromatic properties of neurons subserving foveal and parafoveal vision in rhesus monkey. *Brain Res.* 100: 25–59

Poggio, G. F., Fischer, B. 1977. Binocular interaction and depth sensitivity in striate and prestriate cortex of behaving rhesus monkey. *J. Neurophysiol.* 40: 1392–1405

Pollen, D. A., Ronner, S. F. 1975. Periodic excitability changes across the receptive fields of complex cells in the striate and parastriate cortex of the cat. *J. Physiol. London* 245:667–97

Polyak, S. 1957. *The Vertebrate Visual System,* ed. H. Klüver. Chicago: Univ. Chicago Press. 1390 pp.

Pöppel, E., Held, R., Frost, D. 1973. Residual visual function after brain wounds involving the central visual pathways in man. *Nature* 243:295–96

Pribram, K. H., Spinelli, D. N., Reitz, S. L. 1969. The effects of radical disconnexion of occipital and temporal cortex on visual behaviour of monkeys. *Brain* 92:301–12

Robinson, D. L., Goldberg, M. E. 1977. Functional properties of posterior parietal cortex of the monkey. I. Sensory responses. *Soc. Neurosci.* 3:574 (Abstr.)

Rocha-Miranda, C. E., Bombardieri, R. A. Jr., de Monasterio, F. M., Linden, R. 1973. Receptive fields in the visual cortex of the opossum. *Brain Res.* 63:362–67

Rose, D. 1977. Responses of single units in cat visual cortex to moving bars of light as a function of bar length. *J. Physiol. London* 271:1–23

Rosenquist, A. C., Edwards, S. B., Palmer, L. A. 1974. An autoradiographic study of the projections of the dorsal lateral geniculate nucleus and the posterior nucleus in the cat. *Brain Res.* 80:71–93

Rowe, M. H., Rezak, M., Benevento, L. A. 1977. Geniculate projections to striate cortex (area 17) in the squirrel monkey as demonstrated by transneuronal autoradiography. *Soc. Neurosci.* 3:575 (Abstr.)

Sanides, F. 1970. Functional architecture of motor and sensory cortices in primates in light of a new concept of neocortical evolution. In *The Primate Brain,* ed. C. R. Noback, W. Montangna, pp. 137–208. New York: Appleton. 320 pp.

Sanides, F., Hoffman, J. 1969. Cyto- and myeloarchitecture of the visual cortex of the cat and of the surrounding integration cortices. *J. Hirnforsch.* 11:79–104

Sarmiento, R. F. 1975. The stereoacuity of macaque monkey. *Vision Res.* 15: 493–98

Schilder, P., Pasik, T., Pasik, P. 1971. Extrageniculostriate vision in the monkey. II. Demonstration of brightness discrimination. *Brain Res.* 32:383–98

Schiller, P. H., Finlay, B. L., Volman, S. F. 1976a. Quantitative studies of single-cell properties in monkey striate cortex. I. Spatiotemporal organization of receptive fields. *J. Neurophysiol.* 39:1288–1319

Schiller, P. H., Finlay, B. L., Volman, S. F. 1976b. Quantitative studies of single-cell properties in monkey striate cortex. III. Spatial frequency. *J. Neurophysiol.* 39:1334–51

Schiller, P. H., Malpeli, J. G. 1977. The effect of striate cortex cooling on area 18 cells in the monkey. *Brain Res.* 126:366–69

Schiller, P. H., Stryker, M. 1972. Single-unit recording and stimulation in superior colliculus of the alert rhesus monkey. *J. Neurophysiol.* 35:915–24

Schneider, G. E. 1969. Two visual systems. *Science* 163:895–902

Shatz, C. J., Lindström, S., Wiesel, T. N. 1977. The distribution of afferents representing the right and left eyes in the cat's visual cortex. *Brain Res.* 131:103–16

Singer, W., Tretter, F., Cynader, M. 1975. Organization of cat striate cortex: a correlation of receptive field properties with afferent and efferent connections. *J. Neurophysiol.* 38:1080–98

Sousa, A. P. B., Gattass, R., Oswaldo-Cruz, E. 1978. The projection of the opossum's visual field on the cerebral cortex. *J. Comp. Neurol.* 177:569–88

Spatz, W. B. 1975. An efferent connection of the solitary cells of Meynert. A study with horseradish peroxidase in the marmoset *Callithrix. Brain Res.* 92:450–55

Spatz, W. B. 1977. Topographically organized reciprocal connections between areas 17 and MT (visual area of superior temporal sulcus) in the marmoset (Callithrix jacchus). *Exp. Brain Res.* 27: 559–72

Spatz, W. B., Erdmann, G. 1974. Striate cortex projections to the lateral geniculate and other thalamic nuclei; a study using degeneration and autoradiographic tracing methods in the marmoset *Callithrix. Brain Res.* 82:91–108

Spatz, W. B., Tigges, J. 1972a. Species difference between Old World and New World monkeys in the organization of the striate-prestriate association. *Brain Res.* 43:591–94

Spatz, W. B., Tigges, J. 1972b. Experimental-anatomical studies on the "Middle Temporal Visual Area (MT)" in primates. I. Efferent cortico-cortical connections in the marmoset *Callithrix jacchus. J. Comp. Neurol.* 146:451–64

Spatz, W. B., Tigges, J. 1973. Studies on the visual area MT in primates. II. Projection fibers to subcortical structures. *Brain Res.* 61:374–78

Spatz, W. B., Tigges, J., Tigges, M. 1970. Subcortical projections, cortical associations, and some intrinsic interlaminar connections of the striate cortex in the squirrel monkey (*Saimiri*). *J. Comp. Neurol.* 140:155–74

Spear, P. D., Baumann, T. P. 1975. Receptive-field characteristics of single neurons in lateral suprasylvian visual area of the cat. *J. Neurophysiol.* 38:1403–20

Sprague, J. M., Levy, J., DiBerardino, A., Berlucchi, G. 1977. Visual cortical areas mediating form discrimination in the cat. *J. Comp. Neurol.* 172:441–88

Stensaas, S. S., Eddington, D. K., Dobelle, W. H. 1974. The topography and variability of the primary visual cortex in man. *J. Neurosurg.* 40:747–55

Talbot, S. A. 1942. A lateral localization in cat's visual cortex. *Fed. Proc.* 1:84

Talbot, S. A., Marshall, W. H. 1941. Physiological studies on neural mechanisms of visual localization and discrimination. *Am. J. Ophthalmol.* 24:1255–64

Thompson, J. M., Woolsey, C. N., Talbot, S. A. 1950. Visual areas I and II of cerebral cortex of rabbit. *J. Neurophysiol.* 13:277–88

Tiao, Y.-C., Blakemore, C. 1976. Functional organization in the visual cortex of the golden hamster. *J. Comp. Neurol.* 168:459–82

Tigges, J., Spatz, W. B., Tigges, M. 1973a. Reciprocal point-to-point connections between parastriate and striate cortex in the squirrel monkey (*Saimiri*). *J. Comp. Neurol.* 148:481–90

Tigges, J., Spatz, W. B., Tigges, M. 1974. Efferent cortico-cortical fiber connections of area 18 in the squirrel monkey (*Saimiri*). *J. Comp. Neurol.* 158:219–36

Tigges, J., Tigges, M., Kalaha, C. S. 1973b. Efferent connections of area 17 in *Galago. Am. J. Phys. Anthropol.* 38: 393–98

Tigges, J., Tigges, M., Perachio, A. A. 1977. Complementary laminar terminations of afferents to area 17 originating in area 18 and in the lateral geniculate nucleus in squirrel monkey. *J. Comp. Neurol.* 176:87–100.

Towns, L. C., Giolli, R. A., Haste, D. A. 1977. Corticocortical fiber connections of the rabbit visual cortex: a fiber degeneration study. *J. Comp. Neurol.* 173:537–560

Tretter, F., Cynader, M., Singer, W. 1975. Cat parastriate cortex: a primary or secondary visual area? *J. Neurophysiol.* 38:1099–1113

Trevarthen, C. W. 1968. Two mechanisms of vision in primates. *Psychol. Forsch.* 31:299–377

Trojanowski, J. Q., Jacobson, S. 1976. Areal and laminar distribution of some pulvinar cortical efferents in rhesus monkey. *J. Comp. Neurol.* 169:371–92

Tusa, R. J., Palmer, L. A., Rosenquist, A. C. 1975. The retinotopic organization of the visual cortex in the cat. *Soc. Neurosci.* 1:52 (Abstr.)

Tusa, R. J., Palmer, L. A., Rosenquist, A. C. 1978. The retinotopic organization of area 17 (striate cortex) in the cat. *J. Comp. Neurol.* 177:213–36

Tyler, C. W. 1978. Selectivity for spatial frequency and bar width in cat visual cortex. *Vision Res.* 18:121–22

Ungerleider, L. G., Mishkin, M. 1978. The visual area in the superior temporal sulcus of Macaca mulatta: location and topographic organization. *Anat. Rec.* 190:568

Updyke, B. V. 1977. Topographic organization of the projections from cortical areas 17, 18, and 19 onto the thalamus, pretectum and superior colliculus in the cat. *J. Comp. Neurol.* 173:81–122

Van Essen, D. C., Zeki, S. M. 1978. The topographic organization of rhesus monkey prestriate cortex. *J. Physiol. London* 277:193–226

von Economo, C., Koskinas, G. N. 1925. *Die Cytoarchiteckonik der Hirnrinde des erwachsen Menchen.* Berlin u Wien: Springer. 810 pp.

Wagor, E., Lin, C. S., Kaas, J. H. 1975. Some cortical projections of the dorsomedial visual area (DM) of association cortex in the owl monkey, *Aotus trivirgatus. J. Comp. Neurol.* 163:227–50

Wagor, E., Mangini, N., Pearlman, A. L. 1977. A retinotopic map of mouse extrastriate visual cortex. *Soc. Neurosci.* 3:580 (Abstr.)

Weber, J. T., Casagrande, V. A., Harting, J. K. 1977. Transneuronal transport of [³H] proline within the visual system of the grey squirrel. *Brain Res.* 129:346–52

Weiskrantz, L. 1972. Behavioural analysis of the monkey's visual nervous system. *Proc. R. Soc. London Ser. B* 182:427–55

Weiskrantz, L., Warrington, E. K., Sanders, M. D., Marshall, J. 1974. Visual capacity in the hemianopic field following a restricted occipital ablation. *Brain* 97:709–28

Wilson, J. R., Hendrickson, A. E., Ogren, M. P. 1977. Connections between the dorsal lateral geniculate nucleus and visual cortex in the macaque and squirrel monkey. *Soc. Neurosci.* 3:582 (Abstr.)

Wilson, M. E. 1968. Cortico-cortical connexions of the cat visual areas. *J. Anat. London* 102:375–86

Wilson, M. E., Cragg, B. G. 1967. Projections from the lateral geniculate nucleus in the cat and monkey. *J. Anat. London* 101:677–92

Wong-Riley, M. T. T. 1976. Projections from the dorsal lateral geniculate nucleus to prestriate cortex in the squirrel monkey as demonstrated by retrograde transport of horseradish peroxidase. *Brain Res.* 109:595–600

Woolsey, C. N. 1971. Comparative studies on cortical representation of vision. *Vision Res.* Supp. 3:365–82

Woolsey, C. N., Sitthi-amorn, C., Keesey, V. T., Holub, R. A. 1973. Cortical visual areas of the rabbit. *Ann. Meet. Soc. Neurosci., 3rd, San Diego,* p. 180 (Abstr.)

Wurtz, R. H. 1969. Visual receptive fields of striate cortex neurons in awake monkeys. *J. Neurophysiol.* 32:727–42

Zeki, S. M. 1969. Representation of central visual fields in prestriate cortex of monkey. *Brain Res.* 14:271–91

Zeki, S. M. 1970. Interhemispheric connections of prestriate cortex in monkey. *Brain Res.* 19:63–75

Zeki, S. M. 1971. Cortical projections from two prestriate areas in the monkey. *Brain Res.* 34:19–35

Zeki, S. M. 1973. Colour coding in rhesus monkey prestriate cortex. *Brain Res.* 53:422–27

Zeki, S. M. 1974a. Functional organization of a visual area in the posterior bank of the superior temporal sulcus of the rhesus monkey. *J. Physiol. London* 236:549–73

Zeki, S. M. 1974b. Cells responding to changing image size and disparity in the cortex of the rhesus monkey. *J. Physiol. London* 242:827–41

Zeki, S. M. 1977. Colour coding in the superior temporal sulcus of rhesus monkey visual cortex. *Proc. R. Soc. London Ser. B.* 197:195–223

Zeki, S. M. 1978a. The cortical projections of foveal striate cortex in the rhesus monkey. *J. Physiol London* 277:227–44

Zeki, S. M. 1978b. The third visual complex of rhesus monkey prestriate cortex. *J. Physiol. London* 277:245–72

Zeki, S. M. 1978c. Uniformity and diversity of structure and function in rhesus monkey prestriate visual cortex. *J. Physiol. London* 277:273–90

Zeki, S. M., Sandeman, D. R. 1976. Combined anatomical and electrophysiological studies on the boundary between the second and third visual areas of rhesus monkey cortex. *Proc. R. Soc. London Ser. B* 194:555–62

Ann. Rev. Neurosci. 1979. 2:265–89

VESTIBULAR MECHANISMS ❖11524

Wolfgang Precht

Neurobiologische Abteilung, Max Planck Institut für Hirnforschung,
Frankfurt am Main, West Germany

The progress of research in the vestibular system in the past decade has been characterized by the accumulation of an enormous wealth of new data, particularly in the field of central vestibular mechanisms. Single unit recordings at the extra- and intracellular level performed in various species and in anesthetized, as well as alert, animals have given us a much deeper insight into the neuronal operations performed by this system. It is now possible, for the first time, to correlate neuronal operations with behavior, at least in some of the subfields of vestibular research. In a brief review it is impossible to describe all the new and important findings. Instead, I concentrate on the most recent and important developments in research, specifically as regards the role that the vestibular system plays in the control of eye movements. For the interested reader a comprehensive summary of most aspects of the vestibular system can be found in the *Handbook of Sensory Physiology* (Kornhuber 1974). More recent and specialized reviews are those by Goldberg & Fernández (1975), Precht (1975a,b, 1976, 1978), Wilson & Peterson (1978), and symposium proceedings edited by Baker & Berthoz (1977).

PRIMARY VESTIBULAR NEURONS

An extensive account of the response characteristics of primary afferents has recently been given by Goldberg & Fernández (1975). The present review, therefore, summarizes only those data that have appeared thereafter and are of relevance to the main topic of the review.

Semicircular Canal Afferents

Analyses of the response characteristics of the primary canal neurons of various species in the frequency domain have revealed both an amazing

265

degree of similarity, as well as some major species differences. Figure 1 (upper diagram) shows a comparison between the phase lag of the responses relative to angular acceleration of the head of canal afferents of three widely separated vertebrates, namely the unanesthetized frog (Blanks & Precht 1976), anesthetized cat (Anderson et al 1978), and monkey (Fernández & Goldberg 1971). The shape of the phase plots for the three species is very similar. At frequencies between 0.1 and 1 Hz the primary canal neurons carry essentially a head velocity signal. This means that head acceleration, which represents the adequate stimulus for the canals, has been integrated by peripheral mechanisms. The monkey data show that at frequencies greater than 1 Hz there is, however, progressively less of a phase lag. For the frog one finds smaller phase lags and correspondingly shorter time constants (ca. 3 sec). This species difference is partially due to differences in the physical dimensions of the canals (Melvill Jones & Spells 1963) and/or the greater degree of sensory adaptation in frog afferents. It should be noted that the stimulus-response curves of cat and monkey have been obtained under barbiturate anesthesia, whereas frog afferents were studied in unanesthetized preparations. Apparently, anesthesia does not significantly affect the frequency responses of primary canal neurons, e.g. by blocking efferent control. Recent data obtained in the alert monkey support this notion (Keller 1976). Contrary to the small differences in the phase behavior of different species, more striking differences are seen when the gains are compared (Figure 1, lower diagram). Surprisingly, the gain is lowest in monkey and highest in frog. One of the reasons for the high gain in frog primary afferents may be related to the fact that this animal has a very poorly developed vestibular commissural inhibition (Ozawa et al 1974), which is known to increase the sensitivity of secondary vestibular neurons in higher vertebrates (Shimazu & Precht 1966). Thus, in the frog a high peripheral gain may assure the sensitivity of the vestibular system, which is, indeed, similar to that of other species (Figure 1, lower diagram). In all three species the gain of irregular, adapting units is higher than that of regular, nonadapting neurons. It seems that regular and irregular units are in contact with different types of hair cells (O'Leary et al 1974).

The mean resting activity of primary neurons of frog, cat, and monkey increases from values of less than ten in the frog (Precht et al 1971, Blanks & Precht 1976) to ca. 60 and 90 impulses/sec in cat (Anderson et al 1978, Estes et al 1975) and monkey (Goldberg & Fernández 1971), respectively. At present there exists no reasonable explanation for these large differences. Anesthesia has no significant effects on resting rate in monkey (Keller 1976) and cat (Blanks & Precht, unpublished observations).

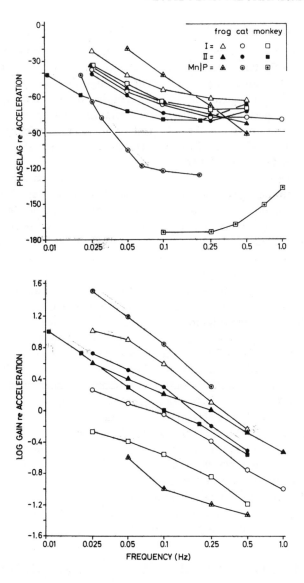

Figure 1 Phases and gains of primary (I), secondary (II) vestibular neurons, abducens motoneurons (Mn) and Purkinje cells (P) (in frog only) during horizontal rotation in the dark (for details and references see text).

Otolith Afferents

Most of the important static and dynamic features of otolithic inputs studied prior to 1975 have been summarized by Goldberg & Fernández (1975). Here I briefly summarize more recent findings that are primarily of a comparative physiological nature. Studies of the comparative physiology of the vertebrate peripheral otolith system comprise work performed in the monkey (Fernández & Goldberg 1976a,b,c), cat (Anderson et al 1978), fish (Macadar et al 1975), and frog (Lowenstein & Saunders 1975, Blanks & Precht 1976, Caston et al 1977).

In all species *tonic, phasic-tonic,* and *phasic* response patterns of otolith units to ramp changes in angular position of the head were noted. *Tonic* neurons are modulated as a function of head position in space (gravity) and are nonadapting. *Phasic-tonic* units respond to gravity and show significant adaptation, i.e. during transitions in head position their firing rate is dependent upon the head velocity. *Phasic* units respond only when the head is moving and have no position sensitivity over wide ranges. Whereas in higher vertebrates the tonic and phasic-tonic response patterns prevail, in lower species (frog, fish) there is a preponderance of phasic-tonic and phasic response types to gravitational stimulation. When measured in the frequency domain, frog phasic and phasic-tonic units show a phase lead of ca. $45 \pm 17°$ at 0.25 Hz with respect to displacement and a gain that increases with frequency (Blanks & Precht 1976, Caston et al 1977). Phasic-tonic units in higher vertebrates (monkey, cat) showed similar phase leads and gain enhancements as a function of frequency (Goldberg & Fernández 1975, Fernández & Goldberg 1976c, Anderson et al 1978). Tonic units, however, show small leads of $0 - 15°$ at 0.025 Hz, which remain constant or decrease to 0 to $-15°$ at 1.0 Hz. Clearly then, the different otolith inputs can convey information to the CNS about both the magnitude and rate of change of gravity and linear accelerations in general.

The considerable species differences in the relative distribution of tonic and phasic units might indicate a greater need for static postural control of head, body, and eyes in higher vertebrates.

VESTIBULAR SYSTEM AND EYE MOVEMENTS

The role of the vestibular system in the control of eye movements has been intensively investigated in the last decade. In this section the major results concerning the functional synaptology of vestibuloocular relations and the response of single units in unanesthetized animals during various kinds of eye movements is described.

Functional Synaptology of the Vestibuloocular Reflex

The pathways mediating the vestibuloocular reflexes (VOR)—that is the pathways conveying vestibular signals to the motoneurons—have classically been divided into direct and indirect pathways (Lorente de Nó 1933, Szentágothai 1950). The *direct* pathway is the three-neuronal arc consisting of primary and secondary vestibular neurons and the ocular motoneurons, and the *indirect* path is formed by polysynaptic routes through the reticular formation. This strict division into two separate systems of pathways has recently been questioned by the finding that the direct path likewise may carry information previously thought to be mediated solely by the indirect route (see below).

DIRECT VESTIBULOOCULAR PROJECTIONS   At the present time detailed knowledge of the synaptology of the vestibuloocular reflex (VOR) is available for the cat (Baker et al 1969, 1973, Maeda et al 1972, Precht & Baker 1972, Precht et al 1967, Richter & Precht 1968, Baker 1977, Schwindt et al 1973, Hwang & Poon 1975) and the rabbit (Highstein 1973a, b, Ito et al 1976a, b, Ghelarducci et al 1977). The results of these physiological studies are summarized in Figure 2 and I do not restate them here (for summary of the anatomy see Tarlov 1970, Graybiel 1977, Büttner-Ennever 1977). Seven points, however, deserve emphasis:

1. There are no significant differences in VOR synaptology between cat and rabbit; this finding may indicate that the functional synaptology of the monkey which has not yet been studied, may be similar.
2. From Figure 2 it is apparent that each canal gives rise to two excitatory and two inhibitory VORs, resulting in 12 reciprocally organized canalocular reflexes. Except for the inhibitory VOR to MR motoneurons (Baker 1977) all reflexes are three-neuronal in nature. The presence of inhibitory VORs implies that with head rotation in the plane of a given canal pair, excitation of ocular motoneurons is achieved by the sum of activation and disinhibition. At the same time the excitability of antagonistic motoneurons is modified by inhibition and disfacilitation.
3. Most inhibitory vestibular relay neurons are located in the superior vestibular nucleus, the exception being the vestibuloocular reflex to the medial rectus, which has its relay in the medial vestibular nucleus.
4. Contrary to some previous reports (Highstein 1973a) the excitatory VORs from the anterior canal to the ipsilateral superior rectus and contralateral inferior oblique muscle have their relay neurons in the dorsal part of the SVN and not the *y*-group (Ghelarducci et al 1977).

5. Besides the medial longitudinal fascicle (MLF), the brachium conjunctivum (BC) and the ascending tract of Deiters (ATD) provide routes for secondary neurons to motoneurons. Whereas the MLF contains the connections of horizontal and posterior canals and their respective motoneurons, the BC path is associated with VOR paths from anterior canals to motoneurons. Furthermore, since the BC path is composed of at least three components, i.e. fibers originating in superior vestibular nucleus, y-group, and a few axons of the p-cell part of the cerebellar dentate nucleus (Ghelarducci et al 1977), it is also involved in otolith-ocular reflexes (see below). The ATD projection to medial rectus motoneurons makes another exceptional linkage from the horizontal canals (Baker 1977).

6. There are still considerable gaps in the knowledge of synaptic arrangement of the otolithic-ocular reflexes. Whereas there is evidence for excitatory three-neuronal connections to some muscles (lateral and medial rectus and superior oblique), inhibitory VORs appear to be established via polysynaptic routes. That such connections must exist has recently been shown by Blanks et al (1978). The utricular-evoked reflexes are presumably mediated through neurons located in the lateral vestibular nucleus (LVN), and the saccular projections may involve both LVN and y-group. In the cat, y-group neurons antidromically driven by stimulation of the IIIrd nucleus are only polysynaptically excited by saccular nerve stimulation (Hwang & Poon 1975); disynaptic projections are established via the LVN. Whether the p-cell part of the cerebellar dentate nucleus is part of the otolith-ocular pathway remains to be clarified; its neurons respond to otolithic stimuli and may be excited mono- and polysynaptically from the vestibular nerve (Ghelarducci et al 1977).

7. Although not included in Figure 2 lower vertebrates such as the frog seem to have a similar arrangement of VOR connections (Precht 1976).

A few studies have been performed concerning the possible nature of the inhibitory transmitter substance released by secondary vestibular neurons of cat (Precht et al 1973) and rabbit (Ito et al 1970). There is good evidence suggesting that GABA is the inhibitory transmitter of the VOR. It is contained in very large boutons located at and near the soma of motoneurons (Bak et al 1976). Nothing is known on the nature of the excitatory transmitter substance in the VOR. Presumed excitatory boutons are predominantly located at the dendrites (Precht & Baker 1972, Bak et al 1976).

INDIRECT VESTIBULOOCULAR PROJECTIONS   In the context of this chapter, indirect projections are those that involve more than the three neurons typical of the direct pathways described above. The most important ones are: (*a*) *intranuclear* and *internuclear* circuits, (*b*) *vestibulo-reticulo-oculomotor* paths, (*c*) *vestibulo-reticulo-vestibulo-oculomotor* connections (reafference), and (*d*) *vestibulo-cerebello-vestibulo-oculomotor* pathways. In the following an attempt is made to correlate recent anatomical and physiological data obtained in this complete field. The functional importance of the indirect pathways compared to the direct routes is discussed.

Intranuclear excitatory polysynaptic circuits were postulated by Precht & Shimazu (1965) to explain the exclusively polysynaptic activation of tonic type I vestibular neurons following weak electrical stimulation of the VIIIth nerve. Some neurons fire at monosynaptic latencies only and others are activated polysynaptically with weak stimuli and, in addition, also monosynaptically when stronger stimuli are used. Finally there are units that

Receptor	Effect	Muscle	Relay-Nucleus	Path-way	Motor Nucleus	Flocc. Stim.
HC	excitation	c - LR	MVN	'MLF'	c - VI	−
		i - MR	LVN	ATD	i - III	+
	inhibition	i - LR	MVN	'MLF'	i - VI	+
		c - MR		extra MLF polysyn.	c - III	−
AC	excitation	i - SR	SVN	BC	c - III	+
		c - IO	SVN	BC	c - III	+
	inhibition	i - IR	SVN	MLF	i - III	+
		c - SO	SVN	MLF	i - IV	+
PC	excitation	c - IR	MVN	MLF	c - III	−
		i - SO	MVN	MLF	c - IV	(+) −
	inhibition	c - SR	SVN	MLF	i - III	−
		i - IO	SVN	MLF	i - III	−
U	excitation	i - LR	LVN*	'MLF'	i - VI	+
		c - MR	LVN*	MLF	c - III	
		i - SO	LVN*	MLF	c - IV	+
S	excitation		y-group	BC		

Figure 2 Summary diagram of the direct (short latency) vestibuloocular reflex paths and their relation to the cerebellar flocculus. Note "MLF" indicates that strictly speaking it is not the MLF but rather a direct projection from the vestibular nuclei. Asterisks denote connections that are highly probable but not proven, and open spaces indicate still unknown connections. Diagram is based on data in cat and rabbit (ref. see text). It should be noted that the indirect pathway to MR through abducens interneurons and MLF has not been included. Receptor organs, HC, AC, PC, U, and S indicate horizontal, anterior, posterior canals, utriculus, and sacculus. Muscles: c and i, are ipsi- and contra-lateral relative to the endorgan; LR, MR, SR, IR, IO, SO indicate lateral, medial, superior, inferior recti, and inferior and superior oblique muscles. Relay: MVN, LVN, SVN medial, lateral, superior vestibular nucleus. Pathway: ATD: ascending tract of Deiters'; MLF: medial longitudinal fasciculus; BC: brachium conjunctivum; Flocc: Flocculus; stim: electrical stimulation of flocculus.

behave opposite to the last group. This latter group undoubtedly receives reafference from the reticular formation. As is shown in the next section, classification of units into mono- and polysynaptic types becomes functionally meaningful when their firing pattern is studied in conjunction with eye movements. Those neurons that are exclusively driven polysynaptically are presumably located within regions of the vestibular nuclei that are free of primary afferents (Brodal et al 1962).

Internuclear or *commissural* pathways mediate crossed inhibition between secondary vestibular neurons of synergistic canal pairs (for ref. see Precht 1975a), which results in an increase in sensitivity of these neurons (Shimazu & Precht 1966). Thus, the direct VORs (see Fig. 2) and the effects brought about by commissural inhibition are mediated by the same vestibular neurons. This is one example of the close interaction between direct and indirect projection and demonstrates the difficulty of functionally separating the two. Indeed, any consideration of VOR dynamics has to take into account commissural mechanisms (Berthoz et al 1973).

Vestibulo-reticulo-oculomotor and vestibulo-reticulo-vestibulo-oculomotor paths were postulated long ago, since lesion of the MLF did not abolish vestibular eye movements and stimulation of the reticular formation, particularly the paramedian pontine reticular formation (PPRF), generated eye movements (for ref. see Cohen 1974). The synaptic circuitry of these complex projections is beginning to show some contours that deserve presentation, although it should be emphasized that we are still at the beginning of the understanding of its functional complexity. To facilitate the description, a schematic diagram (Fig. 3) is presented that shows some of the known direct and indirect (vestibulo-reticulomotor) paths for the motoneurons of the lateral (LR) and medial recti muscles (MR), i.e. for the horizontal system. Starting with the medial rectus (MR) the direct VORs are labeled 1 and 5 (see fig. 2). There are at least two indirect paths: the first pathway involves the interneurons in and around the VIth nuclei, which receive inhibitory (path 2) and excitatory (path 3) disynaptic inputs from V_i and V_c, respectively; their axons, in turn, project via MLF to contralateral MR (Baker & Highstein 1975, Graybiel & Hartwieg 1974, Highstein et al 1976). It is highly probable that the horizontal eye movement–related burst-tonic MLF units (King et al 1976, Pola & Robinson 1978) are, for the most part, axons of the above interneurons, which are part of a horizontal gaze center organizing the joint action of the MR/LR muscle pair. The effects of MLF lesion on horizontal gaze are compatible with this concept (Evinger et al 1977). A second indirect route to MR runs through the *n. prepositus hypoglossi* (paths 6 and 7 in Fig. 3). Neurons in this nucleus receive disynaptic inhibition and excitation after stimulation of ipsi- (V_i) and contralateral (V_c) VIIIth nerves, respectively, and, in turn, project to the contralateral

MR. The opposite pattern was less frequently observed (Baker & Berthoz 1975). These electrophysiological results are supported by a natural stimulation study showing that the majority of the prepositus neurons show type II responses to horizontal rotation (Blanks et al 1977b). It is interesting to note that the phase relationship of prepositus cells to angular acceleration is characterized by a phase lag that is larger than that of most vestibular neurons, which indicates that a partial integration of the velocity signal has already occurred (Blanks et al 1977b). This finding, and the fact that prepositus units fire in relation to eye position and often are of burst-tonic character (Baker et al 1976), strongly suggests that these neurons are involved in the organization of eye movements. Their efferent projection to the cerebellar vermis and afference from the vestibulocerebellum point in this direction as well (Torvik & Brodal 1954). Furthermore, prepositus neurons have visual receptive fields similar to those of collicular neurons (Gresty & Baker 1976) and collicular-evoked horizontal eye movements have been shown to utilize an area caudal to the VIth nucleus, presumably the prepositus nucleus (Hyde & Eliasson 1957).

The lateral rectus (LR) likewise receives direct and indirect paths. Paths 2 and 3 of Figure 3 are the direct inhibitory and excitatory VORs. Indirect projections may reach motoneurons via the prepositus nucleus (paths 6 and 7) or through path 5, which includes an inhibitory burst neuron (B in Fig. 3) located in the medial reticular formation caudal to the VIth nuclei

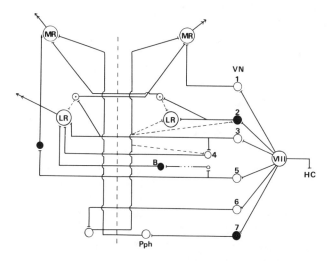

Figure 3   Schematic diagram of the direct and indirect VOR connections to horizontal eye muscles. Abbr.: B, inhibitory burst neurons; HC, horizontal canal; LR, MR lateral and medial rectus motoneurons; Pph, nucl. prepositus hypoglossi; VN, vestibular nuclei. Inhibitory neurons are represented by filled circles. For explanation see text.

(Hikosaka & Kawakami 1977, Hikosaka et al 1977). These inhibitory burst neurons fire in conjunction with the nystagmic, inhibitory quick phase in abducens motoneurons and are silent during the slow phase. No excitatory burst neurons have been found so far. Neurons related to the excitatory quick phase and recorded in the VIth nucleus (neuron 4 in Fig. 3) which are of burst-tonic character, are disynaptically activated from V_c and probably originate from the vestibular nuclei or, less likely, the prepositus nucleus (Hikosaka et al 1978) and abducens interneurons (Baker 1977). Whereas the monosynaptic excitatory path (3 in Fig. 3) shows no eye movement related activity but rather a head velocity signal, the monosynaptic inhibitory path (2 in Fig. 3) and, in particular, the disynaptic excitatory path (4 in Fig. 3, see above), are eye position related (Hikosaka et al 1977). At present it is not known how monosynaptic V_i inhibitory axons receive a weak eye position signal, i.e. an integrated head velocity signal; this may occur via the above mentioned intranuclear chains and/or reafference from reticular integrator circuits (see below).

Just as the caudal PPRF plays an important role in the indirect vestibuloocular paths for the horizontal eye movements, the region of the interstitial nucleus of Cajal (INC) is probably a crucial link for indirect routes to motoneurons organizing vertical and rotatory eye movements (Szentágothai 1943, Hassler & Hess 1954). This nucleus receives disynaptic IPSPs from the ipsilateral and disynaptic EPSPs from the contralateral VIIIth nerves via the MLF (King et al 1978a) and projects monosynaptically to vertical ocular motoneurons (Schwindt et al 1974). The latter projection is both crossed and uncrossed and inhibitory as well as excitatory. Recent anatomical work (Büttner-Ennever 1977, Graybiel 1977) shows that not only the INC but also other mesencephalic reticular nuclei [nucleus of posterior commissure (NPC), nucleus of the pre-rubral field (NPRF)] project to motor nuclei of vertical eye muscles. It is interesting to note, however, that vestibular projections are confined to INC, NPC, and a small part of NPRF. On the other hand, the PPRF projects to the NPRF, but not to INC or NPC. This latter projection may contain, at least in part, the axons of omnipause neurons, which are antidromically activated from this region (King et al 1978b). Recent work in the alert monkey (King 1976) and cat (King et al 1978a) has shown that the mesencephalic reticular region contains burst-tonic and medium lead burst neurons related only to vertical eye movements. Both types of units appear to be missing in the ascending MLF (King et al 1976, Pola & Robinson 1978): MLF axons projecting to motoneurons of vertical eye muscles contain eye position and head velocity signals, and in addition pause, but do not burst, during saccades. Results obtained by bilateral MLF lesions are in agreement with these single unit data: the vertical VOR was abolished, saccades were normal, but excentric fixation positions could not be maintained, i.e. a deficiency of the neural integrator was noted (Evinger et al 1977). On the other hand, lesion of the INC in the cat abolished quick phases of vertical vestibular nystagmus (Szentágothai 1943). Somewhat different results have been obtained in the monkey (Carpenter et al 1970). Given that interstitial neurons also receive synaptic excitation and/or inhibition from the frontal eye fields (King et al 1978a), and presumably from reticular and tectal neurons either directly or indirectly, it appears that this group of neurons is the prime candidate for an immediate premotor nucleus concerned with vertical eye movements. Finally, it should be mentioned that the INC region not only projects directly onto particular ocular motoneurons but also on secondary vestibular neurons (Szentágothai 1943, Markham et al 1966). Possibly, some vestibular neurons receive, by way of interstitial neurons, an eye position signal.

Another known indirect VOR pathway to all vertical eye muscles involves the *prepositus hypoglossi nucleus* (Baker et al 1977) and is exclusively excitatory. The relative importance of this projection is presently not well understood.

Vestibulo-cerebello-vestibular pathways represent yet another indirect route by which vestibular signals may reach motoneurons. They can be divided into a pathway via the cerebellar cortex and a transnuclear route.

The transcortical pathway consists of primary and secondary vestibular neurons reaching the vestibulocerebellum (particularly the flocculus) as mossy fibers, which excite Purkinje cells (Precht & Llinás 1969). The latter, in turn, project their inhibitory axons on certain secondary, vestibular VOR relay neurons (Fukuda et al 1972, Baker et al 1972). In addition to the vestibular input, floccular Purkinje cells receive a visual projection (at least in cat and rabbit) via climbing (Maekawa & Simpson 1973, Simpson et al 1974) and mossy fibers (Maekawa & Takeda 1975). Detailed studies of the inhibitory floccular projections onto the 12 VORs have been performed in the rabbit (Ito et al 1977, Ito 1977a) and are summarized in the last column of Figure 2. As a principle, flocculus Purkinje cells act upon only one, but never both VORs reaching a given muscle. It was also found that certain areas (functional subgroups) in the flocculus are connected to particular VOR pathways in a one-to-one fashion (Ito 1977a). This functional localization is supported by anatomical work as well (Yamamoto & Shimoyama 1977).

It appears that although the nodulus has some weak effects on VORs, in particular those originating from the otoliths, its major targets are relay cells not dealing with vestibuloocular transactions but rather vestibulospinal interactions (Precht et al 1976).

Although a full account of the functional implications of cerebello-vestibular interactions is impossible in the context of this brief review a few major points deserve mention. Ito and his associates (for ref. see Ito 1977b) have elaborated in great detail the intricacies of vestibulocerebellar interactions in the rabbit. Based on the hypothesis that the VOR is an open-loop reflex requiring optokinetic support for proper functioning, (i.e. the labyrinth does not receive a negative feedback signal of eye position) it was assumed that such interaction could occur in the flocculus, which, in turn, via its direct output to the vestibular nuclei, would set the VOR relay neurons in proper functional context. In support of this notion it was found that Purkinje cells of the horizontal zone responded to horizontal rotation in the dark predominantly in the type II (outphase) mode whereas pure optokinetic stimuli evoked similar simple spike responses when the surround was rotated in the opposite direction (complex spike patterns were of opposite polarity). When the animal was rotated against a stable visual world the type II Purkinje cell modulation was augmented. This neuronal behavior is in accordance with the increase of the VOR gain occurring under similar conditions. On the other hand, when the visual world moved with the table at twice the amplitude, the opposite effects were noted at unitary as well as VOR levels. Since floccular lesion abolished the optokinetic influences (Takemori & Cohen 1974, Ito 1977b), a causal relationship between Purkinje cell firing and VOR changes are assumed. Interestingly, in the cat even total cerebellectomy does not abolish the immediate effects of vision on the VOR (Keller & Precht 1978a).

When head rotations were continued for several hours it was found that the VOR in the rabbit underwent adaptive changes: its gain increased in the presence of an earth fixed visual world and decreased when the visual pattern moved at twice the amplitude of the table. Purkinje cells from the horizontal zone sampled in the adapted state showed the corresponding gain increases or decreases. These changes are thought to be causal to the adaptive changes in the VOR. In fact, removal of the flocculus abolished all plastic changes observed in the VOR gain (Ito et al 1974, Robinson 1976). Since the flocculus does not reach motoneurons directly but rather through vestibulooocular relay neurons, it may be expected that type I and type II vestibular neurons are likewise affected by conflicting exposure. That this is the case has recently been shown in the cat by Keller & Precht (1978b); type I and type II neurons show significant changes in phase and gain when compared to normal responses. These changes can in part explain the observed alteration in the VOR.

Whereas in Ito's hypothesis the visual input to the flocculus per se is of crucial importance, Lisberger & Fuchs (1977), in studying the flocculus of the alert, trained monkey arrived at a different conclusion. Purkinje cell discharge was studied during three types of eye movements: VORs evoked by rotation in the dark, smooth pursuit eye movements in the absence of head rotation, and during suppression of VOR by visual fixation in light.

Whereas rotations in darkness caused no modulation of firing, strong modulations of the opposite polarity were observed when studied during smooth pursuit or with the VOR suppressed. During smooth pursuit the amplitude of discharge modulation increased as a function of maximum eye velocity and was not correlated with retinal slip velocity. Parenthetically, it should be added that a small number of Purkinje cells were found whose activity may have been correlated with target slip on the retina, i.e. a true visual signal. Based on the above findings and data obtained from recordings of mossy fibers in the flocculus, it was concluded that the flocculus receives a head velocity signal via primary and secondary vestibular fibers and extracts an eye velocity signal arriving from eye movement centers in the brain stem. Thus, it was suggested that the flocculus does not initiate but rather sustains eye velocity generated by visual and vestibular inputs to extrafloccular pathways. In this context, absence of Purkinje cell modulation during rotation in darkness would be caused by a cancellation of the oppositely directed head and eye velocity signals. The strong Purkinje cell modulation occurring when the VOR is suppressed represents the head velocity signal, which is then fed monosynaptically via inhibitory Purkinje axons onto VOR relay neurons in the vestibular nuclei resulting in suppression of VOR.

Finally, it should be mentioned that both in the rabbit (Llinás et al 1976) and monkey (Noda et al 1977) Purkinje cells may show correlation with saccadic eye movements. Firing may precede eye movement by ca. 11 msec or, more frequently, discharge during the whole saccade. Thus, vestibular VOR relay neurons may also be under cerebellar control during saccades.

Besides the pathways through the vestibulocerebellum, vestibular activity may reach motoneurons via the cerebellar nuclei. One such pathway through the dentate nucleus has already been described above. Another route runs via the fastigial nuclei, which receive monosynaptic as well as polysynaptic (via vestibular nuclei) vestibular inputs (Precht & Llinás 1968, Furuya et al 1975). Natural stimulation provoked horizontal canal responses (type I) in the middle and caudal part (Gardner & Fuchs 1975), and type II canal, as well as otolith, responses in the rostral part (Gardner & Fuchs 1975, Ghelarducci 1973). The caudally located type I neurons were antidromically activated from the contralateral vestibular nuclei (Furuya et al 1975). On anatomical grounds it may be assumed (Walberg et al 1962) that type II neurons and otolithic units (located rostrally) preferentially project to vestibuloocular relay neurons in the vestibular nuclei. Thus, fastigial type II and Purkinje cell type I neurons seem to project to similar areas and it may be assumed that their disfacilitatory and inhibitory actions, respectively, will sum at the level of vestibular neurons.

As for the crossed fastigiovestibular projection, it has been shown that it is involved in transcerebellar crossed inhibition of type I vestibular neurons (Furuya et al 1975). It is very probable that crossed axons of type I fastigial neurons reach type II vestibular neurons, which, in turn, inhibit type I vestibular neurons (Shimazu & Precht 1966, Shimazu & Smith 1971). This transcerebellar inhibition is under the control of the vermal lobuli VI and VII a (Furuya et al 1975). These areas are known to be related to oculomotor function (for ref. see Precht 1975b, 1977) and to receive vestibular inputs as well (Precht et al 1977).

In summary, a combined effort of both anatomists and physiologists during the last decade has brought some order into the complex field of vestibulooccular relationships. The importance of the classical three neuronal arc in the strict sense of the word for vestibuloocular processing has been deemphasized. Instead, the indirect routes from the labyrinth to the eye muscles have been stressed. As also shown in the next section, these indirect routes seem to return signals to the vestibular nuclei so that some of these neurons carry an eye position signal not present in the primary units of the eighth nerve. Furthermore, the perihypoglossal, interstitial nuclei and the internuclear neurons of the abducens nuclei have emerged as very important supranuclear structures for the control of horizontal and vertical eye movements. Finally, the vestibuloocular path through the cerebellum has proven to be of considerable importance for the acute as well as plastic adjustment of vestibuloocular reflexes.

Eye Movement Correlated Firing of Vestibular Neurons

Since the first description of nystagmic modulation of firing of vestibular neurons in the rabbit (Duensing & Schaefer 1958) evidence has been accumulated in various species showing that ca. 50% of the vestibular neurons participate in eye movements; the rest are related to head movements only.

FIRING RELATED TO VESTIBULOOCULAR REFLEX

As shown in the first section, primary vestibular fibers carry a head velocity signal, which was obtained from head acceleration by an integration performed in the cupula endolymph system. To provide the motoneurons with the eye position signal required for the compensatory VORs, a second integration must be performed. Current hypotheses assume that this integration takes place in a "neural integrator" located in the reticular formation (Robinson 1972). If this hypothesis is correct, secondary vestibular neurons may contain an eye position signal only if the integrator returns its output to these neurons. Alternatively, intra- or intervestibular nuclear circuits could perform this integration. This possibility is not disproven, but unlikely.

Vestibular unit behavior during horizontal head rotation is shown in Figure 1 for both cat (Shinoda & Yoshida 1974) and monkey (Buettner et al 1978). Secondary vestibular neurons exhibit a larger phase lag than primary neurons but their phases are still far from those seen in motoneurons. It should be noted that the plots in Figure 1 represent average values and, in the case of the monkey, do not contain the eye position units described by Keller & Kamath 1975, Keller & Daniels 1975, and Fuchs & Kimm, 1975. These are essentially burst-tonic neurons that have phase lags with acceleration of 120–170° at 0.5 Hz, that exhibit modulation during smooth pursuit weaker than that during slow phase vestibular nystagmus, and that cease to be modulated when the VOR is suppressed by fixation (see preceding section). Most interestingly, these eye position related units are driven only disynaptically and never monosynaptically from the VIIIth nerve (Keller & Kamath 1975). There is some disagreement among the authors as to the frequency of occurrence of eye position neurons but this may be related to the rather restricted localization of these units (laterocaudal to the VIth nuclei).

In contrast to the eye position neurons, many vestibular neurons (type I or type II) show phase lags of the order shown in Figure 1 and may be called head velocity units (Keller & Kamath 1975), although among this group there are units that show an eye position signal in the absence of head movement. In the case of fixation suppression of the VOR they continue, however, to exhibit a head velocity signal. They are unmodulated during smooth pursuit movements but do show nystagmic modulation (see below). Very much in contrast to the eye position units, VIIIth nerve stimulation yields monosynaptic activation of these neurons (Keller & Kamath 1975).

In the cat, similar results have been obtained (Shimazu & Precht 1965, Precht & Shimazu 1965, Melvill Jones 1971, Shinoda & Yoshida 1974, Hikosaka et al 1977). In short, when studied during horizontal angular acceleration, vestibular units may be divided into short (ca. 3.0 sec) and long (ca. 7 sec) time constant units having small and large phase lags with acceleration, respectively. In the fully alert cat (Keller & Precht, unpublished observations) slightly longer time constants were noted (3 sec and 11 sec). As in the monkey, short and long time constant units are activated mono- and disynaptically, respectively, by VIIIth nerve stimulation (Precht & Shimazu 1965). Furthermore, head velocity and eye position neurons are found in the alert cat (Keller & Precht, unpublished observations). As mentioned in the previous section there exist in the VIth nucleus monosynaptically activated vestibular axons, which have no relation to eye position (V_c axons of Hikosaka et al 1977). Some V_i axons have a weak eye position relationship. On the other hand, the disynaptically activated axons showed a strong eye position signal that may have, at least in part, originated in the vestibular nuclei.

SACCADIC MODULATION

In addition to the above described eye position neurons, which showed a burst-tonic pattern during saccades, a large number of the head velocity units in the vestibular nuclei also show modulation of firing rate in conjunction with nystagmus (Fuchs & Kimm 1975, Keller & Kamath 1975, Waespe & Henn 1977b, Waespe et al 1977). One group of neurons paused during saccades in all or only one direction, while another group burst during saccades in all or only one direction. Pauses and bursts often started later than or outlasted the saccades.

At present we are far from an understanding of the functional meaning of the complexity of these modulations of vestibular neurons in relation to eye movements. The frequent occurrence of modulation, however, stresses the importance of the vestibuloocular mechanisms even during voluntary saccades and quick phases of vestibular nystagmus.

VESTIBULAR RECEPTOR AND VISUAL-VESTIBULAR CONVERGENCE

As is evident from the discussion of the interactions among the vestibular nuclei, reticular formation, and the cerebellum, the vestibular nuclei are not merely a relay center for vestibuloocular or other vestibular-evoked reflex pathways. Rather these nuclei are complex somatosensory integrating centers as a result of the multiple convergence input from many parts of the brain. I therefore now describe the interactions of vestibular receptors on central neurons as well as visual-vestibular convergence. Other inputs such as those from the spinal cord are not dealt with.

Canal-canal and Canal-otolith Convergence

Since there is no evidence for a significant amount of receptor convergence at the level of primary neurons, information concerning the spatial canal and otolith planes arrives in the central nervous system via separate channels. We now ask: is this spatial separation preserved centrally such that groups of neurons receive only one type of receptor input? The answer is yes and no. When carefully controlled natural stimulation (avoiding pseudoconvergence) is applied one finds that, for example in the frog vestibular nuclei (Precht 1978) and cerebellum (Blanks et al 1977a) only ca. 20% of the units do not show receptor convergence, whereas the rest exhibit various combinations of canal-canal and canal-otolith convergence. Similar findings have been obtained in the vestibular nuclei of the cat (Curthoys & Markham 1971, Markham & Curthoys 1972) and rabbit (Duensing & Schaefer 1959).

The parallel or inplane canal-canal convergence (e.g. left and right horizontal or left posterior and right anterior canals) occurring in conjunction with commissural inhibition (Shimazu & Precht 1966, Kasahara & Uchino 1971) has been shown to increase the sensitivity of vestibular neurons to rotation and may also play a role in compensation following vestibular lesions (Precht et al 1966). Interestingly, when commissural inhibition is not demonstrable, as in the frog (Ozawa et al 1974), a higher vestibular gain is already found in primary neurons (Fig. 1).

The functional meaning of orthogonal canal-canal convergence (e.g. ipsilateral horizontal and posterior canals) also found in many vestibular and cerebellar neurons (Blanks et al 1977a, Precht 1978) is more difficult to interpret. Possibly its function is related to the fact that canal responses are a function of the cosine of the angle between the canal plane and plane of rotation. Such a relationship yields only small differences in responses over a relatively wide range of head positions and may require information from another canal to enable the CNS to precisely determine a particular head axis about which rotation is performed.

Electrophysiological studies are somewhat in disagreement with the results mentioned above in that they indicated very little orthogonal canal convergence (Kasahara & Uchino 1971, Wilson et al 1974). It should be kept in mind, however, that these studies were done in anesthetized animals so that polysynaptic pathways, which are probably mainly responsible for convergence, were less readily activated.

In contrast to the frequent occurrence of canal-canal convergence in the vestibular nuclei and cerebellum, little orthogonal convergence has been observed at the level of ocular motoneurons (Blanks et al 1978). This indicates that the functional synaptology shown in Figure 2 is valid even under natural stimulation conditions and suggests a minor role of orthogonal convergence at the final output stage.

Finally, a very interesting and important case of convergence, namely the canal-otolithic one, deserves mention. That such a convergence occurs in many vestibular (Blanks et al 1978, Duensing & Schaefer 1959), cerebellar (Precht et al 1977), and ocular motoneurons (Anderson et al 1977, Blanks et al 1978) has been well documented. It appears that canal-otolith convergence is particularly strong and ubiquitous in motoneurons of the vertical and oblique eye muscles but is also detectable in the horizontal system (Blanks et al 1978, Anderson et al 1977). The functional meaning of this convergence for the dynamics of the VOR becomes obvious when one compares the phases of responses of vertical motoneurons during a canal input only with those obtained with combined canal-otolithic stimulation (i.e. during roll or pitch rotations). The inherent insufficiency of the canals in the low frequency range in the dark (response lags with acceleration of about 60°) is compensated by the otolithic input yielding a phase lag of 180° (Blanks et al 1978). Interestingly, when one compares the otolith responses of primary afferents (Anderson et al 1978) with those of the motoneurons, a considerable lag is noted suggesting that both the otolith and canal signals are similarly subjected to a neuronal integration. Needless to say, during static head tilt in roll or pitch, it is the otolith input that maintains a compensated eye position. Finally, it should be mentioned that many secondary vestibular fibers terminating in the motor nuclei already contain both canal and otolith signals. This holds for both inhibitory and excitatory axons and indicates that there are excitatory (disynaptic, see Fig. 2) and inhibitory (polysynaptic) maculo-ocular reflexes (Blanks et al 1978).

Visual-Vestibular Convergence

Recent work has demonstrated that an intact vestibular system is necessary for an optimal high velocity gain of optokinetic nystagmus (OKN) and for optokinetic after nystagmus (OKAN). Bilateral labyrinthectomy, in fact, reduces the gain of OKN and abolishes OKAN in rabbits (Collewijn 1976) and monkey (Cohen et al 1973). Similarly, lesions of the vestibular nuclei impair OKN mechanisms (Azzena et al 1974). On the other hand, it was known for a long time that the gain and phase of the VOR were higher when the animal was rotated in the light as compared to the dark (for details see Baarmsa & Collewijn 1974). The poor performance of abducens motoneurons in the dark is shown in the phase diagram of Figure 1, and is particularly dramatic for the cat at low frequencies. Other evidence in support of visual-vestibular interactions was obtained by studying the psychophysics of circular- or linearvection (for ref. see Dichgans & Brandt 1972).

Recently, possible neuronal correlates for these visual-vestibular interactions have been found. Most secondary vestibular neurons (type I and II, eye position and head velocity units) show a sustained and direction specific frequency modulation when a large visual surround is moved around the animal (fish: Dichgans et al 1973, Allum et al 1976; rabbit: Dichgans & Brandt 1972; cat: Dichgans 1977, Keller & Precht 1978a; monkey: Waespe & Henn 1977a). Response magnitude was related to stimulus velocity up to ca. 60°/sec in monkey and 10°/sec in fish and cat, with the important difference that the fish and cat studies were carried out in open-loop condition. It is not quite clear yet whether retinal image velocity or true surround velocity is coded. Most recent studies with the trained monkey seem to indicate that the modulation is not caused by retinal slip but rather represents an internal surround velocity signal (Henn, personal communication). The direction of the visual responses is such that it combines synergistically with the vestibular response: if a neuron is activated by head rotation in darkness to the left, surround rotation to the right also results in activation. Since these two stimuli generate eye movements in the same direction during rotations in the light, vision will improve the vestibular performance, particularly in the low frequency range and/or during constant velocity of rotation. In the latter case, vision can maintain the firing rate of vestibular neurons and signals the vestibular system that the body is still in rotation. It appears that over some range, vestibular and visual effects add linearly (Robinson 1977) and ensure the high fidelity of compensatory eye movements over a wide frequency range.

When the discharge of vestibular unit was studied during OKN and OKAN a linear relation was observed between firing and slow phase nystagmus velocity (Waespe & Henn 1977b). There were, however, several conditions in which a dissociation between the two events occurred (e.g. high velocities and conflict situations). This suggests that the vestibular nuclei, alone, do not control OKN and requires search for other mechanisms.

Finally, the question as to the possible pathways mediating visual effects on vestibular neurons should be considered. Since it has been reported that suppression of the VOR during fixation (Takemori & Cohen 1974) and acute visual effects on the VOR in the rabbit (Ito et al 1974) are impaired or abolished following flocculectomy, an indirect route through the cerebellum has been postulated. However, recent experiments by Keller & Precht (1978a) have demonstrated that visual-vestibular interaction in the cat are still present following flocculectomy as well as after total cerebellectomy, suggesting different or alternate pathways in the brain stem. Clearly more work is needed to resolve this problem.

Interestingly, in the goldfish (Klinke & Schmidt 1970) presumed primary afferents have been shown to be modulated by large field surround motion. Whether primary afferents of higher vertebrates are likewise modulated should be further investigated in alert animals. It is known, however, that monkey afferents do not respond to movement of small targets and striped full field patterns (Keller 1976). Recent work in alert cats has shown that primary afferents do not respond to patterned field surround motion (Blanks & Precht, *Neurosci. Lett.,* 1978, in press).

SUMMARY

It is apparent from this and other reviews of the subject that our knowledge of vestibular function is most complete for the primary canal and otolithic afferents. Relatively little progress has been made in the understanding of receptor mechanisms and the functional importance of the efferent vestibular system. Since most of it has been summarized previously the latter were not considered here.

Considerably more knowledge has accumulated in the field of central vestibular mechanisms, particularly those related to eye movements. Recent advances in functional synaptology of direct and indirect vestibuloocular pathways are described. It appears that the indirect pathways are essential for the central integration of the peripheral head velocity into a central eye position signal. Candidates for the neural integrator are presented and discussed and their connectivity described both for the horizontal and the relatively poorly studied vertical eye movement system. This field will certainly be studied extensively during the next years.

Another interesting field is the role of the cerebellum in the control of the vestibuloocular reflex. Recent data and hypotheses, including the problem of cerebellar plasticity, are summarized and evaluated.

That the vestibular nuclei are by no means a simple relay system for specific vestibular signals destined for other sensory or motor centers is evidenced in this review by the description of multiple canal-canal, canal-otolith, and visual-vestibular convergence at the nuclear level. Canal-otolith and polysensory convergence in vestibular neurons enables them to correct for the inherent inadequacies of the peripheral canal system in the low frequency range. The mechanisms of polysensory interaction in the central vestibular system will undoubtedly be an important and interesting field for future research.

Literature Cited

Allum, J. H. J., Graf, W., Dichgans, J., Schmidt, C. L. 1976. Visual-vestibular interactions in the vestibular nuclei of the goldfish. *Exp. Brain Res.* 26:463–85

Anderson, J. H., Blanks, R. H. I., Precht, W. 1978. Response characteristics of semicircular canal and otolith systems in cat. I. Dynamic responses of primary vestibular fibers. *Exp. Brain Res.* 32:491–507

Anderson, J. H., Precht, W., Blanks, R. H. I. 1977. In *Control of Gaze by Brain Stem Neurons, Developments in Neuroscience,* ed. R. Baker, A. Berthoz, 1:253–60. Amsterdam: Elsevier. 514 pp.

Azzena, G. B., Azzena, M. T., Marini, R. 1974. Optokinetic nystagmus and the vestibular nuclei. *Exp. Neurol.* 42:158–68

Baarsma, E. A., Collewijn, H. 1974. Vestibulo-ocular and optokinetic reactions to rotation and their interaction in the rabbit. *J. Physiol. London* 238:603–25

Bak, I. J., Baker, R., Choi, W. B., Precht, W. 1976. Electron microscope investigation of the vestibular projection to the cat trochlear nuclei. *Neuroscience* 1:477–82

Baker, R. 1977. In *Control of Gaze by Brain Stem Neurons, Developments in Neuroscience,* ed. R. Baker, A. Berthoz, 1:207–22. Amsterdam: Elsevier. 514 pp.

Baker, R., Berthoz, A. 1975. Is the prepositus hypoglossi nucleus the source of another vestibulo-ocular pathway? *Brain Res.* 86:121–27

Baker, R., Berthoz, A., eds. 1977. *Control of Gaze by Brain Stem Neurons.* Vol. 1, *Developments in Neuroscience.* Amsterdam: Elsevier. 514 pp.

Baker, R., Berthoz, A., Delgado-Garcia, J. 1977. Monosynaptic excitation of trochlear motoneurons following electrical stimulation of the prepositus hypoglossi nucleus. *Brain Res.* 121:157–61

Baker, R., Gresty, M., Berthoz, A. 1976. Neuronal activity in the prepositus hypoglossi nucleus correlated with vertical and horizontal eye movement in the cat. *Brain Res.* 101:366–71

Baker, R., Highstein, S. 1975. Physiological identification of interneurons and motoneurons in the abducens nucleus. *Brain Res.* 91:292–98

Baker, R., Mano, N., Shimazu, H. 1969. Postsynaptic potentials in abducens motoneurons induced by vestibular stimulation. *Brain Res.* 15:557–80

Baker, R., Precht, W., Berthoz, A. 1973. Synaptic connections to trochlear motoneurons determined by individual vestibular nerve branch stimulation in the cat. *Brain Res.* 64:402–6

Baker, R., Precht, W., Llinás, R. 1972. Cerebellar modulatory action on the vestibulo-trochlear pathway in the cat. *Exp. Brain Res.* 15:364–85

Berthoz, A., Baker, R., Precht, W. 1973. Labyrinthine control of inferior oblique motoneurons. *Exp. Brain Res.* 18: 225–41

Blanks, R. H. I., Anderson, J. H., Precht, W. 1978. Response characteristics of semicircular canal and otolith systems in cat. II. Responses of trochlear motoneurons. *Exp. Brain Res.* 32: 509–28

Blanks, R. H. I., Precht, W. 1976. Functional characterization of primary vestibular afferents in the frog. *Exp. Brain Res.* 25:369–90

Blanks, R. H. I., Precht, W., Giretti, M. L. 1977a. Response characteristics and vestibular receptor convergence of frog cerebellar Purkinje cell. A natural stimulation study. *Exp. Brain Res.* 27:181–201

Blanks, R. H. I., Volkind, R., Precht, W., Baker, R. 1977b. Responses of cat prepositus hypoglossi neurons to horizontal angular acceleration. *Neuroscience* 2:391–404

Brodal, A., Pompeiano, O., Walberg, F. 1962. *The Vestibular Nuclei and Their Connections: Anatomy and Functional Correlations.* London: Oliver and Boyd. 190 pp.

Buettner, U. W., Büttner, U., Henn, V. 1978. Transfer characteristics of neurons in the vestibular nuclei of alert monkey. *J. Neurophysiol.* In press

Büttner-Ennever, J. A. 1977. In *Control of Gaze by Brain Stem Neurons, Developments in Neuroscience,* ed. R. Baker, A. Berthoz, 1:89–98. Amsterdam: Elsevier. 514 pp.

Carpenter, M. B., Harbison, J. W., Peter, P. 1970. Accessory oculomotor nuclei in the monkey: Projections and effects of discrete lesions. *J. Comp. Neurol.* 140:131–54

Caston, J., Precht, W., Blanks, R. H. I. 1977. Responses of lagena afferents to natural stimuli. *J. Comp. Physiol.* 118:263–89

Cohen, B. 1974. The vestibulo-ocular reflex arc. In *Handbook of Sensory Physiology,* Vol. VI, *Vestibular System,* pp. 477–540, ed. H. H. Kornhuber. Berlin-Heidelberg-New York: Springer. 676 pp.

Cohen, B., Uemura, T., Takemori, S. 1973. Effects of labyrinthectomy on optokinetic nystagmus (OKN) and optokinetic after-nystagmus (OKAN). *Equilibrium Res.* 3:88–93

Collewijn, H. 1976. Impairment of optokinetic (after-) nystagmus by labyrinthectomy in the rabbit. *Exp. Neurol.* 52:146–56

Curthoys, I. S., Markham, C. H. 1971. Convergence of labyrinthine influences on units in the vestibular nuclei of the cat. I. Natural stimulation. *Brain Res.* 35:469–90

Dichgans, J. 1977. In *Control of Gaze by Brain Stem Neurons, Developments in Neuroscience,* ed. R. Baker, A. Berthoz, 1:261–67. Amsterdam: Elsevier. 514 pp.

Dichgans, J., Brandt, T. 1972. Visual-vestibular interaction and motion perception. *Bibl. ophthalmol.* 82:327–38

Dichgans, J., Schmidt, C. L., Graf, W. 1973. Visual input improves the speedometer function of the vestibular nuclei in the goldfish. *Exp. Brain Res.* 18:319–22

Duensing, F., Schaefer, K. P. 1958. Die Aktivität einzelner Neurone im Bereich der Vestibulariskerne bei Horizontalbeschleunigungen unter besonderer Berücksichtigung des vestibulären Nystagmus. *Arch. Psychiatr. Nervenkr.* 198:225–52

Duensing, F., Schaefer, K. P. 1959. Über die Konvergenz verschiedener labyrinthärer Afferenzen auf einzelne Neurone des Vestibulariskerngebietes. *Arch. Psychiatr. Nervenkr.* 199:345–71

Estes, M. S., Blanks, R. H. I., Markham, C. H. 1975. Physiologic characteristics of vestibular first-order canal neurons in the cat. I. Response plane determination and resting discharge characteristics. *J. Neurophysiol.* 38:1232–49

Evinger, C., Fuchs, A. F., Baker, R. 1977. Bilateral lesions of the medial longitudinal fasciculus in monkeys: effects on the horizontal and vertical components of voluntary and vestibular induced eye movements. *Exp. Brain Res.* 28:1–20

Fernández, C., Goldberg, J. M. 1971. Physiology of peripheral neurons innervating semicircular canals of the squirrel monkey. II. Response to sinusoidal stimulation and dynamics of peripheral vestibular system. *J. Neurophysiol.* 34: 661–75

Fernández, C., Goldberg, J. M. 1976a. Physiology of peripheral neurons innervating otolith organs of the squirrel monkey. I. Response to static tilts and to long-duration centrifugal force. *J. Neurophysiol.* 39:970–84

Fernández, C., Goldberg, J. M. 1976b. Physiology of peripheral neurons innervating otolith organs of the squirrel monkey. III. Response dynamics. *J. Neurophysiol.* 39:996–1008

Fernández, D., Goldberg, J. M. 1976c. Physiology of peripheral neurons innervating otolith organs of the squirrel monkey. II. Directional selectivity and force-response relations. *J. Neurophysiol.* 39:985–95

Fuchs, A. F., Kimm, J. 1975. Unit activity in vestibular nucleus of the alert monkey during horizontal angular acceleration and eye movement. *J. Neurophysiol.* 38:1140–61

Fukuda, J., Highstein, S. M., Ito, M. 1972. Cerebellar inhibitory control of the vestibulo-ocular reflex investigated in rabbit IIIrd nucleus. *Exp. Brain Res.* 14:511–26

Furuya, N., Kawano, K., Shimazu, H. 1975. Functional organization of vestibulofastigial projection in the horizontal semicircular canal system in the cat. *Exp. Brain Res.* 24:75–87

Gardner, E. P., Fuchs, A. F. 1975. Single-unit responses to natural vestibular stimuli and eye movements in deep cerebellar nuclei of the alert rhesus monkey. *J. Neurophysiol.* 38:627–49

Ghelarducci, B. 1973. Responses of the cerebellar fastigial neurones to tilt. *Pflügers Arch.* 344:195–206

Ghelarducci, B., Highstein, S. M., Ito, M. 1977. In *Control of Gaze by Brain Stem Neurons, Developments in Neuroscience,* ed. R. Baker, A. Berthoz, 1:167–75. Amsterdam: Elsevier. 514 pp.

Goldberg, J. M., Fernández, C. 1971. Physiology of peripheral neurons innervating semicircular canals of the squirrel monkey. Resting discharge and response to constant angular accelerations. *J. Neurophysiol.* 34:635–60

Goldberg, J. M., Fernández, C. 1975. Vestibular Mechanisms. *Ann. Rev. Physiol.* 37:129–62

Graybiel, A. M. 1977. In *Control of Gaze by Brain Stem Neurons, Developments in Neuroscience,* ed. R. Baker, A. Berthoz, 1:79–88. Amsterdam: Elsevier. 514 pp.

Graybiel, A. M., Hartwieg, E. A. 1974. Some afferent connections of the oculomotor complex in the cat: an experimental study with tracer techniques. *Brain Res.* 81:543–51

Gresty, M., Baker, R. 1976. Neurons with visual receptive field, eye movement and neck displacement sensitivity within and around the nucleus prepositus hypoglossi in the alert cat. *Exp. Brain Res.* 24:429–33

Hassler, R., Hess, W. R. 1954. Experimentelle und anatomische Befunde über die Drehbewegungen und ihre nervösen Apparate. *Arch. Psychiatr. Nervenkr.* 192:488–526

Highstein, S. M. 1973a. The organization of the vestibulooculomotor and trochlear reflex pathways in the rabbit. *Exp. Brain Res.* 17:285–300

Highstein, S. M. 1973b. Synaptic linkage in the vestibuloocular and cerebello-vestibular pathways to the VIth nucleus in the rabbit. *Exp. Brain Res.* 17:301–14

Highstein, S. M., Maekawa, K., Steinacker, A., Cohen, B. 1976. Synaptic input from the pontine reticular nuclei to abducens motoneurons and internuclear neurons in the cat. *Brain Res.* 112: 162–67

Hikosaka, O., Igusa, Y., Imai, H. 1978. Firing pattern of prepositus hypoglossi and adjacent reticular neurons related to vestibular nystagmus in the cat. *Brain Res.* 144:395–403

Hikosaka, O., Kawakami, T. 1977. Inhibitory reticular neurons related to the quick phase of vestibular nystagmus— Their location and projection. *Exp. Brain Res.* 27:377–96

Hikosaka, O., Maeda, M., Nakao, S., Shimazu, H., Shinoda, Y. 1977. Presynaptic impulses in the abducens nucleus and their relation to postsynaptic potentials in motoneurons during vestibular nystagmus. *Exp. Brain Res.* 27:355–76

Hwang, J. C., Poon, W. F. 1975. An electrophysiological study of the sacculo-ocular pathways in cats. *Jpn. J. Physiol.* 25:241–51

Hyde, J. E., Eliasson, S. G. 1957. Brainstem induced eye movements in cats. *J. Comp. Neurol.* 108:139–72

Ito, M. 1977a. In *Control of Gaze by Brain Stem Neurons, Developments in Neuroscience,* ed. R. Baker, A. Berthoz,

1:177–86. Amsterdam: Elsevier. 514 pp.

Ito, M. 1977b. *In Control of Gaze by Brain Stem Neurons, Developments in Neuroscience,* ed. R. Baker, A. Berthoz, 1:391–398. Amsterdam: Elsevier. 514 pp.

Ito, M., Highstein, S. M., Tsuchiya, T. 1970. The postsynaptic inhibition of rabbit oculomotor neurones by secondary vestibular impulses and its blockage by picrotoxin. *Brain Res.* 17:520–23

Ito, M., Nisimaru, N., Yamamoto, M. 1976a. Pathways for the vestibulo-ocular reflex excitation arising from semicircular canals of rabbits. *Exp. Brain Res.* 24:257–71

Ito, M., Nisimaru, N., Yamamoto, M. 1976b. Postsynaptic inhibition of oculomotor neurons involved in vestibuloocular reflexes arising from semicircular canals of rabbits. *Exp. Brain Res.* 24:273–83

Ito, M., Nisimaru, N., Yamamoto, M. 1977. Specific patterns of neuronal connexions involved in the control of the rabbit's vestibuloocular reflexes by the cerebellar flocculus. *J. Physiol.* 265:833–54

Ito, M., Shiida, T., Yagi, N., Yamamoto, M. 1974. Visual influence on rabbit's horizontal vestibulo-ocular reflex that presumably is effected via the cerebellar flocculus. *Brain Res.* 65:170–74

Kasahara, M., Uchino, Y. 1971. Selective mode of commissural inhibition induced by semicircular canal afferents on secondary vestibular neurons in the cat. *Brain res.* 34:366–69

Keller, E. L. 1976. Behavior of horizontal semicircular canal afferents in alert monkey during vestibular and optokinetic stimulation. *Exp. Brain Res.* 24:459–71

Keller, E. L., Daniels, P. D. 1975. Oculomotor related interaction of vestibular and visual stimulation in vestibular nucleus cells in alert monkey. *Exp. Neurol.* 46:187–98

Keller, E. L., Kamath, B. Y. 1975. Characteristics of head rotation and eye movement-related neurons in alert monkey vestibular nucleus. *Brain Res.* 100:182–87

Keller, E. L., Precht, W. 1978a. Persistence of visual response in vestibular nucleus neurons in cerebellectomized cat. *Exp. Brain Res.* 32:591–94

Keller, E. L., Precht, W. 1978b. Firing patterns of central vestibular neurons during adaptive modification of the vestibuloocular reflex by reversing prisms. *Neurosci. Abstr.* In press

King, W. M. 1976. *Quantitative analysis of the activity of neurons in the accessory oculomotor nuclei and the mesencephalic reticular formation of alert monkeys in relation to vertical eye movements induced by visual and vestibular stimulation.* PhD Thesis. Univ. of Washington, Seattle. 113 pp.

King, W. M., Lisberger, S. G., Fuchs, A. F. 1976. Responses of fibers in medial longitudinal fasciculus (MLF) of alert monkeys during horizontal and vertical conjugate eye movements evoked by vestibular or visual stimuli. *J. Neurophysiol.* 39:1135–49

King, W. M., Precht, W., Dieringer, N. 1978a. Convergence of vestibular and frontal cortical input in the region of the interstitial nucleus of Cajal. *Pflügers Arch.,* Suppl. Vol. 373:R72, 259.

King, W. M., Precht, W., Dieringer, N. 1978b. Connections of behaviorally identified cat omnipause neurons. *Exp. Brain Res.* 32:435–38

Klinke, R., Schmidt, C. L. 1970. Efferent influence on the vestibular organ during active movement of the body. *Pflügers Arch. Gesamte Physiol. Menschen Tiere* 318:325–32

Kornhuber, H. H., ed. 1974. *Handbook of Sensory Physiology.* Vol. VI, Pts. 1, 2. Berlin: Springer. 676 pp., 680 pp.

Lisberger, S. G., Fuchs, A. F. 1977. In *Control of Gaze by Brain Stem Neurons, Developments in Neuroscience,* ed. R. Baker, A. Berthoz, 1:381–89. Amsterdam: Elsevier. 514 pp.

Llinás, R., Simpson, J. I., Precht, W. 1976. Nystagmic modulation of neuronal activity in rabbit cerebellar flocculus. *Pflügers Arch.* 367:7–13

Lorente de Nó, R. 1933. Vestibulo-ocular reflex arc. *Arch. Neurol. Psychiatry* 30:245–91

Lowenstein, O., Saunders, R. D. 1975. Otolith-controlled responses from the first-order neurons of the labyrinth of the bullfrog (Rana catesbeiana) to changes in linear acceleration. *Proc. R. Soc. London Ser. B* 191:475–505

Macadar, O., Wolfe, G. E., O'Leary, D. P., Segundo, J. P. 1975. Response of the elasmobranch utricle to maintained spatial orientation, transitions and jitter. *Exp. Brain Res.* 22:1–12

Maeda, M., Shimazu, H., Shinoda, Y. 1972. Nature of synaptic events in cat abducens motoneurons at slow and quick phase of vestibular nystagmus. *J. Neurophysiol.* 35:279–96

Maekawa, K., Simpson, J. I. 1973. Climbing fiber responses evoked in vestibulocerebellum of rabbit from visual system. *J. Neurophysiol.* 36:649–66

Maekawa, K., Takeda, T. 1975. Mossy fiber responses evoked in the cerebellar flocculus of rabbits by stimulation of the optic pathway. *Brain Res.* 98:590–95

Markham, C. H., Curthoys, I. S. 1972. Convergence of labyrinthine influences on units in the vestibular nuclei of the cat. II. Electrical stimulation. *Brain Res.* 43:383–96

Markham, C. H., Precht, W., Shimazu, H. 1966. Effect of stimulation of interstitial nucleus of Cajal on vestibular unit activity in the cat. *J. Neurophysiol.* 29:493–507

Melvill Jones, G. 1971. In *The Control of Eye Movement,* ed. P. Bach-y-Rita, C. C. Collins, pp. 497–518. New York: Academic. 560 pp.

Melvill Jones, G., Spells, K. E. 1963. A theoretical and comparative study of the functional dependence of the semicircular canal upon its physical dimensions. *Proc. R. Soc. London Ser. B* 157:403–19

Noda, H., Asoh, R., Shibagaki, M. 1977. In *Control of Gaze by Brain Stem Neurons, Developments in Neuroscience,* ed. R. Baker, A. Berthoz, 1:371–80. Amsterdam: Elsevier. 514 pp.

O'Leary, D. P., Dunn, R. F., Honrubia, V. 1974. Functional and anatomical correlation of afferent responses from the isolated semicircular canal. *Nature,* 251:225–27

Ozawa, S., Precht, W., Shimazu, H. 1974. Crossed effects on central vestibular neurons in the horizontal canal system of the frog. *Exp. Brain Res.* 19:394–405

Pola, J., Robinson, D. A. 1978. Oculomotor signals in the medial longitudinal fasciculus of the monkey. *J. Neurophysiol.* 41:245–59

Precht, W. 1975a. In *MTP International Review of Sciences, Neurophysiology. Physiol. Series One,* ed. A. C. Guyton, C. C. Hunt, 3:82–149. London, Baltimore: Butterworths Univ. Park Press. 419 pp.

Precht, W. 1975b. In *Basic Mechanisms of ocular motility and their clinical implications,* ed. G. Lennerstrand, P. Bach-y-Rita, pp. 261–80. Oxford and New York: Pergamon. 584 pp.

Precht, W. 1976. In *Frog Neurobiology,* ed. R. Llinás, W. Precht, pp. 481–512. Berlin-Heidelberg-New York: Springer. 1046 pp.

Precht, W. 1977. In *Control of Gaze by Brain Stem Neurons, Developments in Neuroscience,* ed. R. Baker, A. Berthoz, 1:131–41. Amsterdam: Elsevier. 514 pp.

Precht, W. 1978. In *Studies of Brain Function,* ed. V. Braitenberg, Vol. 2. Berlin-Heidelberg-New York: Springer. 226 pp.

Precht, W., Baker, R. 1972. Synaptic organization of the vestibulo-trochlear pathway. *Exp. Brain Res.* 14:158–84

Precht, W., Baker, R., Okada, Y. 1973. Evidence for GABA as the synaptic transmitter of the inhibitory vestibuloocular pathway. *Exp. Brain Res.* 18:415–28

Precht, W., Grippo, J., Richter, A. 1967. Effect of horizontal angular acceleration on neurons in the abducens nucleus. *Brain Res.* 5:527–31

Precht, W., Llinás, R. 1968. *Proc. Int. Union Physiol. Sci.* 7:355

Precht, W., Llinás, R. 1969. Functional organization of the vestibular afferents to the cerebellar cortex of frog and cat. *Exp. Brain Res.* 9:30–52

Precht, W., Llinás, R., Clarke, M. 1971. Physiological responses of frog vestibular fibers to horizontal angular rotation. *Exp. Brain Res.* 13:378–407

Precht, W., Shimazu, H. 1965. Functional connections of tonic and kinetic vestibular neurons with primary vestibular afferents. *J. Neurophysiol.* 28:1014–28

Precht, W., Shimazu, H., Markham, C. H. 1966. A mechanism of central compensation of vestibular function following hemilabyrinthectomy. *J. Neurophysiol.* 29:996–1010

Precht, W., Volkind, R., Blanks, R. H. I. 1977. Functional organization of the vestibular input to the anterior and posterior cerebellar vermis of cat. *Exp. Brain Res.* 27:143–60

Precht, W., Volkind, R., Maeda, M., Giretti, M. L. 1976. The effects of stimulating the cerebellar nodulus in the cat on the responses of vestibular neurons. *Neuroscience* 1:301–312

Richter, A., Precht, W. 1968. Inhibition of abducens motoneurones by vestibular nerve stimulation. *Brain Res.* 11:701–5

Robinson, D. A. 1972. On the nature of visual oculomotor connections. *Invest. Ophthalmol.* 11:497–503

Robinson, D. A. 1976. Adaptive gain control of vestibuloocular reflex by the cerebellum. *J. Neurophysiol.* 39:954–69

Robinson, D. A. 1977. Linear addition of optokinetic and vestibular signals in the vestibular nucleus. *Exp. Brain Res.* 30:447–50

Schwindt, P. C., Precht, W., Richter, A. 1974. Monosynaptic excitatory and inhibitory pathways from medial midbrain nuclei to trochlear motoneurons. *Exp. Brain Res.* 20:223–38

Schwindt, P. C., Richter, A., Precht, W. 1973. Short latency utricular and canal input to ipsilateral abducens motoneurons. *Brain Res.* 60:259–62

Shimazu, H., Precht, W. 1965. Tonic and kinetic responses of cat's vestibular neurons to horizontal angular acceleration. *J. Neurophysiol.* 28:991–1013

Shimazu, H., Precht, W. 1966. Inhibition of central vestibular neurons from the contralateral labyrinth and its mediating pathway. *J. Neurophysiol.* 29:467–92

Shimazu, H., Smith, C. M. 1971. Cerebellar and labyrinthine influences on single vestibular neurons identified by natural stimuli. *J. Neurophysiol.* 34:493–508

Shinoda, Y., Yoshida, K. 1974. Dynamic characteristics of responses to horizontal head angular acceleration in the vestibuloocular pathway in the cat. *J. Neurophysiol.* 37:653–73

Simpson, J. I., Precht, W., Llinás, R. 1974. Sensory separation in climbing and mossy fiber inputs to cat vestibulocerebellum. *Pflügers Arch.* 351:183–93

Szentágothai, J. 1943. Die zentrale Innervation der Augenbewegungen. *Arch. Psychiatr. Nervenkr.* 116:721–60

Szentágothai, J. 1950. The elementary vestibulo-ocular reflex arc. *J. Neurophysiol.* 13:395–407

Takemori, S., Cohen, B. 1974. Loss of visual suppression of vestibular nystagmus after flocculus lesions. *Brain Res.* 72:213–24

Tarlov, E. 1970. Organization of vestibulooculomotor projections in the cat. *Brain Res.* 20:159–79

Torvik, A., Brodal, A. 1954. The cerebellar projection of the perihypoglossal nuclei

(nucleus intercalatus, nucleus prepositus hypoglossi and nucleus of Roller) in the cat. *J. Neuropathol. Exp. Neurol.* 13:515–27

Waespe, W., Henn, V. 1977a. Neuronal activity in the vestibular nuclei of the alert monkey during vestibular and optokinetic stimulation. *Exp. Brain Res.* 27:523–38

Waespe, W., Henn, V. 1977b. Vestibular nuclei activity during optokinetic afternystagmus (OKAN) in the alert monkey. *Exp. Brain Res.* 30:323–30

Waespe, W., Henn, V., Miles, T. S. 1977. In *Control of Gaze by Brain Stem Neurons, Developments in Neuroscience,* ed. R. Baker, A. Berthoz, 1:269–78. Amsterdam: Elsevier. 514 pp.

Walberg, F., Pompeiano, O., Brodal, A., Jansen, J. 1962. The fastigiovestibular projection in the cat. An experimental study with silver impregnation methods. *J. Comp. Neurol.* 118:49–76

Wilson, V. J., Anderson, J. A., Felix, D. 1974. Unit and field potential activity evoked in the pigeon vestibulocerebellum by stimulation of individual semicircular canals. *Exp. Brain Res.* 19:142–57

Wilson, V. J., Peterson, B. W. 1978. Peripheral and central substrates of vestibulospinal reflexes. *Physiol. Rev.* 58:80–105

Yamamoto, M., Shimoyama, I. 1977. Differential localization of rabbit's flocculus Purkinje cells projecting to the medial and superior vestibular nuclei, investigated by means of the horseradish peroxidase retrograde axonal transport. *Neurosci. Lett.* 5:279–83

Ann. Rev. Neurosci. 1979. 2:291–307

THE DEVELOPMENT OF BEHAVIOR IN HUMAN INFANTS, PREMATURE AND NEWBORN

❖11525

Peter H. Wolff and Richard Ferber

Children's Hospital Medical Center, 300 Longwood Avenue,
Boston, Massachusetts 02115

INTRODUCTION

The proliferation of empirical studies on human neonates over the past 20 years can be attributed to at least three motivating factors.

1. A theoretical or ethological orientation that focuses on the human infant, its evolutionary history and species-specific ecology, and analyzes the organization of behavior before extrauterine stimulation has had any significant influence on development.
2. A diagnostic-predictive orientation that focuses on the possible continuities between behavioral adaptation at birth and its developmental outcome. This orientation may be motivated by clinical concerns to identify the neonatal "risk factors" that contribute to deviant development; or by theoretical concerns to examine developmental universals (Piaget 1967) and to determine the etiology of individual variations in human development.
3. Finally, experimental psychology uses the human neonate as a simplified experimental model for testing the boundaries of relevance of general laws in psychology. Its approach to the study of neonatal behavior is "variable-" rather than infant-oriented.

The range of topics that might be considered under the chapter heading is too extensive to permit any exhaustive discussion of all possibly relevant domains of investigation within the allotted space. Therefore, the review is restricted to some major growth points in contemporary behavioral re-

291

0147-006X/79/0315-0291$01.00

search on full-term infants during the first several weeks after birth, and on premature infants for a comparable phase of development. It is further restricted to "observable behavior" in the conventional sense, the data from psychophysiological experiments and polygraphic recordings being considered only as they might clarify mechanisms of observable behavior. Unavoidable omissions of research areas or of critical studies relating to the topic areas discussed therefore reflect the authors' prejudices.

BEHAVIORAL STATE

At present, there is a general consensus that valid and replicable observations of neonatal behavior require careful attention to behavioral state (for example, whether the infant is awake or asleep, what are the characteristics of a sleep state, etc). While a consideration of behavioral state is implicit in all psychological studies, most of which will be discontinued if the subject falls asleep or becomes acutely agitated, state fluctuations in the neonate are more rapid and fluid, and more significantly related to the intensity, frequency, and quality of discrete behavior patterns than they are in the child and adult.

For the study of young infants, behavioral state can be defined as a configuration of functional parameters (e.g. respiration, motility, heart rate, eyes open or closed) that remains relatively stable for predictable periods, occurs repeatedly within the same infant, and is more or less the same across infants. The importance of state resides in the fact that the presence, intensity, and in some cases, the quality of the motor reflexes, the occurrence and characteristics of self-generated behavior and the response thresholds to auditory, tactile, or visual stimulation vary systematically with behavioral state (Wolff 1966, Prechtl et al 1967, Hutt et al 1968, Lenard et al 1968, Eisenberg 1976). The state dependence of behavior patterns is such that states cannot be classified along some linear dimension like responsivity or level of arousal, but must provisionally be treated as qualitatively discrete assemblies of functional parameters or "structures of the whole" (Wolff 1966, Hutt et al 1969). Various schemes proposed for categorizing behavioral states in the full-term newborn infant differ in technical details (Wolff 1966, Prechtl & Beintema 1964, Anders et al 1971, Parmalee 1974). A classification scheme limited to five or six states appears to have greater heuristic validity and utility than schemes that introduce elaborate refinements to account for every minor variation in functional parameters. The available evidence further suggests that a combination of behavioral observations and polygraphic data (of respiration, heart rate, electroencephalogram, electro-oculogram, etc) yields the most reliable set of criteria for categorizing behavioral states (Prechtl et al 1968, Prechtl 1968).

Considerations of behavioral state are of central importance for experiments on stimulus discrimination, learning, conditioning, and habituation in the neonatal period (see below). However, the study of behavioral states as such is also of inherent theoretical interest and clinical importance. The observation and recording of behavioral states and state changes without the confounding influence of adventitious stimulation provide essential information about the neuropsychological organization of the young infant, without which the infant's responses to experimental stimulation cannot be properly interpreted. The sequence of state changes during the neonatal period is sufficiently stable within and across individual infants to warrant the term "state cycles" (Theorell et al 1973, Parmalee 1974). Although not impervious to environmental factors, the sequence of state cycles is at least partially controlled by intrinsic clocking mechanisms that are well established at birth in the healthy full-term infant (Roffwarg et al 1966, Wolff 1973).

The earliest differentiation between quiet and active sleep in preterm infants appears at around 30 weeks of conceptional age; active sleep is fully developed by 35 weeks, and quiet sleep by 36 weeks (Dreyfus-Brisac 1975). Thereafter, the organization, duration, and cycling of sleep and waking states follow a predictable course throughout infancy, which has been summarized by Roffwarg et al (1966), Paul et al (1973), and Parmalee (1975). The organization of behavioral state differs significantly between high risk and normal infants of the same conceptional age (Dreyfus-Brisac 1970, Prechtl et al 1973). Thus the detailed analysis of behavioral states and state changes under unstimulated conditions are a sensitive measure of the young infant's maturational state and central nervous system integrity.

THE BEHAVIORAL REPERTORY OF THE NEWBORN INFANT

The newborn infant does not experience the extrauterine environment as "one great blooming, buzzing confusion" (James 1890), but is born with a remarkably complex repertory of preadapted perceptual and motor mechanisms.

Motor Behavior

The detailed descriptions by Hooker (1952) and Humphrey (1964) on the motility patterns of human fetuses are the only detailed accounts to date of the relation between fetal sensory motor development and morphological maturation. Their findings demonstrate that the human central nervous system is functional by eight and a half weeks after conception. However, the observations were made on aborted fetuses that could be kept alive for only brief periods and under grossly abnormal environmental conditions; therefore, they do not constitute an entirely satisfactory physiological description of sensory motor function during fetal life. By means of ultrasonic radiography it may become possible to investigate prenatal sensorimotor maturation under more physiological conditions. Using this technique, Boddy & Dawes (1975) were able to measure respiratory movements of the human fetus as early as 11 weeks of gestation, and to discriminate between regular and irregular breathing patterns in 20–30 week old fetuses. Although the method was intended primarily as a clinical tool to identify early risk factors in fetal life, it may also prove of great value for studying the intrauterine development of motor patterns and behavioral states in normal fetuses. The sensitivity of the method will be considerably enhanced by recent technical refinements of continuous high speed imaging that permit the visual analysis of discrete motor patterns such as twitches, startles, isolated movements of the limbs and head, etc (Birnholz et al 1978).

The development of motor reflexes from the last trimester of pregnancy until shortly after birth has been used in a clinical context to estimate maturational state of delivered infants independent of chronological age, and to evaluate the neurological integrity of preterm and full-term infants (Robinson 1966, Saint-Anne Dargassies 1966, Dubowitz et al 1970). The systematic comparison of behavior in preterm infants at 40 weeks of conceptional age and full-term infants of the same conceptional age offers an important opportunity to isolate the influence of extrauterine environmental factors on behavioral development. Parmalee (1975) has reviewed a large number of studies that compare the neurological and behavioral development of premature infants at 40 weeks conceptional age and full-term infants of the same gestational age. The majority of findings points to the conclusion that differences in intra- and extrauterine experience do not significantly affect the course or rate of development of sensorimotor functions, although differences in stability of state organization, strength of motor reflexes, visual attention, and general irritability have been reported. Prechtl et al (1975) compared the development of posture, spontaneous movement, and behavioral state in premature infants who had been carefully selected to exclude perinatal risk factors. In keeping with most previous studies, they found that differences between intrauterine and extrauterine stimulus environment had no significant effect on the development of motor functions or behavioral states. The fact that the authors included only low risk premature infants raises the possibility that many of the observed differences between preterm and full-term infants are due to clinical conditions that accompany prematurity, rather than to prematurity as such. The motor reflexes of movement, postural adjustment, locomotion, breathing, sucking, and the like of the full-term newborn infant have been extensively catalogued by Peiper (1963). Similarly detailed information on preterm infants is not available.

Several discrete neonatal motor patterns have been investigated in considerably greater detail. For example, the serial organization, state dependence, and responsiveness to auditory, tactile, and visual stimulation of nutritive and nonnutritive sucking have been carefully documented (see Kessen et al 1970, for review). As one of the few complex motor patterns that is fully organized at birth, readily elicited in most behavioral states, and easily recorded for objective analysis, sucking behavior has also been used as an outcome measure to investigate the newborn infant's capacity for stimulus discrimination, attention, habituation, and learning (see Sameroff 1972, for a recent review). Most, but not all, of the available evidence suggests that sucking behavior cannot be conditioned by classical (Pavlovian) methods in the newborn period, whereas some parameters of nutritive and nonnutritive sucking can be controlled by operant techniques (Sameroff 1972; see also below).

The variety of contradictory results and of competing interpretations concerning the conditionability of motor responses during the neonatal period are reviewed by Sameroff (1972) and Fitzgerald & Brackbill (1976).

Hemispheric lateralization of functions in the newborn period has become an issue of interest as neuropsychological investigations address developmental questions (Wada 1977). Turkewitz (1977) has examined the asymmetry of posture and particularly of head rotation from various perspectives. In the prone position, normal newborn infants show a distinct preference for keeping their head to the right. Without clearly specifying the etiology of this head-right preference in young infants, the author concludes that it contributes to the development of lateralization for perceptual functions. In a review of manual specialization in infancy, Young (1977) summarizes various studies all reporting that there are no consistent manual preferences during the neonatal period.

Perception in the Neonatal Period

Contemporary studies of neonatal perception emphasize the infant's selective responses to biologically meaningful stimulus configurations rather than nonspecific reactions to stimulus parameters that can be dimensionalized precisely, but may have no biological relevance.

HEARING   The healthy full-term infant responds selectively to the pitch, intensity, and quality of sound stimulation, is more sensitive to patterned sounds than pure tones, particularly in the frequency range of the human voice (Eisenberg 1969, Hutt et al 1968). Behavioral state also modifies the kind of motor response observed to high and low frequency sounds. The structure of the sound signal appears to be a more important determinant of response activity than the amount of energy in the signal (see Eisenberg 1976, for a comprehensive review).

Eimas (1974) and Eisenberg (1976) have summarized current research in the infant's capacity for auditory discrimination of speech sounds. On the basis of experiments using habituation and dishabituation of the sucking response as the outcome measure, Eimas concluded that one-month old infants make categorical discriminations between speech sounds along a voicing continuum that is similar to the adults categorical perceptions of speech. The conclusion has been criticized on the grounds that mere difference detection need not imply a discrimination between discrete perceptual events (Trehub 1973). At present, the interesting suggestion that the naive human infant is endowed with "phonetic feature detectors" prior to language exposure cannot be considered as proven (Stevens & Klatt 1974). Molfese et al (1975) and Entus (1977) have presented neurophysiological and neuropsychological evidence suggesting that the left cerebral hemisphere of the full-term infant may be partially specialized for processing speech sounds (phonemes and words), the right hemisphere for processing nonspeech sounds.

VISION   Within the first 24 hours after birth, full-term infants make conjugate pursuit movements to a target in the horizontal or vertical axis (Dayton & Jones 1964, Salapatek 1975), but binocular fixation cannot be clearly demonstrated until after the eighth week (Wickelgren 1967). The newborn does not accommodate to objects at various distances from the retina until at least the eighth week and, until then, appears to have a focal length that is relatively fixed at 19 cm (Haynes et al 1965).

Contemporary vision research on infants emphasizes the capacity for pattern perception and pattern discrimination. Neonates show a distinct visual preference for patterned over nonpatterned surfaces (Fantz et al 1975). The naive infant has only a limited capacity to select for, or discriminate among, visual features, figures, or patterns that are equated in brightness and contour density. However, Fantz & Miranda (1975) observed that neonates show a measurable visual preference for curved over straight contours, and Salapatek (1975) reported that in the first two weeks after birth infants pay more attention to the outer contours of a pattern, whereas two-month old infants concentrate on internal details. There is no persuasive evidence that newborn infants show a visual preference for faces over other geometric patterns, when comparison stimuli are equated for contour density (number or size) (Haaf & Bell 1967). The newborn infant's visual search pattern tends to center on single or limited numbers of features within a figure or pattern (Salapatek 1975). Fantz & Fagan (1975) also reported that conceptional age rather than postnatal age best predicts the developmental progress of visual preference for complex patterns, suggesting that central nervous system maturation, rather than visual experience, determines the development of visual perception during the early weeks after birth.

Bornstein et al (1976) reported the intriguing observation that recovery from habituation after repeated stimulation by the same color stimulus is greater when the comparison stimulus comes from an adjacent hue (as perceived by the adult) than when the comparison stimulus comes from the same adult hue category, although the wavelengths of both comparison stimuli are equally distant from the wavelengths of the habituated stimulus. The experiments were carried out on older rather than newborn infants, but they are one of the few reliable demonstrations that young infants have color vision (see also Peiper 1963), and that they group the visible light spectrum into hue categories much like adults at a stage in development long before linguistic categories for color hue could conceivably be involved.

Bower (1974) demonstrated that by two months infants respond to arrays of visual objects of different sizes, shapes, or distances from the eye, as if they utilize information from binocular parallax, are capable of size constancy (i.e. matching visual objects according to their actual size rather than the visual image projected on the retina), and have a limited capacity for shape constancy (i.e. matching rectangles in different orientations to the line of sight according to the object's real rather than retinal shape). Because of their important implications for contemporary theories of perception, these observations should be replicated on larger samples, and perhaps by different methods. For a recent review of visual perception experiments during the early weeks after birth, see Salapatek (1975).

Learning in the Neonatal Period

An enduring preoccupation of experimental and developmental psychologists is the question how soon after birth human infants can learn, what kind of learning is possible, and how the behavioral repertoire of the neonate is most effectively brought under stimulus control. Autonomic responses (heart rate, pupillary reflex, skin potential) and motor patterns (sucking, head rotation, Babkin reflex, eye blink, and foot withdrawal) have been used as outcome measures, to test for susceptibility to classical and instrumental conditioning, and to determine the infant's capacity for stimulus discrimination as measured by the "orienting response" (Sokolov 1963). The volume of empirical studies on infant learning has grown considerably over the past 15 years, but the theoretical questions, which at first appeared simple and amenable to empirical verification, have become extraordinarily complex. At present, they are confounded by a mass of contradictory results and inconsistent variations of methodological refinement.

From a review of contemporary research on neonatal learning, Sameroff (1972) concludes that there is no clear evidence for classical conditioning in the neonatal period and that many reports claiming such conditioning are confounded by failure to control for behavioral state. On the other hand, he concludes that it is possible to bring behavior patterns with evolutionary survival value (e.g. sucking, head turning) under stimulus control by instrumental conditioning or reinforcement. Fitzgerald & Brackbill (1976) concluded that full-term infants *can* be conditioned by classical (Pavlovian) methods even during the first week, but added the qualification that conditionability is an interactive function between the sensory modality of the conditioning stimulus and the neurological system that controls the response to be conditioned; in other words, that stimulus-response specificity imposes major constraints on neonatal learning, and that behavioral state as well as other biological factors limit the newborn infant's capacity for learning significantly. At present, it is not clear how concepts of behavioral state, of survival value (as a necessary criterion for instrumental conditioning), and of stimulus-response specificity (as a precondition for classical conditioning) can be integrated under a rigorous formulation of either classical or reinforcement conditioning.

On the assumption that the orienting response (Sokolov 1963) is a necessary precondition for learning, and that heart rate deceleration in response to stimulation measures the orienting response (Graham & Clifton 1966), investigators have extensively examined the question whether newborn infants orient to discrete stimuli and, by extrapolation, whether they attend to and "take in" the stimulus world. Most studies concur that young infants show only heart rate acceleration to stimulation (Graham & Jackson 1970), and therefore conclude that infants do not orient or attend to stimulation (see Hirschman & Katkin 1974, for a review). Correspondingly, evidence of heart rate deceleration to low level tactile, auditory, or taste stimulation (Samaroff 1972) is taken as an indication that newborn infants can orient or attend to stimulation; in other words, that the preconditions for learning in the neonatal period are met. Whether heart rate deceleration is a valid indicator of the orienting response in young infants, and whether the orienting response is a necessary precondition for neonatal attending, are issues that have not been persuasively demonstrated, although the relationship is generally assumed. (For more detailed accounts of current theoretical issues and empirical findings on neonatal learning, see recent reviews by Sameroff 1972, Hirschman & Katkin 1974, and Fitzgerald & Brackbill 1976.)

THE NEWBORN INFANT AS SOCIAL PARTNER

The infant's earliest adaptation to the social world and particularly the role of the infant as an active partner in social interchange (Richards 1974, Brazelton et al 1975, Lamb 1977) represents an important growth point of current behavioral research. Interest in the topic stems in part from ethological studies on species-specific behavioral mechanisms controlling the filiative behavior between parent and immature offspring (Lorenz 1971), and in part from the clinical assumption that nonoptimal experience in early life may irreversibly compromise the child's long-term social development (Bowlby 1969).

Among the infant's repertory of social signals or expressions of emotion (Darwin 1873) by which the caretaker gauges the infant's biological or social needs, cry vocalizations probably have the most compelling influence on the social partner. Using sound spectrography and other electronic recording techniques, Wasz-Hockert et al (1968) identified some of the acoustic features by which neonatal vocalizations can be classified as birth, pain, hunger, and pleasure cries (see also Wolff 1969). The mother's biological and social responses may be partially controlled by the type of infant cry perceived, as in other animal species (Wasz-Hockert et al 1968), but they are also significantly modulated by the mother's motivations and prior attitudes to child care (Wolff 1969, Bernal 1972). The spectrographic analysis of infant cry patterns has been proposed as an objective method for diagnosing central nervous system impairment (Karelitz & Fischieli 1962). However, it remains to be demonstrated whether this method is more discriminating than conventional neurological examinations; and for such a demonstration, quantified methods of sound spectrographic analysis are needed.

The pervasive interest in smiling behavior stems from its importance for nonverbal communication in later life, and from its uniqueness to the human species (Darwin 1873). The majority of empirical work on the development and function of smiling has dealt with older infants. Results from these studies have given rise to a number of competing theories about the cognitive, affective, or social-communicative implications of smiling that are reviewed by Sroufe & Waters (1976). Whether smiling serves any developmentally relevant function in the first week after birth remains a matter of controversy. As early as 28 weeks after gestation (Wolff 1963, Emde et al 1976), premature infants display facial grimaces, which are morphologically very similar to smiling in older infants. The full-term newborn smiles almost exclusively during sleep, and particularly during active or light sleep, and drowsiness. Smiling is observed in the absence of known environmental stimuli ("spontaneous" or "endogenous" smiling), as well as in response to gentle tactile or auditory stimuli. Other facial grimaces reminiscent of emotional expressions that are variously labeled as grimaces, frowns, and the like, can be demonstrated in premature as well as full-term infants, particularly during active sleep (Emde et al 1976), but their behavioral function during the neonatal period and their relevance for social development is at best obscure.

The emphasis of recent studies has shifted from documenting isolated affect expressions, to the analysis of social interactions between caretaker and infant, and the mechanism that facilitates such interactions. Again, the majority of studies on mother-child interaction, on "bonding," "attachment," etc, have been conducted with older infants (for a review, see Lamb 1977). In recent years, however, some investigators have focused on social interactions during the infant's first weeks after birth. Klaus & Kennell (1976) summarized a long series of studies carried out by them and their colleagues, many of which could not, for obvious reasons, be carried out under controlled experimental conditions. They reported that physical contact between mother and full-term newborn infant can significantly influence the mother's social response to her child at 3, 14, and 24 months, as measured by the mother's physical contact and verbal interchange with her child. Mothers who have touched, held, and played with their prematurely born infants during the lying-in period were found to be more skillful, more confident, and more stimulating at later stages in the child's development than mothers deprived of such contact (see also Leiderman & Seashore 1975). The most direct effect of increased social contacts between mother and newborn infant appears to be on maternal attitude, which in turn may account for many of the reported variations in developmental outcome in the infant. Thus, it may be the mother rather than the infant who experiences a "sensitive period" for a social attachment during the first few days after the infant's birth (Klaus & Kennell 1976). However, Sander et al (1970), among others, demonstrated that newborn infants are directly sensitive to variations in the social environment, independent of any effect on the mother. Infants cared for by multiple persons with different styles of feeding and handling show more erratic sleep-waking cycles, more fussing, etc than infants cared for by only one person during the same period.

Whether early behavioral differences in response to variations of social stimulation have any direct long-term effects or are easily compensated by later changes in the social environment, and whether stable behavioral effects in older infants are mediated by the caretaker's attitude in response to the baby's behavior, remain unresolved issues. They should be systematically explored before undue importance is attributed to the quality and amount of social interchange between mother and infant during the early days after birth as a cause of deviant social or intellectual development

(Sander et al 1970). In contrast to normal full-term infants, the quality of social stimulation may have a decisive influence on young infants who are already compromised in their developmental potential because of prenatal malnutrition, prematurity and other perinatal risk factors (Scarr-Salapatek & Williams 1973).

While the social relevance of the infant's behavior can obviously be assessed only in relation to the behavior of the caretaker or social partner, current theoretical models, observation techniques, and methods of data analysis may be too limited in conception to capture the essential features of the social transaction. Many current studies infer a causal relation between the infant's and the caretaker's behavior from frequency counts of contiguous events under a Markov process model (Lamb 1977). Yet, there is no reason to assume that temporally contiguous social events are causally related even during the neonatal period (Lewis & Lee-Painter 1974). Instead, a model of hierarchically organized structures of action may be required, which will capture the "syntax" of social interactions regardless of their temporal contiguity.

INDIVIDUAL DIFFERENCES

Full-term newborn infants have been shown to differ from one another with respect to psychophysiological and motor parameters, ease of conditionability to sensory stimuli, stability of behavioral state, amount of sucking and crying, nonspecific motility, and the like (Escalona 1968, Sameroff 1972, Korner 1974, Fitzgerald & Brackbill 1976). Moreover, there is some indication that such individual differences influence the caregiver's action (Korner 1974). More detailed investigations of the issue would probably reveal that newborn infants differ from one another on every conceivable behavioral dimension; whether such differences have any long-term effect for adaptation and behavioral outcome has not been satisfactorily resolved even by extensive longitudinal followup studies (Kagan 1971). A meaningful demonstration of antecedent-consequent relations between individual differences in neonatal behavior and variations of behavioral outcome requires elaborate research designs (see Escalona 1968) that can determine how the infant's individual characteristics and developmental timetable influence the caretaker's behavior, how the caretaker's idiosyncratic style of child care influences the infant's behavior, and how the match between infant and caretaker determines the characteristics of their transactions throughout early ontogenesis.

Individual variations in neonatal behavior have occasionally been investigated in terms of sex differences. The earlier somatic maturation of girls than boys appears to have major consequences for the behavioral development later in life, and is already apparent at birth (Tanner 1974). The importance of such biological sex differences is generally recognized in principle, but empirical evidence on behavioral sex differences at birth is very limited. Female infants are reported to be more sensitive to tactile stimulation than males (Bell & Costello 1964, Rosenblith & DeLucia 1965, Wolff 1969), but the etiology of such differences remains obscure (see Maccoby & Jacklin 1974). The observation that female infants are more sensitive to color hues than males (Peiper 1963) is suggestive because similar sex differences are reported throughout childhood; however, Peiper's original observations must be replicated under psychophysiologically more controlled conditions. Newborn boys and girls have also been reported to differ in stability of state cycles, frequency of unstimulated motor behavior, etc (Korner 1974, Freedman 1974), but the long-term significance of these and other sex differences for behavioral development has not been tested. In view of significant sex differences on biological parameters of birth weight, neonatal maturity, and perinatal risk factors, the lack of systematic information on behavioral sex differences in the newborn infant is somewhat surprising (see Maccoby & Jacklin 1974, for a recent review).

Systematic studies of behavioral differences among infants from different human races or mating groups are very rare. In a frequently cited study, Geber & Dean (1957) examined the motor maturity of Bantu infants by the neurological method of Saint-Anne Dargassies and reported that African infants were motorically precocious in comparison to (French?) white infants. However, no statistical comparisons were reported, no comparison groups were tested, and at the time the method of neurological examination had not been standardized on white infants. Moreover, Warren (1972) re-examined samples from the same populations and failed to find significant differences between African blacks and whites by the same neurological examination.

Freedman (1974) summarized his observations indicating that there are consistent behavioral differences between Caucasian newborn infants and infants from subgroups of the Mongoloid major mating population, African blacks and Australian Aborigines, as measured on items from the Brazelton Infant Assessment Scale. Maternal and fetal nutritional state, perinatal risk factors, and obstetrical practices were partially controlled in the comparison of American-born Chinese and Caucasoid infants, but such confounding variables are not reported for other racial groups. Moreover, many of the items on which differences were reported, imply complex psychological processes ("soothability," "habituation," "defensive reaction") whose functional significance is difficult to interpret as long as neither the method of stimulation nor the behavioral outcome are carefully controlled. Therefore, it would seem premature to draw far-reaching conclusions about racial differences in behavior or about their developmental significance from available evidence.

Wolff (1972) reported major differences between Caucasoid newborns and newborns from various subgroups of the Mongoloid major mating population in the vasomotor flushing response to small quantities of alcohol. The same differences were observed between adults of similar mating populations. While group differences of vasomotor responsiveness to alcohol were large and consistent, the relevance for psychological development remains undetermined.

CONCLUSIONS

Contemporary behavioral research converges on the conclusion that the full-term newborn human infant encounters its extrauterine environment with a remarkably complex array of perceptual and motor mechanisms for adaptation to the physical and social world, and that behavioral organization at birth is neither a simplified replica of psychological processes in human adults, nor an analog of the behavioral organization in phylogenetically less differentiated animal species.

Justified enthusiasm about the remarkable abilities of naive infants, and about the possibilities for the early diagnosis of behavioral risk factors and determinants of individual variations in normal development must, however, be counterbalanced by evolutionary consideration of human neotony (Young 1971). The behavioral capacities of the human newborn are far less differentiated in comparison to long-term developmental outcome than they are in other animal species. Since human infants have a much longer developmental road to travel, their development will be modified more extensively by environmental events and late-maturing functions than that of other animal species.

By virtue of its evolutionary history, the human infant is not a simplified neuronal model that lends itself passively to experimental manipulations, for example, to test the general principles of learning or conditioning. At the same time, the human neonate is not endowed with a miniature replica of psychological abilities observed in children and adults. The neonatal stepping reflex is not a homologue of voluntary walking (Peiper 1963); visual scanning in the newborn period involves different neural mechanisms than the visual search of older infants (Salapatek 1975). "Learning" during early infancy implies different neuropsychological processes than concept formation in older children (Piaget 1967). More generally, human behavioral development does not proceed along linear pathways, but is characterized by regressions (or learning dips), repetitions or recapitulations, and the acquisition of qualitatively different mechanisms to perform the same function at a more differentiated level (Bronson 1965). Analogies of form between behavior patterns of the newborn infant and child therefore do not justify the inference that the behavioral capacities of the adult that resemble neonatal behavior are fixed at birth.

Such considerations may account in part for the fact that contemporary attempts to predict normal and deviant developmental outcome from neurobehavioral status at birth have been generally disappointing. Failures of prediction may be due in part to the inadequate selection of study samples, the failure to standardize examination procedures, and the irrelevance of outcome measures. However, uncontrolled genetic factors, the effect of unsuspected environmental influences, and particularly our lack of knowledge concerning developmental transformations in the covert mechanisms that control manifest behavior severely limit our ability to predict from neonatal behavior to psychological adaptation in the mature person. Such limitations are of more than theoretical interest, as long as the social-psychological consequences of labeling and false assignment of risk for non-optimal development outweigh the potential benefits from currently available methods for behavioral intervention during early infancy.

Literature Cited

Anders, T., Emde, R., Parmalee, A. H., eds. 1971. *A Manual of Standardized Terminology, Techniques and Criteria for Scoring States of Sleep and Wakefulness in Newborn Infants.* Los Angeles: UCLA Brain Information Service.

Bell, R. Q., Costello, N. S. 1964. Three tests for sex differences in tactile sensitivity in the newborn. *Biol. Neonate* 7:335–47

Bernal, J. F. 1972. Crying during the first ten days of life, and maternal responses. *Dev. Med. Child Neurol.* 14:362–72

Birnholz, J. C., Stephens, J. C., Faria, M. 1978. Analysis of fetal movement patterns. *Am. J. Roentgenol.* 130:537–40

Boddy, K., Dawes, G. S. 1975. Fetal breathing. *Br. Med. Bull.* 31:3–7

Bornstein, M. H., Kessen, W., Weiskopf, S. 1976. The categories of hues in infancy. *Science* 191:201–2

Bower, T. G. R. 1974. *Development in Infancy.* San Francisco: Freeman

Bowlby, J. 1969. *Attachment and Loss.* Vol. I. Attachment. London: Hogarth

Brazelton, T. B., Tronick, E., Adamson, L., Als, H., Weise, S. 1975. Early mother infant reciprocity. *Ciba Symp.* 33: 137–68

Bronson, G. 1965. Hierarchical organization of the central nervous system. *Behav. Sci.* 10:7–25

Darwin, C. 1873. *The Expression of Emotions in Man and Animals.* London: Murray

Dayton, G. O., Jones, M. H. 1964. Analysis of characteristics of fixation reflex in infants by use of direct current electro-oculography. *Neurology* 14:1152–56

Dreyfus-Brisac, C. 1970. Sleeping behavior in abnormal newborn infants. *Neuropaediatrie* 3:354–66

Dreyfus-Brisac, C. 1975. Neurophysiological studies in human premates and full-term newborns. *Biol. Psychiatry* 10: 481–96

Dubowitz, L., Dubowitz, V., Goldberg, C. 1970. Clinical assessment of gestational age in the newborn infant. *J. Pediatr.* 77:1–10

Eimas, P. D. 1974. Linguistic processing of speech by young infants. In *Language Perspectives—Acquisition, Retardation and Intervention,* eds. R. L. Schiefelbusch, L. L. Lloyd, pp. 53–73. Baltimore: University Park Press

Eisenberg, R. B. 1969. Auditory behavior in the human neonate. *Int. Audiol.* 4:65–68

Eisenberg, R. B. 1976. *Auditory Competence in early life: The Roots of Communicative Behavior.* Baltimore: Univ. Park Press

Emde, R. N., Gaensbauer, T. J., Harmon, R. J. 1976. Emotional expressions in infancy. *Psychological Issues Monograph Series X.* New York: International Universities Press

Entus, A. K. 1977. Hemispheric asymmetry in processing of dichotically presented speech and non-speech stimuli by infants. In *Language Development and Neurological Theory,* ed. S. Segalowitz, F. Gruber, pp. 63–73. New York: Academic

Escalona, S. K. 1968. *The Roots of Individuality.* Chicago: Aldine

Fantz, R. L., Fagan, J. F. 1975. Visual attention to size and number of pattern details by term and preterm infants. *Child Dev.* 46:3–18

Fantz, R. L., Fagan, J. F., Miranda, S. B. 1975. Early visual selectivity. In *Infant Perception: From Sensation to Cognition,* ed. L. B. Cohen, P. Salapatek, pp. 249–346. New York: Academic

Fantz, R. L., Miranda, S. B. 1975. Newborn attention to form of contour. *Child Dev.* 46:224–28

Fitzgerald, H. E., Brackbill, Y. 1976. Classical conditioning in infancy. *Psychol. Bull.* 83:353–76

Freedman, D. G. 1974. *Human Infancy,* pp. 51–79, 145–76. New York: Wiley

Geber, M., Dean, R. F. A. 1957. The state of development of newborn African children. *Lancet* 1:1216–19

Graham, F. K., Clifton, R. K. 1966. Heart rate changes as a component of the orienting response. *Psychol. Bull.* 65:305–20

Graham, F. K., Jackson, J. C. 1970. Arousal systems and infant heart rate responses. In *Advanced Child Development Behavior,* Vol. 5, ed. H. W. Reese, L. P. Lipsitt, pp. 59–117. New York: Academic

Haaf, R. A., Bell, R. Q. 1967. A facial dimension in visual discrimination by human infants. *Child Dev.* 38:893–99

Haynes, H., White, B. L., Held, R. 1965. Visual accommodation in human infants. *Science* 148:528–30

Hirschman, R., Katkin, E. S. 1974. Psychophysiological functioning, arousal, attention and learning during the first year of life. In *Advanced Child Development Behavior,* Vol. 9, ed. H. W. Reese, pp. 115–150. New York: Academic

Hooker, D. 1952. *The Prenatal Origin of Behavior.* Lawrence, Kansas: Univ. Kansas Press

Humphrey, T. 1964. Some correlations between the appearance of human fetal reflexes and the development of the nervous system. *Prog. Brain Res.* 4:93–117

Hutt, C., Lenard, H. G., von Bernuth, H., Hutt, S. J., Prechtl, H. F. R. 1968. Habituation in relation to state in the human neonate. *Nature* 220:618–20

Hutt, S. J., Lenard, H. G., Prechtl, H. F. R. 1969. Psychophysiological studies in newborn infants. In *Advanced Child Development Behavior,* Vol. 4, ed. H. W. Reese, L. P. Lipsitt, pp. 127–72. New York: Academic

James, W. 1890. *The Principles of Psychology.* New York: Holt

Kagan, J. 1971. *Change and Continuity in Infancy.* New York: Wiley

Karelitz, S., Fischieli, V. R. i962. The cry threshold of normal infants and those with brain damage. *J. Pediatr.* 61:679–85

Kessen, W., Haith, M. M., Salapatek, P. H. 1970. Human infancy. In *Carmichael's Handbook of Child Psychology,* ed. P. H. Mussen, pp. 287–445. New York: Wiley

Klaus, M. H., Kennell, J. H. 1976. *Maternal-Infant Bonding,* pp. 1–15, 38–98, 99–166. St. Louis: Mosby

Korner, A. F. 1974. The effect of the infant's state, level of arousal, sex and ontogenetic stage on the caregiver. In *The Effect of the Infant on its Caregiver,* ed. M. Lewis, L. A. Rosenblum, pp. 105–21. New York: Wiley

Lamb, M. E. 1977. A re-examination of the infant social world. *Human Dev.* 20:65–85

Leiderman, P. H., Seashore, M. J. 1975. Mother-infant neonatal separation. *Ciba Symp.* 33:213–39

Lenard, H. G., von Bernuth, H., Prechtl, H. F. R. 1968. Reflexes and their relationship to behavioral states in the newborn. *Acta Psychiatr. Scand.* 57:177–85

Lewis, M., Lee-Painter, S. 1974. An interactional approach to the mother infant dyad. In *The Effect of the Infant on the Caregiver,* ed. M. Lewis, L. A. Rosenblum, pp. 21–48. New York: Wiley

Lorenz, K. 1971. *Studies in Animal and Human Behavior,* Vol. I. Cambridge: Harvard Univ. Press

Maccoby, E. E., Jacklin, C. N. 1974. *Psychology of Sex Differences,* pp. 17–62. Stanford: Stanford Univ. Press

Molfese, D. L., Freeman, R. B., Palermo, D. S. 1975. The ontogeny of brain lateralization for speech and non-speech stimuli. *Brain and Language,* 2:356–68

Parmalee, A. H. Jr. 1974. Ontogeny of sleep patterns and associated periodicities in infants. *Mod. Probl. Pediatr.* 13:298–311

Parmalee, A. H. Jr. 1975. Neurophysiological and behavioral organization of premature infants in the first months of life. *Biol. Psychiatry* 10:501–12

Paul, K., Dittrichova, J., Pavli Kova, E. 1973. The course of quiet sleep in infants. *Biol. Neonate* 23:78–89

Peiper, A. 1963. Cerebral function in infancy and childhood. New York: Consultants Bureau, pp. 76, 147–247

Piaget, J. 1967. *Biologie et Connaissance.* Paris: Gallimand.

Prechtl, H. F. R. 1968. Polygraphic studies of the full-term newborn infant. II. Computer analysis of recorded data. In *Studies in Infancy,* ed. M. C. O. Bax, R. C. Mackerth, London: Spastics Int. Med. Publ. (SIMP) with Heinemann, pp. 22–40

Prechtl, H. F. R., Beintema, D. 1964. *The Neurological Examination of the Full-term Newborn Infant.* London: SIMP with Heinemann.

Prechtl, H. F. R., Vlach, V., Lenard, H. G., Kerr-grant, D. 1967. Exteroceptive and tendon reflexes in various behavioral states in the newborn infant. *Biol. Neonate* 11:159–75

Prechtl, H. F. R., Akiyama, Y., Zinkin, P., Kerr-grant, D. 1968. Polygraphic studies of the full-term newborn infant. I. Technical aspects and qualitative analysis. In *Studies in Infancy,* ed. M. O. C. Bax, R. C. MacKerth, London: SIMP with Heinemann, pp. 1–21

Prechtl, H. F. R., Theorell, K., Blair, A. W. 1973. Behavioral state cycles in abnormal infants. *Dev. Med. Child Neurol.* 15:606–14

Prechtl, H. F. R., Fargel, J. W., Weinmann, H. M., Bakker, H. H. 1975. Development of motor function and body posture in preterm infants. *Inst. Natl. Santé Rech. Med. (INSERM)* 43:55–66

Richards, M. P. M. 1974. First steps in becoming social. In *The Integration of a Child in a Social World,* ed. M. P. M. Richards. London: Cambridge Univ. Press.

Robinson, R. J. 1966. Assessment of gestational age by neurological examination. *Arch. Dis. Child.* 41:437–47

Roffwarg, H. P., Muzio, J. N., Dement, W. C. 1966. Ontogenetic development of the human sleep-dream cycle. *Science* 152:604–19

Rosenblith, J. F., Delucia, L. A. 1965. Tactile sensitivity and muscular strength in the neonate. *Biol. Neonate* 5:266–82

Saint-Anne Dargassies, A. 1966. Neurological maturation of the premature infant of 28–41 weeks gestational age. In *Human Development,* ed. F. Falkner, pp. 305–26. Philadelphia: Saunders

Salapatek, P. 1975. Pattern perception in early infancy. In *Infant Perception: From Sensation to Cognition,* ed. L. B. Cohen, P. Salapatek, pp. 133–248. New York: Academic

Sameroff, A. J. 1972. Learning and adaptation in infancy. In *Advanced Child Development Behavior,* Vol. 7, ed. H. W. Reese, pp. 169–214. New York: Academic

Sander, L. W., Stechler, G., Burns, P., Julia, H. 1970. Early mother-infant interaction and 24-hour patterns of activity and sleep. *J. Am. Acad. Child Psychiatr.* 9:103–23

Scarr-Salapatek, S., Williams, M. L. 1973. The effects of early stimulation on low-birth-weight infants. *Child Dev.* 44:94–101

Sokolov, E. N. 1963. *Perception and the Conditioned Reflex.* London: Pergamon

Sroufe, L. A., Waters, E. 1976. The ontogenesis of smiling and laughter. *Psychol. Rev.* 83:173–89.

Stevens, K. N., Klatt, D. H. 1974. Role of formant transitions in the voiced-voiceless distinction for stops. *J. Acoust. Soc. Am.* 55:653–59

Tanner, J. M. 1974. Variability of growth and maturity in newborn infants. In *The Effect of the Infant on Its Caregiver,* ed. M. Lewis, L. A. Rosenblum, pp. 77–103. New York: Wiley

Theorell, K., Prechtl, H. F. R., Blair, A. W., Lind, J. 1973. Behavioral state cycles of normal newborn infants. *Dev. Med. Child Neurol.* 15:597–605

Trehub, S. B. 1973. Infants' sensitivity to vowel and tonal contrasts. *Dev. Psychol.* 9:91–96

Turkewitz, G. 1977. The development of lateral differentiation in the human infant. *Ann. N.Y. Acad. Sci.* 299:309–18

Wada, J. 1977. Pre-language and fundamental asymmetry of the infant brain. *Ann. N.Y. Acad. Sci.* 299:370–79

Warren, N. 1972. African infant precocity. *Psychol. Bull.* 78:353–67

Wasz-Hockert, O., Lind, J., Vuorenkoski, V., Partanen, T., Valanne, E. 1968. *The Infant Cry: A Spectrographic and Auditory Analysis.* London: SIMP with Heinemann

Wickelgren, L. W. 1967. Development of convergence. *J. Exp. Child Psychol.* 5:74–85

Wolff, P. H. 1963. Observations on the early development of smiling. In *Determinants of Infant Behaviour, II,* ed. B. Foss, pp. 113–167. London: Methuen

Wolff, P. H. 1966. The causes, controls and organization of behavior in the neonate.

Psych. Issues Monog. Ser. Vol. V, No. 1. New York: Int. Univ. Press

Wolff, P. H. 1969. The natural history of crying and other vocalizations in early infancy. In *Determinants of Infant Behaviour, IV,* ed. B. Foss, pp. 81–109. London: Methuen

Wolff, P. H. 1972. Ethnic differences in alcohol sensitivity. *Science.* 175:449–50

Wolff, P. H. 1973. The organization of behavior in the first three months of life. *Res.*

Publ. Assoc. Res. Nerv. and Ment. Dis. 51:132–53

Young, G. 1977. Manual specialization in infancy. In *Language Development and Neurological Theory,* ed. S. J. Segalowitz, F. A. Gruber, pp. 289–311. New York: Academic

Young, J. Z. 1971. *An Introduction to the Study of Man,* pp. 479–80. New York: Oxford Univ. Press

Ann. Rev. Neurosci. 1979. 2:309–40
Copyright © 1979 by Annual Reviews Inc. All rights reserved

SLOW VIRAL INFECTIONS ❖11526

Benjamin R. Brooks, Burk Jubelt, Jeffrey R. Swarz, and Richard T. Johnson

Department of Neurology, The Johns Hopkins University School of Medicine, Baltimore, Maryland 21205

INTRODUCTION

Slow viral infections of the central nervous system (CNS) are dependent not only on the unique biological properties of viruses, but also on the properties of the neural cells that they infect and on the response of the host to that infection (Johnson & ter Meulen 1978). Knowledge of the structure and biochemical composition of viruses has progressed rapidly in recent years; nevertheless, the nature of the causative agents of two of the most dramatic slow infections of man, kuru and Creutzfeldt-Jakob disease, is so ill-defined that they may not even represent true viruses (Gajdusek 1977). The varied virus-cell interactions are also being determined (Robb 1977), but studies of these relationships in intact hosts move to a new level of complexity because of the mosaic of different cells and the myriad of host responses (Johnson & Griffin 1978). The persistence of viruses within the CNS poses relevant and intriguing problems since, to date, the major slow viral infections uncovered in man have been manifest by chronic neurological disease (ter Meulen & Katz 1977). Furthermore, the pathways by which viruses invade the CNS, the complex interactions of viruses with diverse populations of neural cells, the mechanisms by which viruses can be sequestered in the brain and by which they cause chronic CNS disease, pose tantilizing biological questions. (Thormar, Lin & Karl 1973).

We discuss the nature of viruses and the diversity of virus-host interactions before describing examples of chronic inflammatory, demyelinating, and degenerative diseases of the CNS of man or animals in which some knowledge of pathogenesis has been uncovered.

309

VIRUSES AND THE DIVERSITY OF VIRUS-HOST INTERACTIONS

Nature of Viruses

The definition of a virus is now restricted to . . .

"entities whose genomes are elements of nucleic acid, that replicate inside living cells using the cellular synthetic machinery and causing synthesis of the specialized elements, the virions, that can transfer the genome to other cells" (Luria et al 1978).

Structurally the virus particle, or virion, consists of a single or double strand of a single nucleic acid, either DNA or RNA, encased by, or interwoven with, multiple protein subunits that together form the nucleocapsid. This represents the complete virion in nonenveloped viruses such as polioviruses or papovaviruses. In enveloped viruses, such as herpesviruses, orthomyxoviruses, and paramyxoviruses, the nucleocapsid is surrounded by an envelope of virus-coded proteins and host cell lipids acquired as the nucleocapsid buds through cell membranes (Rifkin & Quigley 1974).

Infection can be effected by even simpler structural agents. Viroids, which are associated with several plant diseases, consist of short pieces of nucleic acid with no protein coat (Diener & Haddi 1977). These botanical curiosities may be relevant to slow infections of the nervous system, because their physicochemical properties resemble those of the undefined agents that cause kuru and Creutzfeldt-Jakob disease (Gajdusek 1977).

Virus-Cell Interaction: Replication of Virus

Virus replication and cellular susceptibility are dependent on the attachment and penetration of the virus and on the degree to which viral nucleic acids and proteins are replicated within the cell (Robb 1977). If a complete cycle of infection occurs within the cell, the following sequence takes place: (*a*) attachment of virus to plasma membrane; (*b*) penetration of the virion or nucleocapsid; (*c*) uncoating of capsid proteins from the nucleic acid; (*d*) replication of the viral genome and synthesis of structural and nonstructural proteins; (*e*) assembly of the nucleocapsid; and (*f*) release of infectious virions (Bablanian 1975).

If receptor sites are not present on the cytoplasmic membrane, virus infection at the cellular level may not occur, and the cell is considered resistant (Lonberg-Holm & Philipson 1974). Absence of systems necessary for virus penetration and uncoating may yield a similar result (Robb & Martin 1972). If the virus is uncoated, then replication may proceed to different degrees depending on the cell type infected (Spring, Roizman & Schwartz 1968). If the viral genome is replicated, in whole or in part, and some or all of the structural proteins (antigens) are not synthesized, then nonproductive (nonpermissive) infection may occur (Sturman & Tamm 1969). The viral genome can be sequestered by integration into the chromosomes of infected cells (Varmus et al 1974, Jaenisch 1976) or be preserved extrachromosomally without integration (Hampar et al 1971, Klein et al 1976). Some structural and/or nonstructural proteins may be generated over long periods of time by nonpermissive infected cells without assembly of mature nucleocapsids and/or virions (Spring & Roizman 1967). Even when nucleocapsids are formed they may not be released from cells as infectious extracellular virus particles (Scheid & Choppin 1976) or they may be released as defective particles without the complete viral genome (Huang & Baltimore 1977). Thus, abortive or nonproductive infection, which fails to generate infectious virus particles, may occur by a variety of mechanisms.

Virus-Cell Interaction: Effect on Cell Function

Cellular cytopathology will depend on the interaction between the virus and host cell. Virus infection may lead to (a) lysis, (b) transformation, (c) altered function or dysfunction, or (d) no detectable pathological effect. Acute cell lysis may result from the turn off of host RNA and protein synthesis as in poliovirus infection (Tamm 1975), direct toxicity by accumulated viral protein as in adenovirus infections (Pereira 1958), or activation of lysosomal enzymes as in murine hepatitis virus infection (Allison & Sandelin 1963).

Transformation may occur with or without infectious virus production (Martin 1970), but probably requires at least partial viral genome integration (Smith, Gelb & Martin 1972). Cell transformation in vitro is manifest by the loss of contact inhibition (Eckhart, Dulbecco & Burger 1971) and may be accompanied in vivo by tumor formation (Ponten 1976).

Alterations of specific cell function or general chronic dysfunction with virus infection is only beginning to be understood. Persistent infection of cells by rubella virus will not cause lysis, but will inhibit mitosis (Rawls & Melnick 1966). This change is present in cloned infected cells, which indicates that mixed cell populations are not the cause of the apparent effect on growth rate (Rawls 1974). Specialized differentiated cellular functions may be affected by persistent virus infections that do not affect overall DNA, RNA, and protein synthesis. Infection of different clones of murine neuroblastoma cells by lymphocytic choriomeningitis virus in vitro will lead to a decrease in choline acetyltransferase activity and acetylcholine esterase activity without changes in cell morphology, cell growth, or total protein synthesis (Oldstone, Holmstoen & Welsh 1977). Thus, specific cellular functions may be affected in persistent noncytolytic viral infections of neurons. Such a mechanism may occur in vivo during rabies virus infection (Johnson 1965).

Virus infection of cells may occur without a direct demonstrable effect upon normal cell activity. Endogenous retrovirus infected cells in mouse lymphoid and gonadal tissue persistently produce infectious virus without any definite pathologic effect (Aaronson & Stephenson 1976). Thus, the virus may enter an "endosymbiotic" relationship with the cell. Such a state may persist unless other factors interrupt this relationship (Johnson 1973). Therefore, the effect of viral infection on cells is dependent on the cell type as well as the virus. For example, simian virus 40 in its natural rhesus monkey host causes persistent noncytopathic infection. However, the same virus causes an acute lytic infection in cell cultures of African green monkey kidneys and transformation in rodent cell cultures (Johnson et al 1977).

Virus-Host Interaction

How a virus produces a disease in animals or man is defined not simply by multiple virus-cell relationships, but by highly complex virus-host relationships (Mims 1964). The virus-host relationship is determined by: (a) which cells are susceptible to viral infection (Hartley, Rowe & Huebner 1970); (b) what types of virus-cell relationship occur in each different cell type (Trentin, Yabe & Taylor 1962); (c) what natural barriers exists to limit infection (Ogra & Karzon 1971); and (d) what types of immune response exist either to limit or cause disease (McFarland 1974).

Most infections of the CNS are preceded by infection of extraneural cells, but disease may be manifest only with the CNS infection. For example, poliovirus infects both cells in the gut and motor neurons in the spinal cord, yet clinical disease and cytopathic changes are limited to the neuronal infection (Sabin & Ward 1941). Even within the CNS the same virus may affect different cells. Some viruses (herpesvirus, reovirus, Sindbis virus) produce lytic infections in neurons, glia, and endothelial cells (Johnson & Mims 1968). Other viruses like poliovirus infect only select neuronal populations (Bodian & Horstmann 1965; Jubelt et al, unpublished data). The same virus can also infect multiple cell types but with different virus-cell interactions. For example, the papovavirus, which causes progressive multifocal leukoencephalopathy, appears to lytically infect oligodendroglia, causing demyelination, and nonpermissively infect astrocytes, altering their growth and morphology (Weiner & Narayan 1974).

The natural barriers that may limit disease in the host include barriers of nonsusceptible cells (Smith 1972) or abortively infected cells (Stevens 1978). Disease in the CNS may result from entry of virus following spread from extraneural sites via peripheral nerves (rabies, herpes virus), the olfactory route (herpes virus), or the hematogenous route (poliovirus, togavirus) (Johnson & Griffin 1978). Endothelial cells of cerebral vessels may allow direct entry to the CNS following lytic infection (Johnson 1965), nonlytic infection (Swarz et al, unpublished data), or transport of infectious virus without infection (Schultz & Fröhlich 1965). Infection of the choroid plexus (Mims 1960), or transport of virus through the choroid plexus (Albrecht 1968), will allow spread via the cerebrospinal fluid (CSF) with possible initial infection of meningeal or ependymal cells (Johnson & Mims 1968). Within the CNS parenchyma, virus may spread through extracellular spaces (Blinzinger & Muller 1971) or via axonal transport (Bak et al 1977).

Both the cellular and humoral immune responses usually serve to limit acute viral infection within the CNS (Griffin & Johnson 1977). However, with lymphocytic choriomeningitis virus infection, in adult mice, the disease is actually the result of an antiviral cellular immune response (Gilden, Cole & Nathanson 1972). This response is mediated by thymic-dependent (T) lymphocytes because anti-theta (T-lymphocyte surface antigen) serum rendered immune donar spleen cells incapable of producing disease despite persistence of viral specific antibody (Cole, Nathanson & Prendergast 1972). Inappropriate cell-mediated immune responses to new or preexisting antigens may occur many weeks after the acute infection of mice with Theiler's virus, a mouse picornavirus (Lipton & Dal Canto 1976, a, b). Immunologically mediated demyelination occurs in a patchy fashion with late onset paralysis in mice who have survived the acute infection with Theiler's virus, a mouse picornavirus (Lipton & Dal Canto 1976a, b). reproducible virus-induced immune-mediated CNS demyelinating disease (Dal Canto & Lipton 1977).

Infection-Disease Relationships

These various properties of the virus, the infected cell, the location of the infected cell within the host, and the response of the host to the infected cell serve to determine the nature of the disease that is produced. Virus interacts with specific host cells to produce acute, chronic, or latent infection. Acute infection may result in acute disease with recovery or acute disease with static or progressive sequelae. Chronic infection may be asymptomatic or lead to progressive disease. Latent infection may be associated with episodic disease, but cumulative effects of episodic activations of permissive infection may give the clinical appearance of progressive disease (Johnson, Narayan & Clements 1978).

Although acute self-limited infection may result in monophasic illness with complete recovery as in acute viral meningitis (Albrecht 1968), it may also cause chronic disease as in the residual paralysis after poliovirus infection (Sabin & Ward 1941). However, such static sequelae may give the appearance of progressive disease in fetal or neonatal infections when normal development unmasks the cell destruction that occurred earlier in development (Kilham & Margolis 1966, Johnson, Narayan & Clements 1978). For example, replication of feline panleukopenia virus, a parvovirus, is limited to mitotic cells. In adult or fetal cats destruction of intestinal and myelopoietic cells occurs, but these cells are rapidly replenished. In the fetal or newborn kitten, infection of the mitotically active external germinal cell layer of the cerebellum results in lysis of a nonregenerating population of cells. These kittens appear normal in infancy but never aquire normal motor control giving the impression of a progressive granuloprival cerebellar degeneration (Kilham & Margolis 1975). Similar static sequelae following arbovirus encephalitis in infants appear as progressive disease because subsequent developmental milestones are not achieved. Nevertheless, virus infection does not persist and further cell destruction does not appear to occur (Finley et al 1967).

Chronic disease may also result from structural progressive sequelae initiated by an acute infection such as hydrocephalus developing after mumps virus infection (Johnson & Johnson 1968). In laboratory animals progressive aqueductal stenosis and hydrocephalus can develop following a variety of viral infections that selectively involve ependymal cells lining the ventricles of the brain (Johnson 1972, 1975a). No acute disease may be observed in infected animals, but ependymal cells are destroyed, infectious virus is cleared, immunity develops, and the inflammatory response resolves. Subsequently, the aqueduct of Sylvius is reduced in size, and may eventually become stenotic or occluded. Progressive obstructive hydrocephalus supervenes with associated clinical signs (Johnson 1975a). While the exact pathogenetic mechanisms are still under investigation, the role of virus-induced aqueductal stenosis in the development of progressive hydrocephalus is well established (Masters, Alpers & Kakulas 1977).

Late progressive sequelae after acute self-limited infections may occur, as in the late onset progressive amyotrophy after poliomyelitis infection (Mulder, Rosenbaum & Layton 1972). In this situation the normal senescence of anterior horn cells may occur, when fewer cells are available. Because of the previous poliovirus-induced attrition of neurons, the decreased number of cells reaches a critical level at which clinical symptoms occur (McComas, Upton & Sica 1973).

Acute infection may be followed by chronic or persistent infection in which virus is continously replicated to varying degrees (Mims 1974). Neonatal infection of mice with lymphocytic choriomeningitis virus produces persistent infection of CNS without progressive pathological changes (Hotchin 1971), whereas hepatitis B virus can lead to self-limited disease with subsequent persistent infection that may or may not lead to immunologically mediated liver damage (Robinson & Lutwick 1976).

Progressive disease associated with persistent infection by rubella virus results from human fetal infection in early pregnancy (Rawls 1968). Progressive and static sequelae of in utero infection include cataracts, microphthalmia, deafness, valvular heart lesions, and microencephaly (Cooper et al 1969). The involved organs in these patients have decreased numbers of cells (Naeye & Blanc 1965) and focal collections of rubella antigen containing cells (Woods et al 1966). Infection of cells in vitro slows the replication rate (Boue & Boue 1969) and leads to chromosome breaks (Nichols 1970). Cloned cell cultures derived from infected human tissues have shorter in vitro life spans than uninfected controls (Rawls & Melnick 1966). Thus, persistent infection may lead to focal or multifocal alterations of in utero fetal growth patterns due to clones of infected cells with mitotic inhibition (Rawls 1974). Virus persists for a variable time postnatally but is eventually cleared, presumably when infected cells become senescent (Rawls 1974). Organs having the slowest cellular rate of senescence (i.e. brain and lens) have the longest persistence of virus (Johnson 1978b).

Latent infections differ from persistent infection in that infectious virus cannot be demonstrated by standard methods of viral cultivation between episodes of acute disease (Stevens 1978). Herpes virus infection of sensory ganglia constitutes the best known example of latent infections that produce episodic disease (Stevens 1975). Latent infections can be established in a variety of nervous tissue, most prominently the sensory ganglia (Cook & Stevens 1976). Schwann cells, satellite cells, and other supporting cells undergo abortive infection, whereas complete virions are replicated in neurons during acute infection (Stevens 1975). Viral DNA has been localized by in situ hybridization only in neurons of latently infected sensory ganglia analyzed directly upon removal (Stevens 1978). Cultivation in vitro "induces" production and release of infectious herpes virus apparently by removing antiviral immunoglobulin G. However, the molecular mechanism by which antibody prevents virus genome replication and antigen synthesis is unknown (Stevens 1978). In addition, the mechanism by which activation in vivo exacerbates latent infections is unclear, although reactivation in mice can occur with depressed cellular immunity despite persistence of viral specific antibody (Openshaw et al 1978).

SLOW VIRAL INFECTIONS

Acute, persistent, and latent infections of the CNS, which produce acute or chronic (static or progressive) disease, are differentiated from those infections defined initially by Sigurdsson as "slow infections" (Sigurdsson 1954). The original characteristics of slow infections have been modified by considerable experimental evidence in the past quarter century (Johnson & ter Meulen 1978). Simply stated slow infections are characterized by:

1. a long incubation period between exposure to the infectious agents and development of clinical disease (months–years),
2. a protracted ingravescent clinical course resulting usually in death (weeks–months),
3. a localization of pathology usually to one organ system.

Slow infections are caused by both conventional viruses and unconventional agents. The clinical and pathological changes in affected animals are directly related to the organ concentration of virus and infectious agent. The prolonged incubation period is probably related to the gradual accumulation of a critical number of virus-affected cells (Brooks et al, submitted for publication).

Conventional viruses such as paramyxoviruses (measles), papovaviruses (JC), and retroviruses (visna) may cause slow encephalitic and demyelinating diseases. Unconventional agents associated with scrapie, transmissible mink encephalopathy, kuru, and Creutzfeldt-Jakob disease cause slowly progressive noninflammatory spongiform encephalopathies in animals and man (Gajdusek 1977). However, a conventional virus (murine neurotropic retrovirus) has been recently shown to cause a similar encephalomyelopathy in mice (Gardner 1978).

How slow infections result in disease provides fascinating insight to biological interactions.

SLOW INFLAMMATORY AND DEMYELINATING DISEASES

During persistent CNS infections, chronic disease may result from nonproductive virus replication in an immunologically intact host, as occurs in subacute sclerosing panencephalitis or from complete replication of virus in an immunologically compromised host as occurs in progressive multifocal leukoencephalopathy. In addition, chronic disease may be the result of immune mediated mechanisms (Theiler's mouse encephalomyelitis virus) or result from other complex virus-host interactions (visna).

Inflammatory Diseases

SUBACUTE SCLEROSING PANENCEPHALITIS (SSPE)   SSPE is an uncommon slowly progressive inflammatory disease of humans initially manifested by behavioral changes and intellectual deterioration. Severe dementia with myoclonus develops and patients usually die within 1–3 years (Freeman 1969).

Pathological changes are limited to the CNS. Cowdry type A eosinophilic intranuclear inclusion bodies are characteristically seen within neurons, astrocytes, and oligodendroglia (Herndon & Rubinstein 1968, ZuRhein & Chou 1968). Both the gray and white matter are involved with diffuse infiltration and perivascular cuffing by mononuclear cells. Glial cell proliferation results in reactive gliosis.

The association of a paramyxovirus with SSPE was first suspected when inclusion bodies composed of tubules resembling paramyxovirus nucleocapsids were demonstrated by electron microscopy (Bouteille et al 1965). High levels of measles antibody are present in both serum and CSF. There is specific staining for measles antigen in brain tissue by immunofluorescence (Connolly et al 1967), but infectious virus could not be isolated from biopsy material by conventional virological methods. Infectious virus was finally isolated from brain biopsies after multiple passages of the patient's brain cells in culture followed by co-cultivation with other tissue culture cells (Horta-Barbosa et al 1969, Payne, Baublis & Itabashi 1969).

The mechanisms by which measles virus causes this slow infection remain unsolved. Several hypotheses have been advanced, including:

1. abnormalities of the host immune response to measles virus;
2. an atypical measles virus causing slow infection in a normal host;
3. interaction of a measles virus with a second agent (Johnson 1970).

An abnormality of the host immune response initially received the most consideration as a possible mechanism for the disease. High levels of measles antibody in both serum and CSF led to the speculation that SSPE might be an immunopathologic process (Kolar 1968). Conversely, Burnet (1968) postulated a specific immunological deficiency of thymus-differentiated cells to measles antigen, since delayed hypersensitivity responses are often depressed during acute measles virus infection (Starr & Berkovich 1964). Even though early studies gave some support to both under- and over-activity of the immune system, more recent studies have supported neither hypothesis (Kreth, Kackell & ter Meulen 1974, 1975). However, in certain cell-mediated assay systems a blocking factor in SSPE serum and CSF was observed, which might explain the failure of the immune system to eliminate the persistent CNS infection (Ahmed et al 1974, Swick et al 1976).

Disease can be induced in hamsters by inoculation of SSPE viruses. In these animals measles virus antigen was demonstrated by immunoperoxidase staining along the inner aspects of cytoplasmic membranes of brain cells, consistent with a blocking factor protecting infected cells from immune surveillance and destruction (Johnson & Swoveland 1977). Another study, however, revealed an inhibitor of antigen-antibody reactions to be nonspecific, occurring also in controls (Karcher, Noppe & Lowenthal 1977). Several studies have also been unable to demonstrate an inhibitor of cell-mediated immune reactions (Perrin, Tishon & Oldstone 1977, Kreth & ter Meulen 1977).

Host immune factors may well be important in the initial induction of persistent measles virus infection but do not appear important for expression of the disease. More than 50% of patients with SSPE have had measles before two years of age suggesting that the pathogenesis may be related either to the presence of maternal antibody or to an age dependent susceptibility of particular cells. For instance, the presence of maternal antibody in neonatal hamsters at the time of intracerebral measles virus inoculation can produce persistent infection and chronic disease when animals are subsequently immunosuppressed (Wear & Rapp 1971). Inoculation of weanling hamsters with SSPE virus strains has resulted in chronic encephalitis with persistent cell-associated virus, while no persistence was found in mature animals, and an acute uniformly fatal encephalitis occurred in newborns (Byington & Johnson 1972). The disappearance of cell-free virus in the weanling hamster correlated with the appearance of measles serum antibody. Thus, the different diseases produced appeared to be due to an age difference in the immune response. Chronic disease has also been produced with intracerebral inoculation of SSPE virus in ferrets (Thormar, Arnesen & Mehta 1977) and rhesus monkeys (Albrecht et al 1977) when these animals had been previously immunized with measles virus or had natural infection.

Therefore, selective pressure exerted by antibody may result in the evolution of a nonlytic, cell-associated mutant measles virus capable of cell-to-cell spread in the tightly packed neuropil of the brain. Such an agent would not be exposed to antibodies present in the extracellular space. Cell culture experiments have demonstrated that the loss of measles virus antigen from the surface of infected cells occurred in the presence of antibodies, yet the cells remained chronically infected (Joseph & Oldstone 1975). This "antigenic modulation" was reversed with full expression of antigenic sites on cell membranes once antibody had been removed from the culture fluid.

Even without the presence of antibody, the cell-associated SSPE viruses appear defective in their ability to replicate complete infectious measles virus virions because of a defect in their assembly. Electron microscopic studies of brain biopsies and cell cultures from patients with SSPE show only nucleocapsids, and enveloped virions are almost nonexistent (Dubois-Dalcq, Coblentz & Fleet, 1974, Dubois-Dalcq et al 1974). In cell cultures the nucleocapsids do not align with the cell membrane, which occurs as prerequisite to budding (Dubois-Dalcq, Coblentz & Fleet 1974, Dubois-Dalcq et al 1976). Laboratory analysis of SSPE viruses have shown minor distinctions from measles viruses with differences in neutralization kinetics

(Payne & Baublis 1973) and RNA genome homology (Hall & ter Meulen 1976). A viral membrane (M) protein of paramyxoviruses appears to stabilize the region of the cell membrane containing virus-specific proteins after it attaches to the inner surface of the plasma membrane (Choppin et al 1971, Nagai et al 1975). The viral nucleocapsid then attaches to the M protein probably by a hydrophobic interaction (Mountcastle et al 1974, McSharry et al 1975). Thus, the interaction between the nucleocapsids and the viral-specific proteins in the cell membrane is necessary for virion assembly, and this interaction is mediated by the M protein. Recent studies have shown that the M proteins of SSPE viruses have higher molecular weights than those of standard wild-type measles viruses (Schluederberg et al 1974, Wechsler & Fields 1978, Hall, Kiessling & ter Meulen 1978), and the mRNA for the SSPE M protein is also larger than that of measles. This probably represents a mutation of the M protein gene (Hall, Kiessling & ter Meulen 1978). These findings lend support to the hypothesis that a mutation involving production of an abnormal M polypeptide may explain the defective infection in the SSPE strains of measles virus.

The possibility that another infection may induce this mutation or precipitate disease expression has been suggested by the preponderence of the disease in rural males, a finding suggesting a zoonotic infection (Brody & Detels 1970). Although one laboratory has observed papovavirus-like particles in brain cell cultures of patients (Koprowski, Barbanti-Brodano & Katz 1970), these findings have not been substantiated by other laboratories, and an infectious papovavrus has not been recovered. Thus, SSPE appears to be caused by a mutant of measles virus that may evolve during the natural course of measles, possibly enhanced by selective antibody pressure. The slow progression of disease can be explained by the cell-associated nature of the infection, but the latency of many years between the initial measles infection and the expression of disease remains unexplained.

OTHER CHRONIC INFLAMMATORY DISEASES   Canine distemper virus (CDV), a paramyxovirus closely related to measles, occasionally produces an SSPE-like disease in dogs. This subacute canine distemper panencephalitis or "old dog encephalitis" occurs years after exposure to CDV or after the acute systemic disease caused by CDV with apparently prolonged persistence of the virus (Appel & Gillespie 1972).

Another rare chronic disease in humans is caused by rubella virus. This has been seen largely in adolescents bearing stigmata of congenital rubella who develop a progressive neurological disease resembling SSPE. Rubella virus has been isolated (Cremer et al 1975) but the pathogenesis of this disease, progressive rubella panencephalitis (PRP), has not yet been established (Townsend et al 1975, Weil et al 1975).

Viral-Induced Demyelinating Diseases

A demyelinating disease is characterized by destruction of myelin sheaths with relative sparing of axons, neurons, and supporting tissues (Raine & Schaumberg 1977). The most common human demyelinating disease is multiple sclerosis. Although a viral etiology for multiple sclerosis has often been postulated, the cause of the disease remains unknown (Johnson 1975b, 1978a). However, several other human and animal demyelinating diseases are caused by virus infections (Weiner, Johnson & Herndon 1973, Lampert 1978), and studies of these diseases have increased our knowledge of the varied mechanisms by which viruses can produce demyelination (Johnson & Herndon 1974). The general mechanisms of viral-induced demyelination are:

1. direct lytic infections of oligodendroglia (Lampert, Sims & Kniazeff 1973);
2. immunopathologic injury to oligodendroglia and/or myelin (Wisniewski 1977).

These immunopathologic mechanisms include (*a*) several types of cell-mediated immune responses directed toward myelin or oligodendroglia (Brostoff 1977), (*b*) induction of antimyelin or antioligodendroglia antibodies (Abramsky et al 1977), (*c*) alteration of myelin membranes by myelinolytic factors released by immune cells in the vicinity of myelin ("bystander" effect) (Wisniewski & Bloom 1975), and (*d*) antigen-antibody complex deposition (Gregson, Kennedy & Leibowitz 1974).

Direct Oligodendroglia Lytic Infection

CORONAVIRUS-INDUCED DEMYELINATION   Lytic infections of oligodendroglia can result in acute or chronic demylinating disease. The JHM strain of mouse-hepatitis virus, an enveloped RNA coronavirus, produces an acute diffuse encephalomyelitis in mice and rats (Cheever et al 1949). The degree of demyelination is dependent on the dose of virus, the route of inoculation, and the age of the animals. Fluorescent antibody staining has shown viral antigen in cells of the white matter (Weiner 1973), and electron microscopy has demonstrated virus in the cytoplasm of oligodendroglia (Lampert, Sims & Kniazeff 1973). Although the major demyelination occurs acutely, animals that recover may develop late demyelinating lesions. JHM virus-like particles in astrocytes and macrophages may be seen by electron microscopy in asymptomatic mice 16 months after the initial intracerebral inoculation (Herndon et al 1975) and a chronic progressive paralytic disease has been seen in some strains of mice 1–3 months after inoculation with the MHV$_3$ strain of mouse hepatitis virus (Le Prévost, Virelizier & Dupuy 1975). The mechanism of this late disease has not been elucidated, but the animals with the chronic paralytic disease did have an associated impairment of delayed hypersensitivity (Dupuy et al 1973).

PROGRESSIVE MULTIFOCAL LEUKOENCEPHALOPATHY (PML)
PML is a chronic demyelinating disease of man caused by nonenveloped DNA papovavirus infection of oligodendroglia. Unlike the acute mouse hepatitis infection, where the immune response is normal and appears to play no role in disease, PML usually occurs as an opportunistic infection in an immunodeficient host (Johnson et al 1977). It is usually seen as a complication of disorders of the reticuloendothelial system such as leukemia, Hodgkin's disease, or sarcoidosis (Richardson 1961), but has also occurred in patients therapeutically immunosuppressed for other diseases (ZuRhein 1969, Sponzilli et al 1975), and for renal transplantation (Manz, Dinsdale & Morrin 1971, ZuRhein & Varakis 1974). PML may occur rarely in patients without underlying disease; however, immunological studies have generally shown a defect in cell-mediated immunity (Weiner & Narayan 1974). The disease follows a progressive course with multifocal neurological deficits, normal CSF, and death within 1 year. Rarely, prolonged survival for years has occurred (Kepes, Chou & Price 1975).

Pathological lesions in PML are limited to the CNS with multifocal, often confluent, areas of demyelination, most prominent in the subcortical white matter. Microscopically, within the demyelinated areas, there is loss of oligodendroglia and myelin with relative sparing of axons. Astrocytes within these areas appear bizarre; they resemble malignant cells being enlarged and having abnormal mitotic figures, nuclear forms, and chromatin patterns. Surrounding these areas, the oligodendroglia are enlarged and usually contain intranuclear inclusions. Inflammation is usually absent.

Despite the paucity of inflammation a viral etiology was postulated because of the presence of intranuclear inclusions and the association with diseases causing immune deficits (Richardson 1961). Subsequent electron microscopic studies showed large numbers of papova-like particles in oligodendroglia and rarely in astrocytes (ZuRhein & Chou 1965). Initial attempts to recover virus using standard laboratory techniques were unsuccessful. A new papovavirus, JC virus, was recovered from the brain of a patient with PML by inoculating brain homogenates into explant cultures of human fetal brain (Padgett et al 1971, 1976). This virus subsequently has been identified in most cases of PML (Narayan et al 1973, Padgett & Walker 1976). Viruses antigenically similar to simian virus 40 (SV40) have also been isolated from two patients with PML (Weiner et al 1972), and spontaneous PML in rhesus monkeys due to SV40 has now been reported (Holmberg et al 1977).

It is not clear whether these viruses remain persistent or latent within the host in the CNS or in extraneural tissues. It is possible that PML represents the first encounter by an immunodeficient host with a common agent. Papovavirus antibodies have been found to be widespread in normal persons (Padgett & Walker 1973, Shah 1972).

There is little doubt, however, that demyelination in this disease is the direct result of a permissive lytic infection of oligodendroglia by an opportunistic papovavirus in an immunodeficient host.

Immune-Mediated Demyelination

Immune-mediated demyelination may occur on both an acute (experimental allergic encephalomyelitis), and chronic (Theiler's murine encephalomyelitis virus), basis in animals. Although evidence remains scanty, an immune-mediated mechanism of demyelination on an acute basis seems likely in post-infectious encephalomyelitis of man, which is similar to the experimental allergic encephalomyelitis animal model (Alvord 1970, 1977). No chronic human demyelinating diseases have been definitely shown to be due to immunopathologic mechanisms.

PICORNAVIRUS-INDUCED DEMYELINATION   Theiler's murine encephalomyelitis virus is a mouse picornavirus that produces two distinct phases of disease, one acute and the other chronic (Lipton 1975). The acute or early disease is a poliomyelitis-like disease with flaccid paralysis occurring in approximately 80% of weanling Swiss mice inoculated intracerebrally. Pathological changes are confined to the gray matter of the spinal cord, brain stem, and thalamus with neuronal degeneration, microglial proliferation, and perivascular mononuclear cell cuffing (Olitsky & Schlesinger 1941). During the acute infection fluorescent antibody staining showed viral antigen in anterior horn cells. Virus growth peaked on day 13 and then declined, but virus persisted for at least 5 months despite rising serum-neutralizing antibody. Thus, the acute disease was caused by a lytic infection of anterior horn cells.

In survivors, a waddling gait often developed 45–60 days after inoculation. At this time an inflammatory response was limited to the leptomeninges and white matter, and demyelination was found with sparing of axons. Viral antigens were found in a few isolated cells of the white matter of the spinal cord (Lipton 1975). Immunosuppression of mice who survived acute disease prevented the late clinical disease and demyelination (Lipton & Dal Canto 1976 a, b). Recently, electron microscopic studies of the acute disease caused by a strain of Theiler's virus, revealed viral crystalline arrays in neurons and, less frequently, oligodendroglia (Penney & Wolinsky 1978). Possibly virus persistence occurs only in oligodendroglia, and virus-induced modifications in their plasma membranes leads to immune-mediated demyelination.

Complex Mechanisms of Virus-Induced Demyelination

VISNA VIRUS LEUKOENCEPHALITIS   Visna is a slow leukoencephalitis of sheep that exemplifies the possible complexities of virus-host interrelationships (Narayan, Griffin & Clements 1978). Visna was a natural disease of Icelandic sheep, which complicated a chronic interstitial viral pneumonia called maedi (Gudnadottir 1974). In the laboratory, visna can be transmitted from sheep to sheep by intracerebral inoculation resulting in a slowly progressive paralytic disease after an incubation period of months to years (Sigurdsson, Palsson & Grimsson 1957, Sigurdsson & Palsson 1958). During the incubation period virus can be recovered consistently from explant cultures of brain and lung and occasionally from lymph nodes, spleen, salivary glands, and peripheral blood leukocytes. Serum and CSF neutralizing antibody and CSF pleocytosis develop over several months, long before the onset of clinical disease (Gudnadottir 1974, Petursson et al 1976).

The pathological changes in the CNS consist of an early, perivascular inflammation followed by patchy demyelination and leukomalcia. Immunosuppression eliminates the inflammatory lesions but does not affect virus growth (Nathanson et al 1976).

The pathogenesis of the viral persistence and the unusual mechanism of the progressive disease are intriguing (Haase 1975). Visna virus is an enveloped RNA virus that resembles the RNA tumor viruses (retroviruses, oncornaviruses) because of the presence of an RNA-dependent DNA polymerase (reverse transcriptase) (Schlom et al 1971, Lin & Thromar 1970). The polymerase transcribes DNA that is complimentary to the viral RNA, i.e. proviral DNA (Haase & Varmus 1973). However, visna virus appears to be acquired by each individual animal (i.e. exogenous), rather than passed from generation to generation with the animal's genetic information (i.e. endogenous) as are many murine retroviruses. In sheep this proviral DNA has been demonstrated by in situ hybridization in foci of cells and probably explains the mechanism of persistent infection (Haase et al 1977). The proviral DNA is probably integrated into the host's chromosomal DNA, and there appears to be a correlation between the amount of infectious virus produced and the quantity of proviral DNA integrated (Clements, Narayan & Griffin 1978). Studies have shown that the viruses recovered from a single sheep prior to and several months after the development of neutralizing antibody were antigenically identical to the parental strain used for inoculation. Subsequent viruses recovered were not neutralized by the animal's serum (Narayan, Griffin & Silverstein 1977). Apparently mutation of the virus had occurred during the prolonged infection as antibody to the parental virus and eventually to subsequent mutants had provided selective pressure for the replication of sequential virulent mutants

(Narayan, Griffin & Clements 1978). This mechanism has also been shown to occur in tissue culture (Narayan, Griffin & Chase 1977). Antigenic drift within the individual host is a new concept in viral disease. The sequential evolution of virulent mutants within the same individual might result in exacerbations of acute pathological inflammation producing cumulative lesions and a clinically progressive disease (Johnson, Narayan & Clements 1978). Thus, in this persistent infection the virus is not defective, and the slow disease may be dependent on a competent host immune system that exerts selective pressure leading to the evolution of virulent mutant viruses with pathological exacerbations resulting in progressive disease.

SLOW SPONGIFORM ENCEPHALOPATHIES IN ANIMALS AND MAN

Unconventional Agents

Slow infections without inflammatory or immunopathologic reactions involving the CNS in animals and man have been shown to be due to transmissible filterable agents that do not behave as classic viruses (Gajdusek 1977). Two of the best studied examples are scrapie in sheep and mink encephalopathy. Pathologic similarities between scrapie and kuru (Hadlow 1959) spurred the initial experiments that led to the elucidation of the etiology of kuru and Creutzfeldt-Jakob disease in man (Gajdusek & Gibbs 1977). The study of scrapie in the natural host and experimental animal provides insight to the pathogenetic mechanisms that lead to disease (Hunter 1974).

SCRAPIE   Scrapie is a chronic sheep disease characterized by progressive ataxia, tremor, pruritus, weakness, and wasting. Scrapie was regarded as a genetic disease for over 100 years before transmission from sheep to sheep with brain homogenate filtrates was demonstrated (Cuillé & Chelle 1936). More recently scrapie has been transmitted to mice and a variety of other animals (Chandler 1961). Passage with serial limiting dilutions has established that the scrapie agent is replicated, but the agent can only be detected by clinical disease or pathological examination, since no immunologic, electron microscopic, or tissue culture systems detect the agent (Dickinson & Fraser 1977).

The pathology of scrapie is limited to CNS tissue. Vacuolization of neuronal and astrocytic processes results in spongiform degeneration of subcortical structures including basal ganglia, hypothalamus, brainstem, cerebellum, and to a lesser degree, spinal cord (Chandler 1963, Hadlow & Eklund 1968, Fraser 1976). Astrocytic hypertrophy with reactive gliosis is seen in areas of neuronal loss. No inflammatory reaction, similar to that in acute or chronic encephalitic disease, is seen (Asher, Gibbs & Gajdusek 1976).

Scrapie has been transmitted to goat, mouse, rat, hamster, gerbil, mink, vole, and to new and old world monkeys, producing similar pathology after varying but prolonged incubation periods (Gajdusek 1977, Gibbs & Gajdusek 1978). It is interesting that the pathology in mice after the first passage from sheep shows an identical subcortical topographic distribution but that further passage in mice produces a more generalized CNS disease (Zlotnik & Rennie 1965). For selected hosts, transmission has also been shown by extraneural inoculation (Dickinson & Fraser 1972), contact (Zlotnik 1968), oral feeding (Gibbs & Gajdusek 1978), and transplacental transmission (Pattison et al 1972).

Pathogenesis studies employing different strains of scrapie agent and different animal hosts (Eklund, Kennedy & Hadlow 1967, Hadlow et al 1974, Dickinson & Fraser 1977) indicated that extraneural inoculation (intraperitoneal or subcutaneous) is followed by early replication in the lymphoreticular system—particularly in spleen (Fraser & Dickinson 1970) —and development of infectivity in many extraneural tissues despite lack of any cytopathology. However, no "viremia" has been demonstrated, so the mechanism of CNS infection is uncertain (Clarke & Haig 1967, Eklund, Kennedy & Hadlow 1967). Sequential clinicopathologic studies in mice have shown that generalized astrocytic hypertrophy is the first demonstrable CNS lesion 5–8 months after intracerebral inoculation and occurs when mice show decreased emergent behavior (Heitzman & Corp 1968). Neuronal vacuolization and loss occur 7–8 months after infection, at a time when more definite clinical signs are present (Suckling et al 1976).

Susceptibility to extraneural, but not intracerebral, inoculation in a given mouse strain varies inversely with age (Outram, Dickinson & Fraser 1973, Hotchin & Buckley 1977). Congenital asplenia (Fraser & Dickinson 1970), splenectomy (Fraser & Dickinson 1970), treatment with prednisone (Outram, Dickinson & Fraser 1974) or immunosuppressant arachis oil (Outram, Dickinson & Fraser 1975) prolongs the incubation period after extraneural, but not intracerebral, inoculation. Scrapie replicates normally in thymectomized (McFarlin et al 1971) and congenitally athymic nude mice (Hotchin & Buckley 1977). In addition, strain differences among scrapie agents and genetic differences among mice play a role in disease expression (Dickinson & Meickle 1971). Infection with scrapie elicits no detectable antibody response (Marsh, Pan & Hanson 1970, Porter, Porter & Cox 1973), cell-mediated immune response (Pattison, Millson & Smith 1964, Asher, Gibbs & Gajdusek 1976), or interferon response (Katz & Koprowski 1968, Worthington 1972). Nevertheless, these responses to other antigens and viral infections are intact in scrapie-infected animals (Gardiner & Marucci 1969, Asher, Gibbs & Gajdusek 1976).

Properties of the transmissible agents that cause human spongiform encephalopathies are probably similar to those of the scrapie agent (Gajdusek & Gibbs 1977, Gibbs & Gajdusek 1978). Scrapie can replicate in serial subculture of brain and spleen tissues from infected animals, but no cytopathology is evident (Clarke & Haig 1970). Infectivity in brain homogenates is resistant to formaldehyde, beta propiolactone, ethylene diamine tetraccetric acid, trypsin, pepsin, ribonuclease A and III, deoxyribonuclease I, heat to 80°C (but not 100°C), and ultrasound (Gibbs & Gajdusek 1978). Resistance to ultraviolet radiation (Latarjet et al 1970) and gamma radiation (Gibbs & Gajdusek 1978) of the scrapie agent indicates a small target size of 150,000 daltons. Purified scrapie is less resistant than scrapie in the crude brain homogenate, which suggests that the presumed parallel in properties between "viroids" and scrapie is untenable (Gajdusek 1977). More direct evidence includes the lack of infectivity of DNA and RNA extracted from infected brain homogenates (Marsh et al 1974, Ward, Porter & Stevens 1974, Hunter et al 1976).

The search for virus particles has been spurred by the observation of 35 nm particles in axons of scrapie-infected sheep brain (Bignami & Parry 1971), 23 nm particles in dendrites of scrapie-infected mouse brain (Baringer & Prusiner 1978), and 14 nm particles in CsC1 gradients of scrapie-infected mouse brain homogenates (Cho & Greig 1975). These 14 nm particles, however, are immunologically related to ferritins and found in normal mouse brain (Cho et al 1977). Analytical differential centrifugation studies indicate a sedimentation coefficient of 400 S for scrapie (Prusiner et al 1977). Although free of plasma membranes, fractions containing the scrapie agent are contaminated by ribosomes (Siakotos et al 1976, Malone et al 1978). Heating the brain homogenates to 80°C, however, disrupts ribosomes and alters the sedimentation coefficient to 500 S, which suggests aggregation (Prusiner et al 1978). It is of interest that previous studies of heated scrapie-infected brain homogenates showed that such treatment altered the topographical distribution of scrapie lesions (Zlotnik & Rennie 1967, Dickinson & Fraser 1969).

TRANSMISSIBLE MINK ENCEPHALOPATHY (TME)   Epidemics of a disease in mink resembling scrapie have occurred after incorporation of animal carcasses into the mink feed (Hartsough & Burger 1965), and it is widely believed that mink encephalopathy is simply scrapie in mink. The pathology in mink is similar to that of scrapie in sheep, except for more involvement of the cortex (Burger & Hartsough 1965). The disease has been transmitted by intracerebral inoculation of infected mink brain homogenates to mink, ferret, goat, hamster, opposum, raccoon, sheep, skunk, new and old world monkeys, but not mice (Gajdusek 1977).

Chediak Higashi mink are deficient in lysozomal enzymes. When they develop TME, spongiform changes are dramatically diminished (Marsh et al 1976). Thus, specific host enzyme systems may be involved in causing the vacuolation characteristic of the spongiform encephalopathies.

KURU   Kuru is the first chronic human neurological disease transmitted to an experimental animal (Gajdusek & Zigas 1957). The disease was endemic to the Fore people of New Guinea, where primarily children and adult females were affected (Hornabrook 1968). Patients presented with ingravescent ataxia followed later by weakness and mental deterioration, leading to death within 1 year. Transmission among humans probably resulted from self-inoculation via cutaneous and conjunctival routes during the cannibalistic mourning ritual (Gajdusek 1973). Pathological changes are similar to those seen in scrapie but the topographical distribution of lesions is different (Klatzo, Gajdusek & Zigas 1959, Kakulas, LeCours & Gajdusek 1967). In addition, 20–60 micron periodic acid Schiff reaction positive, argentophilic plaques are frequently found in the cerebellum (Neuman, Gajdusek & Zigas 1964).

Kuru has been transmitted by intracerebral inoculation of infected human brain homogenate to mink, ferret, a variety of apes, and new and old world monkeys (Gibbs & Gajdusek 1978). The infectious agent has been demonstrated in brain, spleen, lymph node, and liver, but not bodily fluids, or the placenta (Gajdusek et al 1977). The disease produced is strikingly similar to that in man but the topographical distribution is different and the argentophilic plaques are greatly reduced (Beck et al 1966, Lampert et al 1969, Beck et al 1973). No immunological response to the kuru agent has been demonstrated in affected patients or disease-free relatives (Benfante et al 1974).

TRANSMISSIBLE VIRUS DEMENTIA (TVD): CREUTZFELDT-JAKOB DISEASE (CJD)   CJD is a rare, sporadic, subacute neurological disease characterized by progressive dementia, myoclonus, ataxia, and to a lesser degree, rigidity, spasticity, visual loss or lower motor neuron weakness (Roos, Gajdusek & Gibbs 1973). Pathologic changes are similar to those seen in scrapie and kuru though argentophilic plaques are more common in the cerebrum and basal ganglia of patients with Creutzfeldt-Jakob disease as opposed to the cerebellum of patients with Kuru (Hirano et al 1972, Horoupian, Powers & Schaumberg 1972).

CJD has developed after intracerebral inoculation of infected human brain homogenates in cats, ferrets, guinea pigs, hamsters, apes, and a variety of new and old world monkeys (Gajdusek 1977, Manuelides 1975, Manue-lides et al 1977). Disease can also be produced by peripheral inoculation of affected tissues in selected susceptible animals (Gibbs & Gajdusek 1978). The infectious agent is found in brain, spinal cord, cerebrospinal fluid, cornea, spleen, lymph node, and lung of affected patients (Gajdusek et al 1977). Viremia in experimental CJD in guinea pigs has been demonstrated by inoculation of the leukocyte fraction into susceptible animals (Manueli-dis, Gorgacz & Manuelides 1978). No antibody or cell-mediated immune response to the agent has been demonstrated in affected patients or animals (Brown et al 1972, Asher, Gibbs & Gajdusek 1976).

The infectious agent can replicate in vitro in primary and established cell lines from infected human or experimentally infected primate brain tissue but not heterologous non-brain cell cultures (Gibbs & Gajdusek 1978). Infection can be transmitted with cell culture material only if cells are included and not with supernatant fluids alone (Gibbs & Gajdusek 1978).

The first pathologic lesion in monkeys inoculated with infectious material is neuronal vacuolation, as opposed to the proposed astrocytic hypertrophy in scrapie (Masters et al 1976). Astrocytic hypertrophy appears subsequent to neuronal loss.

The mode of transmission of sporadic CJD in man is unknown. Cases have occurred after corneal transplant, intracerebral placement of elec-trodes, and, possibly, intracranial surgery (Traub, Gajdusek & Gibbs 1977, Gajdusek et al 1977). This type of transmission would be equivalent to experimental intracerebral inoculation. No common contact or occupa-tional exposure has otherwise been shown in sporadic CJD (Bobowick et al 1973, Johnson 1977). Nevertheless a higher rate of CJD has been reported among Libyan Jews who have a practice of eating ovine brain and eyes (Herzberg et al 1974).

Familial CJD occurs in 9% of reported cases (Traub, Gajdusek & Gibbs 1977). The genetic analysis is consistent with an autosomal dominant mode of transmission. Thus, a definitely transmissible degenerative neurological disease can appear in a subgroup of patients to be a genetically determined disease. The implication of this observation for other human chronic degen-erative neurological diseases remains to be explored.

Conventional Agents

MURINE RETROVIRUS POLIOENCEPHALOMYELOPATHY   Neuro-tropic murine retroviruses (C type RNA tumor virus) produce late onset progressive paralysis in wild mice that are more than one year old (Gardner 1978). Pathological study of the experimental disease shows spongiform change in the gray matter of the brainstem and spinal cord (Oldstone et al 1977). No inflammatory response is seen. Neonatal animals are more sus-

ceptible to infection than adults. Acute experimental disease after intracerebral inoculation of concentrated virus will produce clinical disease in 3
weeks as opposed to 3 months with unconcentrated virus (Brooks et al,
unplublished data). Complete virus replication is demonstrated in endothelial cells and pericytes without degeneration. Concomitant astrocytic
swelling occurs without complete virus replication. Pathological examination of animals with acute disease shows vacuolation in axons and dendrites
without accumulation of virus in the neurons. In chronic disease aberrent
viral forms are present in the neurons (Andrews & Gardner 1974). Reactive
astrocytosis is common in both forms of disease. Pathogenesis studies comparing acute with chronic disease suggest that primary neuronal infection
is not the cause of the neuronal vacuolation. It is possible, therefore, that
neuronal damage may result from disrupted neuron-glia-endothelial cell
relationships that affect neuronal nutrition. Alternatively, a direct toxic
effect of viral products on the neuron cannot be ruled out (Gardner, submitted for publication).

Thus, a spongiform encephalomyelopathy with clinical and pathological
characteristics similar to those diseases caused by unconventional agents is
now available for study. Since this disease is caused by a conventional virus,
which can be grown in cell culture and assayed in vitro, the mechanism of
spongiform change and its relationship to virus infection of CNS cells can
be approached directly.

CONCLUSIONS

Slow viral infections of the central nervous system highlight fascinating and
complex aspects of virus-host relationships (Kimberlin 1976, ter Meulen &
Katz 1977). We have discussed the diversity of the relationships and presented examples of slow infections where some insights into pathogenesis
have been obtained. It is evident that in almost every case the mechanism
of virus persistence and the development of pathology is different. Thus, in
PML or visna, the viruses are standard mature infectious particles. In other
diseases, such as SSPE, the virus is defective, or in the case of scrapie, kuru,
and CJD the agents may not be true viruses at all. The role of host responses
are equally diverse. In PML, the slow infection represents an opportunistic
infection in an immunodeficient host; in the spongiform encephalopathies
there is no host response; in SSPE a normal host response fails to clear virus;
and in visna the host immune response may play a role in causing chronic
disease by fostering the development of mutant viruses, which may lead to
relapses and progression.

Consequent to these varied interactions many different neurological diseases can result from viral infection. Many viruses, including rubella and
parvoviruses, can lead to congenital malformations and developmental dis-

orders as the result of static sequelae of acute infection. Other viruses, including mumps, influenza virus, and reovirus, however, can cause obstructive hydrocephalus due to progressive sequelae resulting in aqueductal stenosis. Chronic inflammatory diseases develop from infection by measles virus in SSPE and visna virus. Yet, noninflammatory demyelinating disease can result from papovavirus infection in PML and noninflammatory degenerative diseases of the CNS can be caused by the spongiform encephalopathy agents. The diversity of diseases that can result from virus infection lends credence to the speculation that other chronic neurological diseases may be related to viral infection. However, reports of the identification of viral agents in multiple sclerosis, amyotrophic lateral sclerosis, Parkinson's disease, or Alzheimer's disease remain, as yet, unconfirmed (Johnson & Herndon 1974).

One central question remains to be addressed. Why in almost every case do slow infections selectively involve the CNS? The CNS does indeed appear to be a favored target for slow infections, and several unique aspects of CNS structure and function may explain this. 1. The CNS is relatively isolated from the systemic immune system (Rappaport 1975). The blood brain barrier limits the entry of antibody molecules and under normal circumstances prevents the entry of wandering phagocytic cells. Although this provides a barrier to viral invasion, once virus has entered the CNS it may remain relatively isolated from immune clearance. Possibly these factors may explain why the SSPE strains of measles virus are capable of causing chronic disease in the CNS while not causing disease in any other tissue. 2. The CNS represents a highly differentiated cell population functionally connected by distant cell-to-cell contacts (Weiner & Johnson 1977). Such differentiation can mean highly different selective susceptibility. One would assume that all hepatic cells of the liver would contain similar receptor sites, whereas different neurons may show selective vulnerability to specific viruses. It is of interest that both in vivo as well as in vitro the only cell type that has thus far been found to be lytically infected with the JC papovavirus, the cause of PML, is the oligodendrocyte (Weiner et al 1973). 3. Cells of the central nervous system have a high rate of metabolic activity, but turnover of cell populations is minimal and regenerative capacity is poor (Herndon, Price & Weiner 1977). It has already been pointed out that the possible persistence of rubella virus in neural tissue may well be due to the longer life span of neural cells, so that mitotically inhibited rubella-infected cells are not eliminated during the normal turnover of cell populations. In the case of the spongiform encephalopathies, the infectious agent is present in cells in many organs of the body, but clinical and pathological alterations are limited entirely to neural cells. If these agents slowly cause minor dysfunction leading to premature senescence or gradual death of cells, one would expect the disease to be manifested primarily or exclusively in the CNS where target cells can not be replaced. Viral infection can selectively interfere with very specific cell functions as in cell cultures infected with lymphocytic choriomeningitis virus. Although the inhibition of neural transmitter synthesis was referred to as "luxury functions" in the studies in vitro, inhibition of similar function in vivo might have devastating clinical effects even in the absence of any discernible pathological changes in neural tissue.

Thus, in considering the slow viral infections it is also important to consider the biology of neural cells. Possibly the continued investigation of the effects of these infections will provide information not only on the pathogenesis of disease but also on some of the basic biological properties of the CNS.

ACKNOWLEDGEMENTS

Support during the preparation of this review was provided by grants from the Amyotrophic Lateral Sclerosis Society of America (77–4) and the National Institutes for Neurological and Communicative Disorders and Stroke (5-T32-NS 07000).

Literature Cited

Aaronson, S. A., Stephenson, J. R. 1976. Endogenous type-C RNA viruses of mammalian cells. *Biochim. Biophys. Acta* 458:323–54

Abramsky, O., Lisak, R. P., Silberberg, D. H., Pleasure, D. E. 1977. Antibodies to oligodendroglia in patients with multiple sclerosis. *N. Engl. J. Med.* 297:1207–11

Ahmed, A., Strong, D. M., Sell, K. W., Thurman, G. B., Knudsen, R. C., Wistar, R. Jr., Grace, W. R. 1974. Demonstration of a blocking factor in the plasma and spinal fluid of patients with subacute sclerosing panencephalitis. I. Partial characterization. *J. Exp. Med.* 139:902–24

Albrecht, P. 1968. Pathogenesis of neurotropic arbovirus infections. *Curr. Top. Microbiol. Immunol.* 43:44–91

Albrecht, P., Burnstein, T., Klutch, M. J., Hicks, J. T., Ennis, F. A. 1977. Subacute sclerosing panencephalitis: experimental infection in primates. *Science* 195:64–66

Allison, A. C., Sandelin, K. 1963. Activation of lysosomal enzymes in virus-infected cells and its possible relationship to cytopathic effects. *J. Exp. Med.* 116:879–87

Alvord, E. C. Jr. 1970. Acute disseminated encephalomyelitis and "allergic" neuroencephalopathies. In *Multiple Sclerosis and Other Demyelinating Diseases. Handbook of Clinical Neurology,* ed. P. J. Vinken, G. W. Bruyn, 9:500–71. Amsterdam: North-Holland. 706 pp.

Alvord, E. C. Jr. 1977. Demyelination in Experimental Allergic Encephalomyelitis and Multiple Sclerosis. In *Slow Virus Infections of the Central Nervous System.* ed. V. ter Meulen, M. Katz, pp. 166–85. New York: Springer. 258 pp.

Andrews, J. M., Gardner, M. B. 1974. Lower motor neuron degeneration associated with type C RNA virus infection in mice: neuropathological features. *J. Neuropathol. Exp. Neurol.* 33:285–307

Appel, M. J. G., Gillespie, J. H. 1972. Canine Distemper Virus. In *Virology Monographs,* ed. S. Gard, C. Hallaver, K. F. Meyer, 11:1–96. New York: Springer. 153 pp.

Asher, D. M., Gibbs, C. J. Jr., Gajdusek, D. C. 1976. Pathogenesis of subacute spongiform encephalopathies. *Ann. Clin. Lab. Sci.* 6:84–103

Bablanian, R. 1975. Structural and functional alterations in cultured cells infected with cytocidal viruses. *Prog. Med. Virol.* 19:40–83

Bak, I. J., Markham, C. H., Cook, M. L., Stevens, J. G. 1977. Intraaxonal transport of herpes simplex virus in the rat central nervous system. *Brain Res.* 136:415–29

Baringer, J. R., Prusiner, S. B. 1978. Experimental scrapie in mice: ultrastructural observations. *Ann Neurol.* 4:205–11

Beck, E., Daniel, P. M., Alpers, M., Gajdusek, D. C., Gibbs, C. J. Jr. 1966. Experimental "Kuru" in chimpanzees. A pathological report. *Lancet* 2:1056–59

Beck, E., Daniel, P. M., Gajdusek, D. C., Gibbs, C. J. Jr. 1973. Experimental kuru in the chimpanzee: a neuropathological study. *Brain.* 96:441–62

Benfante, R. J., Traub, R. D., Lim, K. A., Hooks, J., Gibbs, C. J. Jr., Gajdusek, D.C. 1974. Immunological reactions in kuru: Attempts to demonstrate serological relationships between kuru and other known infectious agents. *Am. J. Trop. Med. Hyg.* 23:476–88

Bignami, A., Parry, H. B. 1971. Aggregations of 35-nanometer particles associated with neuronal cytopathic changes in natural scrapie. *Science* 171:389–90

Blinzinger, K., Muller, W. 1971. The intercellular gaps of the neuropil as possible pathways for virus spread in viral encephalomyelitis. *Acta Neuropathol.* 17:37–43

Bobowick, A. R., Brody, J. A., Matthews, M. R., Roos, R., Gajdusek, D. C. 1973. Creutzfeldt-Jakob disease: A case-control study. *Am. J. Epidemiol.* 98:381–94

Bodian, D., Horstmann, D. M. 1965. Polioviruses. In *Viral and Rickettsial Infections of Man,* ed. C. F. L. Horsfall, Jr., I. Tamm, pp. 430–73. Philadelphia: Lippincott. 1282 pp. 4th ed.

Boue, A., Boue, J. G. 1969. Effects of rubella virus infection on the division of human cells. *Am. J. Dis. Child.* 118:45–48

Bouteille, M., Fontaine, C., Vedrenne, C., Delarue, J. 1965. Sur un cas d'encephalite subaique a inclusions. Etude anatomo-clinique et ultrastructurale. *Rev. Neurol.* 113:454–58

Brody, J. A., Detels, R. 1970. Subacute sclerosing panencephalitis: A zoonosis following aberrant measles. *Lancet* 2:500–1

Brostoff, S. W. 1977. Immunological responses to myelin and myelin components. In *Myelin,* ed. P. Morell, pp. 415–46. New York: Plenum. 531 pp.

Brown, P., Hooks, J., Roos, R., Gajdusek, D. C., Gibbs, C. J. Jr. 1972. Attempt to identify the agent for Creutzfeldt-Jakob

disease by CF antibody relationship to known viruses. *Nature* 235:149–52

Burger, D., Hartsough, G. R. 1965. Encephalopathy of mink II. Experimental and natural transmission. *J. Infect. Dis.* 115:393–99

Burnet, F. M. 1968. Measles as an index of immunological function. *Lancet* 2: 610–13

Byington, D. P., Johnson, K. P. 1972. Experimental subacute sclerosing panencephalitis in the hamster: correlation of age with chronic inclusion-cell encephalitis. *J. Infect. Dis.* 126:18–26

Chandler, R. L. 1961. Encephalopathy in mice produced by inoculation with scrapie brain material. *Lancet* 1:1378–79

Chandler, R. L. 1963. Experimental scrapie in mouse. *Res. Vet. Sci.* 4:276–85

Cheever, F. S., Daniels, J. B., Pappenheimer, A. M., Bailey, O. T. 1949. A murine virus (JHM) causing disseminated encephalomyelitis with extensive destruction of myelin. I. Isolation and biological properties of the virus. *J. Exp. Med.* 90:181–94

Cho, H. J., Greig, A. S. 1975. Isolation of 14-nm virus-like particles from mouse brain infected with scrapie agent. *Nature* 257:685–86

Cho, H. J., Greig, A. S., Corp, C. R., Kimberlin, R. H., Chandler, R. L., Millson, G. C. 1977. Virus-like particles from both control and scrapie-affected mouse brain. *Nature* 267:459–60

Choppin, P. W., Klenk, H. D., Compans, R. W., Caliguiri, L. A. 1971. The parainfluenza virus SV5 and its relationship to the cell membrane. In *Perspectives in Virology,* ed. M. Pollard, 7:127–58. New York: Academic. 327 pp.

Clarke, M. C., Haig, D. A. 1967. Presence of the transmissible agent of scrapie in the serum of affected mice and rats. *Vet. Rec.* 80:504

Clarke, M. C., Haig, D. A. 1970. Evidence for the multiplication of scrapie agent in cell culture. *Nature* 225:100–1

Clements, J. E., Narayan, O., Griffin, D. E. 1978. The proviral DNA of visna virus: synthesis and physical maps of parental and antigenic mutant DNA. In *Persistent Viruses,* ed. J. G. Stevens, G. Todaro, C. F. Fox. New York: Academic. In press

Cole, G. D., Nathanson, N., Prendergast, R. A. 1972. Requirement for θ bearing cells in LCM virus-induced CNS disease. *Nature* 238:335–37

Connolly, J. H., Allen, I. V., Hurwitz, L. J., Millar, J. H. D. 1967. Measles-virus antibody and antigen in subacute sclerosing panencephalitis. *Lancet* 1:542–44

Cook, M. L., Stevens, J. G. 1976. Latent herpetic infections following experimental viremia. *J. Gen. Virol.* 31:75–80

Cooper, L. Z., Ziring, P. R., Ockerse, A. B., Fedum, B. A., Kiely, B., Krugman, S. 1969. Rubella: Clinical manifestations and management. *Am. J. Dis. Child.* 118:18–29

Cremer, N. E., Oshiro, L. S., Weil, M. L., Lennette, E. H., Itabashi, H. H., Carnay, L. 1975. Isolation of rubella virus from brain in chronic progressive panencephalitis. *J. Gen. Virol.* 29:143–53

Cuillé, J., Chelle, P. L. 1936. La Maladie dite tremblante du mouton est elle inoculable? *C. R. Acad. Sci.* 203:1552–54

Dal Canto, M., Lipton, H. L., 1977. Animal Model: Theiler's virus infection in mice. *Am. J. Pathol.* 88:497–500

Dickinson, A. G., Fraser, H. 1969. Modification of the pathogenesis of scrapie in mice by treatment of the agent. *Nature* 222:892–93

Dickinson, A. G., Fraser, H. 1972. Scrapie: Effect of DH gene on incubation period of extraneurally injected agent. *Heredity* 29:91–93

Dickinson, A. G., Fraser, H. 1977. Scrapie: Pathogenesis in inbred mice: An assessment of host control and response involving many strains of agent. See Alvord, 1977, pp. 3–14

Dickinson, A. G., Meikle, V. M. H. 1971. Host-genotype and agent effects in scrapie incubation: Change in allelic interaction with different strains of agent. *Mol. Gen. Genet.* 112:73–79

Diener, T. O., Haddi, A. 1977. Viroids. In *Comprehensive Virology,* ed. H. Fraenkel-Conrat, R. R. Wagner, 11:285–338. New York: Plenum. 348 pp.

Dubois-Dalcq, M., Barbosa, L. H., Hamilton, R., Sever, J. L. 1974. Comparison between productive and latent subacute sclerosing panencephalitis viral infection in vitro. An electron microscopic and immunoperoxidase study. *Lab. Invest.* 30:241–50

Dubois-Dalcq, M., Coblentz, J. M., Fleet, A. B. 1974. Subacute sclerosing panencephalitis. Unusual nuclear inclusions and lengthy clinical course. *Arch. Neurol. Chicago* 31:355–63

Dubois-Dalcq, M., Reese, T. S., Murphy, M., Fuccillo, D. 1976. Defective bud formation in human cells chronically infected with subacute sclerosing panencephalitis virus. *J. Virol.* 19:579–93

Dupuy, J. M., Le Prevost, C., Levy-Leblonde, E., Virelizier, J. L. 1973. Persistent virus infection with neurological involvement in mice infected with MHV3. *Fed. Proc.* 32:4179

Eckhart, W., Dulbecco, R., Burger, M. 1971. Temperature-dependent surface changes in cells infected or transformed by a thermosensitive mutant of polyoma virus. *Proc. Natl. Acad. Sci. USA* 68:283–86

Eklund, C. M., Kennedy, R. C., Hadlow, W. J. 1967. Pathogenesis of scrapie virus infections in the mouse. *J. Infect. Dis.* 117:15–22

Finley, K. H., Fitzgerald, L. H., Richter, R. W., Riggs, N., Shelton, J. T. 1967. Western encephalitis and cerebral ontogenesis. *Arch. Neurol. Chicago* 16:140–64

Fraser, H. 1976. The Pathology of Natural and Experimental Scrapie. In *Slow Virus Diseases of Animals and Man,* ed. R. H. Kimberlin, pp. 267–305. Amsterdam: Elsevier. 461 pp.

Fraser, H., Dickinson, A. G. 1970. Pathogenesis of scrapie in the mouse: The role of the spleen. *Nature* 226:462–63

Freeman, J. M. 1969. The clinical spectrum and early diagnosis of Dawson's encephalitis. *J. Pediatr.* 75:590–603

Gajdusek, D. C. 1973. Kuru in the New Guinea Highlands. In *Tropical Neurology.* ed. J. D. Spillane, pp. 376–83. London: Oxford Univ. Press. 448 pp.

Gajdusek, D. C. 1977. Unconventional viruses and the origin and disappearance of kuru. *Science* 197:943–60

Gajdusek, D. C., Gibbs, C. J. Jr. 1977. Kuru, Creutzfeldt-Jakob disease and transmissible presenile dementia. See Alvord, 1977, pp. 15–49

Gajdusek, D. C., Gibbs, C. J. Jr., Asher, D. M., Brown, P., Diwan, A., Hoffman, P., Nemo, G., Rohwer, R., White, L. 1977. Precautions in medical care of, and in handling materials from, patients with transmissible virus dementia (Creutzfeldt-Jakob Disease). *N. Engl. J. Med.* 297:1253–58

Gajdusek, D. C., Zigas, V. 1957. Degenerative disease of the central nervous system in New Guinea. The endemic occurrence of "Kuru" in the native population. *N. Engl. J. Med.* 257:974–78

Gardiner, A. C., Marucci, A. A. 1969. Immunological responsiveness of scrapie infected mice. *J. Comp. Pathol.* 79:233–35

Gardner, M. B. 1978. Type C viruses of wild mice: Characterization and natural history of amphotropic, ectropic and xenotropic MuLV. *Curr. Top. Microbiol. Immunol.* 79:215–59

Gibbs, C. J. Jr., Gajdusek, D. C. 1978. Atypical viruses as the cause of sporadic, epidemic, and familial chronic disease in man: Slow viruses and human diseases. In *Perspectives in Virology,* ed. M. Pollard, 10:161–94. New York: Raven. 248 pp.

Gilden, D. H., Cole, C. A., Nathanson, N. 1972. Immunopathogenesis of acute central nervous system disease produced by lymphocytic choriomeningitis virus. II. Adoptive immunization of virus carriers. *J. Exp. Med.* 135:874–89

Gregson, N. A., Kennedy, M., Leibowitz, S. 1974. The specificity of anti-galactocerebroside antibody and its reaction with lysolecithin-solubilized myelin. *Immunology* 26:743–48

Griffin, D. E., Johnson, R. T. 1977. Role of the immune response in recovery from Sindbis virus encephalitis in mice. *J. Immunol.* 118:1070–75

Gudnadottir, M. 1974. Visna-maedi in sheep. *Prog. Med. Virol.* 18:336–49

Haase, A. T. 1975. The slow infection caused by visna virus. *Curr. Top. Microbiol. Immunol.* 72:101–56

Haase, A. T., Stowring, L., Narayan, O., Griffin, D., Price, D. 1977. Slow persistent infection caused by visna virus: role of host restriction. *Science* 195:175–77

Haase, A. T., Varmus, H. E. 1973. Demonstration of a DNA provirus in the lytic growth of visna virus. *Nature* 245:237–39

Hadlow, W. J. 1959. Scrapie and Kuru. *Lancet* 1:289–90

Hadlow, W. J., Eklund, C. M. 1968. Scrapie-A virus induced chronic encephalopathy of sheep. *Res. Publ. Assoc. Res. Nerv. Ment. Dis.* 44:281–306

Hadlow, W. J., Eklund, C. M., Kennedy, R. C., Jackson, T. A., Whitford, H. W., Boyle, C. C. 1974. Course of experimental scrapie virus infection in the goat. *J. Infect. Dis.* 129:559–67

Hall, W. W., Kiessling, W., ter Meulen, V. 1978. Membrane proteins of subacute sclerosing panencephalitis and measles viruses. *Nature* 272:460–62

Hall, W. W., ter Meulen, V. 1976. RNA homology between subacute sclerosing panencephalitis and measles viruses. *Nature* 264:474–77

Hampar, B., Derge, J. G., Martos, L. M., Walker, J. L. 1971. Persistence of a repressed Epstein-Barr virus genome in Burkitt lymphoma cells made resistant to 5-Bromodeoxyuridine. *Proc. Natl. Acad. Sci. USA* 68:3185–89

Hartley, J. W., Rowe, W. P., Huebner, R. J. 1970. Host-range restrictions of murine leukemia viruses in mouse embryo cell cultures. *J. Virol.* 5:221–25

Hartsough, G. R., Burger, D. 1965. Encephalopathy of mink I. Epizootological and clinical observations. *J. Infect. Dis.* 115:387–92

Heitzman, R. J., Corp, C. R. 1968. Behavior in emergence and open-field tests of normal and scrapie mice. *Res. Vet. Sci.* 9:600–1

Herndon, R. M., Griffin, D. E., McCormick, U., Weiner, L. P. 1975. Mouse hepatitis virus-induced recurrent demyelination. *Arch. Neurol. Chicago* 32:32–35

Herndon, R. M., Price, D. L., Weiner, L. P. 1977. Regeneration of oligodendroglia during recovery from demyelinating disease. *Science* 195:693–94

Herndon, R. M., Rubinstein, L. J. 1968. Light and electron microscopy on the development of viral particles in the inclusions of Dawson's encephalitis (subacute sclerosing panencephalitis). *Neurology* 18: Pt. 2, pp. 8–18

Herzberg, L., Herzberg, B. N., Gibbs, C. J. Jr., Sullivan, W., Amyx, H., Gajdusek, D. C. 1974. Creutzfeldt-Jakob disease: Hypothesis for high incidence in Libyan Jews in Israel. *Science* 186:848–49

Hirano, A., Ghatak, N., Johnson, A., Partnow, M. J., Gomori, A. J. 1972. Argentophilic plaques in Creutzfeldt-Jakob disease. *Arch. Neurol. Chicago* 26:530–42

Holmberg, C. A., Gribble, D. H., Takemoto, K. K., Howley, P. M., Espana, C., Osburn, B. I. 1977. Isolation of simian virus 40 from rhesus monkeys (Macaca mulatta) with spontaneous progressive multifocal leukoencephalopathy. *J. Infect. Dis.* 136:593–96

Hornabrook, R. 1968. Kuru—a subacute cerebellar degeneration: The natural history and clinical features. *Brain* 91:53–74

Horoupian, D. S., Powers, J. M., Schaumberg, H. H. 1972. Kuru-like neuropathological changes in a North American. *Arch. Neurol. Chicago* 27:555–61

Horta-Barbosa, L., Fuccillo, D. A., Sever, J. L., Zeman, W. 1969. Subacute sclerosing panencephalitis: Isolation of measles virus from a brain biopsy. *Nature* 221:974

Hotchin, J. 1971. Persistent and Slow Infections. *Monogr. Virol.,* Vol. 3,

Hotchin, J., Buckley, R. 1977. Latent form of scrapie virus: A new factor in slow-virus disease. *Science* 196:668–71

Huang, A. S., Baltimore, D. 1977. Defective interfering animal viruses. In *Comprehensive Virology*, ed. H. Fraenkel-Conrat, R. R. Wagner, 10:73–116. New York: Plenum. 496 pp.

Hunter, G. D. 1974. Scrapie. *Prog. Med. Virol.* 18:289–306

Hunter, G. D., Collis, S. C., Millson, G. C., Kimberlin, R. H. 1976. Search for scrapie-specific RNA and attempts to detect an infectious DNA or RNA. *J. Gen. Virol.* 32:157–62

Jaenisch, R. 1976. Germ line integration and mendelian transmission of the exogenous moloney leukemia virus. *Proc. Natl. Acad. Sci. USA* 73:1260–64

Johnson, K. P., Swoveland, P. 1977. Measles antigen distribution in brains of chronically infected hamsters. An immunoperoxidase study of experimental subacute sclerosing panencephalitis. *Lab. Invest.* 37:459–65

Johnson, R. T. 1965. Experimental rabies. Studies of cellular vulnerability and pathogenesis using fluorescent antibody staining. *J. Neuropath. Exp. Neurol.* 24:662–74

Johnson, R. T. 1970. Subacute sclerosing panencephalitis (Editorial). *J. Infect. Dis.* 121:227–30

Johnson, R. T. 1972. Effects of viral infection on the developing nervous system. *N. Engl. J. Med.* 287:599–604

Johnson, R. T. 1973. Slow infections: Virus-host relationships. In *Slow Virus Diseases*, ed. W. Zeman, E. H. Lennette, pp. 1–9. Baltimore: Williams & Wilkins. 145 pp.

Johnson, R. T. 1975a. Hydrocephalus and viral infections. *Dev. Med. Child Neurol.* 17:807–16

Johnson, R. T. 1975b. The possible viral etiology of multiple sclerosis. *Adv. Neurol.* 13:1–46

Johnson, R. T. 1977. Slow viral diseases of the CNS: Transmissibility vs communicability. *Clin. Neurosurg.* 24:590–99

Johnson, R. T. 1978a. Current knowledge of multiple sclerosis. *South. Med. J.* 71:2–3

Johnson, R. T. 1978b. Teratogenic effects of viruses. In *Infections of the Central Nervous System, Handbook of Clinical Neurology*, ed. P. J. Vinken, G. W. Bruyn, 34:369–89. Amsterdam: North Holland

Johnson, R. T., Griffin, D. E. 1978. Pathogenesis of Viral Infections, See Johnson 1978 b, 34:15–37

Johnson, R. T., Herndon, R. M. 1974. Virologic studies of multiple sclerosis and other chronic and relapsing neurologi-cal diseases. *Prog. Med. Virol.* 18:2144–18

Johnson, R. T., Johnson, K. P. 1968. Hydrocephalus following viral infection: the pathology of aqueductal stenosis developing after experimental mumps virus infection. *J. Neuropath. Exp. Neurol.* 27:591–606

Johnson, R. T., Mims, C. A. 1968. Pathogenesis of viral infections of the nervous system. *N. Engl. J. Med.* 270:23–30, 84–92

Johnson, R. T., Narayan, O., Weiner, L. P., Greenlee, J. E. 1977. Progressive Multifocal Leukoencephalopathy. See Alvord, 1977, pp. 91–100

Johnson, R. T., Narayan, O., Clements, J. 1978. Varied role of viruses in chronic neurologic diseases. See Clements et al 1978, pp.

Johnson, R. T., ter Meulen, V. 1978. Slow infections of the nervous system. *Adv. Intern. Med.* 23:353–83

Joseph, B. S., Oldstone, M. B. A. 1975. Immunologic injury in measles virus infection. II. Suppression of immune injury through antigenic modulation. *J. Exp. Med.* 142:864–76

Kakulas, B., LeCours, A., Gajdusek, D. C. 1967. Further observations of the pathology of kuru. *J. Neuropath. Exp. Neurol.* 26:85–97

Karcher, D., Noppe, M., Lowenthal, A. 1977. A heat stable serum inhibitor of an antigen antibody reaction of subacute sclerosing panencephalitis. *J. Neurol.* 216:51–56

Katz, M., Koprowski, H. 1968. Failure to demonstrate a relationship between scrapie and production of interferon in mice. *Nature* 219:639–40

Kepes, J. H., Chou, S. M., Price, L. W. 1975. Progressive multifocal leukoencephalopathy with 10-year survival in a patient with nontropical sprue. *Neurology* 25:1006–12

Kilham, L., Margolis, G. 1966. Spontaneous hepatitis and cerebellar hypoplasia in suckling rats due to congenital infections with rat virus. *Amer. J. Pathol.* 49:457–75

Kilham, L., Margolis, G. 1975. Problems of human concern arising from animal models of intrauterine and neonatal infections due to viruses: A review. *Prog. Med. Virol.* 20:113–43

Kimberlin, R. H., ed. 1976. *Slow Virus Diseases of Animals and Man.* Amsterdam: Elsevier/North Holland Biomedical Press. 404 pp.

Klatzo, I., Gajdusek, D. C., Zigas, V. 1959.

Pathology of kuru. *Lab. Invest.* 8:799–847

Klein, G., Giovanella, B. C., Singh, S. 1976. Intracellular forms of Epstein-Barr virus DNA in human tumor cells in vitro. *Nature* 260:302–6

Kolar, O. 1968. Immunopathologic aspects of subacute sclerosing panencephalitis. *Neurology* 18: Pt. 2, pp. 107–11

Koprowski, H., Barbanti-Brodano, G., Katz, M. 1970. Interaction between papova-like virus and paramyxovirus in human brain cells. A hypothesis. *Nature* 225:1045–47

Kreth, H. W., Kackell, Y. M., ter Meulen, V. 1974. Cellular immunity in SSPE patients. *Med. Microbiol. Immunol.* 160:191–99

Kreth, H. W., Kackell, M. Y., ter Meulen, V. 1975. Demonstration of in vitro lymphocyte-mediated cytotoxicity against measles virus in SSPE. *J. Immunol.* 114:1042–46

Kreth, H. W., ter Meulen, V. 1977. Cell-mediated cytotoxicity against measles virus in SSPE. I. Enhancement by antibody. *J. Immunol.* 118:291–95

Lampert, P. W. 1978. Autoimmune and virus-induced demyelinating diseases. *Am. J. Pathol.* 91:175–208

Lampert, P. W., Earle, K., Gibbs, C. J. Jr., Gajdusek, D. C. 1969. Experimental kuru encephalopathy in chimpanzees and spider monkeys. Electron microscopic studies. *J. Neuropathol. Exp. Neurol.* 28:353–70

Lampert, P. W., Sims, J. K., Kniazeff, A. J. 1973. Mechanism of demyelination in JHM virus encephalomyelitis. *Acta Neuropathol.* 24:76–85

Latarjet, R., Juel, B., Haig, D. A., Clarke, M. C., Alper, T. 1970. Inactivation of the scrapie agent by near monochromatic ultraviolet light. *Nature* 227:1341–43

Le Prevost, C., Virelizier, J. L., Dupuy, J. M. 1975. Immunopathology of mouse hepatitis virus type 3 infection. III. Clinical and virologic observation of a persistent viral infection. *J. Immunol.* 115:640–43

Lin, F. H., Thormar, H. 1970. Ribonucleic acid dependent deoxyribonucleic acid polymerase in visna virus. *J. Virol.* 6:702–4

Lipton, H. L. 1975. Theiler's virus infection in mice: an unusual biphasic disease process leading to demyelination. *Infect. Immun.* 11:1147–55

Lipton, H. L., Dal Canto, M. C. 1976a. Theiler's virus-induced demyelination: Prevention by immunosuppression. *Science* 192:62–64

Lipton, H. L., Dal Canto, M. C. 1976b. Chronic neurological disease in Theiler's virus infection of SJL/J mice. *J. Neurol. Sci.* 30:201–7

Lonberg-Holm, K., Philipson, L. 1974. *Early Interaction Between Animal Viruses and Cells. Monogr. Virol.,* Vol. 3. Basel: Karger. 148 pp.

Luria, S. E., Darnell, J. E., Baltimore, D., Campbell, A. 1978. *General Virology,* 3rd ed. New York: Wiley. 578 pp.

Malone, T. G., Marsh, R. F., Hanson, R. P., Semancik, J. S. 1978. Membrane free scrapie activity. *J. Virol.* 25:933–35

Manuelidis, E. E. 1975. Transmission of Creutzfeldt-Jakob disease from man to the guinea pig. *Science* 190:571–72

Manuelidis, E. E., Angelo, J. N., Gorgacz, E. J., Manuelidis, L. 1977. Transmission of Creutzfeldt-Jakob disease to Syrian hamsters. *Lancet* 1:479

Manuelidis, E. E., Gorgacz, E. J., Manuelidis, L. 1978. Viremia in experimental Creutzfeldt-Jakob disease. *Science* 200:1069–71

Manz, H. J., Dinsdale, H. B., Morrin, P. A. F. 1971. Progressive multifocal leukoencephalopathy after renal transplantation. Demonstration of papova-like virions. *Ann. Intern. Med.* 75:77–81

Marsh, R. F., Pan, I. C., Hanson, R. P. 1970. Failure to demonstrate specific antibody in transmissible mink encephalopathy. *Infect. Immun.* 2:727–30

Marsh, R. F., Semancik, J. S., Medappa, K. C., Hanson, R. P., Ruechert, R. R. 1974. Scrapie and transmissible mink encephalopathy: Search for infectious nucleic acid. *J. Virol.* 13:993–96

Marsh, R. F., Sipe, J. C., Morse, S. S., Hanson, R. P. 1976. Transmissible mink encephalopathy: Reduced spongiform degeneration in aged mink of the Chediak-Higashi genotype. *Lab. Invest.* 34:381–86

Martin, M. 1970. Characteristics of SV40-DNA transcription during lytic infection, abortive infection and in transformed mouse cells. *Cold Spring Harbor Symp. Quant. Biol.* 35:833–38

Masters, C. L., Alpers, M., Kakulas, B. 1977. Pathogenesis of reovirus type 1 hydrocephalus in mice. *Arch. Neurol.* 34:18–28

Masters, C. L., Kakulas, B. A., Alpers, M. P., Gajdusek, D. C., Gibbs, C. J. Jr. 1976. Preclinical lesions and their progression in the experimental spongiform encephalopathies (Kuru and Creutzfeldt-Jakob disease) in primates. *J. Neuropathol. Exp. Neurol.* 35:593–605

McComas, A. J., Upton, A. R. M., Sica, R. E. P. 1973. Motor neuron disease and ageing. *Lancet* 2:1477–80

McFarland, H. 1974. The role of immunological response in infections of the central nervous system. *Advances in Neurology,* ed. R. A. Thompson, J. R. Green, 6:19–25. New York: Raven. 286 pp.

McFarlin, D. E., Raff, M. C., Simpson, E., Nehlsen, S. H. 1971. Scrapie in immunologically deficient mice. *Nature* 233:336

McSharry, J. J., Compans, R. W., Lackland, H., Choppin, P. W. 1975. Isolation and characterization of the non-glycosylated membrane protein and a nucleocapsid complex from the paramyxovirus SV5. *Virology* 67:365–74

Mims, C. A. 1960. Intracerebral injections and growth of viruses in mouse brain. *Br. J. Exp. Pathol.* 41:52–59

Mims, C. A. 1964. Aspects of the pathogenesis of virus diseases. *Bacteriol. Rev.* 28:30–71

Mims, C. A. 1974. Factors in the mechanism of persistence of viral infections. *Prog. Med. Virol.* 18:1–14

Mountcastle, W. E., Compans, R. W., Lackland, H., Choppin, P. W. 1974. Proteolytic cleavage of subunits of the nucleocapsid of the paramyxovirus simian virus 5. *J. Virol.* 14:1253–61

Mulder, D. W., Rosenbaum, R. A., Layton, D. D. 1972. Late progression of poliomyelitis or form fruste amyotrophic lateral sclerosis. *Mayo Clin. Proc.* 47:756–61

Naeye, R. L., Blanc, W. 1965. Pathogenesis of congenital rubella. *J. Am. Med. Assoc.* 194:1277–83

Nagai, Y., Yoshida, T., Yoshii, S., Maeno, K., Matsumoto, T. 1975. Modification of normal cell surface by smooth membrane preparations from BHK-21 cells infected with Newcastle disease virus. *Med. Microbiol. Immunol.* 161:175–88

Narayan, O., Griffin, D. E., Chase, J. 1977. Antigenic shift of visna virus in persistently infected sheep. *Science* 197:376–78

Narayan, O., Griffin, D. E., Clements, J. E. 1978. Progressive antigenic drift of visna virus in persistently infected sheep. See Clements et al, 1978

Narayan, O., Griffin, D. E., Silverstein, A. 1977. Slow virus infection: Replication and mechanism of persistence of visna virus in sheep. *J. Infect. Dis.* 135:800–6

Narayan, O., Penney, J. B. Jr., Johnson, R. T., Herndon, R. M., Weiner, L. P. 1973. Etiology of progressive multifocal leukoencephalopathy: Identification of papovavirus. *N. Engl. J. Med.* 289:1278–82

Nathanson, N., Panitch, H., Palsson, P. A., Petursson, G., Georgsson, G. 1976. Pathogenesis of visna. II. Effect of immunosuppression upon early central nervous system lesions. *Lab. Invest.* 35:444–51

Neuman, M., Gajdusek, D. C., Zigas, V. 1964. Neuropathologic findings in exotic neurologic disorders among natives of New Guinea. *J. Neuropathol. Exp. Neurol.* 23:486–507

Nichols, W. W. 1970. Virus-induced chromosome abnormalities. *Ann. Rev. Microbiol.* 24:479–500

Ogra, P. L., Karzon, D. T. 1971. Formation and function of poliovirus antibody in different tissues. *Prog. Med. Virol.* 13:156–93

Oldstone, M. B. A., Holmstoen, J., Welsh, R. M. Jr. 1977. Alterations of acetylcholine enzymes in neuroblastoma cells persistently infected with lymphocytic choriomeningitis virus. *J. Cell Physiol.* 91:459–72

Oldstone, M. B. A., Lampert, P. W., Lee, S., Dixon, F. J. 1977. Pathogenesis of the slow disease of the central nervous system associated with WM 1504 E virus I. Relationship of strain susceptibility and replication to disease. *Am. J. Pathol.* 88:193–212

Olitsky, P. K., Schlesinger, R. W. 1941. Histopathology of CNS of mice infected with virus of Theiler's disease (spontaneous encephalomyelitis). *Proc. Soc. Exp. Biol. Med.* 47:79–83

Openshaw, H., Asher, L., Wohlenberg, C., Notkins, A. L. 1978. Latency and reactivation of herpes simplex virus in ganglia: Immune control. *Neurology* 28:370

Outram, G. W., Dickinson, A. G., Fraser, H. 1973. Developmental maturation of susceptibility to scrapie in mice. *Nature* 241:536–37

Outram, G. W., Dickinson, A. G., Fraser, H. 1974. Reduced susceptibility to scrapie in mice after steroid administration. *Nature* 249:855–56

Outram, G. W., Dickinson, A. G., Fraser, H. 1975. Slow encephalopathies, inflammatory responses, and arachis oil. *Lancet* 1:198–200

Padgett, B. L., Walker, D. L. 1973. Prevalence of antibodies in human sera against JC virus, an isolate from a case of progressive multifocal leukoencephalopathy. *J. Infect. Dis.* 127:467–70

Padgett, B. L., Walker, D. L. 1976. New human papovaviruses. *Prog. Med. Virol.* 22:1–35

Padgett, B. L., Walker, D. L., ZuRhein, G. M., Echroade, R. J., Dessel, B. H. 1971. Cultivation of papova-like virus from human brain with progressive multifocal leukoencephalopathy. *Lancet* 1:1256–60

Padgett, B. L., Walker, D. L., ZuRhein, G. M., Hodach, A. E., Chou, S. M. 1976. JC papovirus in progressive multifocal leukoencephalopathy. *J. Infect. Dis.* 133:686

Pattison, I. H., Hoare, M. N., Jebbet, J. N., Watson, W. A. 1972. Spread of scrapie to sheep and goats by oral dosing from scrapie-affected sheep. *Vet. Rec.* 90:465–68

Pattison, I. H., Millson, G. C., Smith, K. 1964. An examination of the action of whole blood, blood cells, or serum on the goat scrapie agent. *Res. Vet. Sci.* 5:116–21

Payne, F. E., Baublis, J. V. 1973. Decreased reactivity of SSPE strains of measles virus with antibody. *J. Infect. Dis.* 127:505–11

Payne, F. E., Baublis, J. V., Itabashi, H. H. 1969. Isolation of measles virus from cell cultures of brain from a patient with subacute sclerosing panencephalitis. *N. Engl. J. Med.* 281:585–89

Penney, J. B. Jr., Wolinsky, J. S. 1978. Neuronal and oligodendroglial infection by the WW strain of Theiler's virus. *Neurology* 28:370

Pereira, H. G. 1958. A protein factor responsible for the early cytopathic effect of adenoviruses. *Virology* 6:601–11

Perrin, L. H., Tishon, A., Oldstone, M. B. A. 1977. Immunologic injury in measles virus infection. III. Presence and characterization of human cytotoxic lymphocytes. *J. Immunol.* 118:282–90

Petursson, G., Nathanson, N., Georgsson, G., Panitch, H., Palsson, P. A. 1976. Pathogenesis of visna. I. Sequential virologic, serologic, and pathologic studies. *Lab. Invest.* 35:402–12

Ponten, J. 1976. The relationship between in vitro transformation and tumor formation in vivo. *Biochim. Biophys. Acta* 458:397–422

Porter, D. D., Porter, H. G., Cox, N. A. 1973. Failure to demonstrate a humoral immune response to scrapie infection in mice. *J. Immunol.* 11:1407–10

Prusiner, S. B., Garfin, D. E., Cochran, S. P., Hooper, C. J., Baringer, J. R., Hadlow, W. J., Eklund, C. M., Race, R. E. 1978. Gradient centrifugation studies of the scrapie agent from murine spleen. *Neurology* 28:369

Prusiner, S. B., Hadlow, W. J., Eklund, C. M., Race, R. E. 1977. Sedimentation properties of the scrapie agent. *Proc. Natl. Acad. Sci. USA* 74:4656–60

Raine, C. S., Schaumberg, H. H. 1977. The neuropathology of myelin diseases. In *Myelin*, ed. P. Morell, pp. 271–324. New York: Plenum. 531 pp.

Rappaport, S. I. 1975. *Blood-brain Barrier in Physiology and Medicine.* New York: Raven. 316 pp.

Rawls, W. E. 1968. Congenital rubella. The significance of virus persistence. *Prog. Med. Virol.* 10:238–85

Rawls, W. E. 1974. Viral persistence in congenital rubella. *Prog. Med. Virol.* 18:273–88

Rawls, W. E., Melnick, J. L. 1966. Rubella virus carrier cultures derived from congenitally infected infants. *J. Exp. Med.* 123:795–816

Richardson, E. P. Jr. 1961. Progressive multifocal leukoencephalopathy. *N. Engl. J. Med.* 265:815–23

Rifkin, D. B., Quigley, J. P. 1974. Virus-induced modification of cellular membranes related to viral structure. *Ann. Rev. Microbiol.* 28:324–51

Robb, J. A. 1977. Virus-cell interactions: A classification for virus-caused human disease. *Prog. Med. Virol.* 33:51–61

Robb, J. A., Martin, R. G. 1972. Genetic analysis of simian virus 40. III. Characterization of a temperature-sensitive mutant blocked at an early stage of productive infection in monkey cells. *J. Virol.* 9:956–68

Robinson, W. S., Lutwick, L. I. 1976. The virus of hepatitis, type B. *N. Engl. J. Med.* 295:1168–75, 1232–36

Roos, R., Gajdusek, D. C., Gibbs, C. J. Jr. 1973. The clinical characteristics of transmissible Creutzfeldt-Jakob disease. *Brain* 96:441–62

Sabin, A. B., Ward, R. 1941. The natural history of poliomyelitis. I. Distribution of virus in nervous and non-nervous tissue. *J. Exp. Med.* 73:771–93

Scheid, A., Choppin, P. W. 1976. Protease activation mutants of Sendai virus: Activation of biological properties by specific proteases. *Virology* 69:265–77

Schlom, J., Harter, D. H., Burry, A., Spiegelman, S. 1971. DNA polymerase activities in virions of visna virus, a causative agent of a "slow" neurological disease. *Proc. Natl. Acad. Sci USA* 68:182–86

Schluederberg, A., Chavanich, S., Lipman, M. B., Carter, C. 1974. Comparative molecular weight estimates of measles and subacute sclerosing panencephalitis virus structural polypeptides by simultaneous electrophoresis in acrylamide gel slabs. *Biochem. Biophys. Res. Commun.* 58:647–51

Schultz, I., Frohlich, E. 1965. Viuria and viraliquoria in dog after intravenous infection of T₅ bacteriophage. *Proc. Soc. Exp. Biol. Med.* 118:136–38

Shah, K. V. 1972. Evidence for an SV40-related papovavirus infection of man. *Am. J. Epidemiol.* 95:199–206

Siakotos, A. N., Bucana, C., Gajdusek, D. C., Gibbs, C. J. Jr., Traub, R. D. 1976. Partial purification of the scrapie agent from mouse brain by pressure disruption and zonal centrifugation in sucrose sodium chloride gradient. *Virology* 70:230–37

Sigurdsson, B. 1954. Rida, a chronic encephalitis of sheep with general remarks on infections which develop slowly and some of their special characteristics. *Br. Vet. J.* 110:341–54

Sigurdsson, B., Palsson, P. A. 1958. Visna of sheep. A slow, demyelinating infection. *Br. J. Exp. Pathol.* 39:519–28

Sigurdsson, B., Palsson, P. A., Grimsson, H. 1957. Visna, a demyelinating transmissible disease of sheep. *J. Neuropathol. Exp. Neurol.* 16:389–403

Smith, H. 1972. Mechanisms of virus pathogenicity. *Bacteriol. Rev.* 36:291–310

Smith, H. S., Gelb, L. D., Martin, M. A. 1972. Detection and quantitation of simian virus 40 genetic material in abortively transformed Balb 3T3. *Proc. Natl. Acad. Sci. USA* 69:152–56

Sponzilli, E. E., Smith, J. K., Malamud, N., McCulloch, J. R. 1975. Progressive multifocal leukoencephalopathy. A complication of immunosuppressive treatment. *Neurology* 25:664–68

Spring, S. B., Roizman, B. 1967. Herpes simplex virus products in productive and abortive infection. I. Stabilization with formaldehyde and preliminary analysis by isopycnic centrifugation in CsCl. *J. Virol.* 1:294–301

Spring, S. B., Roizman, B., Schwartz, J. 1968. Herpes simplex virus products in productive and abortive infection. II. Electron microscopic and immunological evidence for failure of virus envelopment as cause of abortive infection. *J. Virol.* 2:384–92

Starr, S., Berkovich, S. 1964. Effects of measles, gamma-globulin-modified measles and vaccine measles on the tuberculin test. *N. Engl. J. Med.* 270:386–91

Stevens, J. G. 1975. Latent herpes simplex virus and the nervous system. *Curr. Top. Microbiol. Immunol.* 70:31–50

Stevens, J. G. 1978. Latent characteristics of selected herpes viruses. *Adv. Cancer Res.* 26:227–56

Sturman, L. S., Tamm, I. 1969. Formation of viral ribonucleic acid and virus in cells that are permissive or non-permissive for murine encephalomyelitis virus (GDVII). *J. Virol.* 3:8–16

Suckling, A. J., Bateman, S., Waldron, C. B., Webb, H. E., Kimberlin, R. H. 1976. Motor activity changes in scrapie-affected mice. *Brit. J. Exp. Pathol.* 57:742–46

Swick, H. M., Brooks, W. H., Roszman, T. L., Caldwell, D. 1976. A heat-stable blocking factor in the plasma of patients with subacute sclerosing panencephalitis. *Neurology* 26:84–88

Tamm, I. 1975. Cell injury with viruses. *Am. J. Pathol.* 81:163–77

ter Meulen, V., Katz, M., eds. 1977. *Slow Virus Infection of the Central Nervous System.* New York: Springer. 258 pp.

Thormar, H., Arnesen, K., Mehta, P. D. 1977. Encephalitis in ferrets caused by a nonproductive strain of measles virus (D.R) isolated from a patient with subacute sclerosing panencephalitis. *J. Infect. Dis.* 136:229–38

Thormar, H., Lin, F. H., Karl, S. C. 1973. Effects of Viral Infections on the Brain. Chapter 5. In *Biology of Brain Dysfunction,* ed. G. E. Gaull, 1:191–228. New York: Plenum. 459 pp.

Townsend, J. J., Baringer, J. R., Wolinsky, J. S., Malamud, N., Mednick, J. P., Panitch, H. S., Scott, R. A. T., Oshiro, L. S., Cremer, N. E. 1975. Progressive rubella panencephalitis. Late onset after congenital rubella. *N. Engl. J. Med.* 292:990–93

Traub, R., Gajdusek, D. C., Gibbs, C. J. Jr. 1977. Transmissible Virus Dementia: The Relation of Transmissible Spongiform Encephalopathy To Creutzfeldt-Jakob Disease. In *Aging and Dementia,* eds. M. Kinsbourne, L. Smith, pp. 91–146. Flushing, New York: Spectrum. 461 pp.

Trentin, J. J., Yabe, Y., Taylor, B. 1962. The quest for human cancer viruses. *Science* 137:835–41

Varmus, H. E., Guntaka, R. V., Fan, W. J. W., Heasley, S., Bishop, J. M. 1974. Synthesis of viral DNA in the cytoplasm of duck embryo fibroblasts in enucleated cells after infection by avian sarcoma virus. *Proc. Natl. Acad. Sci. USA* 71:3874–78

Ward, R. L., Porter, D. D., Stevens, J. G. 1974. Nature of the scrapie agent: Evidence against a viroid. *J. Virol.* 14:1099–1103

Wear, D. J., Rapp, F. 1971. Latent measles virus infection of the hamster central nervous system. *J. Immunol.* 107:1593–98

Wechsler, S. L., Fields, B. N. 1978. Differences between the intracellular polypeptides of measles and subacute sclerosing panencephalitis virus. *Nature* 272:458–60

Weil, M. L., Itabashi, H. H., Cremer, N. E., Oshiro, L. S., Lennette, E. H., Carnay, L. 1975. Chronic progressive panencephalitis due to rubella virus simulating subacute sclerosing panencephalitis. *N. Engl. J. Med.* 292:994–98

Weiner, L. P. 1973. Pathogenesis of demyelination induced by a mouse hepatitis virus (JHM virus). *Arch. Neurol. Chicago* 28:298–303

Weiner, L. P., Herndon, R. M., Narayan, O., Johnson, R. T., Shah, K., Rubinstein, L. J., Preziosi, T. J., Conley, F. K. 1972. Isolation of virus related to SV40 from patients with progressive multifocal leukoencephalopathy. *N. Engl. J. Med.* 286:385–90

Weiner, L. P., Johnson, R. T. 1977. Virus-Host Cell Interactions in Slow Virus Diseases of the Nervous System. In *Virus Infection and the Cell Surface*, ed. G. Poste, G. L. Nicholson, pp. 195–212. Amsterdam: Elsevier/North Holland Biomedical Press. 342 pp.

Weiner, L. P., Johnson, R. T., Herndon, R. M. 1973. Viral infections and demyelinating diseases. *N. Engl. J. Med.* 288:1103–10

Weiner, L. P., Narayan, O. 1974. Virologic studies of progressive multifocal leuko-

encephalopathy. *Prog. Med. Virol.* 18:229–40

Wisniewski, H. M. 1977. Immunopathology of demyelination in autoimmune diseases and virus infection. *Br. Med. Bull.* 35:54–59

Wisniewski, H. M., Bloom, B. R. 1975. Primary demyelination as a nonspecific consequence of a cell-mediated immune reaction. *J. Exp. Med.* 141:346–59

Woods, W. A., Johnson, R. T., Hostetler, D. D., Lepow, M. L., Robbins, F. L. 1966. Immunofluorescent studies on rubella infected tissue cultures and human tissue. *J. Immunol.* 96:253–60

Worthington, M. 1972. Interferon system in mice infected with the scrapie agent. *Infect. Immun.* 6:643–45

Zlotnik, I. 1968. Spread of scrapie by contact in mice. *J. Comp. Pathol.* 78:19–22

Zlotnik, I., Rennie, J. 1965. Experimental transmission of mouse passaged scrapie to goats, sheep, rats, and hamsters. *J. Comp. Pathol.* 75:147–57

Zlotnik, I., Rennie, J. C. 1967. The effect of heat on the scrapie agent in mouse brain. *Br. J. Exp. Pathol.* 48:171–79

ZuRhein, G. M. 1969. Association of papova-virions with a human demyelinating disease (progressive multifocal leukoencephalopathy). *Prog. Med. Virol.* 11:185–247

ZuRhein, G. M., Chou, S. M. 1965. Particles resembling papovaviruses in human cerebral demyelinating disease. *Science* 148:1477–79

ZuRhein, G. M., Chou, S. M. 1968. Subacute sclerosing panencephalitis. Ultrastructural study of a brain biopsy. *Neurology* 18: Pt. 2, pp. 146–58

ZuRhein, G. M., Varakis, J. 1974. Progressive multifocal leukoencephalopathy in a renal-allograft recipient. *N. Engl. J. Med.* 291:798

Ann. Rev. Neurosci. 1979. 2:341–62

A PHARMACOLOGICAL APPROACH TO THE STRUCTURE OF SODIUM CHANNELS IN MYELINATED AXONS

♦11527

J. M. Ritchie

Department of Pharmacology, Yale University School of Medicine, New Haven, Connecticut 06510

INTRODUCTION

Ever since the papers of Hodgkin & Huxley (1952) on the ionic basis of conduction, the idea has grown that the currents underlying the action potential in excitable tissues, the potassium and sodium currents, flow through discrete channels: the potassium channels and the sodium channels. In the squid giant axon, the potassium current plays an important role in repolarization. However, in frog myelinated nerve, it plays only a minor role: most of the outward current during repolarization is leak current, which is why treatment with TEA, which blocks the potassium channels, seldom lengthens the action potential in frog nerve by more than 50% (see Stämpfli & Hille 1976). In mammalian myelinated fibers the potassium current is virtually absent (Horakova, Nonner & Stämpfli 1968, Nonner & Stämpfli 1969, Chiu et al 1978); and the short duration of the mammalian action potential is due to a very rapid inactivation of the sodium current in these fibers compared with that in the frog (Chiu et al 1978). For these reasons the present review concentrates on the sodium channel in myelinated nerve fibers, and on how pharmacological studies are beginning to elucidate the molecular structure of this component of the membrane that is critical for conduction. Two classes of drugs in particular are considered: first the local anesthetic agents, procaine, lidocaine, cocaine, etc; and secondly, the biotoxins, saxitoxin and tetrodotoxin (Figure 1). These drugs specifically block the sodium currents in myelinated axons with relatively

341

0147-006X/79/0315-0341$01.00

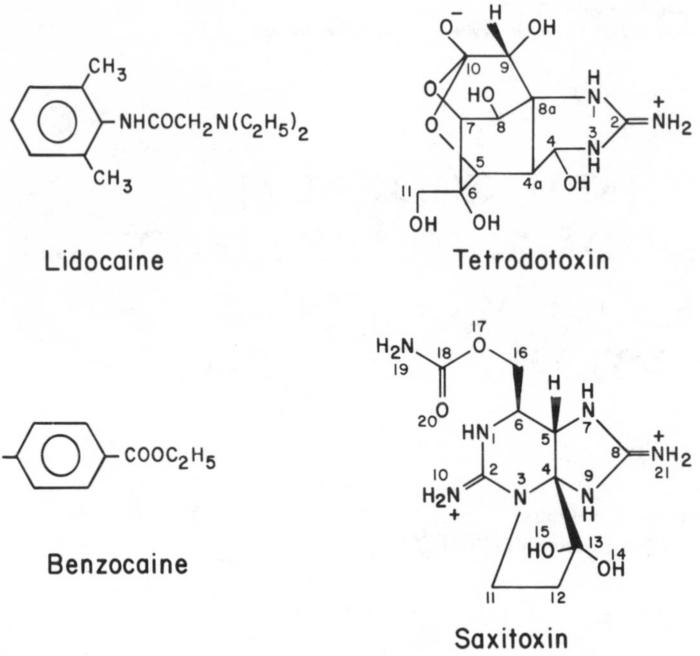

Figure 1   The structure of lidocaine, benzocaine, tetrodotoxin, and saxitoxin.

little effect on the potassium currents (Hille 1968, 1977). How do these agents block conduction, and what does their action tell about the molecular structure of the sodium channel?

LOCAL ANESTHETICS

The Classical View. The Role of the Uncharged Form

All local anesthetics in common use are secondary or tertiary amines; and so they can exist either as uncharged molecules or, in their protonated form, as charged substituted ammonium cations. The balance between the two forms is dependent on pH; the more alkaline the solution, the greater the proportion in the uncharged form. This fact, together with the finding that local anesthetics are more effective when applied to tissues in alkaline solution, led Trevan & Boock in 1927 to conclude that local anesthetic activity resides solely in the uncharged form of the drug. These considerations led in turn to the idea (see Shanes 1958) that local anesthetics act by dissolving in the membrane—somehow entering the region between the sodium channels. As a result, the lateral pressure in the membrane is increased, with a consequent compression of the channels, whose diameter becomes so reduced that sodium ions can no longer pass.

The underlying mechanism is, thus, similar to that proposed for general anesthetic agents, namely that by dissolution in the membrane the uncharged hydrophobic anesthetic molecule expands some microdomain within the membrane that is critical for conduction (see Seeman 1977). Support for this hypothesis comes from the finding of Kendig & Cohen (1977) that high hydrostatic pressure (to 100 atmospheres) reverses the reduction of the compound action potential produced in rat preganglionic sympathetic nerves by the uncharged and potentially uncharged molecules, benzocaine and lidocaine (Figure 1), but not that produced by a permanently charged quaternary analog of lidocaine, QX572, which is too lipophobic to dissolve in the membrane.

The Role of the Charged Form

The action of the local anesthetics described above is presumed to occur primarily at some locus outside the sodium channel. It is clear, however, that there is a second locus of action, namely a receptor site within the sodium channel, at which local anesthetic molecules in their charged form act (Narahashi & Frazier 1971, Ritchie 1975, Hille 1977); indeed, it has also been suggested (see below) that the local anesthetics in their uncharged form may also act here (Hille 1977).

Two lines of evidence established the important role of the cationic form of local anesthetics in producing conduction block. First, when it was realized that the greater efficacy of alkaline solutions of local anesthetics in the Trevan & Boock experiments might just reflect the greater ease of penetration of the uncharged form of the local anesthetic to the nerve membrane, experiments were designed in which the factor of penetration was excluded (Ritchie & Greengard 1961). Thus, in a nerve membrane preloaded with local anesthetic agent, the dependence on pH was shown to be exactly the reverse of that obtained in the Trevan & Boock type of experiment: increasing the pH decreased the effectiveness of the local anesthetic. Second, Narahashi and his colleagues (see Narahashi & Frazier 1971) showed that quaternary forms of local anesthetics (which can exist only as charged substituted cations) are just as effective as the parent amine compounds, provided they are applied to the inside surface of the membrane. In squid giant axon, therefore, they have to be injected internally; and in frog node of Ranvier they have to be allowed to diffuse from the cut end of the fiber along the internodal axoplasm to the inner surface of the node (Koppenhöfer & Vogel 1969). Figure 2a shows diagrammatically how the sodium channel might have been depicted in the early 1970s. The uncharged form of the local anesthetic molecule (U) is important for penetrating any cellular membranes, such as the nerve sheath, to the periaxonal space, and then for penetrating, and crossing, the nerve membrane into the axoplasm. There the local anesthetic would be converted back in part to the

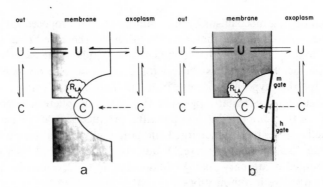

Figure 2 A diagramatic representation of the sodium channel. For explanation see text.

charged form (C), depending on the prevailing intracellular pH, when it would then react (as the cation) with some receptor on the inside surface of the channel to block the sodium currents.

Frequency-Dependence (Use-Dependence) of Local Anesthetic Action

The first important modification of the scheme depicted in Figure 2*a* resulted from the experiments of Strichartz (1973), using quaternary local anesthetic molecules, and of Courtney (1975), using amine local anesthetics. These experiments established two important factors. First, the rate of onset of the inhibition of the sodium currents produced by local anesthetics depends on the degree of opening of both the inactivation (h) and the activation (m) gates. The kinetics of opening and closing of these h and m gates depends both on time and membrane potential, according to the relationships described by Hodgkin & Huxley (1952) for squid axons and modified by Frankenhauser & Huxley (1964) for myelinated fibers. Thus, in voltage-clamped frog nodes of Ranvier exposed internally to a quaternary local anesthetic agent, when the h gates are wide open (by application of hyperpolarizing prepulses) local anesthetic action develops rapidly; on the other hand, when the h gates are partially closed (as a result of depolarizing prepulses) local anesthetic activity develops much more slowly (Strichartz 1973). Open m gates are also needed (Strichartz 1973). For example, if a preparation is left at rest in a solution of a quaternary local anesthetic after cutting the internode (so that the drug has plenty of time to diffuse into, and equilibrate with, the nodal axoplasm), even after 45 min the currents flowing during a small test pulse may be reduced from the control value by only 10%. However, if several large depolarizing pulses are given, so that the m gates are well opened, a tonic inhibition sets in that rapidly may reach a new value (about 90%). Once this tonic state of inhibition of response to the prevailing concentration of local anesthetic in the axoplasm has been reached, it may be reversed partially, but never to the previous small level of inhibition. It seems, therefore, that in the resting state, charged local anesthetic molecules may have little or no access to the site, binding to which results in conduction block. Both the inactivation and activation gates must be open for the anesthetic to reach this site (Figure 2b).

In addition to altering the rate at which equilibrium or steady state conditions are reached, electrical activity affects local anesthetic action in a second way: it seems that even at equilibrium the binding of the local anesthetic molecule to the membrane is itself influenced by previous activity (Strichartz 1973, Courtney 1975). For example, Courtney (1975) exposed nodes of Ranvier to a highly lipid soluble tertiary local anesthetic, GEA 968, long enough to allow equilibration. Because of its high lipid solubility this drug would not be denied access to the sodium channel: it could bypass the h and m gates, if closed, by dissolving in the membrane. The node was allowed to rest for a few hundred milliseconds and then it was subjected to frequent depolarizing test pulses at a rate of 2 sec^{-1}. When this was done, the sodium current obtained in each successive test was found to decrease progressively; after 25 pulses, it was only about one fourth of its initial value. Repetitive use of the sodium channels by short depolarizing voltage pulses thus progressively and dramatically alters the equilibrium between the local anesthetic and the receptor in the channel so that with the same prevailing axoplasmic concentration of drug, more channels are blocked than in the resting membrane. The equilibrium may be restored to its original value either by allowing the fiber to rest, when it recovers with a time constant of the order of 5–10 sec, or by applying hyperpolarizing prepulses.

These experiments thus show that the final equilibrium reached between the drug and its receptor, and the rate at which this equilibrium is approached, are both dependent on previous activity. The reason for the effect on rate is that the local anesthetic (in its cationic form) gains access to, and egress from, the receptor in the sodium channel only when the h and m gates are open. The reason for the effect on the final equilibrium is that the binding of the local anesthetic molecule to the channel is voltage-dependent (see below).

Voltage-Dependent Binding of Local Anesthetic to Sodium Channels

The suggestion that the local anesthetic/receptor interaction is voltage-dependent, first made by Strichartz (1973), is strongly supported by subsequent experiment. A key manifestation of the voltage-dependent nature of the interaction between the local anesthetic molecule and the sodium channel is the finding (Courtney 1975, Hille 1977) that conventional measurements of the steady state sodium inactivation show a large negative voltage shift of the inactivation curve (Figure 3). It seems that when the h gate is open, binding to the receptor is not very firm; but when the h gate is closed, the receptor is modified and the binding stronger (Hille 1977). This mutual interdependence of drug binding and sodium inactivation is equivalent to saying that anesthetic binding increases the probability of the sodium channel being in the inactivated state. And since the process of inactivation is voltage-dependent, so too must be anesthetic binding.

A physical interpretation of the involvement of inactivation in local anesthetic action is shown in Figure 4 (Hille 1977). According to this model, a local anesthetic molecule, having entered the channel past open h and m gates, binds to a site in the sodium channel, and in so doing causes some change, perhaps in the conformation of channel proteins, which enhances the probability of closure of the inactivation gate.

Hille's model makes clear why the action of quaternary analogs or of cationic forms of local anesthetic is use-dependent, and why that of uncharged molecules is not. For charged quaternary molecules gain access to the sodium channel only through a hydrophilic pathway, i.e. when the h and m gates are both open. During a single depolarization/repolarization cycle only a limited amount of drug gets to the receptor from the axoplasm. However, with successive impulses, more and more anesthetic gains access to the sodium channel to be bound to the receptor in it; and more and more of the anesthetic becomes trapped by being bound to the inactivated state for which it has a high affinity. The increased degree of block with repetitive activity is accompanied by a progressive shift of the inactivation curve to

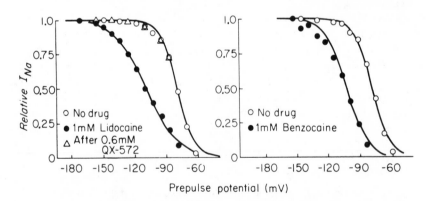

Figure 3   Shift of inactivation curves of voltage-clamped frog nodes of Ranvier by lidocaine (*left*) and by benzocaine (*right*). The holding potential was −90 mV, the final potential −15 mV, and the conditioning prepulses lasted 50 msec. Taken from Hille (1977).

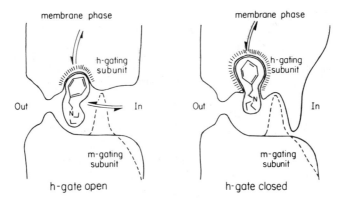

Figure 4   Diagram of local anesthetic molecule binding in the pore of a sodium channel so as to promote sodium inactivation. The molecule can reach the binding site from the intracellular solution if both the inactivation (h) and activation (m) gates are open. It can also reach the site from the membrane phase if it is sufficiently lipophilic. The binding site has an important hydrophobic component (shading) and closure of the inactivation gates enhances the hydrophobic interaction. Taken from Hille (1977).

the left. However, neutral forms of local anesthetics, since they are lipid-soluble, can bypass both the activation and inactivation gates and gain access to the receptor in the sodium channel from within the membrane itself. The availability of uncharged molecules through this route would be so large that a progressive buildup of local anesthetic action would not be expected; nor is it indeed found. The full shift in the inactivation curve should be found in the response to the first stimulus after exposure to such an uncharged molecule. Finally, one would expect with amine local anesthetics, which can exist in both charged and uncharged forms, a spectrum of use-dependent effects that are intermediate between the two extremes obtained with the quaternary and uncharged molecules, depending on the lipid solubility of the anesthetic, the pH of the bathing media, etc. These predictions seem to be well fulfilled (see Table 1 in Hille 1977).

When the h gates are destroyed, for example, by pronase, the frequency-dependence and use-dependence of local anesthetic action disappears (Almers & Cahalan 1977, Yeh & Wu 1978). Furthermore, the voltage-dependence of binding is lost. Both these effects are fully consistent with Hille's scheme.

At How Many Sites Do Local Anesthetics Act?

A major question examined by Hille (1977) was whether the amine local anesthetics and their neutral and quaternary analogs act by common or by different mechanisms. Largely on the basis of the fact that the uncharged benzocaine produces a similar shift in the sodium inactivation curve to that produced by lidocaine, he concluded that, as a first approximation, all inhibitory actions of these drugs are attributable to binding to a single receptor (Hille 1977, Schwarz, Palade & Hille 1977). Slow sodium inactivation (Khodorov et al 1976), cumulative frequency-dependent block (Strichartz 1973, Courtney 1975), and voltage-dependent block (Strichartz 1973) are all seen as manifestations of a single inhibitory mechanism. Although there is no reason to doubt the general validity of Hille's model as far as it applies to quaternary forms of local anesthetics (and protonated forms of the amines), the idea that there is a single mechanism of local anesthetic action has been questioned (Mrose & Ritchie 1978). If there is indeed only a single receptor with which both amine local anesthetics, such as lidocaine, and uncharged molecules, such as benzocaine, interact, mixtures of equiactive solutions of these two types of drug (of lidocaine and benzocaine, for example, chosen to produce the same fall in action potential size at equilibrium) should also produce (at equilibrium) the same response as either solution by itself. However, this is not found experimentally; a mixture of equiactive solutions of lidocaine and benzocaine produces a response that is very much greater than the response produced by either solution alone. Mrose & Ritchie (1978) interpreted this to mean that there must be more than one mechanism of action of local anesthetics. One mechanism would involve the local anesthetic receptor site described by Hille (1977); the other quite likely involves the dissolution of the uncharged local anesthetic molecule in the spaces between the sodium channels, as previously described, and compression of the channel, thus restricting passage of sodium ions through it. Certainly, the fact that benzocaine shifts the inactivation curve in the hyperpolarizing direction, which has been taken as the major evidence for a single site of action of local anesthetic in both muscle and nerve (Hille 1977, Schwarz et al 1977), does not seem to be strong evidence per se for a single site of action. A similar shift in the inactivation curve is obtained with a variety of drugs not commonly thought of as acting at the local anesthetic receptor in the sodium channel: benzyl alcohol on frog node of Ranvier (Chiu & Ritchie, unpublished), trichloroethylene on squid giant axons (Shrivastav et al 1976), tetrodotoxin in heart muscle (Reuter, unpublished), and aconitine, particularly in low concentrations, on frog node of Ranvier (Schmidt & Schmitt 1974).

Noise analysis and a study of membrane selectivity might provide two possible experimental approaches to the problem of the number of sites involved in the action of the local anesthetics. A drug that simply blocked the channel, such as tetrodotoxin or a quaternary local anesthetic, would be expected to reduce the nodal currents during a voltage-clamp depolarizing step, but not the conductance of single unblocked channels; indeed, Sigworth (1977) has shown this to be true for tetrodotoxin. On the other hand, if a restriction of channel diameter were also a factor in producing the block, the single channel conductance should be reduced. The cation selectivity of the channel (Hille 1972) might similarly be expected to remain unaffected in the presence of a drug that merely blocked the channel but to be altered in the presence of a drug that physically distorted the channel.

SAXITOXIN AND TETRODOTOXIN

Uptake of Toxins Determined by Bioassay

Although saxitoxin and tetrodotoxin have been long known, it was not until the mid-1960s that it became clear (Narahashi, Moore & Scott 1964) that these toxins owe their poisonous action solely to a specific blocking of the sodium currents that underly the action potential. This observation soon led to the idea that they physically block the sodium channels; and hence suggested a method for counting the latter by measuring the amount of toxin taken up by a nerve when conduction block had just occurred. Indeed, in a now classical study, Moore, Narahashi & Shaw (1967), using a bioassay method, estimated that the number of sodium channels in lobster nerve was less than $13/\mu m^2$. One problem with the bioassay method is that it is difficult by this means to make the complete binding curve necessary to characterize the toxin-receptor interaction. A substantial advance was therefore made when it was shown (Hafemann 1972) that tetrodotoxin can indeed be tritium-labelled by the Wilzbach method. The problem, however, with both tetrodotoxin and saxitoxin so labelled (Colquhoun et al 1972, Benzer & Raftery 1972, Henderson et al 1973, Almers & Levinson 1975) is that the toxin is exposed to a highly radioactive tritium gaseous atmosphere for several weeks (or for a lesser time, but in the presence of an electric discharge) so that considerable breakdown of the toxin occurs, with the production of radioactive impurities that are subsequently difficult to separate from the toxin itself. For example, in most of the published work on these toxins, the radiochemical purity is only about 30% (see Ritchie & Rogart 1977b). With saxitoxin, however, this problem of purity has been overcome to a large extent by the development of a new method of exchange tritium labelling (Ritchie, Rogart & Strichartz 1976). This yields a relatively stable radioactively labelled marker, which is much purer (radiochemical purity 60–85%) than the Wilzbach labelled toxin and of greater specific activity (about 300 times higher).

Uptake of Exchange-Labelled Saxitoxin by Excitable Tissues

Electrophysiological experiments with saxitoxin and tetrodotoxin indicate that they bind monomolecularly to some receptor in the sodium channel with an extremely small equilibrium dissociation constant (Hille 1968,

Cuervo & Adelman 1970, Colquhoun & Ritchie 1972, Schwarz, Ulbricht & Wagner 1973). This has been confirmed in the experiments on the uptake of exchange-labelled saxitoxin by excitable tissue. First, there is a hyperbolic saturable component of binding of the toxin to a pool of sites presumed to be the sodium channels; and secondly, the concentration of toxin to half-saturate the pool is of the same order of size as that obtained in the electrophysiological experiments. Ritchie & Rogart (1977b) summarized the results obtained with the exchange-labelled toxin on a variety of preparations. In all tissues the equilibrium dissociation constant is 1–10 nM. Furthermore, in all tissues, except myelinated nerve, the site density per unit area is also small. In garfish olfactory nerve, for example, the channel density is $35/\mu m^2$, which corresponds with a mean distance between the sodium channels (if arranged in a regular square array) of 0.2 μm. In squid giant axon and in rat and frog skeletal muscle, the density is ten times higher; nevertheless, in these tissues the channels are still extremely sparsely distributed over the membrane.

The Sodium Channel Density in Myelinated Nerve

In myelinated nerve a special situation would be expected. For the peak sodium current density in voltage-clamp experiments in the frog node of Ranvier is estimated to be more than 100 mA/cm^2 (Nonner, Rojas & Stämpfli 1975), whereas the corresponding value in lobster giant axon is 10 mA/cm^2 (Julian, Moore, & Goldman 1962) and that in squid giant axon is only about 1 mA/cm^2 (Keynes & Rojas 1974). Bearing in mind the greater ionic strength in these latter two marine preparations (so that fewer channels might be needed to pass a given current) one would expect to find, on the basis of the electrophysiological values, a sodium channel density in the frog node that was one to two orders of magnitude greater than in these other preparations (provided, of course, that the single channel conductance in the various species is the same). The first direct test of this prediction was disappointing, because frog sciatic nerve (which consists largely of myelinated fibers) has such large linear components that no saturable component of binding could be distinguished (Strong 1974, Ritchie & Rogart 1977a). For the same reason Colquhoun et al (1972) and Benzer & Raftery (1972) were unable to detect any saturable component of binding to the nodes of Ranvier in garfish trigeminal nerve. However, when special techniques are used (Ritchie 1978), a saturable component of labelled saxitoxin can be seen in frog sciatic nerve. However, the uptake (3.1 f-mole/mg wet) is small, with a considerable scatter in the experimental results. In mammalian sciatic nerve, however, the uptake is much larger, about 20 f-mole/mg wet (Ritchie & Rogart 1977a).

Ritchie & Rogart (1977a) compared the binding of labelled saxitoxin by intact rabbit sciatic nerve with that obtained with *homogenized* rabbit sciatic nerve, arguing that the homogenization procedure would break up the myelin sheath, as well as the nodal and internodal axonal membrane. The surprising finding was that, in spite of there being about 700 times as much internodal membrane as nodal membrane, the maximum saturable uptake in homogenized nerve is virtually the same as in intact nerve. What this seems to mean is that the bulk of the sodium channels are located at the node, with relatively few channels in the internodal region. Thus there seem to be, as Quick & Waxman (1977) have speculated on quite different grounds, structural differences between nodal and internodal axonal membranes, which may have special relevance in the context of demyelinating disease. The fact that nerve fibers that have been demyelinated by experiment or by disease fail to conduct impulses may simply result from an electrophysiological mismatch of the myelinated and demyelinated regions; the safety factor might be too low to allow the last intact node to excite the demyelinated region beyond. However, the binding experiments favor a different explanation, namely that when saltatory conduction fails following demyelination, continuous propagation does not take over because there are insufficient sodium channels in the internodal regions. On the basis that all the channels are located at the node, and using the histological results of Yates, Bouchard & Wherrett (1976) for rabbit sciatic nerve, Ritchie & Rogart (1977a) calculated that the observed maximal saturable uptake of toxin corresponds with a channel density of $12,000/\mu m^2$. A subsequent analysis using a larger number of experiments yielded a somewhat smaller number of $10,000/\mu m^2$ (Ritchie 1978). The sodium channel density in mammalian myelinated fibers is therefore many times greater than in other excitable membranes, which is consistent with the electrophysiological findings of a high current density at the node.

Is the Estimated Channel Density in Myelinated Nerve Too High?

Levinson & Ellory (1973), using X-ray inactivation methods, estimated that the molecular unit forming the sodium channel, if it is a globular protein, has a diameter of 80 Å. With $10,000$ channels/μm^2 this would mean that about 50% of the surface of the mammalian nodal membrane would be covered by sodium channel! Although the absence of space for them would not be serious for potassium channels, which seem to be absent in mammalian nodes (Chiu et al 1978), the high sodium channel density would leave little room for sodium pumps and for other components of the membrane. However, two points should be made in this connection. First, there must be some experimental uncertainty in the determination of the unit diameter (Dr. Clive Ellory, personal communication); if the diameter of the unit forming the sodium channel were only 50 Å rather than 80 Å, then the fraction of the nodal area occupied by the sodium channel would be only

20%. Second, the sodium pumps might well be located not in the nodal membrane but in the Schwann cell immediately surrounding it (see below).

The difference in the maximal saturable binding capacity of rabbit and frog sciatic nerve remains puzzling. One explanation would apparently be that there are fewer channels in the frog node than in the mammalian node. Electronmicroscopic evidence might apparently support this suggestion; for the nodal area in a 15 μm diameter mammalian fiber is about three times that of the corresponding amphibian fiber (Berthold 1968). However, preliminary experiments by Chiu et al (1978) find little or no difference in the maximum sodium conductance of amphibian and mammalian nodes.

The Effective Area of the Node of Ranvier

The conclusion that the sodium channels are concentrated at the nodes of Ranvier in myelinated nerves seems well founded. However, the calculation (Ritchie & Rogart 1977a) of the absolute magnitude of the density is less secure. For example, it assumes (as did Rushton 1951) that the nodal gap has the same width in all fibers. But Berthold (1968) has shown that in cat spinal roots the length of the nodal axon segment increases with increasing fiber size, being 0.9 μm and 1.4 μm for fibers of diameter 5 μm and 15 μm, respectively. Furthermore, Ritchie & Rogart (1977a) assumed, as had Stämpfli (1954), that a 15 μm diameter fiber would have a nodal area of 66 μm^2. Direct measurement gives a much smaller area. Berthold (1968) found a value of about 20 μm^2 for cat nerve; and using Robertson's (1959) electronmicroscopic data for frog node, he calculated that the corresponding area in a frog 15 μm diameter fiber is only about 8 μm^2. The reason for the larger nodal area suggested by Stämpfli (1954) is purely electrophysiological. For measurements of the passive capacitance of a node of Ranvier in a 15 μm diameter fiber yields a value of about 1.6 pF. If the nodal area were only 8 μm^2, this would give a specific capacitance of 20 μF/cm^2, which is felt to be impossibly high. However, if one assumes that the specific nodal capacitance is 3 μF/cm$_2$ (the lowest value in Stämpfli & Hille 1976), with a nodal capacity of 2 pF one obtains a nodal area of 66 μm^2, which is close to that commonly used (e.g. Stämpfli 1954, Ritchie & Rogart 1977a). One could have just as easily assumed, perhaps more plausibly, a more conventional value for the specific capacitance, namely 1 μF/cm^2, in which case the effective nodal area would have been increased to 200 μm^2 and the calculated channel density at the node would decrease. This would allow a more reasonable distribution of the channels over the nodal surface, while still maintaining a value for their density much higher than in other tissues. It would, of course, also reduce the current density in voltage-clamp experiments because they too depend on the assumed value for the nodal area; and it would reduce correspondingly the channel density estimated on the basis of noise measurements. For example, the value of $1000-2000/\mu m^2$ obtained by Conti et al (1976) and by Sigworth (1977) for the frog node may be too high by a factor of 3.

The high value of the density of channels in the node is thus in part a direct result of assuming a value for the membrane capacity per unit area of plasma membrane of $3-7$ $\mu F/cm^2$, i.e. a value that is much higher for the nodal membrane than for other tissues. It is interesting, that although the effective membrane capacity in skeletal muscle is about 6 $\mu F/cm^2$ (referred to unit area of cell membrane), and the corresponding value for many molluscan nerve cells is $5-60$ $\mu F/cm^2$, in both these cases, when the complex morphology is allowed for, the specific capacitance is near to 1 $\mu F/cm^2$ (Hodgkin & Nakajima 1972, Gorman & Mirolli 1972). A similar situation may well exist at the node of Ranvier. As Landon & Hall (1972) point out, the "ideal node of the electrophysiologists bears little resemblance to the morphology of the nodes of the large myelinated peripheral nerve fibers usually employed in their experiments." A nodal capacity of $0.6-1.6$ pF/node (Stämpfli & Hille 1976) would, on the basis of a membrane capacitance of 1 $\mu F/cm^2$, imply an effective area of nodal membrane of $60-160$ μm^2. This value is much larger than that observed directly for frog nodes, which varies from 4 μm^2 in light microscope studies (Hess & Young 1952) to 8 μm^2 in electronmicroscopic studies (Robertson 1959). But so also are the values presently accepted by electrophysiologists, which range from 22 μm^2 (Stämpfli & Hille 1976) to 60 μm^2 (Stämpfli 1954). Thermal studies in myelinated nerve similarly suggest that heat production (and the electrical activity responsible for it), must be occurring over an area of membrane much greater than that suggested even by the capacitance considerations discussed above (Abbott, Hill & Howarth 1968). The suggestion that the effective nodal area is presently grossly underestimated seems further supported by the fact that the resting membrane conductance calculated on the basis of a nodal area of 22 μm^2 for a 15 μm diameter fiber is a good deal higher in this tissue than in all other tissues studied.

The exact location of the putative extra area of membrane involved is unclear. In the absence of any evidence that the nodal membrane is folded (thus increasing its effective area) one might suppose perhaps that some of the axonal membrane in the paranodal or juxtanodal region is involved in the electrical activity of the nerve. An alternative explanation depends on the complex morphology of the node. As Landon & Williams (1963) point out, "although one or two mitochondria are usually visible in the nodal axoplasm they are not impressive as an energy source for so active a region." Furthermore, numerous finger-like filaments of the Schwann cell reach down into the node substance, with the Schwann cell and nodal membranes approaching "to within $2-5$ nm of one another, a considerably smaller gap than is normally seen between two apposed cell membranes" (Landon & Hall 1972). Perhaps, as they suggest, not only is the Schwann cell involved in supplying the energy for the sodium pump at the node, but it might also be involved in the ionic fluxes. Both factors would seemingly imply an effective continuity of Schwann cell and axonal cytoplasm at these regions; and it would be supposed that the nodal collar and the two lateral node-gap walls, which are both part of the Schwann cell, would contribute to the electrical response. The larger surface area would then be adequate to account for the large nodal current, heat production, and binding capacity for saxitoxin. Tight junctions, which might provide such a continuity, have not, however, been described.

In summary, therefore, both the electrophysiological and chemical experiments suggest a high sodium channel density at the nodes of Ranvier. The estimated density for mammalian fibers, of 10,000 channels per μm^2, is essentially founded on a nodal area of 66 μm^2 for a 15 μm diameter fiber. If the effective nodal area, as argued above, is 2–3 times greater than this, clearly the channel density would fall to a value of 3000–5000 channels per μm^2, which would leave much more room for other membrane components to be present.

MOLECULAR STRUCTURE OF THE SODIUM CHANNEL

From considerations of the molecular structure of saxitoxin and tetrodotoxin, Hille (1975) proposed that the sodium channel opens into the extracellular space through an antechamber of cross section 9 Å by 10 Å. Furthermore, these considerations, together with an analysis of the permeability of the voltage-clamped frog node of Ranvier to a variety of organic and inorganic metal cations (Hille 1971, 1972), suggest that deeper in the membrane this antechamber narrows to a pore 3 Å by 5 Å, which is lined by six oxygen atoms and which forms the narrow ionic selectivity filter. These features are illustrated in Figure 5. The two toxins thus block conduction because, as Kao & Nishiyama (1965) first suggested, a guanidinium group in both toxin molecules enters the sodium channel and becomes stuck there because the rest of the molecule is too wide to pass. Within the pore is an ionized carboxylic group that forms a metal cation binding site to which tetrodotoxin and saxitoxin bind as well as a variety of trivalent, divalent, and monovalent cations, including hydrogen ions (Hille 1971, 1972, Henderson et al 1973, 1974, Woodhull 1973). This site in myelinated fibers, postulated on the basis of electrophysiological experiments by Hille (1975), is presumably the same site studied in the binding experiments in a variety of other intact and solubilized tissues (Colquhoun et al 1972, Henderson et al 1973, Reed & Raftery 1976, Catterall 1975a, b).

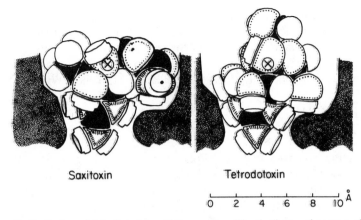

Saxitoxin   Tetrodotoxin

0   2   4   6   8   10 Å

Figure 5   Saxitoxin and tetrodotoxin on their receptors. The shading on the atoms of toxin represent: carbon, black; hydrogen, white; oxygens, dotted margins; nitrogen, dashed margins. The stippled areas represent the receptor in the sagittal section with the narrow selectivity filter below. Most of the receptor is hydrogen bond accepting, and there is a negative charge associated with the selectivity filter. A circled X has been drawn in the same position with respect to the receptor in two cases. The X falls on a hydroxyl group attached to an unusually electropositive carbon. Taken from Hille (1975).

A variety of other drugs affect the sodium channels in myelinated and other nerves (see below). These drugs act in quite small concentrations and presumably act by binding to some site in the membrane rather than by acting in the bulk phase (as the general anesthetics are presumed to do). Although the electrophysiological consequences of exposure to these other drugs are known, there has been as yet relatively little speculation on the nature of their molecular interaction with the membrane. This is why in this review the emphasis is on the local anesthetics and on the two toxins, saxitoxin and tetrodotoxin; only with these two groups of drugs have experiment and speculation advanced sufficiently to permit suggestion on the molecular structure of the membrane. Further study with these other agents, however, it is hoped, will tell us more about the membrane.

LOCATION OF THE *h* GATES

The fact that quaternary forms of local anesthetics are effective only when applied internally, and furthermore seem to gain access to the receptor in the sodium channel only when the electrophysiological conditions are such as to open the *h* gates, provides good evidence that the *h* gate is located on the inside of the nerve membrane. Conclusive proof that this is the case, at least in squid membranes, comes from the finding (Armstrong, Bezanilla & Rojas 1973) that internal, but not external, perfusion of squid axons with the enzyme pronase selectively destroys the inactivation process of the sodium conductance. The turn-off kinetics and the voltage-dependence of the activation m gates are unaffected. As might be expected, tetrodotoxin, which acts from the outside surface, does not protect the inactivation gates, nor does maintained inactivation of the sodium channels during exposure to the enzymes. However, the frequency- and use-dependence of local anesthetic action disappears (Almers & Cahalan 1977, Yeh & Wu 1978).

OTHER DRUGS ACTING ON THE SODIUM CHANNEL

Veratridine, a steroidal alkaloid isolated from the plants of the *Lilliaceae* family (see Ulbricht 1974), batrachotoxin, a steroidal alkaloid with some resemblance to the veratrine alkaloids, isolated from the skin secretions of a small Colombian frog, *Phyllobates aurotaenia* (Khodorov 1978), aconitine, an alkaloid from plants of the *Ranunculaceae* family (Schmidt & Schmitt 1974), and a polypeptide neurotoxin from the venom of the scorpion *Leiurus quinquestriatus* (Koppenhöfer & Schmidt 1968a, b, Okamoto, Takahashi & Yamashita 1977) all slow markedly the process of inactivation. Iodate also inhibits inactivation (Stämpfli 1974). With some of these drugs (e.g. aconitine and scorpion venom) inactivation in addition to being slowed may be incomplete, as if the h gates of some channels had been jammed in the open position. The proteolytic enzyme pronase seems to destroy the inactivation gates, leaving the activation gate and potassium channel unaffected (Armstrong et al 1973). A toxin from the sea anemone *Anemonia sulcata* also seems to prevent the closing of the inactivation gates of the sodium channels (see Abita et al 1977). Similarly, the insecticide DDT prolongs the transient increase of sodium permeability elicited by depolarization of the node, i.e. it slows sodium inactivation (Hille 1968). Other agents, for example, the venom of another scorpion, *Centruroides sculpturatus,* shift the membrane potential dependence of sodium activation by some 40–50 mV in the hyperpolarizing direction (Cahalan 1975). The result is that the m gates, which are normally shut at the resting potential, remain open for hundreds of msec, allowing channels to reopen as recovery from inactivation occurs. Batrachotoxin opens resting sodium channels in squid and frog node, possibly by similarly opening the m gates (Albuquerque 1972, Stämpfli & Hille 1976, Khodorov 1978). The hallucinogenic drug Δ^9tetrahydrocannabinol seems to slow the opening of the m gate (Strichartz, Chiu & Ritchie 1978). The antipsychotic drug haloperidol reduces the sodium currents in myelinated nerve with little effect on sodium activation or, except at high concentrations, on inactivation (Pencek, Schauf & Davis 1978); potassium currents are unaffected. The action of haloperidol, therefore, resembles that of the amine local anesthetics, as might be expected from its structure.

The effects are often not restricted to one parameter. For example, the alkaloid aconitine, in addition to affecting the inactivation process, shifts the m parameter in the hyperpolarizing direction and may double the time constant of activation (Schmidt & Schmitt 1974); it also alters the selectivity of the channel for cations (Mozhayeva et al 1977, Campbell 1978). Batrachotoxin, like aconitine, also alters the selectivity of sodium channels, as if the 3 by 5 Å pore in the sodium channel had been widened (Khodorov 1978). Veratrine alkaloids slow the kinetics of sodium activation by three orders of magnitude. Furthermore, besides acting at the sodium channel, many of these agents, particularly in higher concentrations, also affect the potassium currents. *Leiurus sculpturatus* toxin, for example, markedly reduces the potassium permeability and slows activation of the potassium channel. It is not clear whether these multiple effects indicate the presence of multiple receptor sites, or whether it means that there are multiple responses to occupation of a single site. However, to some extent there is a clear specificity of action within the sodium channel. Thus, *Centruroides* scorpion venom specifically affects the activation process (Cahalan 1975), whereas the *Leiurus* scorpion venom specifically affects the inactivation process (Koppenhöfer & Schmidt 1968a, b). This would seem to argue strongly that there are two independent molecular structures governing the two separate processes.

DO THE VARIOUS AGENTS ACT AT INDEPENDENT SITES IN THE SODIUM CHANNELS?

The metal cation binding site at which the two toxins, saxitoxin and tetrodotoxin, act is clearly different from the site or sites at which the local anesthetics act. First, cationic forms of local anesthetics act only from the inner side of the channel, and are ineffective when applied externally (Narahashi & Frazier 1971, Strichartz 1973). On the other hand, saxitoxin and tetrodotoxin are effective only when applied externally (Koppenhöfer & Vogel 1969). Second, local anesthetics do not interfere with the binding of labelled tetrodotoxin or saxitoxin, nor do batrachotoxin or veratrine (Colquhoun et al 1972, Henderson et al 1973). Third, even in concentration sufficient to modify drastically the inactivation of the sodium channel, veratridine does not affect the kinetics of tetrodotoxin action (Ulbricht 1974).

In studies of the uptake of ^{22}Na by neuroblastoma cells, the three drugs batrachotoxin, veratridine, and aconitine, seem to interact competitively, which suggests that they can all act at the same single site (Catterall 1975a, b), whereas venom from the scorpion *Leiurus sculpturatus* acts at a different site. These two sites are coupled cooperatively in such a way as to suggest that they open the sodium channel by an allosteric mechanism. Occupation of neither of these sites, however, alters the concentration of tetrodotoxin that is required to inhibit uptake; again it seems that tetrodotoxin must act at yet another site.

Thus, at least four independent sites within or near the sodium channel can be identified chemically: one at which saxitoxin and tetrodotoxin act; another at which the local anesthetics act; and a third and fourth at which the venom from the scorpions *Leiurus* and *Centruroides* act (*h* and *m* gates respectively). In addition, there is a membrane locus of action of local anesthetics. Furthermore, the alkaloids veratridine, aconitine, and batrachotoxin, have a complex set of actions which may indicate that, in addition to acting at these sites above, they may also act at some as yet unidentified sites.

SUMMARY

Figure 6 summarizes the present state of our knowledge on the sodium channel in myelinated nerve fibers. Two sites have been discussed in detail: a metal cation binding site accessible by tetrodotoxin and saxitoxin from the outside surface only; and a second site accessible from the inside surface with which local anesthetics combine. Hydrogen ions gain access to this region of the sodium channel (and hence determine the relative local concentration of protonated drug) more readily from the extracellular fluid than from the axoplasm (Schwarz et al 1977). In addition, a variety of other sites have been mentioned, binding of drugs to which alters selectively the

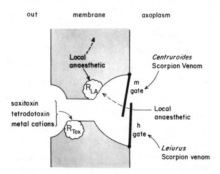

Figure 6   The sodium channel in myelinated axons showing: the receptor site with which saxitoxin, tetrodotoxin, and metal cations interact; the receptor in the channel with which cationic local anesthetics interact; the membrane action of local anesthetics; and the *h* and *m* gates at which various drugs act (*h* gates: aconitine, batrachotoxin, iodate, *Leiurus* scorpion venom, sea-anemone toxin, veratridine; *m* gates: aconitine, batrachotoxin, *Centruroides* scorpion venom, Δ⁹tetrahydrocannabinol).

kinetics of opening and closing of the h and m gates. In myelinated nerve fibers these channels are packed tightly together on the nodal membrane. The highest estimate for the sodium channel density in the mammalian node is $10,000\,\mu m^2$. A re-evaluation of the effective nodal area, however, might reduce this value to $3000-5000/\mu m^2$. This would still leave the nodal membrane rather crowded with sodium channels. Furthermore, the channel density would still be greater than the density of particles, sometimes believed to be sodium channels seen in freeze fracture studies (Rosenbluth 1976). One possibility for resolving this problem is that the units detected by X-ray inactivation (Levinson & Ellory 1973), and those seen in freeze-fracture studies (Rosenbluth 1976) represent not single sodium channels but groups of three. Catterall & Morrow (1978) in a comparison of the binding of saxitoxin and *Leiurus sculpturatus* scorpion toxin venom have concluded that there are three saxitoxin binding sites for each scorpion toxin binding site. On this basis, three saxitoxin molecules might act to block independently each of the three openings of the channels; while the conformational change produced by the scorpion venom molecule would affect the inactivation process of all three channels.

ACKNOWLEDGMENTS

This work was supported in part by grants NS-08304 and NS-12327 from the USPHS and by a grant RG1162 from the U.S. National Multiple Sclerosis Society.

Literature Cited

Abbott, B. C., Hill, A. V., Howarth, J. V. 1968. The positive and negative heat production associated with a single impulse. *Proc. R. Soc. London Ser. B.* 148:149–87

Abita, J. P., Chicheportiche, R., Schweitz, H., Lazdunski, M. 1977. Effects of neurotoxins (veratridine, sea anemone toxin, tetrodotoxin) in transmitter accumulation and release by nerve terminals *in vitro. Biochemistry* 16:1838–44

Albuquerque, E. X. 1972. The mode of action of batrachotoxin. *Fed. Proc.* 31: 1133–38

Almers, W., Cahalan, M. D. 1977. Interaction between a local anesthetic, the sodium channel gates and tetrodotoxin. *Biophys. J.* 17:205a

Almers, W., Levinson, S. R. 1975. Tetrodotoxin binding normal and depolarized frog muscle and the conductance of a single sodium channel. *J. Physiol. London* 247:483–509

Armstrong, C. M., Bezanilla, F., Rojas, E. 1973. Destruction of sodium conductance inactivation in squid axons perfused with pronase. *J. Gen. Physiol.* 62:375–91

Benzer, T. I., Raftery, M. A. 1972. Partial characterization of a tetrodotoxin-binding component from nerve membrane. *Proc. Natl. Acad. Sci. USA* 69:3634–37

Berthold, C.-H. 1968. Ultrastructure of the node-paranode region of mature feline ventral lumbar spinal-root fibres. *Acta Soc. Med. Ups.* 73: Suppl. 9, pp. 37–70

Cahalan, M. 1975. Modification of sodium channel gating by scorpion venom. *J. Physiol. London* 244:511–34

Campbell, D. T. 1978. Aconitine alters selectivity and kinetics of frog muscle sodium channels. *Biophys. J.* 21:42a

Catterall, W. A. 1975a. Cooperative action of action potential Na^+ ionophore by neurotoxins. *Proc. Natl. Acad. Sci. USA* 72:1782–86

Catterall, W. A. 1975b. Activation of the action potential Na⁺ ionophore of culture neuroblastoma cells by veratridine and batrachotoxin. *J. Biol. Chem.* 250: 4053–59

Catterall, W. A., Morrow, C. S. 1978. Binding of saxitoxin to electrically excitable neuroblastoma cells. *Proc. Natl. Acad. Sci USA* 75:218–22

Chiu, S. Y., Ritchie, J. M., Rogart, R. B., Stagg, D. 1978. A quantitative description of membrane currents in mammalian myelinated nerve and its application to excitation. *J. Physiol. London* In press

Colquhoun, D., Henderson, R., Ritchie, J. M. 1972. The binding of labelled tetrodotoxin to non-myelinated nerve fibres. *J. Physiol. London* 277:95–126

Colquhoun, D., Ritchie, J. M. 1972. The interaction of equilibrium between tetrodotoxin and mammalian non-myelinated nerve fibres. *J. Physiol. London* 221:533–53

Conti, F., Hille, B., Neumcke, B., Nonner, W., Stämpfli, R. 1976. Conductance of the sodium channel in myelinated nerve fibres with modified sodium inactivation. *J. Physiol. London* 262:729–42

Courtney, K. R. 1975. Mechanism of frequency-dependent inhibition of sodium currents in frog myelinated nerve by the lidocaine derivative GEA 968. *J. Pharmacol. Exp. Ther.* 195:225–36

Cuervo, L. A., Adelman, W. J. 1970. Equilibrium and kinetic properties of the interaction between tetrodotoxin and the excitable membrane of the squid giant axon. *J. Gen. Physiol.* 55:309–335

Frankenhaeuser, B., Huxley, A. F. 1964. The action potential in the myelinated nerve fibre of *Xenopus laevis* as computed on the basis of voltage clamp data. *J. Physiol. London* 171:302–15

Gorman, A. L. F., Mirolli, M. 1972. The passive electrical properties of the membrane of a molluscan neurone. *J. Physiol. London* 227:35–49

Hafemann, D. R. 1972. Binding of radioactive tetrodotoxin to nerve membrane preparations. *Biochim. Biophys. Acta.* 266:548–56

Henderson, R., Ritchie, J. M., Strichartz, G. 1973. The binding of labelled saxitoxin to the sodium channels in nerve membranes. *J. Physiol. London* 235:783–804

Henderson, R., Ritchie, J. M., Strichartz, G. 1974. Evidence that tetrodotoxin and saxitoxin act at a metal cation binding site in the sodium channels of nerve membrane. *Proc. Natl. Acad. Sci. USA* 71:3936–40

Hess, A., Young, J. Z. 1952. The nodes of Ranvier. *Proc. R. Soc. London Ser. B* 140:301–20

Hille, B. 1968. Pharmacological modifications of the sodium channels of frog nerve. *J. Gen. Physiol.* 51:199–219

Hille, B. 1971. The permeability of the sodium channel to organic cations in myelinated nerve. *J. Gen. Physiol.* 58:599–619

Hille, B. 1972. The permeability of the sodium channel to metal cations in myelinated nerve. *J. Gen. Physiol.* 59:637–58

Hille, B. 1975. The receptor for tetrodotoxin and saxitoxin: a structural hypothesis. *Biophys. J.* 15:615–19

Hille, B. 1977. Local anesthetics: Hydrophilic and hydrophobic pathways for the drug receptor interaction. *J. Gen. Physiol.* 69:497–515

Hodgkin, A. L., Huxley, A. F. 1952. A quantitative description of membrane current and its application to conduction and excitation in nerve. *J. Physiol. London* 117:500–44

Hodgkin, A. L., Nakajima, S. 1972. Analysis of the membrane capacity in frog muscle. *J. Physiol. London* 221:121–36

Horakova, M., Nonner, W., Stämpfli, R. 1968. Action potentials and voltage clamp currents of single rat Ranvier nodes. *Proc. Int. Union Physiol. Sci.* 7:198

Julian, F. J., Moore, J. W., Goldman, D. E. 1962. Current-voltage relations in the lobster giant axon membrane under voltage clamp conditions. *J. Gen. Physiol.* 45:1217–38

Kao, C. Y., Nishiyama, A. 1965. Actions of saxitoxin in peripheral neuromuscular systems. *J. Physiol. London* 180:50–66

Kendig, J. J., Cohen, E. N. 1977. Pressure antagonism to nerve conduction block by anesthetic agents. *Anesthesiology* 47:6–10

Keynes, R. D., Rojas, E. 1974. Kinetic and steady-state properties of the charged system controlling sodium conductance in the squid giant axon. *J. Physiol. London* 239:393–434

Khodorov, B. I. 1978. Chemicals as tools to study nerve fiber sodium channels; effects of batrachotoxin and some local anesthetics. In *Membrane Transport Processes,* ed. D. C. Tosteson, Y. A. Ovchirrikov, R. Latorre, 2:153–74. New York: Raven

Khodorov, B., Shishkova, L., Peganov, E., Revenko, S. 1976. Inhibition of sodium currents in Ranvier node treated with local anesthetics. Role of slow sodium inactivation. *Biochim. Biophys. Acta* 433:409–35

Koppenhöfer, E., Schmidt, H. 1968a. Die Wirkung von Skorpiongift auf die Ionenströme des Ranvierschen Schnürrings. I. Die Permeabilitäten P_{Na} und P_K. *Pfluegers Arch.* 303:133–49

Koppenhöfer, E., Schmidt, H. 1968b. Die Wirkung von Skorpiongift auf die Ionenströme des Ranvierschen Schnürrings. II. Unvollständige Natrium-Inaktivierung. *Pfluegers Arch.* 303: 150–61

Koppenhöfer, E., Vogel, W. 1969. Effects of tetrodotoxin and tetraelthyl-ammonium chloride on the inside of the nodal membrane of *Xenopus laevis. Pfluegers Arch.* 313:361–80

Landon, D. N., Hall, S. 1972. The myelinated nerve fibre. In *The Peripheral Nerve*, ed. D. N. Landon, pp. 1–103. New York: Wiley

Landon, D. N., Williams, P. L. 1963. Ultrastructure of the node of Ranvier. *Nature* 199:575–77

Levinson, S. R., Ellory, J. C. 1973. Molecular size of the tetrodotoxin binding site estimated by irradiation inactivation. *Nature New Biol.* 245:122–23

Moore, J. W., Narahashi, T., Shaw, T. I. 1967. An upper limit to the number of sodium channels in nerve membrane? *J. Physiol. London* 188:99–105

Mozhayeva, G. N., Naumov, A. P., Negulyaev, Y. A., Nosyreva, E. D. 1977. The permeability of aconitine-modified sodium to univalent cations in myelinated nerve. *Biochim. Biophys. Acta* 466:461–73

Mrose, H. E., Ritchie, J. M. 1978. Local anesthetics: Do benzocaine and lidocaine act at the same single site. *J. Gen. Physiol.* 71:223–25

Narahashi, T., Frazier, D. T. 1971. Site of action and active form of local anesthetics. *Neurosci. Res.* 4:65–99

Narahashi, T., Moore, J. W., Scott, W. R. 1964. Tetrodotoxin blockage of sodium conductance increase in lobster giant axons. *J. Gen. Physiol.* 47:965–74

Nonner, W., Rojas, E., Stämpfli, R. 1975. Displacement currents in the node of Ranvier. *Pfluegers Arch.* 354:1–18

Nonner, W., Stämpfli, R. 1969. A new voltage clamp method. In *Laboratory Techniques in Membrane Biophysics*, ed. H. Passow, R. Stämpfli. pp. 171–75. Berlin: Springer

Okamoto, H., Takahashi, K., Yamashita, N. 1977. One-to-one binding of a purified scorpion toxin to Na channels. *Nature* 266:465–68

Pencek, T. L., Schauf, C. L., Davis, F. A. 1978. The effect of haloperidol on the ionic currents in the voltage-clamped node of Ranvier. *J. Pharmacol. Exp. Ther.* 204:400–5

Quick, D. C., Waxman, S. G. 1977. Specific staining of the axon membrane at nodes of Ranvier with ferric ion and ferrocyanide. *J. Neurol. Sci.* 31:1–11

Reed, J. K., Raftery, M. A. 1976. Properties of the tetrodotoxin binding component in plasma membrane isolated from *Electrophorus electricus. Biochemistry* 15:944–53

Ritchie, J. M. 1975. Mechanism of action of local anaesthetic agents and biotoxins. *Br. J. Anesth.* 47:191–98

Ritchie, J. M. 1978. Sodium channel as a drug receptor. In *Cell Membrane Receptors for Drugs and Hormones. A Multidisciplinary Approach*, ed. R. W. Straub, L. Bolis, pp, 227–42. New York: Raven

Ritchie, J. M., Greengard, P. 1961. On the active structure of local anesthetics. *J. Pharmacol. Exp. Ther.* 133:241–45

Ritchie, J. M., Rogart, R. B. 1977a. The density of sodium channels in mammalian myelinated nerve fibers and the nature of the axonal membrane under the myelin sheath. *Proc. Natl. Acad. Sci. USA* 74:211–15

Ritchie, J. M., Rogart, R. B. 1977b. The binding of saxitoxin and tetrodotoxin to excitable tissue. *Rev. Physiol. Biochem. Pharmacol.* 79:1–50

Ritchie, J. M., Rogart, R. B., Strichartz, G. 1976. A new method for labelling saxitoxin and its binding to non-myelinated fibres of the rabbit vagus, lobster walking leg, and garfish olfactory nerve. *J. Physiol. London* 261:477–94

Robertson, J. D. 1959. Preliminary observations on the ultrastructure of nodes of Ranvier. *Z. Zellforsch. Mikrosk. Anat.* 50:553–60

Rosenbluth, J. 1976. Intramembranes particle distribution at the node of Ranvier and adjacent axolemma in myelinated axons of the frog brain. *J. Neurocytol.* 5:731–45

Rushton, W. A. H. 1951. A theory of the effects of fibre size in medullated nerve. *J. Physiol. London* 115:101–22

Schmidt, H., Schmitt, O. 1974. Effect of aconitine on sodium permeability of the node of Ranvier. *Pfluegers Arch.* 349:133–48

Schwarz, J. R., Ulbricht, W., Wagner, H. H. 1973. The rate of action of tetradotoxin on myelinated nerve fibres of *Xenopus laevis* and *Rana esculenta. J. Physiol. London* 233:167–94

Schwarz, W., Palade, P. T., Hille, B. 1977. Effect of pH on use-dependent block of sodium channels in frog muscle. *Biophys. J.* 20:343–68

Seeman, P. 1977. Anesthetics and pressure reversal of anesthesia. *Anesthesiology* 47:1–3

Shanes, A. 1958. Electrochemical aspects of physiological and pharmacological action in excitable cells. *Pharmacol. Rev.* 10:59–271

Shrivastav, B. B., Narahashi, T., Kitz, R. J., Roberts, J. D. 1976. Mode of action of trichloroethylene on squid axon membranes. *J. Pharmacol. Exp. Ther.* 199:179–88

Sigworth, F. J. 1977. Sodium channels in nerve apparently have two conductance states. *Nature* 270:265–67

Stämpfli, R., 1954. Saltatory conduction in nerve. *Physiol. Rev.* 34:101–12

Stämpfli, R. 1974. Intra-axonal iodate inhibits sodium inactivation. *Experientia* 30:505–8

Stämpfli, R., Hille, B. 1976. Electrophysiology of the peripheral myelinated nerve. In *Frog Neurobiology*, ed. R. Llinas, W. Precht, pp. 1–32. Berlin: Springer

Strichartz, G. R. 1973. The inhibition of sodium currents in myelinated nerve by quaternary derivatives of lidocaine. *J. Gen. Physiol.* 62:37–57

Strichartz, G. R., Chiu, S. Y., Ritchie, J. M. 1978. The effect of Δ⁹Tetrahydrocannabinol on the activation of the sodium conductance in node of Ranvier. *J. Pharmacol. Exp. Ther.* In press

Strong, P. W. 1974. *Chemical and biological studies with tetrodotoxin, tritiated tetrodotoxin and radiolabeled succinyl tetrodotoxin.* PhD thesis. Univ. Oregon. 148 pp.

Trevan, J. W., Boock, E. 1927. The relation of hydrogen ion concentration to the action of the local anaesthetics. *Br. J. Exp. Pathol.* 8:307–15

Ulbricht, W. 1974. Drugs to explore the ionic channels in the axon membrane. In *Biochemistry of Sensory Functions*, ed. L. Jaenicke, pp. 351–66. Berlin: Springer

Woodhull, A. 1973. Ionic blockage of sodium channels in nerve. *J. Gen. Physiol.* 61:687–708

Yates, A. J., Bouchard, J.-P., Wherrett, J. R. 1976. Relation of axon membrane to myelin membrane in sciatic nerve during development: comparison of morphological and chemical parameters. *Brain Res.* 104:261–71

Yeh, J. Z., Wu, C. H. 1978. Sodium inactivation modulates local anesthetic block of sodium channels in squid axons. *Biophys. J.* 21:42a

Ann. Rev. Neurosci. 1979. 2:363–97
Copyright © 1979 by Annual Reviews Inc. All rights reserved

ION CHANNELS
IN DEVELOPMENT

Nicholas C. Spitzer

Biology Department, University of California, San Diego, La Jolla, CA 92093

INTRODUCTION

Two of the most prominent features of many excitable cells are their abilities to respond to voltages applied across their membranes, or to chemicals such as neurotransmitters applied to their surface. The response usually consists of a change in membrane permeability, due to the opening or closing of ion channels, which promotes or prevents the flow of ionic currents. These channels are generally quite selective for ions of a particular charge and size. Nerve and muscle cells are attractive for the study of differentiation, since a detailed knowledge about the mature cells is already available and they have highly specialized features, so that one may investigate the time of appearance of specific properties. Further, a great deal can be learned about the qualitative and quantitative properties of the membranes of these cells by impalements with intracellular microelectrodes. Knowledge of the developmental timetable for such properties in these cells may be a useful step toward discovery of the general rules by which their development occurs.

During the last decade it has been shown that the ionic dependence of the responses to electrical stimuli, when they first appear at an early stage of development, is often different from that seen in mature cells. There is a subsequent change in the ionic dependence of the responses during the further course of development, and by inference, a change in the ion channels involved. In contrast, the ionic dependence of the responses to chemical stimuli is generally constant during the same period. These observations raise several questions. What changes in responses occur during development? Are there changes common to many developing cells, so that some general rules emerge? By what mechanisms do the changes occur? What is

363

the role of these changing ionic responses in the growth and maturation of the cells?

This review discusses voltage sensitive and chemosensitive ion channels of embryonic cells. Developmental changes in resting potentials are not discussed; it now appears that a number of the changes reported in the literature were the result of damage to cells by microelectrode impalement at early stages (see Jaffe & Robinson 1978), and that the damage may be reduced as the cells grow older and larger. Measurements of the resting potential need to be supported by information about the specific ionic permeabilities responsible, in order to eliminate the role of a nonspecific leak around the electrode (e.g. Kidokoro 1975a). There do not appear to be any studies of developmental changes in the ionic dependence of generator potentials in neurons. Changes in the ion channels of low resistance intercellular junctions mediating electrical coupling have been covered elsewhere (Furshpan & Potter 1968, DeHaan & Sachs 1972). Studies of the steady ionic currents involved in development have been recently reviewed (Jaffe & Nuccitelli 1977).

The Problems

Several technical problems are common to the studies of ion channels in developing systems. At the embryonic stages of interest, the cells to be examined are generally small and undifferentiated anatomically, and are thus difficult to locate and identify unequivocally. Furthermore they are often fragile, and easily and rapidly damaged by impalement with electrodes, so that reliable data can be obtained only for periods of seconds or minutes. Low resting potentials cause inactivation of voltage dependent channels that may then be erroneously interpreted to be absent. One solution of these problems has been the selection of experimental preparations with relatively large and accessible cells, such as the tunicate embryo, with a mosaic development that permits identification of cell lineages. Another solution has been to prepare explant or dissociated primary cell cultures which can facilitate location, identification, and experimental manipulation of the cells. Cultured cells often develop in vitro as they do in vivo. Clonal cell lines have been derived from spontaneously occurring tumors, from deliberately transformed cells, or from fusion of a clonal line with other cells. Cultures of clonal lines yield large homogeneous populations of cells, but they may have chromosomal instabilities or demonstrate abnormal development (Fischbach & Nelson 1977).

The Approaches

There have been four major approaches to the problem of assaying qualitatively and quantitatively the kind of ion channels present during development. The first has been to voltage clamp the cells studied, and examine the currents that flow as a result of the activation of the ion channels. Ionic substitutions and specific pharmacological blocking agents allow analysis of the ionic composition of the currents. This technique has been very successful with large cells. The second and somewhat simpler approach has been to record the change in membrane voltage that results from the ionic currents. This is less direct than the first method since the voltage measured is not a direct reflection of the current unless the membrane resistance is constant. However this method is valuable in the study of small cells. The third approach has been to employ agents that hold open specific ion channels, such as veratridine for the voltage dependent Na^+ channel. The consequences of the ensuing ion fluxes can be detected by measuring the accumulation of radioisotopes, sampling the steady membrane potential, or monitoring some other parameter. This technique is particularly useful when the sensitivity of the channel is constant during development. The fourth approach has been to measure the specific binding of a probe with a high affinity for the ion channel or receptor in mature cells. Toxins and antibodies, tagged with fluorescent compounds, heavy metals, or radioisotopes have been used successfully. This method is attractive because it does not require impalement with a microelectrode, and may yield information about the number and distribution of ion channels. However, binding cannot be equated with the ability of the channels to permit ion fluxes. Furthermore, changes in the binding site as a function of development may occur independently of changes in the permeability of the channel. The general utility of this approach is impaired by the fact that there are a number of channels for which no adequate probe has been discovered (e.g. the voltage dependent calcium channel).

CHANGES IN ION CHANNELS: ELECTRICALLY EXCITABLE CELLS

Neurons

ROHON-BEARD NEURONS   Rohon-Beard cells are found in the spinal cord of lamprey, elasmobranch, teleost, and amphibian embryos, and are likely to be primary sensory neurons. They are present very early in development; in *Xenopus* the majority of these cells have their last round of DNA synthesis (birthdate) during the gastrula stage (approximately 15 hr after fertilization of the egg; Spitzer & Spitzer 1975), prior to the appearance of the neural plate. The cells can be directly viewed as early as stage 18, an early neurula (Nieuwkoop & Faber 1956), a few hours later, and recognized on the basis of their large size (~ 25 μm) and the mediolateral position and dorsal location of the cell body. The physiological development of these cells has been studied by intracellular recordings (Spitzer & Baccaglini 1975, 1976, Baccaglini & Spitzer 1977), and the ionic dependence of electrical excitability between 19 hrs and 2½ wks of development has been examined by ion substitution and the application of various pharmacological blocking agents.

The egg and blastula cells of *Xenopus* have been found incapable of making action potentials (Palmer & Slack 1970, Slack & Warner 1975). Similar results have been obtained for presumptive neural cells of *Ambystoma* (Warner 1973). At stages 18–20 (late neurula, 19–21 hrs after fertilization) the Rohon-Beard cells appear electrically inexcitable. Depolarizing current pulses fail to elicit action potentials. Hyperpolarizing current pulses of long duration occasionally evoke responses that may be early signs of excitability produced by anode break excitation (Figure 1A).

Rohon-Beard neurons often produce action potentials when examined at stage 20. The action potentials consist of an overshooting plateau of long duration, often several hundred msec. Although the cells have processes by stage 22 (Muntz 1964), it seems likely that these responses are generated by the membrane of the cell soma. Cells of the embryonic spinal cord are loosely adhesive; impaled cells are frequently removed from the tissue without their processes, and continue to produce typical action potentials. During stages 20–25 (early tailbud) the inward current is carried largely by Ca^{2+}, as indicated by the following lines of evidence:

1. *Ionic dependence:* The overshoot varies with the logarithm of the external concentration of Ca^{2+}, as predicted by the Nernst equation. The overshoot is unaffected by removal of either Na^+ or Cl^-. Ba^{2+} substitutes for Ca^{2+}, producing action potentials many seconds in duration.
2. *Pharmacology:* These action potentials are blocked by small amounts of La^{3+}, Co^{2+}, or Mn^{2+}, agents that block Ca^{2+} currents in other cells. They are unaffected by moderate levels of tetrodotoxin (TTX, 10^{-6} gm/ml), which selectively blocks voltage dependent Na^+ channels in many adult cells.
3. *Conductance:* The conductance increase at the onset of these action potentials, measured by square current pulse analysis, is unaffected by removal of Na^+ or addition of tetraethylammonium (TEA) which blocks outward K^+ current and prolongs the action potentials. Ca^{2+} is the remaining ion distributed with the appropriate concentration gradient. Furthermore, the measured values of conductance agree with those calculated from simple theory, again suggesting that the increase in conductance is due chiefly to Ca^{2+}.

4. *Activation:* The action potentials can often be produced in cells with small resting potentials at which most Na^+ channels are inactivated, although the voltage dependence of Na^+ inactivation during development has not been studied. This feature is common to many Ca^{2+} channels in mature cells.

5. *Kinetics:* Although not studied extensively, the rate of rise of these action potentials is slow ($\sim$ 10 V/sec), like that of other action potentials generated by Ca^{2+} channels.

Typical records are illustrated in Figure 1B. From the mean calcium conductance (G_{Ca}, 2.6 X 10^{-4} mho/cm^2) at the onset of the plateau of the action potential, the peak calcium current entering the cells has been estimated (I_{Ca}, 3.1 X 10^{-5} A/cm^2). Calculations indicate that a single action potential could raise the internal concentration of free Ca^{2+} from 10^{-7} to 2.8 X 10^{-5} M (Baccaglini & Spitzer 1977). Since the overshoot does not change with repetitive firing at low frequencies (0.5 Hz), the calcium influx must be rapidly sequestered. The action potentials become shorter in duration with repetitive stimulation, however, probably as a result of changes in either the calcium or potassium conductance.

At intermediate stages of development (from early tailbud to early larva, stages 25–40, 28 hr–2½ days), Rohon-Beard cells produce action potentials that consist of a spike followed by a plateau, and are several tens of msec in duration. The spike is produced by an influx of Na^+, since it is selectively blocked by replacement of this ion with relatively impermeant ions such as Tris or choline, while the plateau is spared. In contrast, the plateau is due to a Ca^{2+} current since it is blocked by La^{3+}, Co^{2+}, or Mn^{2+}, leaving the spike (Figure 1C).

At larval stages (stages 40–51; 2½ days–2½ weeks) Rohon-Beard cells produce brief action potentials, roughly one msec in duration, which are Na^+ dependent by the following criteria:

1. *Ionic dependence:* The overshoot varies with the logarithm of the external concentration of Na^+, as predicted by the Nernst equation, and is abolished by removal of Na^+.

2. *Pharmacology:* TTX abolishes the spike, while La^{3+}, Co^{2+}, and Mn^{2+} do not affect the overshoot and duration.

3. *Inactivation and kinetics:* These action potentials cannot be elicited from cells with low resting potentials, which suggests an inactivation process similar to that seen for Na^+ channels in other cells. Furthermore these action potentials have a rapid rate of rise, often several hundred V/sec, reminiscent of Na^+ channels seen elsewhere (Figure 1D).

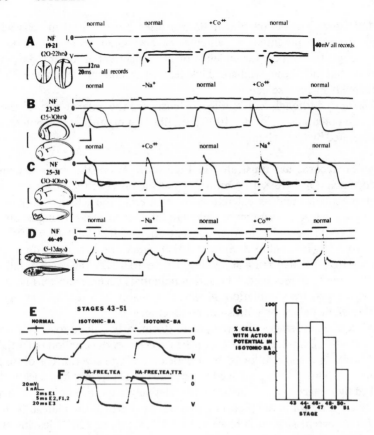

Figure 1 Electrical excitability of *Xenopus* Rohon-Beard neurons; records from 6 cells, typical of the developmental stages indicated below. Size of embryo indicated in mm. *A* Nieuwkoop and Faber stages 19–21: depolarizing current usually does not produce an action potential. A small hump (dot) is occasionally seen during repolarization; hyperpolarizing current may give a response (arrows) that is reversibly blocked by Co^{2+}. *B* Stages 23–25: depolarizing current evokes a Ca^{2+} action potential when threshold is reached; it is largely unaffected by replacement of Na^+ with Tris, but abolished by Co^{2+}. *C* Stages 25–31: the Na^+–Ca^{2+} action potential consists of a spike followed by a plateau. The plateau is eliminated by Co^{2+}, and the spike is abolished by removal of Na^+. *D* Stages 46–49: the Na^+ action potential at these late stages is blocked by removal of Na^+; Co^{2+} has little effect. *E* Stages 43–51: the action potential depends on Na^+ in normal saline (E1), but Ca^{2+} channels are still present: replacement of Na^+ and Ca^{2+} by Ba^{2+} yields action potentials of long duration (E2, 3). *F* Stages 43–51: similar to *E*, long duration action potentials are seen upon replacement of Na^+ by Tris and TEA (F1); unlike the Na^+ spikes these action potentials are not blocked by TTX (F2). *G* The percentage of cells capable of making long duration action potentials in isotonic $-Ba^{2+}$ decreases as a function of developmental age; 150 cells tested, from animals at stages 43–51 (3½–17 days). (From Spitzer & Baccaglini 1976, Baccaglini & Spitzer 1977.)

Some and perhaps all of the Rohon-Beard cells gradually lose their Ca^{2+} channels. At stage 43, when they make Na^+ dependent action potentials in normal saline, they can all make divalent cation dependent action potentials of long duration, when Na^+ is removed and outward K^+ current channels are blocked (Figure 1E, F). The percent of cells able to make divalent cation dependent action potentials declines with age, as illustrated in Figure 1G. By stage 51 two thirds of cells have lost this ability. Rohon-Beard cells are also dying during this period; the decline in cell number makes it difficult to find them beyond two weeks.

Thus the ionic dependence of the inward current of the action potential of these cells changes with age, from Ca^{2+} to Ca^{2+}–Na^+ to Na^+ alone. At the same time, the duration of the action potential becomes shorter and the input resistance (R_{in}) decreases. These processes seem continuously graded. There is some variability in the timing of the changes (as much as a few hours), which could be due to differences either in the time of induction of the cells or in the rate of later development. No anteroposterior gradient in the development of the action potential has been seen. In contrast to the changes seen in the inward current, no qualitative change has been seen in the outward current during the period studied. The outward current channels are present from the time that Ca^{2+} action potentials are first recorded. The rapid repolarization and hyperpolarization of cells are reduced by TEA, which prolongs the action potential. These results are consistent with the observation that cells in the neural plate of *Ambystoma* embryos acquire voltage dependent outward current channels (delayed rectification) around the time of closure of the neural tube (Warner 1973). Outward current channels seem to appear before inward current channels in these cells.

DORSAL ROOT GANGLION CELLS Dorsal root ganglion cells are the major class of peripheral sensory neurons in vertebrates, and are thought to take over the sensory function of the disappearing Rohon-Beard neurons. There may be as many as 40 pairs of dorsal root ganglia per tadpole in *Xenopus* (Kent 1969). The anterior ganglia are first recognizable at stage 39 (2½ days) (Nieuwkoop & Faber 1956). There appears to have been no investigation of the birthdate of these cells. The cells clearly do not develop synchronously, and proliferation, differentiation, and cell death are probably all occurring simultaneously in the dorsal root ganglia of a given tadpole.

The ionic basis of the action potentials of dorsal root ganglion cells in embryos 4½ to 51 days old has been studied by the same techniques employed in the study of Rohon-Beard neurons (Baccaglini 1978). Diameters of impaled cells ranged from about 7 to $70\,\mu m$. The spectrum of ionic dependence of the inward current of the action potentials seen in these cells is very similar to that found in the cell bodies of Rohon-Beard neurons. Ca^{2+} action potentials were found only in small cells. $Ca^{2+}-Na^+$ action potentials could be divided into two classes; the first type could make a Ca^{2+} action potential in the absence of Na^+, while the second type could not. The first type was present in cells of small and intermediate size, while the second type was present in cells of all sizes. Na^+ action potentials also fell in two classes, both present in cells of all sizes. The first type consisted of a typical, short duration spike, like the one in $Ca^{2+}-Na^+$ action potentials. The second type consisted of a spike with an inflection on the falling phase; this spike was prolonged by Co^{2+} or La^{3+}, and unaffected by moderate levels of TTX. One possible explanation is that a Ca^{2+} current is blocked, which normally activates a K^+ conductance (Meech & Standen 1975). Cells with a type II Na^+ action potential had a smaller resting potential and larger input resistance than cells with a type I Na^+ action potential. This Na^+ type II action potential is illustrated in Figure 2A.

It seems likely that the different kinds of action potentials recorded from these cells represent stages in developmental sequence like that reported for Rohon-Beard cells (Baccaglini 1978). The absence of information about the age of the cells (e.g. time since birthdate) makes it difficult to prove this point. However assuming that size is in some way proportional to age, the results suggest that some young cells make Ca^{2+} action potentials and older ones make Na^+ action potentials. Other cells may undergo differentiation and changes in the ionic basis of their action potentials without growth in size. This developmental interpretation receives support from examination of the frequency of appearance of the different kinds of action potentials as a function of the age of the tadpole (Figure 2B). It is evident that the percent of cells with Ca^{2+} and $Ca^{2+}-Na^+$ action potentials decreases with age, whereas the number of Na^+ type I action potentials increases. Na^+ type II action potentials are seen only at later stages. Evidence of an outward (K^+) current was present in all cells tested, and no difference in sensitivity of different action potentials to TEA was detected. The cells with Na^+ type II action potentials are similar to denervated muscle fibers, with TTX insensitive spikes, low resting potentials, and high input resistance (Purves 1976). These action potentials are seen at about the time that the first degenerating dorsal root ganglion cells have been reported (Prestige 1965). It may be that this kind of electrical excitability is an early sign of denervation and possibly a prelude to cell death.

NEURONS IN DISSOCIATED CELL CULTURES   Cultures of cells dissociated from the posterior neural plate and underlying mesoderm of stage 15 *Xenopus* embryos have permitted the study of the development of electrical excitability in amphibian neurons in vitro (Spitzer & Lambor-

ghini 1976, Willard 1977 and in preparation). The tissue is dissociated in
Ca^{2+}–Mg^{2+}-free saline containing EDTA, and cells from a single embryo
are deposited on the bottom of a tissue culture plastic Petri dish. The cells
attach and differentiate morphologically at room temperature in a medium
consisting of Steinberg's salt solution and 0.1% bovine serum albumin.

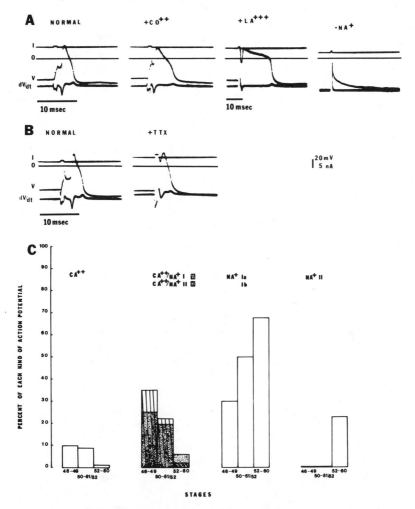

Figure 2 Action potentials from *Xenopus* dorsal root ganglion neurons. *A, B:* Na^+ type II
action potentials from 2 cells. The inflection on the falling phase is prolonged by Co^{2+} or
La^{3+}; the action potential is abolished by replacement of Na^+ with Tris (*A,* stage 53). The
action potential is not blocked by TTX (*B,* stage 54). *C:* The percentage of cells capable of
giving each kind of action potential changes as a function of developmental age. 227 cells were
tested in animals at stages 46–60 (4½–46 days). (From Baccaglini 1978.)

There are as many as 30 neurons and several hundred muscle cells per dish in these low density cultures; the cells survive for periods up to 5 days as they metabolize their stores of yolk and lipid. They can be impaled with intracellular microelectrodes (Figure 3) and perfused with salines of various compositions.

The cultured neurons produce action potentials in their somata when depolarized by current injected through the recording electrode. These action potentials depend on Ca^{2+} when cells are examined as early as 8 hrs after plating. Nerve cells from 20 hr cultures have $Ca^{2+}-Na^+$ action potentials. Neurons in 3 and 4 day cultures have Na^+ action potentials (see Figure 4). Sibling reference embryos at the same age as cells in culture have Rohon-Beard neurons with the same properties. It appears that the neurons in culture go through the same sequence of stages in excitability along roughly the same time course as the Rohon-Beard neurons in vivo. Birth-dating indicates that nearly 80% of the cultured neurons are born by stage 13 (Spitzer & Lamborghini 1976), and are thus either Rohon-Beard neurons, extramedullary neurons, or large ventral neurons (Spitzer & Spitzer 1975). Accordingly, some of the cells studied electrophysiologically are likely to be Rohon-Beard neurons. In addition, functional neuromuscular synapses are observed in these cultures. One possibility is that this developmental sequence of changes in electrical excitability is found in motoneurons (probably the large ventral neurons) as well. The 20% of cells with birthdates after stage 13 are of unknown identity, but follow the same developmental sequence.

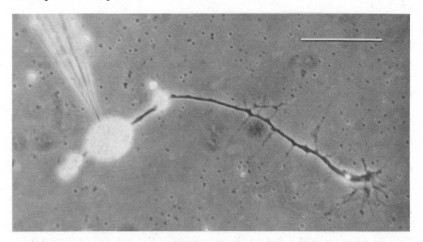

Figure 3 Neuron from dissociated *Xenopus* neural plate, grown in culture for 16 hrs and viewed with phase contrast optics. Note phase dark process and growth cone; phase bright yolk granules obscure the nucleus in the cell body, indicated by the microelectrode. Calibration is 50 μm.

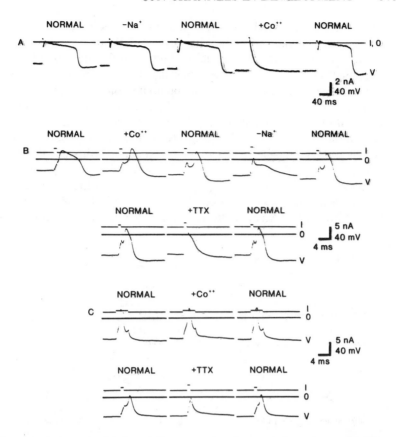

Figure 4 Action potentials from 4 *Xenopus* neurons in neural plate cultures of 3 different ages. *A* Depolarizing current pulses elicit Ca²⁺ action potentials at 8 hrs in culture (stage 24/25); these are unaffected by replacement of Na⁺ with Tris and are blocked by Co²⁺. *B* Na⁺–Ca²⁺ action potentials at 21 hrs in culture (stage 32) consist of a spike followed by a plateau. The plateau is removed by Co²⁺ and the spike disappears upon removal of Na⁺ or addition of TTX. *C* Na⁺ action potentials at 3 days in culture (stage 44) are not greatly affected by Co²⁺, but are blocked by TTX (in another cell). (From Spitzer & Lamborghini 1976.)

The dissociated amphibian cell cultures have made possible a detailed study of the development of electrical excitability in the processes of the neurons (Willard 1977 and in preparation). The lack of an adequate technique for visualization of the neurites has so far prohibited such experiments in vivo. The cultured neurons send out neurites that are often several hundred micra in length. Cells were stimulated either intracellularly, through the recording electrode, or extracellularly, with a glass pipet placed against the neurite. The ionic basis of the action potentials produced in the soma or in the process (and recorded in the soma) was analyzed by ion substitutions and the application of various blocking agents.

The action potentials elicited by stimulation of the process appear as small responses in the cell body, after a delay reflecting the conduction velocity. Occasionally they bring the cell body to threshold and initiate an action potential in the soma. Intracellular injection of current into the cell body is capable of producing an action potential in the neurite, or the large somatic action potential at higher current strengths. Confirmation of the localization of the action potential in the neurite was achieved by cutting it off with the extracellular stimulating electrode, leaving the action potential in the cell body unaffected. Between 6 and about 11 hrs in culture the action potentials in the neurites depend on Ca^{2+}, and are blocked by Co^{2+} but unaffected by removal of Na^+ (Figure 5A). From 11 hrs on, the action potentials in the neurites depend on Na^+; they are blocked by removal of Na^+ or addition of TTX but unaffected by Co^{2+} (Figure 5B). Neurites with

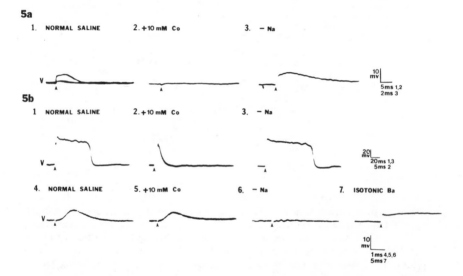

Figure 5 Action potentials from the somata and processes of two neurons from *Xenopus* neural plate cultures, at 8 hrs (*a*) and 11.5 hrs (*b*) in vitro. Brief stimuli are indicated by arrows. *a* Depolarization of the neurite produces a Ca^{2+} action potential (1–3), recorded in the cell body as a small event that fails to overshoot the zero potential; it is blocked by Co^{2+} but unaffected by substitution of choline for Na^+. Depolarization of the cell body gives a longer, overshooting Ca^{2+} action potential (not shown). *b* Depolarization of the cell body still yields an overshooting Ca^{2+} action potential (1–3) blocked by Co^{2+} and unaffected by replacement of Na^+ with choline. In contrast, depolarization of the neurite produces a Na^+ action potential (4–6), recorded as a small event that does not overshoot the zero potential; it is unaffected by the presence of Co^{2+} and blocked by the removal of Na^+. The neurite still has Ca^{2+} channels at this time, since substitution of Ba^{2+} for Na^+ and Ca^{2+} leads to action potentials of long duration (7); these channels are absent from the neurites of cells in slightly older cultures (From Willard, in preparation.)

Na^+ action potentials are able to make action potentials in isotonic Ba^{2+} for a brief period during their development. This capability is rapidly lost, suggesting that the Ca^{2+} channels disappear. Of particular interest is a period of roughly 3 hrs between 11 and 14 hrs in culture, when the action potential in the neurite depends on Na^+ while that of the cell body still depends on Ca^{2+}. There appears to be a high degree of synchrony in the differentiation of these neurons, perhaps due to the fact that the majority of cells have the same birthdate. These results indicate that the ion channels of the membrane of the cell soma and the neurite may be separately regulated. It would be of interest to know the distribution of voltage dependent channels out at the distal end of the neurite, where new membrane is added (Bray 1970). This question is not easily addressed electrophysiologically however, as the cable properties of the neurite will make the distinction between passive and active membrane difficult over short distances. A proximodistal decrease in the number of intramembranous particles in the neurites of cultured cells has been seen by freeze-fracture techniques (Pfenninger & Bunge 1974).

Since the development of electrical excitability in cell bodies of dissociated cells in vitro proceeds as it does in the somata of neurons in vivo, there was reason to expect that the neurites of the cultured cells would undergo normal development as well. Kullberg, Lentz & Cohen (1977) observed that extracellular stimulation of the spinal cord of *Xenopus* embryos evoked endplate potentials that were blocked by TTX at stage 24. These findings are consistent with the results obtained from the studies in vitro, which indicate that there are Na^+ action potentials in the processes of neurites at a time when there are Ca^{2+} action potentials in the cell body.

Most other studies of neurons in vitro have been carried out on older cultures, often at periods of weeks after plating. Chick spinal cord cells in these cultures have brief Na^+ action potentials (Fischbach & Dichter 1974). Dorsal root ganglion cells of mouse and chick have Ca^{2+}–Na^+ action potentials (Matsuda, Yoshida & Yonezawa 1976, Ransom & Holz 1977, Dichter & Fischbach 1977), as do rat superior cervical ganglion neurons (O'Lague, Potter & Furshpan 1978). It is presently unknown whether these cells go through other stages in their development of electrical excitability at earlier times. If they do, it appears that dorsal root ganglion and sympathetic ganglion cells may be arrested at the stage of the Ca^{2+}–Na^+ action potential.

CLONAL NEURONAL CELL LINES IN CULTURE   The electrical excitability of neuroblastoma cells has been studied in several laboratories, and differentiation of this membrane property appears to occur under some conditions. Several initial studies of mouse C1300 neuroblastoma cells revealed that passive responses, delayed rectification, and action potentials that were often overshooting could be elicited in response to depolarizing current pulses (Nelson, Ruffner & Nirenberg 1969). Evidence for voltage dependent Ca^{2+} and Na^+ currents was obtained (Harris & Dennis 1970, Moolenaar & Spector 1977). It was suggested that these different responses represent stages of neuronal differentiation of the cells (Nelson, Ruffner & Nirenberg 1969), although they were apparently not found in sequence. Hybrids of neuroblastoma and clonal fibroblast cells (L cells) exhibited a similar spectrum of responses (Nelson 1973). In contrast, other investigators found that neuroblastoma cells never gave completely passive responses (Harris & Dennis 1970). These results were also obtained upon examination of neuroblastoma grown for at least 20 generations in suspension culture in which attachment and subsequent differentiation were inhibited (Schubert et al 1973). The cells were transferred from suspension to tissue culture dishes and studied immediately, before any morphological differentiation occurred. Round and apparently undifferentiated cells were capable of producing action potentials. These findings have been recently confirmed, and it has been shown that action potentials produced in small round cells depend on Ca^{2+} (Miyake 1978). The early reports of passive cells may have been due to low resting potentials resulting from damage upon impalement with the microelectrode. Furthermore under culture conditions permitting morphological differentiation, these cells exhibit action potentials with separate Na^+ and Ca^{2+} dependent peaks. Delayed rectification is first detected at this time. After selection for nondividing cells, a Na^+–Ca^{2+} action potential and a Na^+ action potential with a single peak can be recorded in different cells. The overshoots for the peaks of these action potentials vary with the external concentrations of Na^+ and Ca^{2+} as expected. Repeated selection for mature cells results in an increase in the percentage of Na^+ action potentials recorded. These results suggest a sequence of differentiation similar to that reported for amphibian neurons and chick muscle cells in vivo.

A clonal line of rat pheochromocytoma cells (PC12) grows in size in response to nerve growth factor (NGF; Dichter, Tischler & Greene 1977), but it is not clear that NGF is triggering the differentiation of electrical excitability in these cells (Patrick, Heinemann & Schubert 1978).

DORSAL UNPAIRED MEDIAN (DUM) NEURONS   Embryos of the grasshopper, *Schistocerca nitens,* are a convenient preparation for the study of neuronal development (Goodman 1979, Goodman & Spitzer, in preparation). The time from egg laying to hatching is 21 days at 34°C, and developmental stages can be defined reproducibly. The embryos are larger than those of many other insects, with relatively large cells. The pattern of neurogenesis has been understood from anatomical studies (Wheeler 1893,

Roonwal 1937, Bate 1976). Within the first few days neuroblasts become arrayed in regular rows and columns on the ventral surface of the primordial ganglia. There are about 30 on each side of the midline for each ganglion; by asymmetric mitoses they give rise to most of the roughly 3000 neurons in each thoracic ganglion and the 500 neurons in each abdominal ganglion (Gymer & Edwards 1967, Sbrenna 1971). A single neuroblast is found on the dorsal surface of each primordial ganglion, in the midline at the posterior margin [first called the median cord neuroblast (Wheeler 1893)]. Like the ventral neuroblast, it gives rise to mitotic daughters. Unlike the ventral daughters, which develop in many pillars projecting through the ganglion from the ventral to the dorsal surface, the dorsal daughters develop in several chains lying on the dorsal surface of the ganglion and projecting anteriorly.

The dorsal neuroblast and its descendents can be visualized in dissected preparations, and are easily accessible for intracellular recording at early stages of development. These descendent cells have now been shown to become the DUM neurons described in the adult nervous system (Hoyle et al 1974). The asymmetric divisions of the neuroblasts (25–30 μm in diameter) give rise to ganglion mother cells (8–10μm) as early as 8 days of incubation; these divide again to yield ganglion cells of similar size. Mitotic metaphase plates can be clearly seen. The ganglion cells then develop into morphologically recognizable neurons as they extend processes and grow in size. The first two chains of cells elongate during an extended period of embryonic development. It is possible to recognize uniquely identified DUM neurons at the ends of the chains in ganglion after ganglion by dye injections, and to identify a cell reliably on the basis of its distance from the end of the left or right chain (Goodman & Spitzer, in preparation). By day 13 those at the end of the chain become detached, however, and can no longer be recognized by the position of their cell bodies. Most neuroblasts die by 14 days of incubation.

The development of electrical excitability in the mitotic daughters has been studied (Spitzer & Goodman, in preparation). On day 12 the neuroblast and the immediately adjacent ganglion mother cells have resting potentials of −70 to −80 mV. They can be fully depolarized by the injection of square current pulses without producing any action potential, and the current-voltage relationship is linear for ± 50 mV polarization (see Figure 6A1). These cells are strongly electrically coupled; furthermore Lucifer Yellow dye (450 MW; a generous gift of Walter Stewart, NIH; see Stewart 1978) injected into a single cell spreads rapidly to cells to which it is electrically coupled (Goodman & Spitzer, in preparation). It seemed possible that voltage dependent channels were not revealed because of the long time constant of the coupled network of cells, which prevented depolarization to threshold before channel inactivation occurred. This possibility was tested by exposing the cells to 10^{-5}gm/ml veratridine, which specifically depolarizes cells possessing voltage dependent Na^+ channels (Narahashi 1974). The treatment has no effect on the resting potentials of these cells, in contrast to those further along the chain. These cells thus appear electrically inexcitable.

Ganglion cells further anterior from the neuroblast in 12 day embryos have resting potentials of –40 to –55 mV, and are no longer electrically coupled to the neuroblast and ganglion mother cells. When these cells are depolarized they produce action potentials that fail to overshoot the zero potential and have a characteristic double-humped appearance (Figure 6A2). These action potentials are likely to be produced in the neurites, since this response has been shown in adult DUM neurons to be the sum of action potentials arising from two symmetrical processes distal to a bifurcation (Heitler & Goodman 1978). Dye injections of Lucifer Yellow do not result in spread to other cells, and confirm the existence of short processes. These responses depend chiefly on Na^+, since they are rapidly abolished by replacement of this ion or the addition of 10^{-9} gm/ml TTX. Furthermore these cells, like those further anterior along the chain, are rapidly depolarized by more than 25 mV by exposure to veratridine in normal saline; the prior removal of Na^+ or addition of TTX prevents this depolarization. These cells thus have electrically excitable processes.

Cells at the ends of the chains in 12 day embryos have –40 to –55 mV resting potentials and produce overshooting action potentials, 2–4 msec in duration (Figure 6A3). The large size and rapid rate of rise suggests that they are produced in the cell body. The cells fire repetitively to long current pulses. Frequently cells with large resting potentials will produce only a few action potentials (often not overshooting) before failing permanently and revealing the action potentials in the process. These cells may be making the transition from inexcitable to excitable cell bodies. An extensive arborization of processes is revealed by dye injection. These action potentials depend on both Na^+ and Ca^{2+}; they persist in either the absence of Na^+ or the presence of Co^{2+} (Figure 6A4), although they are blocked by both treatments applied simultaneously. There is no evidence for a long duration Ca^{2+} action potential. The action potentials of mature DUM cells require the presence of both Na^+ and Ca^{2+}; either ion alone is insufficient (Goodman & Heitler, in preparation) (Figure 6B). One possibility is that the inward Ca^{2+} current is diminished during the further development of these cells as in the other neurons discussed above. However a decrease in the membrane resistance during development could also account for these results. Further work is needed to sort out the alternatives.

The results indicate that the neuronal differentiation of cells increases with distance from the dorsal neuroblast. Cells at the ends of the chains show the same sequence of differentiated states when examined in progres-

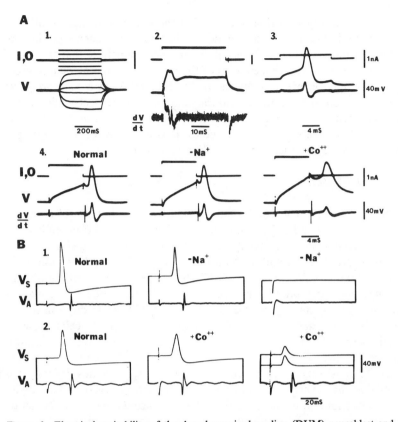

Figure 6   Electrical excitability of the dorsal unpaired median (DUM) neuroblast and its descendent cells (including DUM neurons) in embryonic grasshoppers, compared to excitability of a mature DUM neuron (DUMETi) in adults. (*A*1–3) Records from 3 cells at different stages of development in a 12 day embryo. *A*1 The DUM neuroblast and electrically coupled daughter cells exhibit passive responses to current pulses. *A*2 Uncoupled cells further from the DUM neuroblast have begun to grow axons which often bifurcate; depolarization of these cells produces small double-humped action potentials, in which each hump probably arises from a separate site of spike initiation distal to the bifurcation, as demonstrated for cells in the adult. The inward current is carried largely by Na^+ (not shown). *A*3 Older cells, most distant from the DUM neuroblast, produce overshooting Na^+–Ca^{2+} action potentials in their cell body: these action potentials persist in the absence of Na^+ or in the presence of Co^{2+} (*A*4), but are blocked by simultaneous application of both treatments. (from Spitzer & Goodman, in preparation) *B* Ionic dependence of soma and axon spikes of DUMETi in an adult grasshopper. Recordings are made from the cell body (V_s) and right peripheral axon (V_a), while stimulating the left peripheral axon. In contrast with embryonic cells, the soma spike is gradually abolished by either removal of Na^+ or addition of Co^{2+}. The axon spike is slowly eliminated by removal of Na^+, and unaffected by Co^{2+}, as in embryonic cells. Soma spikes could not be elicited by direct depolarization of the soma in the absence of Na^+ (not shown). Similarly, soma spikes could not be produced in the presence of Co^{2+}, either by antidromic stimulation superimposed on steady depolarization of the soma (bottom right panel) or by direct depolarization of the soma (not shown). (From Goodman & Heitler, in preparation.)

sively older embryos. It thus appears that the spatial sequence of differentiated states in the chains is a reflection of the temporal sequence of differentiation for individual cells. These observations demonstrate that the DUM neurons are the clonal descendents of the dorsal unpaired median neuroblast. Dividing cells are initially electrically coupled. They extend processes after they become postmitotic, and the neurons become electrically uncoupled somewhat later. Electrical excitability first develops in the processes and shortly thereafter in the cell bodies. It is not known if uncoupling is causally related to other aspects of differentiation.

Striated Muscle

CHICK MUSCLE   The development of electrical excitability of chick leg muscle cells in vivo has been studied for a 3 week period from embryonic day 13 through hatching (day 21) to several weeks posthatch (Kano 1975). Myoblast proliferation and fusion occur during this time (Herrmann, Heywood & Marchok 1970, Williams & Goldspink 1971). Differentiation of the sarcoplasmic reticulum and transverse tubular system seems to begin at the earliest myotube stage (Ezerman & Ishikawa 1967). At 13 days most cells give a passive response to depolarization, although some cells show an afterhyperpolarization suggestive of delayed rectification; these cells show no sign of inward currents. At 16–19 days most cells produce an action potential consisting of a plateau that is several seconds in duration; it is likely that the inward current in these cells is carried largely by Ca^{2+}, with a Na^+ contribution in some cells. At days 19 through hatching the action potential is composed of a spike due to Na^+, followed by a plateau due to Ca^{2+}. Within the first few days posthatch the plateau component disappears, and the action potential consists of a Na^+ spike only a few msec in duration (see Figure 7). The plateau is not elicited by the addition of Ba^{2+} to the oldest cells studied, which suggests that they have lost the Ca^{2+} channels. Recorded values of the resting potentials were often large even at the early stages, but increased during this period; it seems unlikely that these changes explain the variations in the responses. Thus the cells differentiate through a sequence in which they are initially inexcitable; when the action potential is produced, the inward current is carried by Ca^{2+}, then Na^+–Ca^{2+}, and finally Na^+ only. Delayed rectification seems to be present at very early stages. These results are derived from a population of cells, to which new ones may be continually added; it is not clear that the sequence requires 3 weeks for completion in any given cell. The first Na^+ action potentials without a plateau are recorded at day 18; thus full differentiation might occur in as little as 5 days.

The onset of excitability of chick muscle has also been studied in cell cultures. Myoblasts from leg muscle of 12 day embryos form myotubes in

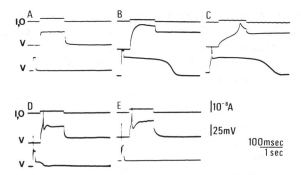

Figure 7 Electrical excitability of chick striated muscle; records from 5 cells typical of the developmental stages indicated below. The bottom trace for each cell is a slow sweep display of the voltage recorded on the middle trace. *A* Passive response from a 13 day embryo. *B* Plateau response from a 17 day embryo. *C* Spike plus plateau response from a 19 day embryo. *D* Spike plus plateau response from a 20 day embryo. *E* Spike response from a chick 8 days after hatching. See text for the ionic basis of these responses. (From Kano 1975.)

2 to 3 days, and most exhibit cross-striations by a week in culture (Kano, Shimada & Ishikawa 1972, Kano & Shimada 1973). The same kinds of action potentials are recorded from the muscle cells in cultures. The type of response generally is not correlated with the age of the culture, perhaps due to asynchrony in the fusion of myoblasts (see Spector & Prives 1977). The size and rate of rise of the TTX sensitive spike increases from day 4 to 14 days in culture, however (Kano & Yamamoto 1977). It seems likely that development is occurring in vitro through the same stages seen in vivo.

TUNICATE MUSCLE   The development of electrical excitability in tunicate muscle has been the object of detailed study (Takahashi, Miyazaki & Kidokoro 1971, Miyazaki, Takahashi & Tsuda 1972). The large egg undergoes a mosaic development and the cells destined to give rise to muscle cells can be followed and impaled at all stages. Cells are tightly electrically coupled at the blastula and gastrula stages. Morphological differentiation yields 6 chains of 7–8 mononucleate muscle cells each in the larval tail. The time of development from first cleavage to hatching is 3 days, and muscles become innervated at 2¼ days (Ohmori & Sasaki 1977). At early stages, from the mature unfertilized egg through blastula, no active response is seen to depolarizing currents at the low resting potentials. However anode break stimulation yields a response that overshoots the resting potential, presumably by removing inactivation of voltage dependent channels (Figure 8A1, 2). This action potential depends on Na^+ and Ca^{2+} (Miyazaki, Takahashi, & Tsuda 1974). Voltage clamp experiments have shown that the maximum Na^+ current is about $10^{-5}A/cm^2$ while the Ca^{++} current is about $5 \times 10^{-7}A/cm^2$ in artificial sea water. These currents flow through channels that are very similar to the Na^+ and Ca^{2+} channels in other excitable cell membranes; the Na^+ channel is not blocked by TTX (Okamoto, Takahashi & Yoshii 1976a, b). In later stages, from early gastrula to a tadpole with roughly half its ultimate tail length, the resting potential becomes much larger; this is the result of the disappearance of the resting permeability to Na^+ present at early stages. A large increase in depolarization occurs with a small increase in current intensity above a threshold, and the depolarization outlasts the current pulse by several seconds (Figure 8B). The current-voltage relation shows that there has been a large increase in the inward (anomalous K^+) rectification, which is greater at this stage than at any other. These responses thus depend on the voltage dependent decreases in K^+ conductance as the cells are depolarized, and an influx of Na^+ and Ca^{2+} (Miyazaki et al 1974). At a still later stage, from young tadpole to hatching, the action potential consists of an overshooting spike followed by a plateau that shortens from seconds to about 100 msec in duration during this period (Figure 8C). Most of the inward current is carried by Ca^{2+}, although there may be a small Na^+ component in the later phase of the plateau. This action potential is followed by a long-lasting depolarizing afterpotential. Finally at a relatively mature stage, from hatched tadpole to a time just prior to metamorphosis, the action potential is a spike less than 50 msec in duration (Figure 8D). The current-voltage

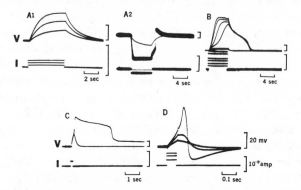

Figure 8 Action potentials in the egg and developing striated muscle cells of the tunicate; records from 5 cells typical of the stages indicated below. *A* From the egg to the 128 cell stage, only passive responses are elicited with depolarizing current pulses (*A*1). Hyperpolarizing currents yield small active responses (*A*2). *B* From early gastrula to young tadpole larva, when the resting potential is larger, depolarizing currents yield active responses that outlast the current pulse by more than a second. *C* From young larva to hatching, an overshooting action potential is recorded. *D* From hatching to just before metamorphosis, a smaller and briefer action potential is elicited. See text for the ionic dependence of these responses. (From Takahashi, Miyazaki & Kidokoro 1971, Miyazaki, Takahashi & Tsuda 1972.) Copyright by the American Association for the Advancement of Science.

relationship indicates that a substantial increase in delayed rectification has appeared at this stage. The inward current now seems to be carried almost exclusively by Ca^{2+}. These action potentials are unaffected by high levels of TTX. Thus the ionic dependence of the inward current changes from Ca^{2+}–Na^+ to Ca^{2+}; the outward (K^+) current channel becomes significant relatively late in development.

RAT MUSCLE   Action potentials have been recorded from the diaphragm muscle of 17 day rat fetuses, although their ionic basis was not studied (Diamond & Miledi 1962). Recordings from the diaphragm at the first day postpartum indicate that action potentials are resistant to moderate levels of TTX (Harris & Marshall 1973), suggesting the presence of either Ca^{2+} channels or TTX insensitive Na^+ channels. The action potentials become TTX sensitive by 20 days. The ionic dependence of action potentials in embryonic muscle fibers is not known, but action potentials in the striated muscles of newborn rats are prolonged by the presence of small amounts of Ba^{2+}, compatible with the presence of Ca^{2+} channels (S. Bevan, personal communication). No Ca^{2+} component has been detected in the action potentials in adult soleus muscle (Kidokoro 1975b). It may be that the early stages of development of electrical excitability have already occurred by the time of birth.

CLONAL RAT MUSCLE CELL LINE   There is a clear differentiation of electrical excitability in a clonal line of developing rat skeletal muscle cells (L6) (Kidokoro 1973, 1975a, b, Land, Sastre & Podleski 1973). The myoblasts are electrically excitable and produce brief, small Na^+ action potentials. After fusion of these cells the multinucleate myotubes have overshooting Na^+–Ca^{2+} action potentials with two components. The initial spike is due to Na^+; the plateau on the repolarizing phase is due largely to Ca^{2+}. The current-voltage relationship indicates that delayed rectification is now present. A transverse tubular system develops in the myotubes (Klier, Schubert & Heinemann 1977, Kidokoro 1975a), as in primary cultures (Schiaffino, Cantini & Sartore 1977). Kidokoro has reported that cross-striated fibers have action potentials like those in myotubes, although the plateau is now shorter in duration; the Ca^{2+} component seems to persist in these cells. Land et al (1973) found Na^+ action potentials that appeared to lack a Ca^{2+} component in some myotubes; often these were blocked by moderate levels of TTX, but occasionally they were not. The Na^+ action potentials were prolonged by the addition of Mn^{2+} or La^{3+}, like the action potentials in some dorsal root ganglion neurons in vivo (Baccaglini 1978). It is not clear that L6 muscle cells reach the same endpoint as rat skeletal muscle developing in vivo. The variation in results could be due to genetic drift of the cell line or to different culture conditions (e.g. density of plating). It is not known whether the clonal muscle line is genetically defective or growing in an environment that is inadequate to support normal differentiation. An L6 mutant (MK_1) resistant to the toxic effects of high concentrations of K^+ has myotube action potentials that are not abolished by elevated K^+ as is normal L6, although the cells are depolarized normally (Gartner, Land & Podleski 1976).

Cardiac Muscle

The action potentials of the chick heart ventricle depend on Ca^{2+} and Na^+ at 2–4 days of embryonic development, and on Na^+ by day 6. A similar change has been observed for heart cells grown in culture under appropriate conditions. The first autonomic innervation in vivo occurs later, at about day 12. A comparable shift in the ionic dependence of the action potentials of the rat heart occurs between 10 and 21 days of gestation. The transverse tubular system appears between 6 and 10 days after birth in these cells. This work has been recently reviewed (Lieberman & Sano 1976).

Egg Cells

The ability of egg membranes to make action potentials was first discovered in the tunicate (Miyazaki, Takahashi & Tsuda 1972). Since then electrical excitability has been found in coelenterate, echiuroid, annelid, echinoderm, and vertebrate eggs. It appears to be absent in some arthropod and chordate eggs, however. These results raise the possibility that development involves loss, in addition to acquisition, of ion channels. Knowledge of the role of these action potentials in the development of eggs may advance understanding of their role in developing nerve and muscle cells. The activation potential that occurs at fertilization is a separate, although related, event. Several features are common to the action potentials recorded from eggs. First, a pattern of ionic specificity emerges since the voltage dependent permeabilities are either to Ca^{2+} or to Na^+ and Ca^{2+}; no purely Na^+ action potentials have been described. Second, there is pharmacological uniformity in that the Ca^{2+} component is blocked by Co^{2+}, and the Na^+ component, when present, is insensitive to TTX. Finally, the inward current channels show inactivation in those cases in which they have been examined. This field has been recently reviewed (Hagiwara & Miyazaki 1977, Hagiwara & Jaffe 1979).

CHANGES IN ION CHANNELS: CHEMICALLY EXCITABLE CELLS

The development of electrical excitability occurs at very early stages. The discussion of chemically excitable cells is restricted to those examined at comparably early times. Most nerve and muscle cells appear to show no change in the ionic basis of their response to a neurotransmitter. Attention is directed to the magnitude and polarity of the membrane potential changes resulting from stimulation of presynaptic nerves or from bath application or iontophoresis of chemicals onto cells in developing systems. In a few cases the reversal potentials and the ionic permeabilities have been identified. The changes in localization of transmitter receptors and their ion channels to particular regions of the cell membrane have been reviewed elsewhere (Fambrough 1976, Lømo 1976, Purves 1976, Patrick et al 1978).

Neurons

The time of onset of chemosensitivity has apparently not been determined for any neurons in vivo. Cells have already developed their adult sensitivity characteristics by the stages at which they have been examined. Cerebellar Purkinje cells of rats in the first postnatal week, a time prior to the onset of synaptogenesis, are as sensitive to the iontophoretic application of several putative neurotransmitters as cells in the adult (Woodward et al 1971). Rohon-Beard neurons in *Xenopus* embryos, which are depolarized by gamma-aminobutyric acid (GABA), are sensitive to iontophoretic application of this chemical when they have just become excitable and make Ca^{2+} action potentials (Spitzer 1976). It is not known if they are sensitive at an earlier stage. The sensitivity does not appear to change during the subsequent period of development in which the ionic basis of the action potential shifts from Ca^{2+} to Na^+.

There seems to be no case in which cultured nerve cells have been studied prior to the appearance of chemosensitivity. Dissociated cell cultures from the neural plate of stage 15 *Xenopus* embryos (17 hrs after fertilization) already contain at least two classes of neurons by 24 hrs in culture (Spitzer 1976). The first class is depolarized by GABA, and unaffected by acetylcholine (ACh), norepinephrine, serotonin, dopamine, glycine, and glutamate. These may be Rohon-Beard neurons. The second class is depolarized by glutamate, hyperpolarized by glycine and GABA, and unaffected by the other transmitters tested; these may be motoneurons. No changes in these responses have been detected with further time in culture. The chemosensitivity of neurons dissociated from 12–14 day mouse embryo spinal cords has been examined in cultures three weeks after plating, by which time many synapses are observed (Ransom, Bullock & Nelson 1977). Cells show hyperpolarizing responses to iontophoretically applied glycine and GABA, with reversal potentials similar to one another and to that of the inhibitory postsynaptic potentials. Cells also show depolarizing responses to glutamate, with reversal potentials more negative than those of the excitatory postsynaptic potentials. Dissociated superior cervical ganglion cells from day old rats are depolarized by iontophoresis of ACh and GABA at 7–30 days in culture, and give no response to norepinephrine, dopamine, and glutamate (Obata 1974). The reversal potential for GABA lies between 0 and –20 mV. Recordings have been made from the larger dorsal root ganglion cells as early as two days in culture. These cells are depolarized by GABA, with a reversal potential similar to that determined for the superior cervical ganglion cells; they are not sensitive to ACh, norepinephrine, and glutamate.

Mouse neuroblastoma cells are sensitive to iontophoretically applied ACh (Harris & Dennis 1970). Although they give no response in the morphologically undifferentiated state (Schubert et al 1973), they exhibit depolarizing or hyperpolarizing responses after they have grown processes; often both responses can be elicited from the same cell (Nelson, Peacock & Amano 1971, Peacock & Nelson 1973). The timing of appearance of ACh sensitivity has not been pinpointed. The depolarizing response has a short latency, rapid rate of rise, and is blocked by curare. It is accompanied by an increase in conductance and shows a reversal potential close to zero. The hyperpolarizing response has a longer latency, a slower rate of rise, and is blocked by atropine. It seems to be associated with a conductance change and to have a reversal potential at –80 mV. Some cells are hyperpolarized by dopamine; the basis of this response has not been characterized. There is considerable variability in the expression of ACh sensitivity; many cells in culture give no response although they are morphologically differentiated and electrically excitable.

Striated Muscle

The time of appearance of sensitivity to ACh in striated muscle has been determined in several embryonic preparations. The anterior trunk musculature of *Xenopus* embryos becomes sensitive to bath application of ACh as early as Nieuwkoop and Faber stage 19–20, 21–22 hrs after fertilization of the egg and about an hour before synaptic activity is detected (Kullberg 1974, Cohen & Kullberg 1974, Blackshaw & Warner 1976, Kullberg, Lentz & Cohen 1977). No fusion of these mononucleate cells occurs (Muntz 1975). The muscle cells are depolarized to –5 to –10 mV, which does not appear to be significantly different from the reversal potential recorded in adult frog muscle (–15 mV; Takeuchi 1963). The development of ACh sensitivity is paralleled by the binding of I^{125} α-bungarotoxin (α-BGT) (Kullberg 1974), a neurotoxin that has a specific affinity for ACh receptors in adult and neonatal muscle (Miledi & Potter 1971, Berg et al 1972). The muscles in the tail of the larval tunicate become sensitive to bath or iontophoretically applied ACh at a little over 1½ days after fertilization of the egg, and 15 hrs prior to the appearance of synaptic activity (Ohmori & Sasaki 1977). The sensitivity increases rapidly over the next two hours. The reversal potential of this response was determined within a few hours of its appearance by measuring currents under voltage clamp conditions, and found to be about −10 mV. The same value was obtained for both the ACh response and the excitatory junctional potential of the older free swimming larva. Thus the reversal potential does not change during development and is similar to that found at the vertebrate end plate.

The chemosensitivity of developing striated muscle has also been studied in culture. Responses to iontophoretic application of ACh were investigated in cells dissociated from leg muscles of 16–18 day old embryonic rats and cultured for 2–7 days (Fambrough & Rash 1971, Ritchie & Fambrough 1975). Short bipolar cells, presumably myoblasts, show no sensitivity; long bipolar cells and myotubes are depolarized by ACh. Sensitivity is not dependent on cell fusion since it is present in some mononucleate cells and appears in all cells even when fusion is blocked by lowering the Ca^{2+} concentration in the medium. The mean reversal potential of the response is −3 mV, independent of the age of the culture (Ritchie & Fambrough 1975). The permeability changes, to Na^+ and K^+, are very similar to those at the vertebrate end plate in the adult (Takeuchi 1963); the small effect of Ca^{2+} seen in the adult is not present. Similar results have been obtained from cultures of chick myotubes derived from striated muscle. Explants of thigh muscle from 11–13 day embryos were studied 3–20 days later by iontophoretic application of ACh. The reversal potential of the response lies between 0 and +10 mV (Harris, Marshall & Wilson 1973). Pectoral myotubes from 11 day embryos grown with spinal cord neurons were investigated at 8–30 days in culture. The postsynaptic potentials of myotubes are blocked by curare and have reversal potentials between −5 and −15 mV (Fischbach 1972). The electrical noise produced during application of ACh to cultured 12 day chick pectoral myotubes has been analyzed (Sachs & Lecar 1973). The value of the elementary conductance is close to that of adult frog muscle ($\sim 10^{-10} \ \Omega^{-1}$; Katz & Miledi 1972). The time of appearance of ACh receptors in chick myotubes in culture has been determined by binding of α-BGT and normally occurs after myoblast fusion (Prives & Paterson 1974). ACh receptors can also appear in the absence of fusion, as in the rat.

The chemosensitivity of a cultured clonal line of rat striated muscle cells (L6) has been investigated (Steinbach 1973, 1975). Some mononucleate myoblasts show a slow hyperpolarizing response to iontophoretically applied ACh in which the membrane potential is driven towards −60 mV. This response is not blocked by curare or α-cobratoxin, and may not be due to a conductance change. Myoblasts were never seen to give a depolarizing response to ACh (Harris et al 1971). After fusion, the multinucleate myotubes show a fast depolarizing response, with a reversal potential at −3 mV.

It is blocked by curare and α-cobratoxin and depends on an increase in permeability to Na^+, K^+, perhaps Ca^{2+}, but not Cl^-. The long-lasting hyperpolarizing response is gradually lost, but some cells initially show both responses. The time course of toxin binding site appearance in the cultures parallels cell fusion (Patrick et al 1972). L6 cells depolarized by elevated concentrations of K^+ are not able to make action potentials, but will do so in the presence of curare or α-BGT (Gartner, Land & Podleski 1976). These results suggest some interaction between the ACh receptor and the voltage dependent ion channels in these cells.

Cardiac and Smooth Muscle

Beating embryonic chick hearts at 44 hrs–3 weeks of age are arrested by bath application of ACh; this effect is blocked by atropine. From 44 hrs to 4 days intracellular recordings indicate that ACh causes a depolarization of the mean resting potential of atrial muscle and a slower rate of diastolic depolarization; the action potential duration is unchanged (Pappano 1972). The depolarization depends on an increase in Na^+ conductance, since it is blocked by replacement of this ion with Tris. In the absence of Na^+, ACh causes a hyperpolarization, which suggests that ACh increases K^+ conductance as well. By 6 days of development ACh causes hyperpolarization of atrial muscle, a slower rate of diastolic depolarization, and a decrease in duration of the action potential, consistent with an increase in K^+ conductance alone. The effect of ACh on adult vertebrate atrial muscle is to increase conductance to K^+, which yields a shortening of the action potential and hyperpolarization of the cells (Trautwein 1963). Thus it appears that the effect of ACh on the early embryonic heart is different from that on adult heart. The change in ionic dependence of the response occurs well in advance of innervation (see above). Ventricular cells are less affected by ACh than atrial cells, both in vivo and in vitro (Nakanishi & Takeda 1969); these results parallel the future normal distribution of cholinergic innervation. The rate of the chick heartbeat is increased by epinephrine between 40 hrs and 9 days of development (Fingl, Woodbury & Hecht 1952, Jones 1958). The effect of epinephrine on adult heart muscle is to decrease K^+ conductance and increase Ca^{2+} and possibly resting Na^+ conductance (Trautwein 1963, Hauswirth, Noble & Tsien 1968). It is not known if the effect on embryonic heart has the same mechanism. The embryonic rat heart has been less extensively studied (Hall 1957, Pager, Bernard & Gargouil 1965).

Smooth muscle cells cultured from taenia coli of newborn guinea pigs are depolarized by iontophoretic application of ACh, which produces large increases in membrane conductance (Purves 1974). The reversal potential for the response is −12 to −22 mV, compatible with values previously reported for intestinal smooth muscle in adults (−9 mV, Bolton 1972; −26 mV, Bennett 1966).

Egg Cells

Xenopus oocytes have been shown to be sensitive to neurotransmitters. Depolarizing, hyperpolarizing, and mixed response are recorded to the iontophoretic application of ACh; epinephrine, serotonin, and dopamine evoke membrane hyperpolarizations (Kusano, Miledi & Stinnakre 1977). The chemosensitivity of cells at early cleavage stages has not yet been reported.

MECHANISMS FOR APPEARANCE AND DISAPPEARANCE OF ION CHANNELS IN DEVELOPMENT

The changes in the ionic dependence of action potentials and neurotransmitter responses are likely to be due to changes in the channels responsible for specific ionic conductances, although changes in the ion concentration gradients driving the currents have not been excluded. The mechanisms by which such changes occur are difficult to study, in part because biochemical isolation of a functional ion channel has been hard to achieve. Rhodopsin can be incorporated into artificial membranes and appears to function as a light-activated ion channel (Montal, Darszon & Trissl 1977). Although the ACh receptor has been purified, its function as an ion channel is uncertain (Sobel et al 1978). Several mechanisms can be considered. First, there could be unequal partitioning of different channels in the membrane of dividing cells during development. Channels would appear or disappear by concentration or dilution. This process could entail premitotic partitioning between the membranes of the two future daughter cells, or postmitotic partitioning to different regions of the membrane of a single daughter. The former mechanism has been proposed for development of tunicate muscle (Okamoto, Takahashi & Yoshii 1976a). Second, modification of existing channels could occur through changes in the local ionic environment altering the tertiary structure of a molecule, or covalent modification, by such changes as phosphorylation, glycosylation, or peptide bond cleavage, altering the primary structure. Increases in intracellular Ca^{2+} can acutely increase the K^+ conductance of many neurons (e.g. Meech & Standen 1975), or reduce the voltage dependent Na^+ conductance in the tunicate egg (Takahashi & Yoshii 1978). The intracellular Ca^{2+} concentration is strongly buffered over the long term, however. Sustained changes in anaerobic glycolysis make some inexcitable adult muscles electrically excitable, probably by a change in intracellular pH (Moody 1978). In these cases the rapid time course of the change makes protein synthesis seem an unlikely explanation. Third, new channel proteins could be synthesized and inserted in the membrane. This mechanism is compatible with the demonstration of synthesis of ACh receptors in denervated striated muscle (Brockes & Hall 1975) and in myotube cultures (Devreotes & Fambrough 1976), and increased K^+ dependence of the resting potential of cells acquiring new membrane during cleavage of the *Xenopus* egg (deLaat & Bluemink 1974, but see DiCaprio, French & Sanders 1976). Finally, loss of channels could occur by degradation and removal from the membrane. Such a process has been demonstrated for the ACh receptors of denervated muscle (Berg & Hall 1975, Chang & Huang 1975) and in myotube cultures (Devreotes & Fambrough 1976, Merlie, Changeux & Gros 1976). Techniques for isolation of ion channels will be helpful in evaluating these possibilities.

POSSIBLE ROLES OF CA^{2+} ACTION POTENTIALS IN DEVELOPMENT OF NERVE AND MUSCLE

Although many nerve and muscle cells can make Ca^{2+} action potentials for a brief period early in development, their normal frequency in vivo is unknown at present and their function is still to be determined. One possibility is that they are important for the signalling ability of cells. The Na^+ gradient for some cells may be insufficient to produce an action potential because of large intracellular concentrations (Slack, Warner & Warren 1973, Slack & Warner 1975), and reliance on a steeper Ca^{2+} gradient may have developed. A second possibility is that the prolonged reversal of membrane potential is necessary for some other membrane event. Prolonged membrane depolarization during activation of sea urchin and *Urechis* eggs has been shown to block polyspermic fertilization (Jaffe 1976, Gould-Somero & Jaffe 1977), perhaps by blocking fusion of sperm and egg membranes. It may be that insertion of membrane proteins such as neurotransmitter receptors is facilitated by transient reduction or reversal of membrane potential (see Ohmori & Sasaki 1977). A third possibility is that Ca^{2+} is needed inside the cell for some aspect of development. The activation of a particular K^+ conductance depends on intracellular Ca^{2+} in many cells (Meech & Standen 1975), and could promote the progressive shortening of the action potential during development. Ca^{2+} could activate the contractile apparatus that may be involved in the elongation of neural processes (Bray & Bunge 1973) or the shortening of muscle sarcomeres. Alternatively, Ca^{2+} might be necessary for exocytotic secretion (Katz & Miledi 1965, Vacquier 1975). A fourth possibility follows from studies of the activation of eggs, that have shown that a Ca^{2+} influx leads to a release of intracellular Ca^{2+} and an efflux of H^+ (Hagiwara & Jaffe 1979 for review, see also Meech & Thomas 1977). Changing intracellular pH could have a broad effect on cellular metabolism. Recent studies indicate that Ca^{2+} entry may be important for the determination of neurotransmitter synthesis (Walicke, Campenot & Patterson, 1977). The consequences of Ca^{2+} influx into embryonic nerve and muscle cells can be further studied in culture, where it is possible to increase it with artificial ionophores or block it with various pharmacological agents.

CONCLUSIONS

There is a change in the ionic dependence and duration of action potentials during development, in neurons, striated muscle, and cardiac muscle that have been examined at early stages. In a number of cases inward current is carried in large part by Ca^{2+} at first, a Na^+ current appears later, and in some instances the Ca^{2+} current disappears at late stages of development. The change in ionic dependence is paralleled by a decrease in duration of the action potential.

There appears to be no developmental change in the ionic dependence of the response of striated muscle cells to neurotransmitters, when it has been examined from the time of first appearance. This conclusion must be considered tentative because most of the data indicate only a constancy of reversal potentials. When several conductances are involved, changes in one could be obscured by compensating changes in others, or masked by changes in ion gradients. In contrast, there appears to be a change in the ionic dependence of the response of chick cardiac muscle to a neurotransmitter. No conclusion can be reached with respect to the responses of neurons to neurotransmitters: although no developmental changes have been seen, the time of onset of the responses is unknown and studies to date may have been conducted too late.

Changes in chemosensitivity may be coincident with changes in electrical excitability. The onset of chemosensitivity in tunicate, chick, and rat (L6) skeletal muscle is paralleled by the appearance of chiefly Ca^{2+} dependent action potentials. The loss of Na^+ dependence of the ACh response of chick cardiac muscle occurs at about the same time as the appearance of the fast Na^+ component of the cardiac action potential. Further study is needed to determine if this correlation holds for neuronal development.

Innervation is not necessary for the development of electrical excitability of nerve or striated muscle, since the changes occur in dissociated cell cultures on a normal time course in several instances. Similarly the acquisiton of chemosensitivity by striated muscle in culture without neurons exhibits the time course and ionic dependence of cells in vivo. Changes in excitability and chemosensitivity of chick cardiac muscle occur prior to functional innervation. The rat heart may become innervated at the time of changes in excitability and chemosensitivity. Innervation does regulate the number and distribution of neurotransmitter receptors in the embryo and in the adult, however.

Cell fusion during myogenesis is not required for the development of some membrane properties, since myoblasts become chemosensitive even when fusion is prevented. It is not known if the normal development of electrical excitability depends on fusion in those cases in which it occurs.

The disappearance of the Ca^{2+} component of the action potential is not well correlated with the development of a transverse tubular system and sarcoplasmic reticulum for the intracellular storage and release of Ca^{2+} in skeletal and cardiac muscle.

The mechanisms by which the developmental changes occur, as well as the role of Ca^{2+} action potentials in development, remain to be elucidated.

ACKNOWLEDGEMENTS

I thank Corey Goodman and Alan Willard for permission to include unpublished results, and am especially grateful to them and to Kate Barald, Darwin Berg, and Janet Lamborghini for their critical reviews of the manuscript. I have profited greatly from valuable discussions with all the Spitzbergen community. I thank Jim Coulombe and Marsha Revenaugh for preparation of the figures and Mary Lou Space for typing the manuscript. Supported by grants from the NSF and the NIH.

Literature Cited

Baccaglini, P. I. 1978. Action potentials of embryonic dorsal root ganglion neurones in *Xenopus* tadpoles. *J. Physiol. London* In press

Baccaglini, P. I., Spitzer, N. C. 1977. Developmental changes in the inward current of the action potential of Rohon-Beard neurones. *J. Physiol. London* 271:93–117

Bate, C. M. 1976. Embryogenesis of an insect nervous system. I. A map of the thoracic and abdominal neuroblasts in *Locusta migratoria. J. Embryol. Exp. Morphol.* 35:107–23

Bennett, M. R. 1966. Model of the membrane of smooth muscle cell of the guinea pig taenia coli muscle during transmission from inhibitory and excitatory nerves. *Nature* 211:1149–52

Berg, D. K., Hall, Z. W. 1975. Loss of α-bungarotoxin from junctional and ex-

trajunctional acetylcholine receptors in rat diaphragm muscle *in vivo* and in organ culture. *J. Physiol. London* 252: 771–89

Berg, D. K., Kelly, R. B., Sargent, P. B., Williamson, P., Hall, Z. W. 1972. Binding of α-bungarotoxin to acetylcholine receptors in mammalian muscle. *Proc. Natl. Acad. Sci. USA* 69:147–51

Blackshaw, S., Warner, A. 1976. Onset of acetylcholine sensitivity and endplate activity in developing myotome muscles of *Xenopus. Nature* 262:217–18

Bolton, T. B. 1972. The depolarizing action of acetylcholine or carbachol in intestinal smooth muscle. *J. Physiol. London* 220:647–71

Bray, D. 1970. Surface movements during the growth of single explanted neurons. *Proc. Natl. Acad. Sci. USA* 65:905–10

Bray, D., Bunge, M. B. 1973. The growth cone in neurite extension. In *Locomotion of Tissue Cells. Ciba Foundation Symposium.* Amsterdam: Elsevier. 14: 195–209

Brockes, J. P., Hall, Z. W. 1975. Synthesis of acetylcholine receptor by denervated rat diaphragm muscle. *Proc. Natl. Acad. Sci. USA* 72:1368–72

Chang, C. C., Huang, M. C. 1975. Turnover of junctional and extrajunctional acetylcholine receptors of the rat diaphragm. *Nature* 253:643–44

Cohen, M. W., Kullberg, R. W. 1974. Temporal relationship between innervation and appearance of acetylcholine receptors in embryonic amphibian muscle. *Proc. Can. Fed. Biol. Soc.* 17:176

DeHaan, R. L., Sachs, H. G. 1972. Cell coupling in developing systems: The heart-cell paradigm. In *Current Topics in Developmental Biology,* ed. A. A. Moscona, A. Monroy, 7:193–228. New York: Academic.

deLaat, S. W., Bluemink, J. G. 1974. New membrane formation during cytokinesis in normal and cytochalasin B-treated eggs of *Xenopus laevis.* II. Electrophysiological observations. *J. Cell Biol.* 60:529–40

Devreotes, P. N., Fambrough, D. M. 1976. Synthesis of acetylcholine receptors by cultured chick myotubes and denervated mouse extensor digitorum longus muscles. *Proc. Natl. Acad. Sci. USA* 73:161–64

Diamond, J., Miledi, R. 1962. A study of foetal and new-born rat muscle fibres. *J. Physiol. London* 162:393–408

DiCaprio, R. A., French, A. S., Sanders, E. J. 1976. On the mechanism of electrical coupling between cells of early *Xenopus* embryos. *J. Membr. Biol.* 27:393–408

Dichter, M. A., Fischbach, G. D. 1977. The action potential of chick dorsal root ganglion neurones maintained in cell culture. *J. Physiol. London* 267:281–98

Dichter, M. A., Tischler, A. S., Greene, L. A. 1977. Nerve growth factor-induced increase in electrical excitability and acetylcholine sensitivity of a rat pheochromocytoma cell line. *Nature* 268: 501–4

Ezerman, E. B., Ishikawa, H. 1967. Differentiation of the sarcoplasmic reticulum and T system in developing chick skeletal muscle *in vitro. J. Cell Biol.* 35:405–20

Fambrough, D. M. 1976. Development of cholinergic innervation of skeletal, cardiac, and smooth muscle. In *Biology of Cholinergic Function,* ed. A. M. Gold-berg, I. Hanin, pp. 101–60. New York: Raven.

Fambrough, D., Rash, J. E. 1971. Development of acetylcholine sensitivity during myogenesis. *Dev. Biol.* 26:55–68

Fingl, E., Woodbury, L. A., Hecht, H. H. 1952. Effects of innervation and drugs upon direct membrane potentials of embryonic chick myocardium. *J. Pharmacol. Exp. Ther.* 104:103–14

Fischbach, G. D. 1972. Synapse formation between dissociated nerve and muscle cells in low density cell cultures. *Dev. Biol.* 28:407–29

Fischbach, G. D., Dichter, M. A. 1974. Electrophysiologic and morphologic properties of neurons in dissociated chick spinal cord cultures. *Dev. Biol.* 37:100–16

Fischbach, G. D., Nelson, P. G. 1977. Cell culture in neurobiology. In *Handbook of Physiology Section 1: The Nervous System, Vol. 1, Part 1.* ed. J. M. Brookhart, V. B. Mountcastle, E. R. Kandel, pp. 719–74. Baltimore: Williams & Wilkins.

Furshpan, E. J., Potter, D. D. 1968. Low-resistance junctions between cells in embryos and tissue culture. In *Current Topics in Developmental Biology,* ed. A. A. Moscona, A. Monroy 3:95–127. New York: Academic.

Gartner, T. K., Land, B., Podleski, T. R. 1976. Genetic and physiological evidence concerning the development of chemically sensitive voltage-dependent ionophores in L6 cells. *J. Neurobiol.* 7:537–49

Goodman, C. G. 1979. Isogenic grasshoppers: Genetic variability and development of identified neurons. In *Topics in Neurogenetics,* ed. X. Breakefield. New York: Elsevier North Holland. In press

Gould-Somero, M., Jaffe, L. A. 1977. Electrically mediated fast polyspermy block in eggs of the marine worm *Urechis caupo. J. Cell. Biol.* 75:37a

Gymer, A., Edwards, J. S. 1967. The development of the insect nervous system. I. An analysis of postembryonic growth in the terminal ganglion of *Acheta domesticus. J. Morphol.* 123:191–97

Hagiwara, S., Jaffe, L. A. 1979. Electrical properties of egg cell membranes. *Ann. Rev. Biophys. Bioeng.* 8 In press

Hagiwara, S., Miyazaki, S. 1977. Ca and Na spikes in egg cell membrane. In *Cellular Neurobiology,* ed. Z. Hall, R. Kelly, C. F. Fox, 15:147–58. New York: Liss.

Hall, E. K. 1957. Acetylcholine and epinephrine effects on the embryonic rat heart. *J. Cell Comp. Physiol.* 49:187–200

Harris, A. J., Dennis, M. J. 1970. Acetylcholine sensitivity and distribution on mouse neuroblastoma. *Science* 167:1253–55

Harris, A. J., Heinemann, S., Schubert, D., Tarakis, H. 1971. Trophic interaction between cloned tissue culture lines of nerve and muscle. *Nature* 231:296–301

Harris, J. B., Marshall, M. W. 1973. Tetrodotoxin-resistant action potentials in newborn rat muscle. *Nature New Biol.* 243:191–92

Harris, J. B., Marshall, M. W., Wilson, P. 1973. A physiological study of chick myotubes grown in tissue culture. *J. Physiol. London* 229:751–66

Hauswirth, O., Nobel, D., Tsien, R. W. 1968. Adrenaline: Mechanism of action on the pacemaker potential in cardiac Purkinje fibers. *Science* 162:916–17

Heitler, W. J., Goodman, C. G. 1978. Multiple sites of spike initiation in a bifurcating locust neurone. *J. Exp. Biol.* In press

Herrmann, H., Heywood, S. M., Marchok, A. P. 1970. Reconstruction of muscle development as a sequence of macromolecular syntheses. In *Current Topics in Developmental Biology*, ed. A. A. Moscona, A. Monroy 5:181–234. New York: Academic

Hoyle, G., Dagan, D., Moberly, B., Colquhoun, W. 1974. Dorsal unpaired median insect neurons make neurosecretory endings on skeletal muscle. *J. Exp. Zool.* 187:159–65

Jaffe, L. A. 1976. Fast block to polyspermy in sea urchin eggs is electrically mediated. *Nature* 261:68–71

Jaffe, L. A., Robinson, K. R. 1978. Membrane potential of the unfertilized sea urchin egg. *Dev. Biol.* 62:215–28

Jaffe, L. F., Nuccitelli, R. 1977. Electrical controls of development. *Ann. Rev. Biophys. Bioeng.* 6:445–76

Jones, D. S. 1958. Effects of acetylcholine and adrenalin on the experimentally uninnervated heart of the chick embryo. *Anat. Rec.* 130:253–59

Kano, M. 1975. Development of excitability in embryonic chick skeletal muscle cells. *J. Cell Physiol.* 86:503–10

Kano, M., Shimada, Y. 1973. Tetrodotoxin-resistant electric activity in chick skeletal muscle cells differentiated *in vitro*. *J. Cell Physiol.* 81:85–90

Kano, M., Shimada, Y., Ishikawa, K. 1972. Electrogenesis of embryonic chick skeletal muscle cells differentiated *in vitro*. *J. Cell Physiol.* 79:363–66

Kano, M., Yamamoto, M. 1977. Development of spike potentials in skeletal muscle cells differentiated *in vitro* from chick embryo. *J. Cell Physiol.* 90:439–44

Katz, B., Miledi, R. 1965. The effect of calcium on acetylcholine release from motor nerve terminals. *Proc. R. Soc. London Ser. B* 161:496–503

Katz, B., Miledi, R. 1972. The statistical nature of the acetylcholine potential and its molecular components. *J. Physiol. London* 224:665–99

Kent, G. C. 1969. *Comparative anatomy of the vertebrates.* St. Louis: Mosby. 2nd ed. 437 pp.

Kidokoro, Y. 1973. Development of action potentials in a clonal rat skeletal muscle cell line. *Nature New Biol.* 241:158–59

Kidokoro, Y. 1975a. Developmental changes of membrane electrical properties in a rat skeletal muscle cell line. *J. Physiol. London* 244:129–43

Kidokoro, Y. 1975b. Sodium and calcium components of the action potential in a developing skeletal muscle cell line. *J. Physiol. London* 244:145–59

Klier, F. G., Schubert, D., Heinemann, S. 1977. The ultrastructural differentiation of the clonal myogenic cell line L6 in normal and high K^+ medium. *Dev. Biol.* 57:440–49

Kullberg, R. W. 1974. *Onset and development of synaptic activity at an amphibian neuromuscular junction.* PhD thesis. McGill Univ. Montreal. 173 pp.

Kullberg, R. W., Lentz, T. L., Cohen, M. W. 1977. Development of the myotomal neuromuscular junction in *Xenopus laevis:* An electrophysiological and fine-structural study. *Dev. Biol.* 60:101–29

Kusano, K., Miledi, R., Stinnakre, J. 1977. Acetylcholine receptors in the oocyte membrane. *Nature* 270:739–41

Land, B. R., Sastre, A., Podleski, T. R. 1973. Tetrodotoxin-sensitive and -insensitive action potentials in myotubes. *J. Cell Physiol.* 82:497–510

Lieberman, M., Sano, T. 1976. *Developmental and Physiological Correlates of Cardiac Muscle.* New York: Raven. 322 pp.

Lømo, T. 1976. The role of activity in the control of membrane and contractile properties of skeletal muscle. In *Motor Innervation of Muscle,* ed. S. Thesleff, pp. 289–331. London: Academic

Matsuda, Y., Yoshida, S., Yonezawa, T. 1976. A Ca-dependent regenerative response in rodent dorsal root ganglion cells cultured *in vitro*. *Brain Res.* 115:334–38

Meech, R. W., Standen, N. B. 1975. Potassium activation in *Helix aspersa* neurones under voltage clamp: A component mediated by calcium influx. *J. Physiol. London* 249:211–39

Meech, R. W., Thomas, R. C. 1977. The effect of calcium injection on the intracellular sodium and pH of snail neurones. *J. Physiol. London* 265:867–79

Merlie, J. P., Changeux, J. P., Gros, F. 1976. Acetylcholine receptor degradation measured by pulse chase labelling. *Nature* 264:74–76

Miledi, R., Potter, L. T. 1971. Acetylcholine receptors in muscle fibers. *Nature* 233:599–603

Miyake, M. 1978. The development of action potential mechanism in a mouse neuronal cell line *in vitro. Brain Res.* 143:349–54

Miyazaki, S., Takahashi, K., Tsuda, K. 1972. Calcium and sodium contributions to regenerative responses in the embryonic excitable cell membrane. *Science* 176:1441–43

Miyazaki, S., Takahashi, K., Tsuda, K. 1974. Electrical excitability in the egg cell membrane of the tunicate. *J. Physiol. London* 238:37–54

Miyazaki, S., Takahashi, K., Tsuda, K., Yoshii, M. 1974. Analysis of non-linearity observed in the current-voltage relation of the tunicate embryo. *J. Physiol. London* 238:55–77

Montal, M., Darszon, A., Trissl, H. W. 1977. Transmembrane channel formation in rhodopsin-containing bilayer membranes. *Nature* 267:221–25

Moody, W. Jr. 1978. Gradual increase in the electrical excitability of crayfish slow muscle fibers produced by anoxia or uncouplers of oxidative phosphorylation. *J. Comp. Physiol.* In press

Moolenaar, W. H., Spector, I. 1977. Membrane currents examined under voltage clamp in cultured neuroblastoma cells. *Science* 196:331–33

Muntz, L. 1964. *Neuromuscular foundations of behaviour in embryonic and larval stages of the anuran, Xenopus laevis.* PhD. thesis. Univ. Bristol. 97 pp.

Muntz, L. 1975. Myogenesis in the trunk and leg during development of the tadpole of *Xenopus laevis* (Daudin 1802). *J. Embryol. Exp. Morphol.* 33:757–74

Nakanishi, H., Takeda, H. 1969. Effect of acetylcholine on the electrical activity of cultured chick embryonic heart. *Jpn. J. Pharmacol.* 19:543–50

Narahashi, T. 1974. Chemicals as tools in the study of excitable membranes. *Physiol. Rev.* 54:813–89

Nelson, P. G. 1973. Electrophysiological studies of normal and neoplastic cells in tissue culture. In *Tissue Culture of the Nervous System,* ed. G. Sato, pp. 135–160. New York: Plenum

Nelson, P. G., Peacock, J. H., Amano, T. 1971. Responses of neuroblastoma cells to iontophoretically applied acetylcholine. *J. Cell Physiol.* 77:353–62

Nelson, P., Ruffner, W., Nirenberg, M. 1969. Neuronal tumor cells with excitable membranes grown *in vitro. Proc. Natl. Acad. Sci. USA* 64:1004–10

Nieuwkoop, P. D., Faber, J. 1956. *Normal Table of Xenopus laevis (Daudin).* Amsterdam: North Holland. 252 pp.

Obata, K. 1974. Transmitter sensitivities of some nerve and muscle cells in culture. *Brain Res.* 73:71–88

Ohmori, H., Sasaki, S. 1977. Development of neuromuscular transmission in a larval tunicate. *J. Physiol. London* 269:221–54

Okamoto, H., Takahashi, K., Yoshii, M. 1976a. Membrane currents of the tunicate egg under the voltage-clamp condition. *J. Physiol. London* 254:607–38

Okamoto, H., Takahashi, K., Yoshii, M. 1976b. Two components of the calcium current in the egg cell membrane of the tunicate. *J. Physiol. London* 255:527–61

O'Lague, P. H., Potter, D. D., Furshpan, E. J. 1978. Studies on rat sympathetic neurons developing in cell culture. I. Growth characteristics and electrophysiological properties. *Dev. Biol.* In press

Pager, J., Bernard, C., Gargouil, Y.-M. 1965. Evolution, au cours de la croissance foetale, des effets de l'acétylcholine au niveau de l'oreillette du rat. *C. R. Seances Soc. Biol. Paris* 159:2470–75

Palmer, J. F., Slack, C. 1970. Some bio-electric parameters of early *Xenopus* embryos. *J. Embryol. Exp. Morphol.* 24:535–54

Pappano, A. J. 1972. Sodium-dependent depolarization of noninnervated embryonic chick heart by acetylcholine. *J. Pharmacol. Exp. Ther.* 180:340–50

Patrick, J., Heinemann, S. F., Lindstrom, J., Schubert, D., Steinbach, J. H. 1972. Appearance of acetylcholine receptors during differentiation of a myogenic cell line. *Proc. Natl. Acad. Sci. USA* 69:2762–66

Patrick, J., Heinemann, S., Schubert, D. 1978. Biology of cultured nerve and muscle. *Ann. Rev. Neurosci.* 1:417–43

Peacock, J. H., Nelson, P. G. 1973. Chemosensitivity of mouse neuroblastoma cells *in vitro. J. Neurobiol.* 4:363–74

Pfenninger, K. H., Bunge, R. P. 1974. Freeze-fracturing of nerve growth cones and young fibers. A study of developing plasma membrane. *J. Cell Biol.* 63:180–96

Prestige, M. C. 1965. Cell turnover in the spinal ganglia of *Xenopus laevis* tadpoles. *J. Embryol. Exp. Morphol.* 13:63–72

Prives, J. M., Paterson, B. M. 1974. Differentiation of cell membranes in cultures of embryonic chick breast muscle. *Proc. Natl. Acad. Sci. USA* 71:3208–11

Purves, D. 1976. Long-term regulation in the vertebrate peripheral nervous system. In *International Review of Physiology, Neurophysiology II,* 10:125–177. Baltimore: Univ. Park Press

Purves, R. D. 1974. Muscarinic excitation: A microelectrophoretic study on cultured smooth muscle cells. *Br. J. Pharmacol.* 52:77–86

Ransom, B. R., Bullock, P. N., Nelson, P. G. 1977. Mouse spinal cord in cell culture. III. Neuronal chemosensitivity and its relationship to synaptic activity. *J. Neurophysiol.* 40:1163–77

Ransom, B. R., Holz, R. W. 1977. Ionic determinants of excitability in cultured mouse dorsal root ganglion and spinal cord cells. *Brain Res.* 136:445–53

Ritchie, A. K., Fambrough, D. M. 1975. Ionic properties of the acetylcholine receptor in cultured rat myotubes. *J. Gen. Physiol.* 65:751–67

Roonwal, M. L. 1937. Studies on the embryology of the African migratory locust, *Locusta migratoria migratoroides. Philos. Trans. R. Soc. London Ser. B* 227:175–224

Sachs, F., Lecar, H. 1973. Acetylcholine noise in tissue culture muscle cells. *Nature New Biol.* 246:214–16

Sbrenna, G. 1971. Postembryonic growth of the ventral nerve cord in *Schistocerca gregaria* Forst (Orthoptera: Acrididae). *Boll. Zool.* 38:49–74

Schiaffino, S., Cantini, M., Sartore, S. 1977. T-system formation in cultured rat skeletal muscle. *Tissue Cell* 9:437–46

Schubert, D., Harris, A. J., Heinemann, S., Kidokoro, Y., Patrick, J., Steinbach, J. H. 1973. Differentiation and interaction of clonal cell lines of nerve and muscle. In *Tissue Culture of the Nervous System,* ed. G. Sato, pp. 55–86. New York: Plenum

Slack, C., Warner, A. E. 1975. Properties of surface and junctional membranes of embryonic cells isolated from blastula stages of *Xenopus laevis. J. Physiol. London* 248:97–120

Slack, C., Warner, A. E., Warren, R. L. 1973. The distribution of sodium and potassium in amphibian embryos during early development. *J. Physiol. London* 232:297–312

Sobel, A., Heidmann, T., Hofler, J., Changeux, J.-P. 1978. Distinct protein components from *Torpedo marmorata* membranes carry the acetylcholine receptor site and the binding site for local anesthetics and histrionicotoxin. *Proc. Natl. Acad. Sci. USA* 75:510–14

Spector, I., Prives, J. M. 1977. Development of electrophysiological and biochemical membrane properties during differentiation of embryonic skeletal muscle in culture. *Proc. Natl. Acad. Sci. USA* 74:5166–70

Spitzer, N. C. 1976. Chemosensitivity of embryonic amphibian neurons *in vivo* and *in vitro. Abstr. Soc. Neurosci. 6th Ann. Meet.* p. 204

Spitzer, N. C., Baccaglini, P. I. 1975. Changes in the ionic basis of the action potential in Rohon-Beard neurons during development. *Abstr. Soc. Neurosci. 5th Ann. Meet.* p. 782

Spitzer, N. C., Baccaglini, P. I. 1976. Development of the action potential in embryonic amphibian neurons *in vivo. Brain Res.* 107:610–16

Spitzer, N. C., Lamborghini, J. E. 1976. The development of the action potential mechanism of amphibian neurons isolated in culture. *Proc. Natl. Acad. Sci. USA* 73:1641–45

Spitzer, N. C., Spitzer, J. L. 1975. Time of origin of Rohon-Beard neurons in spinal cord of *Xenopus laevis. Am. Zool.* 15:781

Steinbach, J. H. 1973. *Nerve-muscle interaction in vitro: A study of some requirements for localization of acetylcholine sensitivity.* PhD thesis. Univ. California, San Diego. 141 pp.

Steinbach, J. H. 1975. Acetylcholine responses in clonal myogenic cells *in vitro. J. Physiol. London* 247:393–405

Stewart, W. W. 1978. Functional connections between cells as revealed by dye-coupling with a highly fluorescent naphthalimide tracer. *Cell* 14:741–59

Takahashi, K., Miyazaki, S., Kidokoro, Y. 1971. Development of excitability in embryonic muscle cell membranes in certain tunicates. *Science* 171:415–18

Takahashi, K., Yoshii, M. 1978. Effects of internal free Ca upon the Na and Ca channels in the tunicate egg analysed by the internal perfusion technique. *J. Physiol. London* 279:519–49

Takeuchi, N. 1963. Some properties of conductance changes at the end-plate membrane during the action of acetylcholine. *J. Physiol. London* 167:128–40

Trautwein, W. 1963. Generation and conduction of impulses in the heart as affected by drugs. *Pharmacol. Rev.* 15:277–332

Vacquier, V. D. 1975. The isolation of intact cortical granules from sea urchin eggs: Calcium ions trigger granule discharge. *Dev. Biol.* 43:62–74

Walicke, P. A., Campenot, R. B., Patterson, P. H. 1977. Determination of transmitter function by neuronal activity. *Proc. Natl. Acad. Sci. USA* 74:5767–71

Warner, A. E. 1973. The electrical properties of the ectoderm in the amphibian embryo during induction and early development of the nervous system. *J. Physiol. London* 235:267–86

Wheeler, W. M. 1893. A contribution to insect embryology. *J. Morphol.* 8:1–160

Willard, A. L. 1977. Action potential mechanisms in processes of amphibian neurons developing in culture. *Abstr. Soc. Neurosci. 7th Ann. Meet.* p. 123

Williams, P. E., Goldspink, G. 1971. Longitudinal growth of striated muscle fibers. *J. Cell Sci.* 9:751–67

Woodward, D. J., Hoffer, B. J., Siggins, G. R., Bloom, F. E. 1971. The ontogenetic development of synaptic junctions, synaptic activation and responsiveness to neurotransmitter substances in rat cerebellum. *Brain Res.* 34:73–97.

Ann. Rev. Neurosci. 1979. 2:399–446

BIOCHEMISTRY OF NEUROTRANSMITTER RELEASE

♦11529

Regis B. Kelly, James W. Deutsch, Steven S. Carlson, and John A. Wagner

Department of Biochemistry and Biophysics, University of California, San Francisco, California 94143

INTRODUCTION

Thanks to the precision of electrophysiological measurements a detailed analysis of the extent and time course of transmitter release is available. The invasion of a nerve terminal by an action potential opens voltage-dependent calcium channels transiently. The brief influx of calcium rapidly stimulates release of quanta of transmitter, which in turn produce postsynaptic change in conductance. Anatomical studies of the morphology of nerve terminals have correlated the quantum of transmitter release with exocytosis of the contents of a single synaptic vesicle. The exocytotic event occurs at morphologically specialized regions of the nerve terminal. By comparison with the wealth of physiological and anatomical data, our knowledge of the molecular events that occur at the nerve terminal is meager indeed. However, recent developments in membrane biochemistry, in our understanding of non-neuronal secretory systems, and in the purification and characterization of nerve terminal components encourage the hope that in the next few years the biochemical basis of transmitter release will be unraveled.

We review here what is presently known of the molecular interactions that take place during the release of neurotransmitter from nerve terminals. We discuss 1. some of the more recent anatomical and electrophysiological observations on neurotransmitter release that either suggest molecular mechanisms or place restrictions on what molecular models may be constructed; 2. the various models suggested by the anatomical and electrophysiological observations, in the light of our current knowledge of non-neural secretory cells and model membrane systems; and 3. the extent

399

0147-006X/79/0315-0399$01.00

to which these models can be verified by data from experiments involving (a) the isolation and identification of the components of the nerve terminal; (b) the reconstruction of the secretory process in vitro; and (c) the inhibition of transmitter release by specific presynaptic neurotoxins. Space limitations oblige us to refer to reviews rather than original references wherever possible.

MORPHOLOGICAL AND ELECTROPHYSIOLOGICAL FINDINGS

The Active Zone

The close correlation between the quantal content measured electrophysiologically and the number of exocytotic sites detected by freeze fracture is strong evidence that transmitter release involves exocytosis (Heuser et al 1978). Important clues to the molecular basis of transmitter release are given by the morphological specializations in nerve terminals (Jones 1975, Heuser & Reese 1977, Holtzman 1977). The most characteristic feature of the nerve terminal is the synaptic vesicle, which corresponds to the quantum of transmitter release (Katz 1969). Synaptic vesicles are not uniformly distributed in the nerve terminal, but are clustered around dense projections at the presynaptic plasma membrane, which stain heavily with phosphotungstic acid and bismuth iodide (Birks et al 1960, Bloom & Aghajanian 1968, Akert et al 1972). Whether or not the vesicle clustering is due to a matrix holding them together or to their exclusion from the filament-containing region inside nerve terminals (Gray 1975) is not clear. The structure of the active zone is most easily interpreted at the frog neuromuscular junction. A bar, 40 nm wide, concave towards the cytoplasm, traverses the nerve terminal perpendicular to its long axis. The densely staining material is found on the cytoplasmic side of the bar, while on each side of it run two rows of intramembranous particles, which could be calcium channels (Heuser et al 1974). Along each side of the dense material is a single row of vesicles, separated from the presynaptic membrane by a gap of less than 50 Å. There are approximately 50 closely opposed vesicles per terminal bar, and about 500 such bars per neuromuscular junction (Heuser 1976, 1978). Very similar presynaptic morphologies are found at synapses in the central nervous system (Akert et al 1975). These morphological specializations of the nerve terminal have been called active zones (Couteaux & Pecot-Dechavassine 1974).

If nerve terminals are stimulated continuously during fixation (Couteaux & Pecot-Dechavassine 1970, Dreyer et al 1973, Heuser et al 1974), or are quick frozen immediately after stimulation (Heuser 1977, Heuser et al 1978), vesicles fuse with the presynaptic membrane exclusively at the active zone. If the quantal content is raised to unphysiologically high levels either by blocking the voltage-dependent potassium channels with 4-aminopyridine (Lundh & Thesleff 1977, Heuser et al 1978) or by electrotonic depolarization (Katz & Miledi 1977b), more than one vesicle is released per stimulus per active zone. At maximal stimulation, 10 to 20 of the 50 vesicles per active zone fuse with the membrane (Heuser et al 1978). It appears that the closely apposed vesicles constitute a readily releasable pool. Even when release occurs in the absence of extracellular calcium as, for example, in the presence of Brown Widow Spider venom, most of the release occurs at the active zone, although a small fraction of exocytosis may occur at other sites (Pumplin & Reese 1977). The anatomical results suggest, therefore, that the active zone is a region of the nerve terminal specialized for exocytosis, either because it is the location of the exocytotic machinery, or of the Ca^{++} entry sites, or both.

Membrane recycling after transmitter release has been extensively reviewed (Heuser & Reese 1977, Holtzman 1977). Anatomical studies point to the presence of large intramembranous particles that are associated with synaptic vesicles at rest, with the plasma membrane during exocytosis, and with the coated vesicles during endocytosis (Heuser & Reese 1975). From the biochemical point of view it is important to know which membrane components of the synaptic vesicles, plasma membranes, coated vesicles, and cisternae are common to all, and which are unique.

Role of Calcium in Transmitter Release

Calcium entry into neurons, its role in transmitter release, and its removal from the intraterminal cytoplasm have been recently reviewed (Llinas 1977, Parsegian 1977, Rahamimoff 1976, Llinas & Heuser 1977, Blaustein et al 1977). Evidence that release is due to elevated intracellular calcium stems mainly from microinjection studies (Miledi 1973) and from light emission measurements of aequorin injected into nerve terminals (Llinas et al 1972, Llinas & Nicholson 1975). There is little question that elevated intraterminal calcium results from the opening of voltage-sensitive channels in the nerve terminal membrane that are specific for calcium, and, to a lesser extent, for strontium and barium (Hubbard et al 1968a, b). External magnesium competes with external calcium, although it facilitates release in Ca^{++}-free media (Hurlbut et al 1971). Any discussion of the role of membrane fusion in exocytosis must consider to what extent magnesium could substitute for calcium in this process, assuming it could enter the terminal. Although Miledi (1973) reported that Mg^{++} was "practically ineffective" in evoking release after intraterminal injection, no data were presented. A more quantitative determination of the ability of intraterminal magnesium to cause exocytosis in vivo would be of great value.

CALCIUM INFLUX   Llinas and collaborators (Llinas 1977) have measured calcium currents in voltage-clamped squid stellate ganglia that have their Na^+ and K^+ channels blocked. From measurements of the delay between the onset of the voltage clamp and the calcium current, they showed that about 80% of the synaptic delay is due to the time required to activate calcium channels. Therefore less than 200 μsec remains for exocytosis induced by internal calcium, for transmitter diffusion across the cleft, and for activation of the postsynaptic channels. The time for exocytosis itself must be of the order of 100 μsec or less and is probably even shorter at the vertebrate neuromuscular junction. A second conclusion from their work is that the postsynaptic depolarization is observed for only a few msec after the calcium current ceases, which provides a measure of the efficiency with which intraterminal calcium must fall to normal levels. Thirdly, during the "off-response," the burst of transmitter release that follows the removal of the depolarizing voltage, transmitter is released even when the nerve terminal is no longer depolarized. This makes it improbable that membrane depolarization plays an important role in the release process (Remler 1973), beyond its role in activating calcium entry. A final point is that the rate of release appears to be linearly related to the inward calcium current, a result difficult to reconcile with the nonlinear dependence of release on extracellular calcium (Dodge & Rahamimoff 1967, Hubbard et al 1968b, Katz & Miledi 1970, Cooke et al 1973).

The question of whether the release process is linear or nonlinear with intracellular calcium is an important one for biochemical identification of the Ca^{++}-binding site. From the electrophysiological data alone, a nonlinear response to calcium is expected. Blocking the voltage dependent K^+ channels with 4-aminopyridine can increase the rate of transmitter release after nerve stimulation to approximately 10^3 quanta/msec, which is about 10^6 times higher than the rate of spontaneous release (Heuser et al 1978). If the release rates were linearly related to the internal calcium concentration and the initial internal calcium concentration were 5×10^{-8}M (Dipolo et al 1976) then a 10^6-fold increase would require the internal calcium to reach 5×10^{-2}M. If instead a third power dependence on free internal calcium were assumed, then the internal concentration would only have to reach 5×10^{-6}M. Since a release mechanism linearly dependent on internal calcium is rather unlikely, linearity with calcium current into the terminal (Llinas 1977) must be explained. The discrepancy could be resolved if the relationship between the internal calcium concentration and neurotransmitter release were sigmoidal. Experiments done at low internal calcium concentrations would be in an exponential region of the curve, while experiments that produce a high internal calcium concentration (i.e. those done in the presence of 4-aminopyridine and 40 mM external calcium) could be in a more linear range. Alternatively, even if the measured currents are truly calcium currents, since the internal calcium depends on the rate of calcium removal as well as rate of calcium entry, the relationship between calcium currents and internal calcium concentrations can be complex.

CALCIUM REMOVAL   As mentioned earlier, reduction of Ca^{++} entry into the nerve terminal is quickly followed by reduction in transmitter release. It is probably valid to conclude that the calcium that enters is rapidly scavenged by calcium-buffering systems, which could include mitochondria (Bygrave 1978, Rahamimoff 1976), synaptic or coated vesicles (Breer et al 1977, Blitz et al 1977), intraterminal membranes (Kendrick et al 1977), and the presynaptic plasma membrane itself (Babel-Guerin et al 1977). Since the relative merits of each of these systems has been reviewed recently (Blaustein et al 1977, Llinas & Heuser 1977), we restrict ourselves to emphasizing that there may be more than one site of calcium removal. It has been known for some time that after a stimulus the probability of release decays not as a simple exponential but as a sum of exponentials (Magleby & Zengel 1975, 1976). Several rates of decay could mean there are independent mechanisms of calcium removal, possessing different affinities for calcium. Support for this hypothesis comes from the work of Zengel & Magleby (1977) who have shown that two of the decay rates have different selectivities for divalent cations.

CA^{++} INDEPENDENT RELEASE   Although transmitter release can occur in the absence of external calcium, it is common to assume that calcium is still required but comes from intraterminal storage sites (Lowe et al 1976). Evidence for this is scant. The best evidence for calcium involvement in Ca^{++} independent release is the demonstration that if the extracellular calcium level is sufficiently low, an increase in calcium permeability can cause the rate of release to drop (Rotshenker et al 1976, Shimoni et al 1977). Calcium chelators that penetrate membranes have also been shown to reduce Ca^{++} independent release rates (Ornberg & Smyth 1976).

Conclusions

The anatomical and physiological observations suggest molecular models that must be tested biochemically. How the morphological components of the release site, the bar, the dense material, the vesicles, and the rows of intramembranous particles are involved in exocytosis would be easier to understand if we knew their composition. The electrophysiological observations suggest that calcium "effectors" be sought. Furthermore, the precise kinetic measurements that are possible by electrophysiological techniques place limits on what types of molecular models should be expected. This point is taken up in the next section.

MODELS OF EXOCYTOSIS

The anatomical and electrophysiological results can be summarized by dividing the release process into two stages (Fig. 1). The first stage is assembly of an active zone at the nerve terminal membrane. As illustrated in Figure 1A, the sequence of steps in this stage are filling the vesicle with transmitter, movement of the vesicle up to the active zone, and recognition of the active zone region by the vesicle (Holtzman 1977, Heuser & Reese 1977). Movement of vesicles to the active zone could occur by association with filamentous structures (Gray 1975) or by random motion. Since the time for a vesicle to diffuse one vesicle diameter is about 200 μsec (Parsegian 1977) and nerve terminals are rarely stimulated faster than once every 10 msec, diffusion of vesicles alone would be adequate to refill the sites. The striking order of vesicles at the active zone make it likely that the third step involves recognition of some distinct feature of the vesicle by a component of the active zone, such as the dense material or the plasma membrane, or both. The second stage of transmitter release (Fig. 1B) is the exocytotic event itself triggered by the influx of calcium. Exocytosis must involve movement of the vesicle from a distance of about 50 Å to about 5 to 10 Å from the plasma membrane followed by membrane fusion. Note that in the figure we suggest that separate machinery may exist for exocytosis and for recognition.

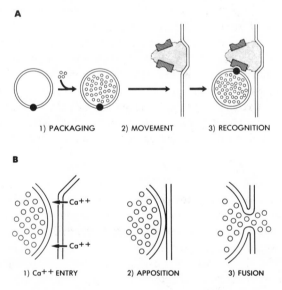

A

1) PACKAGING   2) MOVEMENT   3) RECOGNITION

B

1) Ca++ ENTRY   2) APPOSITION   3) FUSION

Figure 1   Partial reactions during exocytosis. *A* The three steps involved in bringing a filled vesicle to the active site. *B* The three steps involved in release of transmitter.

In this review we focus primarily on exocytosis and discuss the mechanism of assembly of the active zone only when pertinent. Electrophysiological studies suggest that calcium triggers exocytosis by combining with an effector. This effector may be protein or phospholipid. In this section we look for clues to help us differentiate between protein and phospholipid mechanisms 1. by comparing the calcium-binding affinities of phospholipid and protein to the estimated intraterminal calcium concentration, 2. by assessing protein requirements for membrane apposition and fusion, and 3. by exploring the capacities of protein and phospholipid models to show nonlinear dependence on calcium concentration and to show divalent cation selectivity.

Intraterminal Ca^{++} Concentration During Exocytosis

From the data of Llinas and collaborators, Parsegian has argued that the calcium entering in a few hundred μsec can only diffuse 1000–2000 Å from the entry site. This would dictate that only vesicles in close proximity to calcium entry sites could undergo exocytosis, a conclusion that fits well with the suggestion of Heuser et al (1974) that the rows of particles along active zones are in fact calcium channels. Interestingly, intramembranous particles found near sites of exocytosis in unicellular organisms (Satir et al 1973), could also be calcium channels. Satir & Oberg (1977) showed that a nonsecretory mutant of Tetrahymena did not have the "rosette" of particles at the known release site, but that discharge in this mutant could be stimulated by calcium and a calcium ionophore.

If it is assumed that the calcium that enters during nerve stimulation is uniformly distributed throughout the 2000 Å layer of cytoplasm closest to the presynaptic membrane in the squid stellate ganglion, the internal calcium concentration at the release site can be calculated to reach about 10^{-5}M (Parsegian 1977; corrected in Llinas & Heuser 1977). This number is close to that expected for release that is dependent on the third power of the internal calcium, as discussed above. Note that the estimated value would be higher if (a) the calcium diffusion constant were lower than 6×10^{-7} cm^{2}sec^{-1} in the nerve terminal, (b) calcium entered only at active zones, and (c) the internal calcium were not uniformly distributed, but preferentially associated with negatively charged membrane surfaces. The concentration of divalent cations in the vicinity of a negatively charged surface can be considerably higher than in bulk solution (McLaughlin et al 1971, Muller & Finkelstein 1974, Hauser et al 1976), even at physiological ionic strength. Because of these unknown parameters, the most reasonable conclusion is that internal calcium concentrations in the region of the exocytotic site reach between 10^{-6} and 10^{-3}M during the peak of transmitter release.

The approximate calcium concentration at the active zone can also be estimated on kinetic grounds, assuming that about half the effector sites have to be filled during evoked release of transmitter. This assumption is not unreasonable for cooperative processes. If half or more of the binding sites have to be occupied in 100 μsec, then if k_1 is the bimolecular association rate constant for Ca^{++} binding to the effector, k_2 is the corresponding dissociation rate constant and $[Ca^{++}]_{int}$ is the internal calcium concentration, $k_1[Ca^{++}]_{int} + k_2 > 0.7 \times 10^4$ sec^{-1}. If Ca^{++} stays bound for a millisecond or longer, then k_2 can be ignored. The largest possible value of k_1, about 10^9 M^{-1} sec^{-1}, is for a diffusion-controlled reaction between small molecules. Thus $[Ca^{++}]_{int}$ has to be at least 10^{-5}M for half-saturation to occur or 10^{-6}M for 5% binding to occur. It is unlikely that $[Ca^{++}]_{int}$ can be greater than 10^{-3}M, which corresponds to a k_1 of 10^7 M^{-1} sec^{-1} for half-saturation. We conclude that whatever the nature of the calcium binding site, if more than 5% of the sites are occupied during exocytosis, the internal calcium concentration must be in the range of 10^{-6} to 10^{-3}M, and, if calcium remains bound for approximately 1 msec, the dissociation constant of the binding site (k_2/k_1) is in approximately the same range.

Apposition of Synaptic Vesicle and Plasma Membrane

At rest, the shortest distance between the vesicle and the plasma membrane is about 50 Å. Since it is unlikely that fusion of synaptic vesicle and plasma membranes can occur at these distances, it is reasonable to propose that the first event in exocytosis must involve movement of the vesicle into close apposition with the plasma membrane. In non-neural secretory cells, very extensive apposition may represent a distinct morphological step in exocytosis (Chi et al 1976, Lawson et al 1977, Peixoto de Menezes & Pinto da Silva 1978). Nerve terminals soaked in high concentrations of divalent cations also show vesicles in extremely extensive apposition to the plasma membrane (Heuser et al 1971, Boyne et al 1974, Llinas & Heuser, 1977). In both of these situations intramembranous particles move away from the sites of apposition to give bald patches. Since regions free of intramembranous particles are also seen in erythrocytes in the absence of apposition when the internal calcium concentration is high (Volsky & Loyter 1977), we cannot be certain whether apposition causes bald patches, or bald patches produce apposition. At present, no evidence is available indicating that bald patch formation is an intermediate morphological step at the synapse during transmitter release (Llinas & Heuser 1977).

Before asking what brings the membranes together it is appropriate to ask what keeps them apart. One suggestion is that separation is due to an electrostatic repulsion between negatively charged membranes (VanDerKloot & Kita 1973, Dean 1975). Calculation of the energy barrier for apposition involves knowledge of the magnitude of the fixed surface charges that give rise to electrostatic repulsion and of the attractive forces of the London–van der Waals type. With appropriate assumptions it can be shown that the height of the energy barrier for a 40 nm vesicle is approximately 8 kT where k is Boltzmann's constant and T the absolute temperature (Dean 1975).

It was postulated some time ago that the true energy barrier to membrane apposition is a hydration barrier (Bass & Moore 1966, Remler 1973). Experimental support for the role of hydration in separating phosphatidylcholine bilayers was obtained by measuring lamellar spacing while reducing the activity of water (LeNeveu et al 1977). Based on these data, a 15 kT hydration energy barrier must be overcome to bring a vesicle to within 15 Å of the plasma membrane (Parsegian 1977). Unfortunately, the equivalent hydration barrier has not been calculated for phosphatidylethanolamine, phosphatidylserine, and other common membrane components. Phosphatidylcholine is well known to be hygroscopic, whereas phosphatidylethanolamine and phosphatidylserine have a lower affinity for water (Finer & Darke 1974), especially in the presence of calcium (Hauser et al 1975). It would be surprising if the hydration barrier were as large for these phospholipids as it is for phosphatidylcholine, although there is a preliminary claim that the hydration barrier is comparable for charged and neutral phospholipids (Cowley et al 1977).

Whether the barrier is due to electrostatic or hydration energy, energy must be expended to cross it. Of course, energy barriers due to steric hindrance by protein or carbohydrate may also have to be overcome. This energy could come from a contractile apparatus (Berl et al 1973), from removing hydrophilic groups (Parsegian 1977), from calcium charge screening (Dean 1975), from calcium crossbridging of either phospholipids (Lansman & Haynes 1975) or of specialized calcium binding proteins (Katz 1969, Morris & Schober 1977, Creutz et al 1978) located on both the synaptic vesicle and the synaptic plasma membrane. Finally, apposition could be due to the tight binding of a group on the vesicle surface to a complementary site on the cytoplasmic face of the plasma membrane. The presence of carbohydrate on the surface of chromaffin granules (Meyer & Burger 1976), for example, suggests the possibility of a lectin-carbohydrate mechanism of binding.

In the evaluation of each mechanism it must be borne in mind that the time available for vesicle apposition is of the order of 100 μsec, which is too short for extensive enzymatic catalysis to be involved. An enzyme with a turnover time of 10^4/sec, for instance, could only catalyze one chemical event. Several phospholipases, for example, have a substrate turnover number of approximately 10^2/sec (Strong et al 1976). Parsegian (1977) has argued that the conventional actomyosin system could cause only a 7 Å movement in 100 μsec. Furthermore, there is not much space for an actomyosin system within a region of apposition (Trifaro 1978).

Apposition of membranes could come about by calcium crossbridging. From electrophoretic measurements made at physiological ionic strengths, binding of calcium to synaptic vesicles or to secretory granules is associated with a free energy change of -5.5 Kcal/mole (Dean 1975, Matthews & Nordman 1976). If part of this energy could be utilized for the apposition step, perhaps only a few calciums need be bound to overcome a 15 kT barrier. Calcium-crossbridging of membranes has been shown to occur in vitro when phosphatidic acid liposomes are rapidly mixed with calcium (Lansman & Haynes 1975, Haynes 1977).

Morris & Schober (1977) have presented evidence that calcium binds to specific sites on secretory granules, rather than to negatively charged phospholipids. The trivalent ion Tb^{3+} aggregates chromaffin granules and also gives electron dense patches on the granule membrane in the presence of phosphates. The electron dense material increases with Tb^{3+} concentration until the entire granule surface is covered. At 0.5 mM Tb^{3+} Morris and Schober see spots in a trigonal array separated by about 50 nm, which they claim are calcium binding proteins. Calcium and magnesium compete for Tb^{3+} binding only when their concentrations are in the mM range. Unfortunately, binding was measured at low ionic strengths and was uncorrected for surface charge effects.

Fusion of Membranes

No matter what the barrier to vesicle apposition, or the source of energy for moving the vesicle to the plasma membrane, we still have to deal with what could cause membrane fusion. Fortunately, fusion of pure phospholipid bilayers is readily achieved in vitro, and offers a good model of in vivo events, even if the protein model holds.

When experiments are designed to detect fusion and not exchange of phospholipids (Martin & MacDonald 1976, Papahadjopoulos et al 1976), it is found that fusion requires divalent cations, negatively charged phospholipids, and a temperature such that the divalent cations induce a membrane phase change from a fluid to a solid state (Papahadjopoulos et al 1977, Miller & Racker 1976). Surprisingly, more extensive fusion is obtained if there is an osmotic gradient across the liposome membrane, a result interpreted as due to increased surface energy (Miller et al 1976). Pollard and his co-workers have also proposed that osmotic pressure is involved in release, but in their model the buildup of osmotic pressure begins only after vesicle-membrane apposition (Pollard et al 1977).

We have considered the requirements for in vitro membrane fusion, but have so far avoided discussion of the molecular nature of the fusion event itself. Based on the association of fusion and membrane phase transition, it has been suggested that hydrophobic regions are exposed at phase boundaries, and that apposition of these hydrophobic regions acts to nucleate membrane fusion (Fig. 2A) (Papahadjopoulos et al 1977, Van Der Bosch & McConnell 1975). The phase boundary could be that between the bulk phospholipid membrane and the lipid annulus surrounding membrane proteins (Jost et al 1973, Hesketh et al 1976). It has also been suggested on the basis of morphological studies that an intermediate in fusion is a single phospholipid bilayer separating vesicle contents from the external space (Fig. 2B) (Satir et al 1973, Pinto da Silva & Nogueira 1977). Such a membrane arrangement could result from a precursor such as that in Figure 2A. An alternative possibility is shown in Figure 2C. It is well known that both phosphatidylethanolamine and cardiolipin lamellar structures are in equilibrium with hexagonal 2 and cubic phases at low water content (Reiss-Husson 1967, Rand & Sengupta 1972). Since apposition of vesicle and plasma membrane gives rise to separations of only a few angstroms, the environment of the phospholipids is analogous to that in which such phase equilibria occur. In Figure 2C two lamellae have undergone a transition to a hexagonal 2 or cubic arrangement. If they revert to a lamellar structure at right angles to the initial plane of apposition then fusion will result. Cleavage of membranes (dashed line) is required in the plane perpendicular

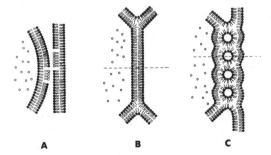

A   **B**   **C**

Figure 2   Possible phospholipid rearrangements during fusion. *A* Apposition of boundary regions between "fluid" and "solid" phases. A could be a precursor of B and C. *B* Fusion of two bilayers to form a single bilayer diaphragm. Note that the phospholipid monolayers facing the cytoplasm are excluded from the diaphragm. *C* Region of apposition of two phospholipid lamellae undergoes a transition to a hexagonal 2 or cubic phase.

In both *B* and *C,* the intermediate must split along the dashed line to give exocytosis. In *C,* if the hexagonal phase reverts to lamellae at right angles to the initial plane of apposition, exocytosis will occur.

to the plane of apposition. Preferential cleavage in the required plane could result from surface energy in the vesicle, resulting, for example, from an osmotic gradient (Miller et al 1976).

Protein Versus Phospholipid Mediated Exocytosis

With this background we return to consider the molecular nature of the "calcium effector." In particular, we ask whether both apposition and fusion can be explained by molecular models involving only phospholipids or whether more sophisticated mechanisms such as specific binding proteins or enzymatic catalysis must be invoked. Since little is known about the molecular events during exocytosis except that calcium is required, any mechanism that is to be seriously considered must show calcium selectivity, nonlinear dependence on calcium, and must operate at concentrations in the range of 10^{-3} to 10^{-6}M calcium (See above). Surprisingly, these three criteria can be satisfied by a purely phospholipid model, without invoking specialized protein machinery.

If synaptic vesicle-plasma membrane apposition involves only phospholipids, then the dissociation constant for calcium binding to phospholipid would have to be 10^{-3} to 10^{-6}M. Data on calcium binding to phosphatidylserine monolayers (Hauser et al 1976) showed dissociation constants as low as 10^{-6}M but only in very low ionic strengths. This is because the fixed negative charge on the membrane surface increases surface calcium concentrations to values that are orders of magnitude above the bulk calcium concentration (Hauser et al 1976, McLaughlin et al 1971, Muller & Finkelstein 1974). The true dissociation constant, corrected for surface charge, is 10^{-4}M. Calcium binds to phosphatidic acid vesicles with an association rate constant of 1.9×10^7 M^{-1} sec^{-1} and with a dissociation constant of 3.6×10^{-4}M, but again measurements were made in low ionic strength medium (Haynes 1977). In general, the concentration of calcium required to half-saturate negatively charged phospholipids is about 10^{-4}M.

The interactions of phospholipids with divalent cations show nonlinearity with cation concentrations. Calcium screening of surface charge does not depend linearly on calcium concentration (Dean 1975) and phospholipid vesicle aggregation and fusion, and isothermal lateral phase separation all show marked nonlinearities with calcium (Lansman & Haynes 1975, Papahadjopoulos et al 1977, Ohnishi & Ito 1974, MacDonald et al 1976). Another phospholipid interaction that is nonlinear with calcium and may be a useful model of the fusion process is the activation of prothrombin by negatively charged phospholipids, where Hill coefficients of 3 to 4 have been measured (Nelsestuen 1976).

The other characteristic of transmitter release is the specificity for calcium over magnesium. If calcium were involved in a selective binding interaction, for example, at a site on a nerve terminal protein, the observed competition with magnesium could be explained. A less obvious but real possibility is that the specificity for calcium is due to its selective interaction with phospholipids. In model studies, for example, calcium binds more strongly than magnesium to a phosphatidylserine monolayer (Hauser & Dawson 1967). Calcium is also more effective than magnesium in eliminating phosphatidylserine and phosphatidylglycerol phase transitions (Jacobson & Papahadjopoulos 1975, Papahadjopoulos et al 1977), in causing lateral phase separation in phosphatidylserine-containing membranes (Ohnishi & Ito 1974), and in bringing about lamellar to hexagonal phase transitions in water-cardiolipid mixtures (Rand & Sengupta 1972). It has been suggested that since the binding of calcium to the phosphate of phosphatidylserine involves removal of bound water, selectivity for calcium might result from the ease with which calcium loses its hydration shell (Hauser et al 1975). Gomperts (1976) has suggested that Ca^{++}/Mg^{++} selectivity arises because of the ability of calcium to form polydentate complexes.

Since apposition of phospholipid bilayers and fusion of phospholipid bilayers can occur at reasonable calcium concentrations, can be nonlinear with calcium concentration, and can show calcium selectivity, we come to the surprising conclusion that the exocytotic event need not involve proteins. Proteins may be involved in recognizing a vesicle and holding it in the vicinity of the plasma membranes (Fig. 1A). Local calcium influx could then trigger apposition, phospholipid rearrangement, and fusion. Although the affinity of phospholipids for calcium is close to the lower limit of calculated values, it should be remembered that the vesicle membrane, the plasma membrane, and the calcium entry site could all be within 50 Å of each other. Of course, the fact that fusion can occur without proteins does not prove that they are not involved in exocytosis.

EXPERIMENTAL APPROACHES

To distinguish between protein and phospholipid models it is necessary to examine the molecular composition of synaptic vesicle membrane and presynaptic plasma membrane. We would like to pay special attention to the presence of charged phospholipids, phase boundaries, the fluidity of hydrocarbon tails, and interactions with calcium. We are also interested in proteins on cytoplasmic faces that could be involved in recognition, apposition, or fusion. In the next section we discuss attempts to purify these membranes and analyze their composition. After isolating the components, the next step for the biochemist is to attempt to reconstitute a secretory system in vitro with the properties of in vivo release. In the third section preliminary attempts to develop in vitro secretory systems are described. Finally, in many cases in which purification and reconstitution are not yet possible, useful biochemical information can be derived by the use of specific biochemical inhibitors. In the fourth section, we discuss the modification of transmitter release by inhibitors, especially the presynaptic neurotoxins.

Purification of Nerve Terminal Components

To answer some of the questions about neurotransmitter release raised in the previous section it is necessary to purify synaptic vesicles, presynaptic plasma membranes, and nerve terminals and to determine their molecular composition. The purification of nerve terminal components, however, raises special problems, especially because of the heterogeneity of the components. Heterogeneity can be biological in origin because in brain tissue there are many types of cells and many types of synapses. It can also be artifactually introduced because synaptosomes and synaptic plasma membranes are not unique entities like proteins but are structures produced by homogenization. Such nerve terminal components therefore have a heterogeneous size distribution that reflects in part the rigor of the homogenization procedure. In addition, homogenization produces membrane fragments that tend to vesiculate. Such artificially formed vesicles are unwelcome because they can resemble natural synaptic vesicles in the electron microscope, and can trap cytoplasmic elements inside. In addition, the buoyant density of vesiculated structures on iso-osmotic sucrose or Ficoll density gradients depends in part on the volume of solution trapped inside (Wagner et al 1978). Thus even techniques such as sucrose and Ficoll gradients that are commonly thought to fractionate membranes independently of size in fact separate vesiculated structures on the basis of their size as well as their density.

The second problem is that there is no specific assay for some nerve terminal components. The synaptic plasma membrane in particular has no distinctive biochemical or morphological marker. Although various biochemical markers including Na^+K^+-ATPase, acetylcholinesterase, 5'-nucleotidase, and ganglioside sialic acid have been used to identify presynaptic plasma membrane (Jones 1975, Mahler 1977, Morgan 1976), none of these markers are found exclusively on presynaptic plasma membranes. In the electron microscope, the presence of synaptic junctional complexes can be used to identify presynaptic plasma membranes but usually only a small fraction of the membrane structures have recognizable junctional complexes (Jones & Matus 1974).

Finally, even when an assay is available, it is often difficult to assess the purity of the preparation even though such an assessment is crucial to evaluating composition. The most commonly used criterion of purity is the absence of contaminants determined morphologically and by enzyme markers. The problems associated with morphological criteria of purity have been discussed elsewhere (DePierre & Karnovsky 1973, Laduron 1977). Even if these criticisms are overlooked, morphological criteria are not sufficient for the biochemist wishing a quantitative estimate of contamination since soluble components are ignored. The classic biochemical approach to tissue fractionation uses enzyme markers to detect the presence of contaminating membranes. The pitfalls of this approach have also been thoroughly documented (DePierre & Karnovsky 1973, Barondes 1974, Morgan 1976, Laduron 1977) and do not need reiteration. While the enzyme marker approach is useful in correlating enzyme activities with each other, it is less useful in estimating the level of protein or lipid contamination. For example, in one case, although a synaptic plasma membrane fraction was ten times richer in Na^+K^+-ATPase than the microsomal fraction, both had identical polypeptide chain compositions on gel electrophoresis (Gurd et al 1974). If soluble proteins such as tubulin monomers stuck to membranes during homogenization at low ionic strength, they will not be detected by the marker method.

In protein and virus purification, purity can also be assessed using relative specific activity if an assay is available. The specific activity of the nerve terminal component should increase at each step of purification until it reaches a maximum value that cannot be increased by further purification steps. At present, although this is the method of choice, the usefulness of this approach is restricted by the limited number of well-established purification procedures available to the membrane biochemist. In addition, since synaptosomes and synaptic plasma membranes are intrinsically heterogeneous fractions, even the available procedures are sometimes of little use. Because the problems associated with the purification of intrinsically heterogenous components are quite different from those associated with homogeneous ones, we have divided the subsequent discussion into two parts, one dealing with the purification of structures that are formed on homogenization, the other with purification of preexisting structures such as synaptic vesicles and coated vesicles.

STRUCTURES FORMED BY HOMOGENIZATION

Synaptosomes   The isolation and characterization of synaptosomes has been recently reviewed (Jones 1975, Morgan 1976, Morgan & Gombos 1975, Whittaker 1969). Synaptosomes are morphologically defined as a membrane "bag" containing many synaptic vesicles. They are intrinsically heterogeneous: in size as observed by the electron microscope (Whittaker 1968), in sedimentation rate (Cotman et al 1970), and in density (Whittaker 1968). The variation in physical properties of these particles greatly increases the difficulty of their purification.

The purity of the preparations has been studied by electron microscopic examination of pelleted and sectioned material. The fraction of structures with synaptosomal morphology has been estimated to be about 50% (Cotman 1974, Joo & Karnushina 1975, Morgan 1976). To estimate the degree of purity biochemically, it is informative to calculate what degree of purification would be required to give pure synaptosomes. If there are 10^{11} synapses/g cortex (Gurd et al 1974) and the volume per nerve terminal is $0.1~\mu m^3$ (Whittaker 1968) then about 1% of the volume of cortex is in nerve terminals. Thus, to obtain pure nerve terminals 99% of the cortical material must be removed, which requires a hundredfold purification. The specific activity of acetylcholine and choline acetyltransferase (markers for nerve terminals) in isolated synaptosome fractions is about 2 to 5 fold higher than the homogenate (De Robertis et al 1963). In addition, about 10% of the protein of the cortex is recovered in the synaptosomal fraction (Cotman 1974), which is very high, if only 1% of the cortex is nerve terminals. Although these calculations are unquestionably inexact, they do imply that the biochemical purity of synaptosome preparations is considerably lower than the morphological purity. If a significant fraction of the mass of these nerve terminal preparations is contaminating material, then the detailed biochemical compositions summarized by Whittaker (1969) are not likely to be meaningful.

Synaptosomes have recently been isolated from the electric organs of *Torpedo* (Morel et al 1977, Michaelson & Sokolovsky 1978). Purification increased the specific activity of acetylcholine 10 fold and of choline acetyltransferase about 5 fold. Purity was assessed both morphologically and with the enzyme marker method. Since 4% of the electric organ is nerve terminals (Morel et al 1977), by the criterion of relative specific activity, the synaptosomes purified from *Torpedo* are about 20–40% pure. These are without doubt the best synaptosome preparations currently available.

Presynaptic plasma membranes   Synaptic plasma membranes are prepared from synaptosomes by osmotic lysis of the isolated nerve terminal preparation followed by equilibrium density centrifugation (Jones 1975). By morphological criteria, the membranes are 80% pure, but this value rests on the assumption that all vesiculated structures of 0.5—1.2 μm diameter vesicles are exclusively derived from presynaptic plasma membrane (Jones & Matus 1974). Since the starting synaptosomes are at least 50% contaminated and presynaptic plasma membranes do not have an unusual density on equilibrium centrifugation (Gurd et al 1974), it is difficult to feel confident that the estimate of 80% purity is a valid one. Using Whittaker's (1968) estimate that 8% of the volume of the nerve terminal is in presynaptic membrane, we calculate that a 50–200 fold purification is required. Consequently the characterizations of plasma membrane preparations (Levitan et al 1972, Morgan 1976, Mahler 1977, Morgan & Gombos 1975, Gurd 1977, Kelly & Cotman 1977) should be viewed with some caution.

The synaptic junctional complex and the postsynaptic densities are derivatives of synaptic plasma membranes with characteristic morphologies in the electron microscope (Jones 1975, Cotman & Taylor 1972, Therien & Mushynski 1976, Cotman et al 1974). In the electron microscope 20% of the structures in preparations of synaptic junctional complexes have pre- and postsynaptic plasma membrane elements joined by synaptic cleft material; the remaining 80% do not have presynaptic membranes (Cotman & Taylor 1972). Since these preparations are largely postsynaptic, we will not discuss them further.

PREEXISTING STRUCTURES

Synaptic vesicles   Unlike the synaptosome and synaptic plasma membrane, synaptic vesicles have been successfully purified. These organelles are unique structures, homogeneous in size, which can be assayed by their content of transmitter. There are two major difficulties in synaptic vesicle isolation, however, both of which arise from vesiculation of other membranes during homogenization. Vesiculation can trap cytoplasmic elements such as transmitters and so produce "pseudosynaptic vesicles." Furthermore, vesiculation produces membrane bags of a range of diameters, including some that have the same dimensions as the synaptic vesicle. These empty bags can have the same morphology as synaptic vesicles, similar sedimentation rates, and will presumably copurify on permeation chromatography. Osmotically shocked liver membranes give structures resembling synaptic vesicles (Laduron 1977).

Cholinergic synaptic vesicles were originally obtained in large quantities from the electric organs of elasmobranchs (Whittaker et al 1972), using the techniques of differential centrifugation followed by equilibrium density centrifugation on sucrose gradients. Application of these techniques to tissue from *Torpedo marmorata* yielded a preparation whose specific activity was 566 nmoles acetylcholine/mg protein. After removal of contaminating acetylcholine esterase activity by permeation chromatography on glass bead columns (Morris 1973), the specific activity was increased to 790 nmoles acetylcholine/mg protein (Nagy et al 1976). The criteria for purity were the reduction of contaminating structures in the electron microscope and of contaminating cytoplasmic and microsomal enzyme markers (Nagy et al 1976, Whittaker et al 1972, Morris 1973).

Improved procedures have resulted in preparations from *Narcine brasiliensis* with specific activities of 8,000 nmoles acetylcholine/mg protein. A three-hundred-fold purification was required, after which the specific activity across the vesicle peak was constant and not increased by other fractionation procedures. The vesicle preparations were estimated to be at least 90% pure by the following criteria: (*a*) Light-scattering material and the vesicle contents cosedimented in analytical ultracentrifugation. (*b*) The protein, lipid, and vesicle contents had the same density on glycerol density gradients, cosedimented during preparative velocity sedimentation and comigrated during electrophoresis. (*c*) The vesicles purified from *Narcine* not only had a higher specific activity than those obtained by previous workers, they also had a 10-fold higher lipid to protein ratio (Wagner et al 1978). Since the buoyant density (which is a measure of the protein to lipid ratio) is the same for both vesicle preparations, the lower specific activity preparations presumably contain a soluble protein or protein-rich membrane contaminants, or both.

The purification of synaptic vesicles from brain has not yet been as successful as that from electric organ, in part because brain contains vesicles of many different types. Isolation procedures involving osmotic shock of synaptosomes followed by equilibrium density centrifugation in sucrose yielded a fraction enriched in acetylcholine and appropriately sized structures (Whittaker & Sheridan 1965, Sheridan & Whittaker 1964). Introduction of a Millipore filtration step into the purification markedly reduces the level of contaminants (Morgan et al 1973a) but since no assay for vesicles was used in these studies, it is uncertain whether the number of vesicles is equally reduced.

Although synaptic vesicles purified from lysed synaptosomes by sucrose gradient sedimentation have been used widely, for example to study vesicle composition (Morgan et al 1973a, b, Breckenridge et al 1972a, b, 1973) and transmitter uptake (see below), the vesicles are not pure. The specific activity of the sucrose gradient fraction can be increased ten times by permeation chromatography on a glass bead column (Nagy et al 1976). The highest specific activity that has been obtained is 14 nmoles ACh/mg protein or 12 nmoles/μmole lipid phosphorus. This is lower than that for *Narcine* vesicles, but the latter have diameters of 800 Å. A pure phospholipid vesicle with a 250 Å radius containing 6000 acetylcholine molecules would have a specific activity of about 300 nmoles acetylcholine/μmole phospholipid (assuming an area per phospholipid of 60 Å^2 and a membrane thickness of 50 Å). The presence of cholesterol and membrane protein would increase the specific activity. Either only a small fraction of the brain synaptic vesicles are cholinergic, or the vesicles are contaminated by other membrane fragments. The latter seems to us most reasonable since there was no peak of protein corresponding to the peak of vesicles after permeation chromatography, the last step in the isolation (Nagy et al 1976). Catecholamine- and acetylcholine-containing vesicles could be separated (Nagy et al 1977), but the catecholamine-containing vesicles had an even lower specific activity.

Other attempts to purify brain synaptic vesicles have also had limited success (Kadota & Kadota 1973, Tsudzuki et al 1977). The problem of tissue heterogeneity can be avoided by use of superior cervical ganglia as source of vesicles (Wilson et al 1973). Unfortunately, the specific activity obtained in this case (4.4 nmoles/mg protein) is not much higher than that obtained from brain tissue.

Catecholamine-containing vesicles can be isolated from sympathetic nerves. Two types of vesicles are commonly seen in thin sections of post-ganglionic sympathetic neurons and nerve terminals: small dense core vesicles (300–600 Å diameter), and large dense core vesicles (600–1500 Å) (Geffen & Livitt 1971). Since large dense core vesicles have a higher buoyant density than the small vesicles, they can be separated by sucrose density gradient equilibrium centrifugation (Chubb et al 1970, Bisby & Fillenz 1971, Nelson & Molinoff 1976, Yen et al 1973). The properties of the large vesicle preparations have been reviewed by Lagercrantz (1976). These vesicles have a specific activity of about 100 nmol noradrenaline/mg protein, which is considerably lower than that of chromaffin granules, 2500 nmol/mg, but higher than the purest brain synaptic vesicles (Nagy et al 1976). The purity of these vesicles must be relatively high, however, since proteins characteristic of adrenergic vesicles (Lagercrantz 1976) can be detected by SDS gel electrophoresis.

The small dense core noradrenergic vesicles have not been purified significantly. These small vesicles appear to contain little dopamine β-hydroxylase, although the question is not completely resolved (DePotter & Chubb 1977, Kirksey et al 1978, Nelson & Molinoff 1976).

Coated vesicle isolation   Coated vesicles are directly implicated as the organelle involved in membrane recycling at the nerve terminal by the studies of Heuser & Reese (1973, 1977). These vesicles are thought to carry membrane, extracellular protein, and lipoproteins to the inside of the cell (Pearse 1976, Silverstein et al 1977). They were first isolated by Pearse (1975), who took advantage of their high buoyant density and their distinctive morphology. In the electron microscope the coated vesicles from brain have diameters of about 700 Å (Pearse 1976). The coats are made up of an 180,000 mw protein called clathrin (Pearse 1975) that forms polyhedral lattices constructed from 12 pentagons plus a variable number of hexagons (Crowther et al 1976). Coated vesicles have subsequently been isolated from a number of cell types including chromaffin cells (Pearse 1976), a nonsecretory lymphoma cell line (Pearse 1976), and the human placenta (Ockleford et al 1977). It remains to be shown that vesicles from brain are homogeneous and not derived from glial as well as neuronal cells.

Endocytotic vesicles from electric organ   Zimmermann & Whittaker (1977) and Zimmermann & Denston (1977a, b) have reported the isolation of endocytotic vesicles from electric organ using an original procedure. Tissue blocks were perfused with a dense solution during stimulation. Endocytotic vesicles formed during perfusion were filled with the dense solution and so could be isolated by their unusual density on a sucrose density gradient. The size of the vesicles isolated in this way are smaller than synaptic vesicles but equal to that of vesicles found in the nerve terminal after stimulation. They do not have the characteristic coat of coated vesicles, but do become filled with newly synthesized acetylcholine.

CONCLUSIONS   Before turning to the molecular characterization of the components of nerve terminals it is worth reiterating that only synaptic vesicles from the electric organ of elasmobranchs can be confidently assumed to approach biochemical purity. The properties of synaptic vesicles are considered therefore in some detail whereas other components are described more briefly. The need for new and higher resolution techniques with which to fractionate and purify presynaptic components, and for specific markers of these components, is obvious.

Characterization of Nerve Terminal Components

SYNAPTIC VESICLES   Purified elasmobranch synaptic vesicles contain 40,000 (Wagner et al 1978) to 80,000 (Ohsawa et al 1976) acetylcholine molecules, a number that is consistent with that estimated from electrophysiological measurements (Fletcher & Forrester 1975, Kuffler & Yoshikami 1975) if the difference in vesicle diameter is taken into account (Wagner et al 1978). The vesicles contain ATP at a ratio of acetylcholine to ATP of 2:1 (Carlson et al 1978) or 5:1 (Dowdall et al 1974), but they also have small amounts of GTP and ADP (Wagner et al 1978). Originally, it was thought that most of the vesicular protein was internal, and that 36% of the total protein was a 10,000 dalton protein named "vesiculin" (Whittaker et al 1972, 1974). However, when purer vesicle preparations were examined, no vesiculin was found and at least 90% of the vesicular protein was determined to be in the vesicle membrane (Wagner et al 1978). The phospholipid to protein ration of the latter vesicles is unusually high (5:1 by weight), even higher than that of myelin. The vesicle phospholipid composition (J. Deutsch, unpublished observations) is very similar to that published for less pure vesicles (Nagy et al 1976), and to the phospholipid composition of the entire electric organ. The vesicles contain negatively charged phospholipids, as required by the phospholipid model of fusion, but it is not known if they are on the cytoplasmic surface of the vesicle. Purified synaptic vesicles also have a net negative charge (Carlson et al 1978) as required if calcium-screening or -binding is to be relevant (Dean 1975), but the number of calcium binding sites, or their affinity is not known. Although purified synaptic vesicles have been reported to contain 8 major polypeptide chains (Wagner et al 1978), it is not known which, if any, correspond to the large intramembranous particles (Heuser & Reese 1975) or to the postulated recognition site (Fig. 1A). An ATPase activated by acetylcholine has been reported in *Torpedo* synaptic vesicle preparations (Breer et al 1977).

It is enlightening to compare the composition of the well-characterized chromaffin granule to that of the synaptic vesicle. The chromaffin granule also has a high lipid to protein ratio (Winkler et al 1974), a net negative charge (Dean & Matthews 1974), and contains cholesterol (Blaschko et al 1967). The phospholipids in chromaffin granules are known from fluorescence polarization (Bashford et al 1976) and spin resonance order parameter measurements (Marsh et al 1976) to be in a fluid state. The order parameter is increased in the presence of calcium, which could be attributed to the calcium-induced membrane phase change required for phospholipid fusion. Chromomembrin b and complex carbohydrates have been identified on the cytoplasmic surface (Konig et al 1976, Meyer & Burger 1976) and so are candidates for fusion or recognition components.

The phospholipid, fatty acid, and protein compositions of brain synaptic vesicle preparations have been published (Morgan et al 1973a, b, Breckenridge et al 1973) and the protein composition of synaptic plasma membranes has been recently reviewed (Mahler 1977), but these must be viewed with caution until the purity of the vesicles is more rigorously established.

One potentially useful way to identify nerve terminal components is to use immunological techniques to generate antibodies to a specific nerve terminal antigen. A synaptic vesicle fraction, free of contaminating enzyme markers (Reeber et al 1978) was used to immunize rabbits (Bock et al 1975) and the antiserum analyzed by quantitative immunoelectrophoresis. A protein was identified ("synaptin") that was specific for nervous tissue (Bock et al 1975). The vesicle fraction contained at least 10 times more synaptin than did brain homogenates (Bock & Jorgenson 1975). Using adsorption, Bock & Helle (1977) reported synaptin to be on the outside of synaptic vesicles and on the inside of the synaptosomal plasma membrane. Antisynaptin antibodies also cross-react with chromaffin granule membranes (Bock & Helle 1977). Synaptin is reported to be a glycoprotein of 45,000 MW as determined by SDS polyacrylamide gel electrophoresis (Bock 1978).

Unfortunately the questionable purity of the vesicle preparation weakens the interpretation of these intriguing results. Localization by immunocyto-chemistry of the sites in nerve terminals to which antisynaptin antibodies bind would be valuable.

A few studies of synaptic vesicle or secretory granule composition have been motivated by a search for fusogens, agents that promote cell fusion (Ahkong et al 1975). Chromaffin granules contain high amounts (as much as 20 mol % of the total phospholipid) of lysophosphatidylcholine (Blasch-hko et al 1967), but it may be exclusively on the interior half of the bilayer (Voyta et al 1978), and it is questionable whether it could be involved in fusion. Baker et al (1975) found that 1 mol% or less of the phospholipid in guinea pig brain and *Torpedo marmorata* synaptic vesicle preparations was lysophosphatidylcholine, whether or not the synaptosomes or electric organ had been stimulated. Pickard & Hawthorne (1978) suggested that there is an increase in hydrolysis of phosphatidylinositol to diacylglycerol (a fusogen) in brain synaptic vesicles when synaptosomes are stimulated to release transmitter. The validity of these studies is limited by the extent of contamination of vesicles with nonvesicle membrane. Fusogen-induced fusion of erythrocytes may be due to the direct or indirect action of fusogen on membrane. Volsky & Loyter (1977) have shown that fusogens can increase intracellular Ca^{++} in erythrocytes, which in turn brings about the appearance of smooth membrane patches in freeze fracture. If these bald patches are precursors to fusion, the fusogens may be operating indirectly by allowing Ca^{++} to enter the erythrocyte.

COATED VESICLES   Coated vesicle preparations from pig brain contain 75% protein and 25% phospholipid by weight. The lipid and cholesterol compositions are known. Of the total protein 70–90% was clathrin (Pearse 1976), which is a good indication of purity. Two of the minor proteins, representing about 10% of the total protein, appear to be very similar to the two major proteins of the sarcoplasmic reticulum Ca^{++} dependent ATPase. Since coated vesicle preparations accumulate calcium, they may be involved in regulating calcium levels in nerve terminals (Blitz et al 1977). Once again it is important to show that the Ca^{++}-ATPase and calcium uptake are associated with vesicles that have coats and not with minor contaminants.

SYNAPTOSOMES AND SYNAPTIC PLASMA MEMBRANES   The identifi-cation of the molecular components of the presynaptic plasma membrane, which should include the active zone structures, has been hampered by lack of a convincing morphological or biochemical marker for these structures.

It has been suggested that they are rich in gangliosides (Breckenridge et al 1972b) and that they have an unusually large amount of polyunsaturated fatty acids (Breckenridge et al 1972a, b) but once again the questionable purity of the membrane fraction limits its usefulness. Brain synaptic vesicle proteins have been compared with the proteins of the presynaptic plasma membranes and few if any differences have been seen (Morgan et al 1973a, b, Breckenridge & Morgan 1972). Antibodies against synaptic vesicles cross-react with synaptic plasma membranes and vice versa (Howe et al 1977). The lack of difference may be misleading since the major proteins of these fractions comigrate with the major fibrous proteins tubulin, myosin, actin, and tropomyosin (Blitz & Fine 1974, Mahler et al 1975, Mahler 1977). Any differences might be concealed by adventitious sticking of these proteins to membranes during purification at low ionic strength. The association of choline acetyltransferase with synaptic vesicles (Fonnum 1968) or α-actinin with chromaffin granules (Jockusch et al 1977) is a function of ionic strength. Differences are found between the protein composition of secretory granule membranes of the pancreas and adrenal medulla and their corresponding plasma membranes (Meldosi 1974, Winkler et al 1974, Hortnagl 1976, Castle et al 1975).

As with synaptic vesicles, investigators have used immunological techniques in attempts to develop markers specific for the nerve terminal. Both synaptosomal and synaptic membrane fractions have been used as antigens (Morgan & Gombos 1975). The specificity of the antisera has been measured by reaction with other membrane fractions using complement fixation (Morgan & Gombos 1975), or crossed immunoelectrophoresis (Bock et al 1975, Jorgensen 1976). Although it has been established that these antibodies are brain specific, they were produced by injection of impure fractions and assayed on the same fractions. Consequently, the antigens need not necessarily be exclusively in nerve terminals, but could be in axons, dendrites, neuronal soma, or glial cells. More attractive evidence for specificity is the selective labeling of cortical layers in brain sections using fluorescent antibodies (Kornguth et al 1969, Livett et al 1974). Immunocytochemical detection of labeled antibodies in the electron microscope is required to verify that the antibodies indeed bind exclusively to nerve terminals.

CONCLUSIONS   Although much remains to be known, some important facts about transmitter release have emerged from the characterization of nerve terminal components. The content of acetylcholine in the synaptic vesicle is in reasonable agreement with values calculated from electrophysiology, but the function of the nucleotides is unknown. Calcium uptake by coated vesicles preparations suggests that they may be involved in calcium buffering in the nerve terminal. The phospholipid composition and the net charge of the synaptic vesicle are compatible with a phospholipid model of exocytosis.

In Vitro Release of Neurotransmitter

To determine the molecular events that occur during transmitter release it would be highly desirable to have an in vitro system that accurately reproduced the properties of synapses in vivo. Ultimately, mixing purified vesicles and purified presynaptic membranes in the presence of calcium and known cofactors ought to give transmitter release by exocytosis. What has been achieved so far is much more limited. Synaptosome preparations retain the ability to take up transmitters and to release them. They can be considered as a partial realization of an in vitro system, simpler than intact tissue but more complex than a true reconstituted one. In addition, very preliminary attempts have been made to look for factors that cause vesicles to lose their contents. These two approaches to in vitro reconstitution will be discussed, but first we must deal with the problem of nonexocytotic release.

NONEXOCYTOTIC RELEASE   For electrophysiologists and anatomists the arguments in favor of exocytosis of transmitter are almost overwhelming. Biochemists, however, have consistently felt less convinced (Marchbanks 1975). Here we attempt to reconcile the biochemical findings that have been used to support nonexocytotic release with the hypothesis of vesicular exocytosis.

An estimate of the contribution of nonquantal release to the release of neurotransmitter at the neuromuscular junction can be made in two ways. From the hyperpolarization response induced by curare (Katz & Miledi 1977a, Vyskocil & Illes 1977), it can be estimated that the steady-state concentration of acetylcholine in the synaptic cleft is about 10^{-8}M in the absence of evoked release. From this figure it is possible to calculate that the leakage rate from the terminal is 10^{-18} moles/sec/terminal (Katz & Miledi 1977a). However, only about 1% of this rate of leakage can be accounted for by spontaneous quantal release; the remaining 99% must be ascribed to nonquantal leakage across the nerve terminal membrane. Even during stimulation a large fraction of transmitter efflux must be due to nonquantal leakage. For example, if the nerve is stimulated at a frequency of 1/sec, and about 400 quanta are released per stimulus, then the evoked release would be only about four times the leakage rate.

Alternatively, acetylcholine release can be measured directly by the accumulation of acetylcholine when hydrolysis is inhibited by an antiesterase. Fletcher & Forrester (1975) have used this approach to determine the rate of spontaneous (8×10^{-19} moles/sec/terminal) and evoked (4×10^{-18} moles/

sec/terminal at 2 Hz) release of acetylcholine from the rat hemidiaphragm. Again, only about 2% of the spontaneous release can be accounted for by spontaneous quantal release, leading to the conclusion that 98% of this release is probably due to leakage across the nerve terminal. This work confirmed the earlier study of Potter (1970) who demonstrated that the leakage rate of acetylcholine was significant (0.76% of total ACh per minute) and that the release rate could only be increased by a factor of 5 to 10 over a range of stimulation frequencies (1 to 20 Hz). These values are in adequate agreement with those of other workers using similar techniques (Krnjevic & Mitchell 1961, Mitchell & Silver 1963, Krnjevic & Straughan 1964) and the estimates of Katz & Miledi (1977a) cited earlier. Studies on the release of acetylcholine from superior cervical ganglion (reviewed by MacIntosh & Collier 1976) have led to a similar conclusion about synapses other than the neuromuscular junction.

The data from both the biochemical and the electrophysiological studies have led to the same conclusion, namely, that the rate of leakage of transmitter from nerve terminals is of the same order of magnitude as the rate of release by exocytosis after nerve stimulation. Whereas the electrophysiologist can easily discriminate the two forms of release, this is not a simple task for the biochemist. Although it is sometimes assumed that quantal release, but not leakage, is calcium dependent, the dependence of leakage rate on calcium or on K^+ depolarization has rarely been measured. Simpson (1974) showed that the rate of neurotransmitter release from an unstimulated nerve was 1.2 pmoles/min in the presence of calcium (a value comparable to Potter's estimate of 0.74 pmoles/min), but fell to a value of only 0.16 pmoles/min in the absence of calcium. It would be of great interest to know if the hyperpolarization response observed by Katz & Miledi (1977a) and Vyskocil & Illes (1977) is calcium dependent, and if this leakage is increased by depolarization of the nerve terminal.

The difficulty in discriminating exocytotic release from simple leakage of transmitter in biochemical experiments may account for some of the apparent conflict between biochemical observations and electrophysiological and anatomical ones. For example, in biochemical experiments release comes from two pools, one cytoplasmic and the other vesicular, whereas normally an electrophysiologist only deals with the vesicular pool.

EVALUATION OF SYNAPTOSOMES AS MODELS OF TRANSMITTER RELEASE   Synaptosomal preparations can release transmitter if they are depolarized by elevated K^+, veratridine, or electrical stimulation in the presence of calcium. There is now an extensive literature detailing the release from synaptosomes of a variety of substances: norepinephrine (Blaustein 1975, Levi et al 1976, Mulder et al 1975), acetylcholine (Blaustein 1975, Murrin et al 1977, Wonnacott & Marchbanks 1976), amino acids (Osborne & Bradford 1975, De Belleroche & Bradford 1977), dopamine (Andrews et al 1978, Patrick & Barchus 1976, Mulder et al 1975), ATP (White 1977, 1978), GABA (Levi et al 1976, Olsen et al 1977, Tapia & Meza-Ruiz 1977, Levy et al 1974), and substance P (Schenker et al 1976, Lembeck et al 1977). Earlier literature has been reviewed by Bradford (1975) and Levi & Raiteri (1976). In nearly all cases, depolarization in the presence of calcium causes an increase in the rate of efflux that is between 50% and 200% of the spontaneous efflux rate although in a few cases, larger increases have been reported. During depolarization the efflux depletes between 1 and 10% of the initial stores/min.

Synaptosomal preparations contain many types of subcellular structures other than synaptosomes. For such preparations to be of value in the study of transmitter release it is important that transmitters are released only by the synaptosomes and not by the contaminants. We know of no autoradiographic evidence that transmitter release in vitro is exclusively from structures with the morphology of synaptosomes. This is important to establish since glial cells have been shown to release transmitter (Minchin & Iversen 1974, Dennis & Miledi 1974).

The following arguments have been presented to show that release of transmitters from synaptosomal preparations resembles the release from nerve terminals in vivo: 1. conditions that should depolarize synaptosomes increase the rate of calcium uptake (Blaustein 1975); 2. these conditions also increase the rate of neurotransmitter release (Bradford 1975); and 3. neurotransmitter release is calcium dependent, and is inhibited by magnesium (Levy et al 1974). Although these observations suggest that exocytosis may be involved, they are not compelling, as we shall discuss. Elevated calcium uptake on depolarization could be due to reduction of calcium efflux by internal sodium. Such a mechanism has been invoked to explain elevated neurotransmitter release in the presence of ouabain (Baker & Crawford 1975). In addition, since other organelles, particularly mitochondria, can actively scavenge calcium it must be convincingly demonstrated that calcium uptake during depolarization is exclusively into synaptosomes. Comparison of the rate of Ca^{++} uptake into squid nerve terminals during stimulation (35 $\mu A/cm^2$, Llinas 1977) with the rate of Ca^{++} uptake into synaptosomes (0.024 $\mu A/cm^2$, Blaustein 1975) illustrates how profoundly different the two systems may be.

The second argument that synaptosomes are useful models is that they release transmitter in response to stimuli that increase Ca^{++} uptake. While transmitter release has been reported in all the synaptosome studies described earlier it is important to bear in mind that (a) the stimulated release is only a small fraction of the stores and is usually only a small increment over background; (b) other substances, such as amino acids, not thought to be neurotransmitters, are released by depolarization (Osborne & Bradford 1975); (c) efflux via exchange sites could well increase during depolarization (although release was still found in the presence of the GABA exchange inhibitor diaminobutyric acid (Levi et al 1976)); and (d) release could result from an enhanced lysis rate of depolarized synaptosomes. Lysis of synaptosomes during incubation has been detected by a loss of lactic dehydrogenase (Sperk & Baldessarini 1977).

The third argument is that the calcium dependence of transmitter release from synaptosomes and its inhibition by magnesium suggest that exocytosis may be involved. Calcium dependence is not a sufficient criterion, because 1. the rate of release is not sigmoidally dependent on extracellular calcium (Levy et al 1974); 2. glial cells also show a calcium dependent release of acetylcholine (Minchin & Iversen 1974); 3. Calcium is a ubiquitous regulator of cell function, and could, for example, cause an increase in the rate of norepinephrine uptake (Levi et al 1976). Conceivably, it could enhance synaptosome lysis by activating calcium dependent phospholipases. Since stimulation in the presence of calcium releases such a small fraction of the total transmitter store, it would be difficult to detect calcium dependent lysis by release of a cytoplasmic enzyme. 4. As discussed earlier, in addition to exocytotic release, nerve terminals leak neurotransmitter to an extent barely detectable electrophysiologically, but which constitutes a large fraction of the neurotransmitter released. This leakage rate may be calcium dependent (Simpson 1974). A large fraction of transmitter release from synaptosomes could be leakage. In some experiments addition of EGTA causes about the same fractional reduction in the spontaneous release and in release from a depolarized synaptosome preparation (De Belleroche & Bradford 1972, Murrin et al 1977). This result could be interpreted either as exocytosis occurring during "spontaneous" release, or leakage stimulated by K^+ depolarization. If synaptosomes are to have value as an in vitro model of exocytotic release rather than leakage, it is important that more stringent criteria for exocytosis be imposed, such as: 1. transmitter, ATP, and intravesicular protein should be released together (Schneider et al 1967); 2. a vesicle marker, such as a specific antigen, should be transferred to the synaptosomal plasma membrane (Howe et al 1977).

The heterogeneity of brain synaptosomes imposes strict limitations on their use in investigation of the molecular mechanism of transmitter release. A more promising source of more uniform nerve ending preparations is the homogeneous tissue of the elasmobranch electric organ. Recently, Michaelson & Sokolovsky (1978) demonstrated release of ACh from a synaptosome-enriched fraction from Torpedo. Release was elicited in the presence of calcium by elevated K^+ or the ionophore A23187. It is not clear whether the release was linearly dependent on calcium. For reasons that are not apparent, generally only about 30% of the initial ACh could be released by the various treatments. The authors claim a correlation between in vitro release and a decrease in the number of synaptic vesicles seen in electron micrographs. This would be an important contribution if verified.

It may be valid to question whether the synaptosome preparation is generally of much value in the study of exocytosis when pure cell lines are available that show calcium dependent release of a single transmitter type and avoid some of the difficulties with homogenization discussed above (Greene & Rein 1977, Schubert & Klier 1977, McGee et al 1978). Indeed despite the vast amount of work on synaptosomes it is difficult to ascertain the novel contributions these studies have made to our biochemical understanding of transmitter release at the nerve terminal.

RECONSTITUTION OF THE RELEASE PROCESS   Reconstitution in vitro of all the partial reactions described in Figure 1 with purified components is the eventual goal of the biochemist interested in neurotransmitter release. In this section we review the success to date in packaging transmitter into synaptic vesicles in vitro, in observing selective binding of vesicle to plasma membrane, in observing apposition and, finally, in detecting the fusion event itself. Because data on neural tissue are unfortunately very scarce, we will allude to non-neural secretory systems when they present a useful model.

Packaging   While ATP-dependent uptake of catecholamines into chromaffin granules has been studied extensively, little is known about uptake into synaptic vesicles (Holtzman 1977). The active uptake of catecholamine and ATP into chromaffin granules appears to be driven by a proton-motive force (Johnson et al 1978, Casey et al 1977). The uptake sites for catecholamine and nucleotide can be inhibited by different pharmacological agents (Kostron et al 1977). Because of their weakly acidic nature, catecholamines are taken up by phospholipid liposomes if there is a transmembrane pH gradient (Nichols & Deamer 1976).

Less is known about uptake into synaptic vesicles. Purified *Torpedo* synaptic vesicles do not accumulate acetylcholine to a level higher than extravesicular acetylcholine (Marchbanks 1975, Suszkiw 1976). A less pure fraction accumulated only about 50 molecules/vesicle (Mattson & O'Brien 1976). As we have discussed earlier, no convincing purification of brain synaptic vesicles has been published. Using synaptic vesicle preparations prepared by differential centrifugation or equilibrium density centrifugation of lysed synaptosomes, various workers have detected uptake of dopamine, noradrenaline, serotonin, GABA, and histamine (Matthaei et al 1976, Philippu & Beyer 1973, Philippu & Matthaei 1975, Lentzen & Philippu 1977,

Tanaka et al 1976, Toll et al 1977). The ATP dependence of this uptake encourages the hope that it is specific. However, convincing evidence that the observed uptake truly represents packaging into vesicles would require 1. further purification of the vesicles. For example, the specific activity (dopamine/mg protein) of the vesicle preparation used by Philippu & Beyer (1973) was only 5 times greater than the homogenate; 2. evidence that uptake is truly into synaptic vesicles and not into contaminating vesicles; 3. measurement of the number of molecules packaged per vesicle; 4. evidence that uptake and not exchange is measured; 5. in the case of the weakly acidic transmitters evidence the "uptake" does not reflect passive uptake in the presence of a pH gradient (Nichols & Deamer 1976).

Apposition   The second step in exocytosis (Fig. 1B) of neurotransmitter is the apposition of the synaptic vesicle and the release site. As far as we can tell this step has not been examined using neural tissue, although there are some intriguing experiments involving non-neural secretory systems. Several workers have looked at calcium dependent aggregation of secretory granules, although this is a questionable model for exocytosis. Regions of apposition were devoid of intramembranous particles. The calcium concentrations in which aggregation was observed ranged from 10^{-5}M (Dahl & Gratzl 1976) to 10^{-2}M (Schober et al 1977), but the relevance of this value is once again questionable since the experiments were done at unphysiologically low ionic strength. No aggregation of purified *Narcine* (Kelly, unpublished observations) or brain (Nagy et al 1977) synaptic vesicles has been observed even at high calcium concentrations. Recently, a protein (synexin) has been detected in the adrenal medulla that causes a calcium dependent turbidity increase when incubated with chromaffin granules (Creutz et al 1978). Synexin binds to chromaffin granules if calcium is present, and causes the turbidity change at Ca^{++} concentrations of less than 100 μM (at low ionic strength). Despite the difficulty of correlating the turbidity change with morphological aggregation, and the lack of evidence for a role in vivo, this is an intriguing result.

A more interesting model would involve vesicle-plasma membrane interaction, rather than vesicle aggregation. Milutinovic et al (1977) report a calcium dependent binding of zymogen granules to a radioactively labeled plasma membrane fraction from cat pancreas. They demonstrate binding specificity, but can find no evidence for exocytosis.

Release of vesicle contents   The final reaction is the fusion of the vesicle to the plasma membrane with release of vesicle contents. Several experimenters have tried to isolate fusion factors by measuring their ability to release vesicle contents. Izumi et al (1975) showed that an ammonium sulfate fraction of cytoplasm from adrenal medulla increases the release of catecholamine from chromaffin granules as much as six fold in the presence of Mg^{++} and adenine nucleotide. EGTA reduces the release rate approximately 50%. Hata et al (1976) detected a small increase in the release of ACh from rat brain synaptic vesicles when they were incubated with a soluble fraction obtained from synaptosome fractions. The maximal effect was obtained in Mg-ATP and 10^{-4}M calcium. Because the factor is small and is susceptible to some proteases, it may be a small polypeptide. Michaelson et al (1978) have found that a high speed supernatant from *Torpedo* electric organ homogenate increases release of ACh from *Torpedo* synaptic vesicles. The stimulation does not require ATP, but does require calcium (half-maximal at 50 μM calcium). It has been suggested that ATP-dependent lysis of chromaffin granules is a model for secretion in vivo (e.g. Pollard et al 1977).

Exocytotic release requires the interaction of vesicle and plasma membrane (Fig. 1B). The facilitation of release by soluble factors may represent a partial reaction in the release process or it may be artifactual. More convincing would be the demonstration of a role of plasma membrane in transmitter release in vitro. Recently, Izumi et al (1977) have shown that both salt and detergent extracts of adrenal medullary microsomes stimulate catecholamine release in a fashion similar to that of their previously described releasing factor (Izumi et al 1975), but the two factors may not be identical. Michaelson et al (1978) showed that less pure vesicles release ACh 30% faster than more pure vesicles in the presence of the high speed supernatant. They interpret this result as evidence for plasma membrane involvement. Davis & Lazarus (1976) have demonstrated release of insulin from a mixture of β granules and plasma membranes, which again is membrane specific and which mimicks in vivo insulin release in its sensitivity to drugs.

These studies represent the beginning of what will undoubtedly be an intense search to identify the components involved in transmitter release by exocytosis. Since vesicles can lose their contents by lysis as well as fusion we look forward to a clear demonstration that the synaptic vesicle membrane actually fuses with the plasma membrane in an in vitro reconstruction.

The Use of Presynaptic Neurotoxins

Neurotoxins that interfere with the release of neurotransmitter have been studied with two hopes. The first is to unravel the molecular mechanisms of transmitter release by the classic biochemical approach of selective inhibition of a key step in a pathway. The second is to use toxin binding to identify and purify the macromolecular components of nerve terminals.

Although neither of these hopes has yet been completely fulfilled, much valuable information is now available and our thinking is certainly much clearer than it was several years ago. It is now generally accepted, for example, that the bacterial neurotoxins, botulinum and tetanus toxin, are similar in mode of action (Dasgupta & Sugiyama 1977), but different from the snake toxins, β-bungarotoxin, taipoxin, crotoxin, and notexin, whose properties are almost identical (Chang et al 1977). The venon of the Black and Brown Widow Spider exhibits yet a third mechanism of action (Frontali et al 1976). The mechanism of action of these three toxin classes has been recently reviewed (Howard 1978). We try here to pay more attention to what we have learned about the mechanism of transmitter release from the study of the neurotoxins, reviewing each class of neurotoxin in turn.

BACTERIAL TOXINS

Mechanism of inhibition   Electrophysiological observations on botulinum intoxication have narrowed the range of mechanisms that can be invoked. The action potential reaches the nerve terminal (Harris & Miledi 1971) and quantal release can occur (Boroff et al 1974, Cull-Candy et al 1976a), which suggests that the point of inhibition must be somewhere in between. It is unlikely that the calcium channels are blocked since both release evoked by intracellular calcium and release evoked by calcium ionophores and external calcium are reduced (Cull-Candy et al 1976a, Kao et al 1976). Therefore, the calcium sensitivity of the release apparatus is altered or the exocytotic mechanism itself is impaired. Since both the spontaneous release rate (m.e.p.p. frequency) and the evoked release are inhibited by tetanus or botulinum toxin (Duchen & Tonge 1973, Harris & Miledi 1971, Boroff et al 1974) it can be concluded that both forms of release share a common pathway. This point will be pertinent later in our discussions of the spider venoms.

Binding studies   As mentioned above, some recognition mechanism must make neuronal tissue the preferred site of toxin binding. Were the toxin molecules to bind randomly to tissue, it is difficult to see how such a small number of molecules (10^6) could kill a mouse. Autoradiographic evidence has suggested that radioactive botulinum toxin (Hirokawa & Kitamura 1975) and tetanus toxin (Wernig et al 1977, Price et al 1977) bind to neuromuscular junctions. The binding is not extremely specific, with the ratio of counts/mg muscle being only twice as high in endplate-containing regions (Wernig et al 1977). There is considerable evidence that botulinum toxin (Ledeen & Mellanby 1977) and tetanus toxin bind to gangliosides (Ledley et al 1977, Helting et al 1977). The most convincing demonstration, perhaps, that long chain gangliosides are involved in tetanus binding, was that hybrid (neuroblastoma X glioma) cells, unable to bind the toxin, acquired toxin-binding capacity after exposure to gangliosides (Dimpfel et al 1977). Whether or not ganglioside binding is physiologically relevant is not clear from these results. Zimmermann & Piffaretti (1977) have shown that tetanus toxin produces morphological changes in nondividing neuroblastoma C1300 cells but not in dividing cells. By immonofluorescence procedures, toxin binds to both dividing and nondividing cells. They argue that binding to gangliosides does not necessarily correlate with the toxin's biological activity. However, as pointed out by Ledeen & Mellanby (1977), the ratio of cholera toxin-binding gangliosides to the number of cholera molecules required to kill a cell is around 10^9. Considering the vast excess of tetanus toxin used in most binding studies, the majority of the binding observed thus may be physiologically unnecessary.

The bacterial neurotoxins deserve more attention from biochemists. Since only about 10^6 molecules are required to kill a mouse, and death is not instantaneous, it is unlikely that the binding event itself is lethal, but rather that binding triggers an enzymatic response. Comparison with other bacterial toxins suggests that an ADP-ribosylation of a key protein could be involved (Chung & Collier 1977, van Heyningen 1977), and indeed preliminary pharmacological observations support such a hypothesis (Howard et al 1976). It has recently been reported that tetanus toxin, like diphtheria toxin, turns off protein synthesis (Dimpfel 1977). The bacterial neurotoxins also are the most promising for binding studies, since specificity of binding can be shown. Finally, they seem to act at a more specific site than the other presynaptic neurotoxins, as we shall see.

α-LATROTOXIN (BLACK AND BROWN WIDOW SPIDER TOXINS)

Mechanism of inhibition   Much of the earlier work on the venoms of these spiders involved homogenates of the whole venom gland. Since the venom contains several toxic elements (Frontali et al 1976, Grasso 1976), many of the early electrophysiological and anatomical studies were difficult to interpret. With the recent purification from other toxic polypeptides of α-latrotoxin, the 103,000 dalton polypeptide that blocks the vertebrate neuromuscular junction (Frontali et al 1976, Ornberg & Smyth 1976, Grasso 1976), the situation has become clearer. Fortunately, the more significant earlier findings have been confirmed using the purified protein.

A strong clue to the molecular mechanism of action of this toxin is the finding of Finkelstein et al (1976) that α-latrotoxin causes cation selective channels to form in black lipid membranes. Since channel opening and closing can be detected in this system it was possible to demonstrate that each channel has a conductance of 3.6×10^{-10} reciprocal ohms, about 4 times the conductance of the acetylcholine receptor channel (Katz & Miledi 1972). The channel is relatively nondiscriminating for cations, allowing even calcium to permeate. It was also reported that the toxin can bind gangliosides.

From electrophysiological studies it was learned that the toxin brings on a dramatic increase in miniature e.p.p. frequency at the neuromuscular junction even in the absence of extracellular calcium (Longenecker et al 1970). When calcium is present, bursts of quantal release are seen (Del Castillo & Pumplin 1975). After the release of a number of quanta approximately equal to the average number of vesicles in a nerve terminal, synaptic transmission is blocked (Longenecker et al 1970). In low external sodium the response to the toxin is slowed; conversely ouabain accelerates it (Ornberg & Smyth 1976). One interpretation of these results is that the toxin is indeed a sodium ionophore that causes a depolarization of the nerve terminal and consequently the release of transmitter. Since the opening of a single channel could give significant depolarization in a small region of the terminal, the bursts of release in calcium-containing medium could be due to the opening of a single channel. (Del Castillo & Pumplin 1975). To explain the release in the absence of external calcium, it is worth remembering that increasing internal sodium either by blocking the sodium pump or by removing extracellular K^+ causes release in the absence of extracellular calcium, presumably by mobilizing intraterminal calcium stores (Baker & Crawford 1975).

Morphological changes   Anatomical results support the conclusions of electrophysiology by showing that prolonged exposure to the toxin results in the disappearance of vesicles and swelling of the nerve terminal (Clark et al 1972). Obviously the toxin causes all the transmitter to be released and, at the same time, prevents recovery of nerve terminal membrane by endocytosis. Examination of the nerve terminal by freeze fracture during the peak of toxin-induced release shows that, in calcium-containing medium, exocytosis occurs at the normal active zone regions. In calcium-free medium, however, there is evidence for exocytosis at regions outside this normal active zone (Pumplin & Reese 1977). This is an important result if we accept the hypothesis that Black Widow Spider venon (BWSV) can mobilize intracellular calcium stores in calcium-free medium. The conclusion drawn is that release occurs at active zones because calcium enters there; release of calcium from internal sites still gives a preponderance of exocytosis at active zones, but allows exocytosis at more distant regions (Pumplin & Reese 1977). Exocytotic events that do not occur close to receptors in the postsynaptic membrane could give rise to small quantal events such as are found in botulinum-treated preparations. Were these "small" m.e.p.p.'s occurring in BWSV-intoxicated preparations, even in the

absence of calcium they would not be detected because of the high frequency of "normal" m.e.p.p.'s.

An intriguing aspect of BWSV action is that it can reverse the inhibition
due to botulinum toxin (Del Castillo & Pumplin 1975, Kao et al 1976,
Pumplin & Reese 1977), allowing the rapid release of normal sized quanta
of transmitter. Freeze-fracture studies have suggested that release at the
active zones is inhibited by botulinum toxin to a greater extent than release
at regions outside the active zones (Pumplin & Reese 1977). Perhaps selective inhibition of release at active zones gives rise to the "log jams" of
vesicles observed by Kao et al (1976), as well as to the small amplitude
m.e.p.p.'s found in botulinum-poisoned preparations. It is impossible at
present to decide if botulinum toxin blocks the calcium sensitivity of the
active zone, or the exocytotic apparatus itself, if indeed these are separable
sites.

THE PRESYNAPTIC SNAKE NEUROTOXINS

Mechanism of inhibition   The biochemical properties of the snake venon
toxins are well characterized. All four have a phospholipase A activity
residing in a basic 13,500 molecular weight polypeptide chain that has
sequence homology with other phospholipases (Eaker et al 1976, Kondo et
al 1978). The phospholipase activity is calcium dependent, is of the A2 type
(Kelly et al 1975, Strong et al 1976), and works best on membranes at their
phase transition temperatures (Strong & Kelly 1977). Cholesterol, which
eliminates the phase transition, inhibits phospholipid hydrolysis in artificial
bilayers. Hence, the sensitivity of bacterial mitochondrial and sarcoplasmic
reticulum membranes to the toxin could be due to their cholesterol content
(Strong & Kelly 1977).

Conceivably, the phospholipase activity could be irrelevant to the toxin's
biological activity. The arguments that do favor the involvement of phospholipase activity in toxicity are: (*a*) Strontium, a competitive inhibitor of
the phospholipase activity, protects the neuromuscular junction from toxin
inhibition (Strong et al 1976, Kelly et al 1975, Chang et al 1977), (*b*)
chemical modification of the toxin eliminates neurotoxicity and phospholipase activity in parallel (Halpert et al 1976, Strong et al 1977); and (*c*)
prolonged exposure to the toxin damages the plasma membrane of the nerve
terminal (Strong et al 1977). Complexes of horseradish peroxidase and
β-bungarotoxin bind to cytoplasmic elements inaccessible before the prolonged exposure (Strong et al 1977).

The problem of specificity   Although the phospholipase activity of the
snake toxins is thought to cause toxicity, normal phospholipases are not as
effective as the toxins in blocking neuromuscular transmission. One expla-

nation is that presynaptic plasma membranes are a preferred substrate for the enzymatic activity. Since β-bungarotoxin preferentially hydrolyzes membranes at their phase transitions (Strong & Kelly 1977), the possibility could be entertained that nerve terminal membranes are rich in phase boundaries. Unfortunately, pancreatic phospholipase A2, which has the same phospholipid specificity (Op den Kamp et al 1975), is not toxic. Alternatively, the nerve terminal could contain a type of phospholipid exquisitely sensitive to the toxins's phospholipase activity. No evidence in support of such a specificity has been found (Strong et al 1977, Howard 1978). Finally, the presynaptic snake toxins could have, in addition to their enzymatic site, a recognition site that recognizes and binds to a specific component of the nerve terminal. Binding would increase the local concentration of toxin and give rise to preferential hydrolysis of nerve terminal membranes (Oberg & Kelly 1976a, Abe et al 1977, Chang et al 1977, Howard & Truog 1977). While this hypothesis is quite plausible, the evidence in favor of it is meager. Selective binding to the endplate regions of neuromuscular junctions has not been shown and the relevance of synaptosomal toxin binding (Oberg & Kelly 1976a) to the mechanism of action is still questionable. Chemical modification of β-bungarotoxin with ethoxyformic anhydride has indicated the presence of a site required for toxicity that appears distinct from the catalytic site (Howard & Truog 1977). Evidence supporting a limited number of toxin binding sites at the nerve terminal is that toxin chemically modified to lose its phospholipase activity and toxicity can protect a neuromuscular junction against inhibition by native toxin (Abe et al 1977, Strong et al 1977). Such biological measures of binding are no substitute for physical measurements, especially since pretreatment of a preparation with bee venom phospholipase A2 also can protect a preparation from at least one of the symptoms of β-bungarotoxin intoxication (Abe & Miledi 1978).

These toxins can also modify transmitter uptake into synaptosomes (Howard 1978, Dowdall et al 1977), even when the phospholipase activity is inhibited. Since the toxins are highly basic proteins and since protamine, a basic polypeptide, can also inhibit choline uptake (Dowdall et al 1977), inhibition in this case may be by electrostatic binding of protein.

When one examines the electrophysiological response of the neuromuscular junction to the snake presynaptic neurotoxins, the overwhelming conclusion is that almost every step in transmitter release is affected. After exposure to toxin, evoked and spontaneous release fall, rise, and fall again (Abe et al 1976). Nerve terminal spikes disappear (Abe et al 1976, Dennis & Kelly, unpublished observations). Delayed release increases (Kamenskaya & Thesleff 1974, Oberg & Kelly 1976b). Although the nerve terminals still contain acetylcholine, K^+ depolarization does not cause release (Oberg & Kelly 1976b). Bursts of transmitter release have also been reported (Abe et al 1976). Anatomical observations of in vitro poisoned terminals show that vesicles are still present, but that large numbers of endocytotic (coated) vesicles accumulate (Strong et al 1977, Abe et al 1976, Cull-Candy et al 1976b). The accumulation is due to block of endocytosis rather than to an enhanced rate of endocytosis since the uptake of horseradish peroxidase is blocked (Lassignal & Heuser 1977). Such a wide range of physiological effects is perhaps to be expected since hydrolysis of membrane phospholipids could result in membrane damage leading to an irreversible perturbance of metabolite uptake, ion pumping, ion permeability, the fusion mechanism, or combinations of these. When the polycationic nature of these toxins is also taken into account the plethora of physiological responses is no longer remarkable.

That toxicity is due to blockage of transmitter release does not necessarily mean that a toxin acts exclusively on nerve terminals. For example, nerve terminals might be more susceptible to phospholipase-catalyzed hydrolysis than other parts of cells nearer to the cell soma. Binding studies with labeled β-bungarotoxin (Oberg & Kelly 1976a) and crotoxin (Jeng et al 1978) have shown extensive binding to non-neural tissue. In fact it has been postulated that the function of the enzymatically inactive B chain of crotoxin and the gamma chain of taipoxin is to act as a "chaperone" minimizing nonspecific binding (Jeng et al 1978, Fohlman et al 1977).

CONCLUSION   The hope motivating the study of presynaptic neurotoxins has been to elucidate the mechanism of transmitter release and to identify toxin binding macromolecules on nerve terminal membranes. The first goal has been attained, at least partially. Botulinum and Black Widow Spider venom, have been especially useful in verifying exocytosis, in pointing out sites of calcium entry, etc. The second goal has been unfulfilled. The bacterial toxins appear to bind to relatively ubiquitous gangliosides, the spider venoms may be channel formers, and the snake toxins are phospholipases. It is tempting to make the gloomy conclusion that no distinctive macromolecules are present on nerve terminal membranes, and that, since all the important machinery is intracellular, the toxins must penetrate the membrane either directly or by making the membrane leaky, or by hydrolyzing it. Hopefully this prediction is incorrect and toxins that specifically block choline uptake, the calcium channel, and the exocytotic site will soon be available.

CONCLUSIONS

Exocytotic release of neurotransmitter, at specialized regions of the nerve terminals called active zones, involves the apposition and fusion of vesicle membranes with the nerve terminal plasma membrane. Fusion between artificial phospholipid bilayers requires divalent cations, charged phospholipids, and "fluid" membrane interiors. If the phospholipid bilayer studies are good models for in vivo events, then they suggest a composition and structure for the synaptic vesicle and the presynaptic plasma membrane. The suggestions can be investigated by purifying and characterizing these membranes. Unfortunately, purification has been only partially successful because of the heterogeneity of the tissue, lack of adequate purification procedures, and lack of specific markers. The synaptic vesicles from elasmobranch electric organ have been the most successfully purified and preliminary characterization is consistent with, but does not verify, a phospholipid model of fusion.

Ultimately, elucidation of the fusion mechanisms responsible for neurotransmitter release will require reconstitution of the exocytotic event in vitro. A problem that biochemists face in studying transmitter release is that nerve terminals are constantly leaking neurotransmitter. Release of transmitter from synaptosomes can theoretically yield information on the release process, but results have so far been meager, largely because of the difficulty of distinguishing exocytotic release from leakage. The more direct approach of fusing synaptic vesicles to plasma membranes is beginning to look promising.

An alternative source of biochemical information about transmitter release is the interaction of presynaptic neurotoxins with the nerve terminal. Presynaptic neurotoxins fall into three classes: bacterial toxins, which bind to gangliosides and inhibit the release mechanism; the spider venom toxins, which are thought to raise intraterminal calcium and so deplete the terminal of vesicles; and the snake neurotoxins, which have a phospholipase activity that preferentially hydrolyzes nerve terminal phospholipids. These three classes of toxin are not only of value as specific inhibitors, they may also become useful as markers of presynaptic terminals.

Literature Cited

Abe, T., Alema, S., Miledi, R. 1977. Isolation and characterization of presynaptically acting neurotoxins from the venom of *Bungarus* snakes. *Eur. J. Biochem.* 80:1–12

Abe, T., Limbrick, A. R., Miledi, R. 1976. Acute muscle denervation induced by β-bungarotoxin. *Proc. R. Soc. London Ser. B* 194:545–53

Abe, T., Miledi, R. 1978. Inhibition of β-bungarotoxin action by bee venom phospholipase A2. *Proc. Roy. Soc. Lond. B* 200:225–230

Ahkong, Q. F., Fisher, D., Tampion, W., Lucy, J. A. 1975. Mechanisms of cell fusion. *Nature* 253:194–95

Akert, K., Pfenninger, K., Sandri, C., Moor, H. 1972. Freeze-etching and cytochemistry of vesicles and membrane complexes in synapses of the central nervous system. *Structure and Function of Synapses*, ed. G. D. Pappas, D. P. Purpura, pp. 67–86. New York: Raven. 308 pp.

Akert, K., Peper, K., Sandri, C. 1975. Structural organization of motor end plate and central synapses. *Cholinergic Mech-*

anisms, ed. P. G. Waser, pp. 43–52. New York: Raven. 555 pp.

Andrews, D. W., Patrick, R. L., Barchas, J. D. 1978. The effects of 5-hydroxtryptophan and 5-hydroxytryptamine on dopamine synthesis and release in rat brain striatal synaptosomes. J. Neurochem. 30:465–70

Babel-Guerin, E., Boyenval, J., Droz, B., Dunant, Y., Hassig, R. 1977. Accumulation of calcium in cholinergic axon terminals after nerve activity. Localization by electron microscope radioautography at the nerve-electroplaque junction of Torpedo. Brain Res. 121:348–52

Baker, P. F., Crawford, A. C. 1975. A note on the mechanism by which inhibitors of the sodium pump accelerate spontaneous release of transmitter from motor nerve terminals. J. Physiol. London 247:209–26

Baker, R. R., Dowdall, M. J., Whittaker, V. P. 1975. The involvement of lysophosphoglycerides in neurotransmitter release: The composition and turnover of phospholipids of synaptic vesicles of guinea pig cerebral cortex and Torpedo electric organ and the effect of stimulation. Brain Res. 100:629–44

Barondes, S. H. 1974. Synaptic macromolecules: Identification and metabolism. Ann. Rev. Biochem. 43:147–68

Bashford, C. L., Johnson, L. N., Radda, G. K., Ritchie, G. A. 1976. Lipid ordering and enzymatic activities in chromaffin granule membranes. Eur. J. Biochem. 67:105–14

Bass, L., Moore, W. J. 1966. Electrokinetic mechanism of miniature post synaptic potentials. Proc. Natl. Acad. Sci. USA 55:1214–17.

Berl, S., Puszkin, S., Nicklas, W. J. 1973. Actomyosin-like protein in brain. Science 179:441–46.

Birks, R. I., Huxley, H. E., Katz, B. 1960. The fine structure of the neuromuscular junction of the frog. J. Physiol. London 150:134–44

Bisby, M. A., Fillenz, M. 1971. The storage of endogenous noradrenaline in sympathetic nerve terminals. J. Physiol. London 215:163–79

Blaschko, H., Firemark, H., Smith, A. D., Winkler, H. 1967. Lipids of the adrenal medulla. Lysolecithin, a characteristic constituent of chromaffin granules. Biochem. J. 104:545–49

Blaustein, M. P. 1975. Effects of potassium, veratridine and scorpion venom on calcium accumulation and transmitter release by nerve terminals in vitro. J. Physiol. London 247:617–55

Blaustein, M. P., Kendrick, N. C., Fried, R. C., Ratzlaff, R. W. 1977. Calcium metabolism at the mammalian presynaptic nerve terminal: Lessons from the synaptosome. In Society for Neuroscience Symposia, Vol. 2, ed. W. M. Cowan, J. A. Ferrendelli, pp. 172–94. Bethesda, Maryland: Soc. Neurosci. 461 pp.

Blitz, A. L., Fine, R. E. 1974. Muscle-like contractile proteins and tubulin in synaptosomes. Proc. Natl. Acad. Sci. USA 71:4472–76

Blitz, A. L., Fine, R. E., Toselli, P. A. 1977. Evidence that coated vesicles isolated from brain are calcium-sequestering organelles resembling sarcoplasmic reticulum. J. Cell Biol. 75:135–47

Bloom, F. E., Aghajanian, G. K. 1968. Fine structure and cytochemical analysis of the staining of synaptic junctions with phosphotungstic acid. J. Ultrastruct. Res. 22:361–75

Bock, E. 1978. Nervous system specific proteins. J. Neurochem. 30:7–14

Bock, E., Helle, K. B. 1977. Localization of synaptin on synaptic vesicle membranes, synaptosomal plasma membranes and chromaffin granule membranes. FEBS Lett. 82:175–78

Bock, E., Jorgensen, O. S. 1975. Rat brain synaptic vesicles and synaptic plasma membranes compared by cross electrophoresis. FEBS Lett. 52:37–39

Bock, E., Jorgensen, O. S., Dittmann, L., Eng, L. F. 1975. Determination of brain-specific antigens in short term cultivated rat astroglial cells and in rat synaptosomes. J. Neurochem. 25:867–70

Boroff, D. A., Del Castillo, J., Evoy, W. H., Steinhardt, R. A. 1974. Observations on the action of Type A botulinum toxin on frog neuromuscular junctions. J. Physiol. London 240:227–53

Boyne, A. F., Bohan, T. P., Williams, T. H. 1974. Effects of calcium containing fixation solutions on cholinergic synaptic vesicles. J. Cell Biol. 63:780–95

Bradford, H. F. 1975. Isolated nerve terminals as an in vitro preparation for the study of dynamic aspects of transmitter metabolism and release. In Handbook of Psychopharmacology, Vol. 1, ed. L. Iversen, S. Iversen, S. Synder, pp. 191–252. New York and London: Plenum. 298 pp.

Breckenridge, W. C., Gombos, G., Morgan, I. G. 1972a. The decosahexaenoic acid of the phospholipids of synaptic membranes, vesicles and mitochondria. Brain Res. 33:581–83

Breckenridge, W. C., Gombos, G., Morgan, I. G. 1972b. The lipid composition of adult rat brain synaptosomal plasma membranes. *Biochim. Biophys. Acta* 266:695–707

Breckenridge, W. C., Morgan, I. G. 1972. Common glycoproteins of synaptic vesicles and the synaptosomal plasma membrane. *FEBS Lett.* 22:253–56

Breckenridge, W. C., Morgan, I. G., Zanetta, J. P., Vincendon, G. 1973. Adult rat brain synaptic vesicles. II. Lipid composition. *Biochim. Biophys. Acta* 320:681–86

Breer, H., Morris, S. J., Whittaker, V. P. 1977. Adenosine triphosphatase activity associated with purified cholinergic synaptic vesicles of *Torpedo marmorata*. *Eur. J. Biochem.* 80:313–18

Bygrave, F. L. 1978. Mitochondria and the control of intracellular calcium. *Biol. Rev. Cambridge Philos. Soc.* 53:43–79

Carlson, S. S., Wagner, J. A., Kelly, R. B. 1978. Purification of synaptic vesicles from elasmobranch electric organ and the use of biophysical criteria to demonstrate purity. *Biochemistry* 17:1188–99

Casey, R. P., Njus, D., Radda, G. K., Sehr, P. A. 1977. Active proton uptake by chromaffin granules: Observation by amine distribution and phosphorus-31 nuclear magnetic resonance techniques. *Biochemistry* 16:972–77

Castle, J. D., Jamieson, J. D., Palade, G. E. 1975. Secretion granules of the rabbit parotid gland. Isolation, subfractionation and characterization of the membrane and content subfractions. *J. Cell Biol.* 64:182–210

Chang, C. C., Jai-Su, M., Lee, J. D., Eaker, D. 1977. Effects of Sr^{2+} and Mg^{2+} on the phospholipase A and the presynaptic neuromuscular blocking actions of β-bungarotoxin, crotoxin and taipoxin. *Naunyn-Schmiedebergs Arch. Pharmakol.* 299:155–61

Chi, E. Y., Lagunoff, D., Koehler, J. K. 1976. Freeze-fracture study of mast cell secretion. *Proc. Natl. Acad. Sci. USA* 73:2823–27

Chubb, I. W., De Potter, W. P., De Schaepdryver, A. F. 1970. Evidence for two types of noradrenaline storage particles in dog spleen. *Nature* 228:1203–4

Chung, D. W., Collier, R. J. 1977. The mechanism of ADP-ribosylation of elongation factor 2 catalyzed by fragment A from diphtheria toxin. *Biochim. Biophys. Acta* 483:248–57

Clark, A. W., Hurlbut, W. P., Mauro, A. 1972. Changes in the fine structure of the neuromuscular junction of the frog caused by black widow spider venom. *J. Cell Biol.* 52:1–14

Cooke, J. D., Okamoto, K., Quastel, D. M. J. 1973. The role of calcium in depolarization-secretion coupling at the motor nerve terminal. *J. Physiol. London* 228:459–97

Cotman, C. W. 1974. Isolation of synaptosomal, synaptic plasma membrane fractions. *Methods in Enzymol.* 31A:445–52

Cotman, C. W., Banker, G., Churchill, L., Taylor, D. 1974. Isolation of postsynaptic densities from rat brain. *J. Cell Biol.* 63:441–55

Cotman, C., Brown, D. H., Harrell, B. W., Anderson, N. G. 1970. Analytical differential centrifugation: An analysis of the sedimentation properties of synaptosomes, mitochondria and lysosomes from rat brain homogenates. *Arch. Biochem. Biophys.* 136:436–47

Cotman, C. W., Taylor, D. 1972. Isolation and structural studies on synaptic complexes from rat brain. *J. Cell Biol.* 55:696–711

Couteaux, R., Pecot-Dechavassine, M. 1970. Vesicules synaptiques et poches au niveau des zones actives de la jonction neuromusculaire. *C R Acad. Sci. Ser. D* 271:2346–49

Couteaux, R., Pecot-Dechavassine, M. 1974. Les zones specialisees des membranes presynaptiques. *C R Acad. Sci. Ser. D* 278:291–93

Cowley, A. C., Fuller, N., Rand, R. P., Parsegian, V. A. 1977. Measurement of repulsion between charged phospholipid bilayers. *Biophys. J.* 17:85A

Creutz, C. E., Pazoles, C. J., Pollard, H. B. 1978. Identification and purification of an adrenal medullary protein (synexin) that causes calcium-dependent aggregation of isolated chromaffin granules. *J. Biol. Chem.* 253:2858–66

Crowther, R. A., Finch, J. T., Pearse, B. M. F. 1976. On the structure of coated vesicles. *J. Mol. Biol.* 103:785–98

Cull-Candy, S. G., Lundh, H., Thesleff, S. 1976a. Effects of botulinum toxin on neuromuscular transmission in the rat. *J. Physiol. London* 260:177–203

Cull-Candy, S. G., Fohlman, J., Gustavsson, D., Lullman-Rauch, R., Thesleff, S. 1976b. The effects of taipoxin and notexin on the function and fine structure of the murine neuromuscular junction. *Neuroscience* 1:175–80

Dahl, G., Gratzl, M. 1976. Calcium-induced fusion of isolated secretory vesicles from the islet of Langerhans. *Cytobiologie Z. Exp. Zellforsch.* 12:344–55

DasGupta, B. R., Sugiyama, H. 1977. Biochemistry and pharmacology of botulinum and tetanus toxins. In *Perspectives in Toxinology,* ed. A. W. Bernheimer, Ch. 5, pp. 87–119. New York: Wiley. 204 pp.

Davis, B., Lazarus, N. R. 1976. An in vitro system for studying insulin release caused by secretory granules–plasma membrane interaction: Definition of the system. *J. Physiol. London* 256:709–29

Dean, P. M. 1975. Exocytosis modelling: An electrostatic function for calcium in stimulus-secretion coupling. *J. Theor. Biol.* 54:289–308

Dean, P. M., Matthews, E. K. 1974. Calcium-ion binding to the chromaffin-granule surface. *Biochem. J.* 142:637–40

De Belleroche, J. S., Bradford, H. F. 1972. The stimulus-induced release of acetylcholine from synaptosome beds and its calcium dependence. *J. Neurochem.* 19:1817–19

De Belleroche, J. S., Bradford, H. F. 1977. On the site of the origin of transmitter amino acids released by depolarization of nerve terminals in vitro. *J. Neurochem.* 29:335–43

Del Castillo, J., Pumplin, D. W. 1975. Discrete and discontinuous action of brown widow spider venom on the presynaptic nerve terminals of frog muscle. *J. Physiol. London* 252:491–508

Dennis, M. J., Miledi, R. 1974. Electrically induced release of acetylcholine from denervated Schwann cells. *J. Physiol. London* 237:431–52

DePierre, J. W., Karnovsky, M. L. 1973. Plasma membranes of mammalian cells. A review of methods for their characterization and isolation. *J. Cell Biol.* 56:275–303

De Potter, W. P., Chubb, I. W. 1977. Biochemical observations on the formation of small noradrenergic vesicles in the splenic nerve of the dog. *Neuroscience* 2:167–74

De Robertis, E., De Lores Arnaiz, G. R., Salganicoff, L., De Iraldi, A. P., Zieher, L. M. 1963. Isolation of synaptic vesicles and structural organization of the acetylcholine system within brain nerve endings. *J. Neurochem.* 10:225–35

Dimpfel, W. 1977. Effect of tetanus toxin on protein synthesis in neuronal cultures. *Naunyn-Schmiedebergs Arch. Pharmakol.* 297(Suppl. 1):R58

Dimpfel, W., Huang, R. T. C., Habermann, E. 1977. Gangliosides in nervous tissue cultures and binding of ^{125}I-labeled tetanus toxin, a neuronal marker. *J. Neurochem.* 29:329–34

Dipolo, R., Requena, J., Brinley, F. J. Jr., Mullins, L. J., Scarpa, A., Tiffert, T. 1976. Ionized calcium concentrations in squid axons. *J. Gen. Physiol.* 67:433–67

Dodge, F. A., Rahamimoff, R. 1967. Cooperative action of calcium ions in transmitter release at the neuromuscular junction. *J. Physiol. London* 193:419–32

Dowdall, M. J., Boyne, A. F., Whittaker, V. P. 1974. Adenosine triphosphate; A constituent of cholinergic synaptic vesicles. *Biochem. J.* 140:1–12

Dowdall, M. J., Fohlman, J. P., Eaker, D. 1977. Inhibition of high-affinity choline transport in peripheral cholinergic endings by presynaptic snake venom neurotoxins. *Nature* 269:700–2

Dreyer, F., Peper, K., Akert, K., Sandri, C., Moor, H. 1973. Ultrastructure of the active zone in the frog neuromuscular junction. *Brain Res.* 62:373–80

Duchen, L. W., Tonge, D. A. 1973. The effects of tetanus toxin on neuromuscular transmission and on the morphology of motor endplates in slow and fast skeletal muscle of the mouse. *J. Physiol. London* 228:157–72

Eaker, D., Halpert, J., Fohlman, J., Karlsson, E. 1976. Structural nature of presynaptic neurotoxins from the venoms of the Australian Tiger Snake, *Notechis scutatus scutatus* and the Taipan *Oxyuranus scutellatus scutellatus.* In *International Symposium on Animal, Plant and Microbial Toxins,* ed. A. Ohsaka, K. Hayashi, Y. Sawai, Vol. 2, pp. 27–45. New York: Plenum. 562 pp.

Finer, E. G., Darke, A. 1974. Phospholipid hydration studied by deuteron magnetic resonance spectroscopy. *Chem. Phys. Lipids* 12:1–16

Finkelstein, A., Rubin, L. L., Tzeng, M. C. 1976. Black widow spider venom: Effect of purified toxin on lipid bilayer membranes. *Science* 193:1009–11

Fletcher, P., Forrester, T. 1975. The effect of curare on the release of acetylcholine from mammalian motor nerve terminals and an estimate of quantum content. *J. Physiol. London* 251:131–44

Fohlman, J., Lind, P., Eaker, D. 1977. Taipoxin, an extremely potent presynaptic snake venom neurotoxin. Elucidation of the primary structure of the acidic carbohydrate-containing taipoxin-subunit, prophospholipase homolog. *FEBS Lett.* 84:367–71

Fonnum, F. 1968. Choline acetyltransferase binding to and release from membranes. *Biochem. J.* 109:389–98

Frontali, N., Ceccarelli, B., Gorio, A., Mauro, A., Siekevitz, P., Tzeng, M.,

Hurlbut, W. P. 1976. Purification from black widow spider venom of a protein factor causing the depletion of synaptic vesicles at neuromuscular junction. *J. Cell Biol.* 68:462–79

Geffen, L. B., Livett, B. G. 1971. Synaptic vesicles in sympathetic neurons. *Physiol. Rev.* 51:99–157

Gomperts, R. D. 1976. Calcium and cell activation. In *Receptors and Recognition,* ed. P. Cuatrecasas, M. F. Greaves. Series A, Vol. 2, pp. 43–102. New York: Chapman & Hall. 229 pp.

Grasso, A. 1976. Preparation and properties of a neurotoxin purified from the venom of black widow spider (*latrodectus mactans tredecimguttatus*). *Biochim. Biophys. Acta* 439:406–12

Gray, E. G. 1975. Presynaptic microtubules and their association with synaptic vesicles. *Proc. R. Soc. London Ser. B* 190:369–72

Greene, L. A., Rein, G. 1977. Release, storage and uptake of catecholamines by a clonal cell line of nerve growth factor (NGF) responsive pheochromocytoma cells. *Brain Res.* 129:247–63

Gurd, J. W. 1977. Synaptic plasma membrane glycoproteins: Molecular identification of lectin receptors. *Biochemistry* 16:369–74

Gurd, J. W., Jones, L. R., Mahler, H. R., Moore, W. J. 1974. Isolation and partial characterization of rat brain synaptic plasma membranes. *J. Neurochem.* 22:281–90

Halpert, J., Eaker, D., Karlsson, E. 1976. The role of phospholipase activity in the action of a presynaptic neurotoxin from the venom of *Notechis scutatus scutatus* (Australian Tiger Snake). *FEBS Lett.* 61:72–76

Harris, A. J., Miledi, R. 1971. The effect of type D botulinum toxin on frog neuromuscular junctions. *J. Physiol. London* 217:497–515

Hata, F., Kuo, H., Matsuda, T., Yoshida, H., 1976. Factors required for calcium sensitive acetylcholine release from crude synaptic vesicles. *J. Neurochem.* 27: 139–44

Hauser, H., Darke, A., Phillips, M. C. 1976. Ion-binding to phospholipids, Interaction of calcium with phosphatidylserine. *Eur. J. Biochem.* 62:335–44

Hauser, H., Dawson, R. M. C. 1967. The binding of calcium at lipid-water interfaces. *Eur. J. Biochem.* 1:61–69

Hauser, H., Phillips, M. C., Barratt, M. D. 1975. Differences in the interaction of inorganic and organic (hydrophobic) cations with phosphatidylserine membranes. *Biochim. Biophys. Acta* 413: 341–53

Haynes, D. H. 1977. Divalent cation-ligand interactions of phospholipid membranes: Equilibria and kinetics. In *Metal-Ligand Interactions in Organic Chemistry and Biochemistry,* part 2, pp. 189–212. Dordrecht, Holland: Reidel.

Helting, T. B., Zwisler, O., Wiegandt, H. 1977. Structure of tetanus toxin. II. Toxin binding to ganglioside. *J. Biol. Chem.* 252:194–98

Hesketh, T. R., Smith, G. A., Houslay, M. D., McGill, K. A., Birdsall, N. J. M., Metcalfe, J. C., Warren, G. B. 1976. Annular lipids determine the ATPase activity of a calcium transport protein complexed with dipalmitoyllecithin. *Biochemistry* 15:4145–51

Heuser, J. E. 1976. Morphology of synaptic vesicle discharge and reformation at the frog neuromuscular junction. In *Motor Innervation of Muscle,* ed. S. Thesleff, pp. 51–115. London: Academic. 351 pp.

Heuser, J. E. 1977. Synaptic vesicle exocytosis revealed in quick-frozen frog neuromuscular junctions treated with 4-aminopyridine and given a single electrical shock. In *Society for Neuroscience Symposia,* ed. W. M. Cowan, J. A. Ferrendelli, pp. 215–239, Vol. 2 Bethesda, Md: Soc. Neurosci. 461 pp.

Heuser, J. E. 1978. Synaptic vesicle exocytosis and recycling during transmitter discharge from the neuromuscular junction. *Proc. Dahlem Conf., Berlin.* In press

Heuser, J. E., Katz, B., Miledi, R. 1971. Structural and functional changes of frog neuromuscular junctions in high calcium solutions. *Proc. R. Soc. London Ser. B.* 178:407–15

Heuser, J. E., Reese, T. S. 1973. Evidence for recycling of synaptic vesicle membrane during transmitter release at frog neuromuscular junction. *J. Cell Biol.* 57: 315–44

Heuser, J. E., Reese, T. S. 1975. Redistribution of intramembranous particles from synaptic vesicles: Direct evidence for vesicle recycling. *Anat. Rec.* 181:374

Heuser, J. E., Reese, T. S. 1977. Structure of the synapse. In *Handbook of Physiology, The Nervous System* I. pp. 261–94. Washington, D.C: Am. Physiol. Soc.

Heuser, J. E., Reese, T. S., Dennis, M. J., Jan, Y., Jan, L. 1978. Exocytosis of synaptic vesicles captured by quick freezing. *J. Cell Biol.* In press

Heuser, J. E., Reese, T. S., Landis, D. M. D. 1974. Functional changes in frog neuromuscular junction studied with freeze-fracture. *J. Neurocytol.* 3:109–131

Hirokawa, N., Kitamura, M., 1975. Localization of radioactive ^{125}I-labeled botulinus toxin at the neuromuscular junction of mouse diaphragm. *Naunyn-Schmiedeberg's Arch. Pharmacol.* 287: 107–110

Holtzman, E. 1977. The origin and fate of secretory packages, especially synaptic vesicles. *Neuroscience* 2:327–355

Hortnagl, H. 1976. Membranes of the adrenal medulla: A comparison of membranes of chromaffin granules with those of endoplasmic reticulum. *Neuroscience* 1:9–18

Howard, B. 1978. Effects of polypeptide neurotoxins on acetylcholine storage and release. In *Cholinergic Mechanisms and Psychopharmacology,* ed. D. J. Jenden, pp. 565–585. New York: Plenum Press. 885 pp.

Howard, B. D., Truog, R. 1977. Relationship between the neurotoxicity and phospholipase A activity of β-bungarotoxin. *Biochemistry* 16:122–125

Howard, B. D., Wu, W. C. S., Gundersen, C. B. 1976. Antagonism of botulinum toxin by theophylline. *Biochem. Biophys. Res. Commun.* 71:413–415

Howe, P. R. C., Fenwick, E. M., Rostas, J. A. P., Livett, B. G. 1977. Immunochemical comparison of synaptic plasma membrane and synaptic vesicle antigens. *J. Neurocytol.* 6:339–352

Hubbard, J. I., Jones, S. F., Landau, E. M. 1968a. On the mechanism by which calcium and magnesium affect the spontaneous release of transmitter from mammalian motor nerve terminals. *J. Physiol. London* 194:355–80

Hubbard, J. I., Jones, S. F., Landau, E. M. 1968b. On the mechanism by which calcium and magnesium affect the release of transmitter by nerve impulses. *J. Physiol. London* 196:75–86

Hurlbut, W. P., Longenecker, H. E. Jr., Mauro, A. 1971. Effects of calcium and magnesium on the frequency of miniature end-plate potentials during prolonged tetanization. *J. Physiol. London* 219:17–38

Izumi, F., Kashimoto, T., Miyashita, T., Wada, A. 1977. Involvement of membrane associated protein in ADP-induced lysis of chromaffin granules. *FEBS Lett.* 78:177–80

Izumi, F., Oka, M., Morita, K., Azuma, H. 1975. Catecholamine releasing factor in bovine adrenal medulla. *FEBS Lett.* 56:73–76

Jacobson, K., Papahadjopoulos, D. 1975. Phase transitions and phase separations in phospholipid membranes induced by changes in temperature, pH, and concentration of bivalent cations. *Biochemistry* 14:152–61

Jeng, T. W., Hendon, R. A., Fraenkel-Conrat, H. 1978. Search for relationships among the hemolytic, phospholipolytic and neurotoxic activities of snake venoms. *Proc. Natl. Acad. Sci. USA* 75:600–4

Jockusch, B. M., Burger, M. M., DaPrada, M., Richards, J. G., Chaponnier, C., Gabbiani, G. 1977. α-Actinin attached to membrane of secretory vesicles. *Nature* 270:628–29

Johnson, R. G., Carlson, N. J., Scarpa, A. 1978. ΔpH and catecholamine distribution in isolated chromaffin granules. *J. Biol. Chem.* 253:1512–21

Jones, D. G. 1975. *Synapses and Synaptosomes, Morphological Aspects.* London: Chapman and Hall. 258 pp.

Jones, D. H., Matus, A. I. 1974. Isolation of synaptic plasma membrane from brain by combined flotation-sedimentation density gradient centrifugation. *Biochim. Biophys. Acta* 356:276–87

Joo, F., Karnushina, I. 1975. Morpholometric assessment of the composition of the synaptosomal fractions obtained by the use of Ficoll gradients. *J. Neurochem.* 24:839–40

Jorgensen, O. S. 1976. Localization of the antigens D1, D2, and D3 in the rat brain synaptic membrane. *J. Neurochem.* 27:1223–27

Jost, P. C., Griffith, O. H., Capaldi, R. A., Vanderkooi, G. 1973. Evidence for boundary lipid in membranes. *Proc. Natl. Acad. Sci. USA* 70:480–84

Kadota, K., Kadota, T. 1973. Isolation of coated vesicles, plain synaptic vesicles and flocculent material from a crude synaptosome fraction of guinea pig whole brain. *J. Cell Biol.* 58:135–51

Kamenskaya, M. A., Thesleff, S. 1974. The neuromuscular blocking action of an isolated toxin from the Elapid (*oxyuranus scutellactus*). *Acta Physiol. Scand.* 90:716–24

Kao, I., Drachman, D. B., Price, D. L. 1976. Botulinum toxin: Mechanism of presynaptic blockade. *Science* 193:1256–58

Katz, B. 1969. *The Release of Neural Transmitter Substances.* Liverpool: Univ. Press. 60 pp.

Katz, B., Miledi, R. 1970. Further study of the role of calcium in synaptic transmission. *J. Physiol. London* 207:789–801

Katz, B., Miledi, R. 1972. The statistical nature of the acetylcholine potential and its molecular components. *J. Physiol. London* 224:665–99

Katz, B., Miledi, R. 1977a. Transmitter leakage from motor nerve terminals. *Proc. R. Soc. London Ser. B* 196:59–72

Katz, B., Miledi, R. 1977b. Suppression of transmitter release at the neuromuscular junction. *Proc. R. Soc. London Ser. B* 196:465–69

Kelly, P. T., Cotman, C. W. 1977. Identification of glycoproteins and proteins at synapses in the central nervous system. *J. Biol. Chem.* 252:786–93

Kelly, R. B., Oberg, S. G., Strong, P. N., Wagner, G. M. 1975. β-bungarotoxin. A phospholipase which stimulates transmitter release. *Cold Spring Harbor Symp. Quant. Biol.* 40:117–25

Kendrick, N. C., Blaustein, M. P., Fried, R. C., Ratzlaff, R. W. 1977. ATP-dependent calcium storage in presynaptic nerve terminals. *Nature* 265:246–48

Kirksey, D. F., Klein, R. L., Baggett, J. McC., Gasparis, M. S. 1978. Evidence that most of the dopamine β-hydroxylase is not membrane bound in purified large dense cored noradrenergic vesicles. *Neuroscience* 3:71–81

Kondo, K., Narita, K., Lee, C. Y. 1978. Amino acid sequences of the two polypeptide chains in β₁-bungarotoxin from the venom of *Bungarus multicinctus*. *J. Biochem.* 83:101–15

Konig, P., Hortnagl, H., Kostron, H., Sapinsky, H., Winkler, H. 1976. The arrangement of dopamine β-hydroxylase (EC 1.14.2.1) and chromomembrin B in the membrane of chromaffin granules. *J. Neurochem.* 27:1539–41

Kornguth, S. E., Anderson, J. W., Scott, G. 1969. Isolation of synaptic complexes in a cesium chloride density gradient: Electron miscroscopic and immunohistochemical studies. *J. Neurochem.* 16:1017–24

Kostron, H., Winkler, H., Peer, L. J., Konig, P. 1977. Uptake of adenosine triphosphate by isolated adrenal chromaffin granules: A carrier mediated transport. *Neuroscience* 2:159–66

Krnjevic, K., Mitchell, J. F. 1961. The release of acetylcholine in the isolated rat diaphragm. *J. Physiol. London* 155:246–62

Krnjevic, K., Straughan, D. W. 1964. The release of acetylcholine from the denervated rat diaphragm. *J. Physiol. London* 170:371–78

Kuffler, S. W., Yoshikami, D. 1975. The number of transmitter molecules in a quantum: An estimate from iontophoretic application of acetylcholine at the neuromuscular synapse. *J. Physiol. London* 251:465–82

Laduron, P. 1977. Tissue fractionation in neurobiochemistry—Analytical tool or a source of artifacts. In *International Review of Neurobiology*, ed. J. R. Smythies, R. J. Bradley, vol. 20, pp. 251–83. New York: Academic. 351 pp.

Lagercrantz, H. 1976. On the composition and function of large dense-cored vesicles in sympathetic nerves. *Neuroscience* 1:81–92

Lansman, J., Haynes, D. H. 1975. Kinetics of a Ca⁺⁺-triggered membrane aggregation reaction of phospholipid membranes. *Biochim. Biophys. Acta* 394:335–47

Lassignal, N. L., Heuser, J. E. 1977. Evidence that β-bungarotoxin arrests synaptic vesicle recycling by blocking coated vesicle formation. *Neurosci. Abstr.* 3:373

Lawson, D., Raff, M. C., Gomperts, B., Fewtrell, C., Gilula, N. B. 1977. Molecular events during membrane fusion. A study of exocytosis in rat peritoneal mast cells. *J. Cell Biol.* 72:242–59

Ledeen, R. W., Mellanby, J. 1977. Gangliosides as receptors for bacterial toxins. In *Perspectives in Toxinology*, ed. A. W. Bernheimer, pp. 15–42. New York: Wiley. 204 pp.

Ledley, F. D., Lee, G., Kohn, L. D., Habig, W. H., Hardegree, M. C. 1977. Tetanus toxin interactions with thyroid plasma membranes. Implications for structure and function of tetanus toxin receptors and potential pathophysiological significance. *J. Biol. Chem.* 252:4049–55

Lembeck, F., Mayer, N., Schindler, G. 1977. Substance P in rat brain synaptosomes. *Naunyn-Schmiedebergs Arch. Pharmakol.* 301:17–22

LeNeveu, D. M., Rand, R. P., Parsegian, V. A., Gingell, D. 1977. Measurement and modification of forces between lecithin bilayers. *Biophys. J.* 18:209–30

Lentzen, H., Philippu, A. 1977. Uptake of tyramine into synaptic vesicles of the caudate nucleus. *Naunyn-Schmiedebergs Arch. Pharmakol.* 300:25–30

Levi, G., Raiteri, M. 1976. Synaptosomal transport processes. In *Int. Rev. Neurobiol.* vol. 19, ed. J. R. Smithies, R. J. Bradley, pp. 51–74, New York: Academic. 323 pp.

Levi, G., Rusca, G., Raiteri, M. 1976. Diaminobutyric acid: A tool for discriminating between carrier-mediated and non-carrier-mediated release of

GABA from synaptosomes? *Neurochem. Res.* 1:581–90

Levitan, I. B., Mushynski, W. E., Ramirez, G. 1972. Highly purified synaptosomal membranes from rat brain. Preparation and characterization. *J. Biol. Chem.* 247:5376–81

Levy, W. B., Haycock, J. W., Cotman, C. W. 1974. Effects of polyvalent cations on stimulus-coupled secretion of [¹⁴C]-γ-aminobutyric acid from isolated brain synaptosomes. *Mol. Pharmacol.* 10:438–49

Livett, B. G., Rostas, J. A. P., Jeffrey, P. L., Austin, L. 1974. Antigenicity of isolated synaptic membranes. *Exp. Neurol.* 43:330–38

Llinas, R. R. 1977. Calcium and transmitter release in squid synapse. In *Society for Neuroscience Symposia,* vol. 2, ed. W. M. Cowan, J. A. Ferrendelli, pp. 139–60. Bethesda, Md: Soc. Neurosci. 461 pp.

Llinas, R., Blinks, J. R., Nicholson, C. 1972. Calcium transient in presynaptic terminal of squid giant synapse: Detection with aequorin. *Science* 176:1127–29

Llinas, R. R., Heuser, J. E. 1977. Depolarization-release coupling systems in neurons. *Neurosci. Res. Prog. Bull.* 15:557–687

Llinas, R., Nicholson, C. 1975. Calcium role in depolarization-secretion coupling: An aequorin study in squid giant synapse. *Proc. Natl. Acad. Sci. USA* 72:187–90

Longenecker, H. E., Hurlbut, W. P., Mauro, A., Clark, A. W. 1970. Effects of black widow spider venom on the frog neuromuscular junction. *Nature* 225:701–3

Lowe, D. A., Richardson, B. P., Taylor, P., Donatsch, P. 1976. Increasing intracellular sodium triggers calcium release from bound pools. *Nature* 260:337–38

Lundh, H., Thesleff, S. 1977. The mode of action of 4-aminopyridine and guanidine on transmitter release from motor nerve terminals. *Eur. J. Pharmacol.* 42:411–12

MacDonald, R. C., Simon, S. A., Baer, E. 1976. Ionic influences on the phase transition of dipalmitylphosphatidylserine. *Biochemistry* 15:885–91

MacIntosh, F. C., Collier, B. 1976. Neurochemistry of cholinergic terminals. *Handbuch der experimentellen Pharmakologie* 42:99–228

Magleby, K. L., Zengel, J. E. 1975. A dual effect of repetitive stimulation on posttetanic potentiation of transmitter release at the frog neuromuscular junction. *J. Physiol. London* 245:163–82

Magleby, K. L., Zengel, J. E. 1976. Augmentation: A process that acts to increase transmitter release at the frog neuromuscular junction. *J. Physiol. London* 257:449–70

Mahler, H. R. 1977. Proteins of the synaptic membrane. *Neurochem. Res.* 2:119–47

Mahler, H. R., Gurd, J. W., Wang, Y. J. 1975. Molecular topography of the synapse. In *The Nervous System,* vol. 1, ed. D. B. Tower, pp. 455–66. New York: Raven. 685 pp.

Marchbanks, R. M. 1975. Biochemistry of cholinergic neurons. In *Handbook of Psychopharmacology,* vol. 3, ed. L. Iversen, S. Iversen, S. Snyder, pp. 247–326. New York and London: Plenum. 486 pp.

Marsh, D., Radda, G. K., Ritchie, G. A. 1976. A spin-label study of the chromaffin granule membrane. *Eur. J. Biochem.* 71:53–61

Martin, F. J., MacDonald, R. C. 1976. Phospholipid exchange between bilayer membrane vesicles. *Biochemistry* 15:321–27

Matthaei, H., Lentzen, H., Philippu, A. 1976. Competition of some biogenic amines for uptake into synaptic vesicles of the striatum. *Naunyn-Schmiedebergs Arch. Pharmakol.* 293:89–96

Matthews, E. K., Nordman, J. J. 1976. The synaptic vesicle: Calcium ion binding to the vesicle membrane and its modification by drug action. *Molec. Pharmacol.* 12:778–88

Mattson, M. E., O'Brien, R. D. 1976. The binding of acetylcholine to synaptic vesicles from *Torpedo californica* electroplax. *J. Neurochem.* 27:867–72

McGee, R., Simpson, P., Christian, C., Mata, M., Nelson, P., Nirenberg, M. 1978. Regulation of acetylcholine release from neuroblastoma x glioma hybrid cells. *Proc. Natl. Acad. Sci. USA* 75:1314–18

McLaughlin, S. G. A., Szabo, G., Eisenman, G. 1971. Divalent ions and the surface potential of charged phospholipid membranes. *J. Gen. Physiol.* 58:667–87

Meldosi, J. 1974. Secretory mechanisms in pancreatic acinar cells. Role of cytoplasmic membrane. In *Adv. Cytopharmacol.* 2:71–85

Meyer, D. I., Burger, M. M. 1976. The chromaffin granule surface. Localization of carbohydrate on the cytoplasmic surface of an intracellular organelle. *Biochim. Biophys. Acta* 443:428–36

Michaelson, D. M., Pinchasi, I., Sokolovsky, M. 1978. Factors required for calcium dependent acetylcholine release from isolated *Torpedo* synaptic vesicles. *Biochem. Biophys. Res. Commun.* 80: 547–53

Michaelson, D. M., Sokolovsky, M. 1978. Induced acetylcholine release from active purely cholinergic *Torpedo* synaptosomes. *J. Neurochem.* 30:217–31

Miledi, R. 1973. Transmitter release induced by injection of calcium ions into nerve terminals. *Proc. R. Soc. London Ser. B* 183:421–25

Miller, C., Arvan, P., Telford, J. N., Racker, E. 1976. Ca^{++}-induced fusion of proteoliposomes: Dependence on transmembrane osmotic gradient. *J. Membr. Biol.* 30:271–82

Miller, C., Racker, E. 1976. Fusion of phospholipid vesicles reconstituted with cytochrome *c* oxidase and mitochondrial hydrophobic protein. *J. Membr. Biol.* 26:319–33

Milutinovic, S., Argent, B. E., Schulz, I., Sachs, G. 1977. Studies on isolated subcellular components of cat pancreas. II. A Ca^{++}-dependent interaction between membranes and zymogen granules of cat pancreas. *J. Membr. Biol.* 36: 281–95

Minchin, M. C. W., Iversen, L. L. 1974. Release of [³H]gamma-aminobutyric acid from glial cells in rat dorsal root ganglia. *J. Neurochem.* 23:533–40

Mitchell, J. F., Silver, A. 1963. The spontaneous release of acetylcholine from the denervated hemidiaphragm of the rat. *J. Physiol. London* 165:117–29

Morel, N., Israel, M., Manaranche, R., Mastour-Frachon, P. 1977. Isolation of pure cholinergic nerve endings from *Torpedo* electric organ. Evaluation of their metabolic properties. *J. Cell. Biol.* 75:43–55

Morgan, I. G. 1976. Synaptosomes and cell separation. *Neuroscience* 1:159–65

Morgan, I., Gombos, G. 1975. Biochemical studies of synaptic macromolecules: Are there specific synaptic components. In *Neuronal Recognition,* ed. S. H. Barondes, pp. 179–202. New York: Plenum. 367 pp.

Morgan, I. G., Vincendon, G., Gombos, G. 1973a. Adult rat brain synaptic vesicles. I. Isolation and characterization. *Biochim. Biophys. Acta* 320:671–80

Morgan, I. G., Wolfe, L. S., Mandel, P., Gombos, G. 1971. Isolation of plasma membranes from rat brain. *Biochim. Biophys. Acta* 241:737–51

Morgan, I. G., Zanetta, J. P., Breckenridge, W. C., Vincendon, G., Gombos, G. 1973b. The chemical structure of synaptic membranes. *Brain Res.* 62:405–11

Morris, S. J. 1973. Removal of residual amounts of acetylcholinesterase and membrane contamination from synaptic vesicles isolated from electric organ of Torpedo. *J. Neurochem.* 21:713–15

Morris, S. J., Schober, R. 1977. Demonstration of binding sites for divalent and trivalent ions on the outer surface of chromaffin-granule membranes. *Eur. J. Biochem.* 75:1–12

Mulder, A. H., Van den Berg, W. B., Stoof, J. C. 1975. Calcium-dependent release of radiolabeled catecholamines and serotonin from rat brain synaptosomes in a superperfusion system. *Brain Res.* 99:419–24

Muller, R. U., Finkelstein, A. 1974. The electrostatic basis of Mg^{++} inhibition of transmitter release. *Proc. Natl. Acad. Sci. USA* 71:923–26

Murrin, L. C., De Haven, R. N., Kuhar, M. J. 1977. On the relationship between [³H]choline uptake activation and [³H]acetylcholine release. *J. Neurochem.* 29:681–87

Nagy, A., Baker, R. R., Morris, S. J., Whittaker, V. P. 1976. The preparation and characterization of synaptic vesicles of high purity. *Brain Res.* 109:285–309

Nagy, A., Varady, G., Joo, F., Rakonczay, Z., Pilc, A. 1977. Separation of acetylcholine and catecholamine containing synaptic vesicles from brain cortex. *J. Neurochem.* 29:449–59

Nelsestuen, G. L. 1976. Role of γ-carboxyglutamic acid. An unusual protein transition required for the calcium-dependent binding of prothrombin to phospholipid. *J. Biol. Chem.* 251:5648–56

Nelson, D. L., Molinoff, P. B. 1976. Distribution and properties of adrenergic storage vesicles in nerve terminals. *J. Pharmacol. Exp. Ther.* 196:346–59

Nichols, J. W., Deamer, D. W. 1976. Catecholamine uptake and concentration by liposomes maintaining pH gradients. *Biochim. Biophys. Acta* 455:269–71

Oberg, S. G., Kelly, R. B. 1976a. Saturable binding to cell membranes of the presynaptic neurotoxin, β-bungarotoxin. *Biochim. Biophys. Acta* 433:662–73

Oberg, S. G., Kelly, R. B. 1976b. The mechanism of β-bungarotoxin action. I. Modification of transmitter release at the neuromuscular junction. *J. Neurobiol.* 7:129–41

Ockleford, C. D., Whyte, A., Bowyer, D. E. 1977. Variation in the volume of coated vesicles isolated from human placenta. *Cell Biol. Int. Rep.* 1:137–46

Ohnishi, S., Ito, T. 1974. Calcium-induced phase separations in phosphatidylserine-phosphatidylcholine membranes. *Biochemistry* 13:881–87

Ohsawa, K., Dowe, G. H. C., Morris, S. J., Whittaker, V. P. 1976. Preparation of ultra-pure synaptic vesicles from the electric organ of *Torpedo marmorata* by porous glass bead chromatography and estimation of their acetylcholine content. *Exp. Brain Res.* (Abstr) 24 (5):19

Olsen, R. W., Ticku, M. K., van Ness, P. C., Greenlee, D. 1977. Effects of drugs on gamma-aminobutyric acid receptors, uptake release and synthesis in vitro. *Brain Res.* 139:277–95

Op Den Kamp, J. A. F., Kauerz, M. Th., van Deenen, L. L. M. 1975. Action of pancreatic phospholipase A2 on phosphatidylcholine bilayers in different physical states. *Biochim. Biophys. Acta* 406:169–77

Ornberg, R. L., Smyth, T. Jr. 1976. The physiological mode of action of a presynaptic neurotoxin isolated from black widow spider venom. *Neurosci. Abstr.* 2:1009

Osborne, R. H., Bradford, H. F. 1975. The influence of sodium, potassium and lanthanum on amino acid release from spinal medullary synaptosomes. *J. Neurochem.* 25:35–41

Papahadjopoulos, D., Hui, S., Vail, W. J., Poste, G. 1976. Studies on membrane fusion. I. Interactions of pure phospholipid membranes and the effect of myristic acid, lysolecithin, proteins and dimethylsulfoxide. *Biochim. Biophys. Acta* 448:245–64

Papahadjopoulos, D., Vail, W. J., Newton, C., Nir, S., Jacobson, K., Poste, G., Lazo, R. 1977. Studies on membrane fusion. III. The role of calcium-induced phase changes. *Biochim. Biophys. Acta* 465:579–99

Parsegian, V. A. 1977. Considerations in determining the mode of influence of calcium on vesicle-membrane interaction. In *Society for Neuroscience Symposia*, ed. W. M. Cowan, J. A. Ferrendelli, Vol. 2, pp. 161–71. Bethesda, Md: Soc. Neurosci. 461 pp.

Patrick, R. L., Barchas, J. D. 1976. Dopamine synthesis in rat brain striatal synaptosomes. I. Correlations between ultratridine-induced synthesis stimulation and endogenous dopamine release. *J. Pharm. Exp. Therp.* 197:89–96

Pearse, B. M. F. 1975. Coated vesicles from pig brain; purification and biochemical characterization. *J. Mol. Biol.* 97:93–98

Pearse, B. M. F. 1976. Clathrin: A unique protein associated with intracellular transfer of membrane by coated vesicles. *Proc. Natl. Acad. Sci. USA* 73:1255–59

Peixoto de Menezes, A., Pinto da Silva, P. 1978. Freeze-fracture observations of the lactating rat mammary gland. Membrane events during milk fat secretion. *J. Cell Biol.* 76:767–78

Philippu, A., Beyer, J. 1973. Dopamine and noradrenaline transport into subcellular vesicles of the striatum. *Naunyn-Schmiedebergs Arch. Pharmakol.* 278:387–402

Philippu, A., Matthaei, H. 1975. Uptake of serotonin, gamma-aminobutyric acid and histamine into synaptic vesicles of the pig caudate nucleus. *Naunyn-Schmiedebergs Arch. Pharmakol.* 287:191–204

Pickard, M. R., Hawthorne, J. N. 1978. The labeling of nerve ending phospholipids in guinea pig brain in vivo and the effect of electrical stimulation on phosphatidylinositol metabolism in prelabeled synaptosomes. *J. Neurochem.* 30:145–57

Pinto da Silva, P., Nogueira, M. L. 1977. Membrane fusion during secretion. A hypothesis based on electronmicroscope observation of Phytophthora Palmivora Zoospores during encystment. *J. Cell Biol.* 73:161–81

Pollard, H. B., Pazoles, C. J., Hoffman, P. G., Zinder, O., Nikodijevik, O. 1977. Regulation of release from isolated adrenergic secretory vesicles by ATP-mediated changes in transmembrane potential and anion permeability. In *Cellular Neurobiology*, ed. Z. Hall, R. Kelly, C. F. Fox, pp. 259–66. New York: Allan R. Liss. 320 pp.

Potter, L. T. 1970. Synthesis, storage and release of [^{14}C]acetylcholine in isolated rat diaphragm muscles. *J. Physiol. London* 206:145–66

Price, D. L., Griffin, J. W., Peck, K. 1977. Tetanus toxin: Evidence for binding at presynaptic nerve endings. *Brain Res.* 121:379–84

Pumplin, D. W., Reese, T. S. 1977. Action of brown widow spider venom and botulinum toxin on the frog neuromuscular junction examined with the freeze-fracture technique. *J. Physiol. London* 273:443–59

Rahamimoff, R. 1976. The role of calcium in transmitter release at the neuromuscular junction. In *Motor Innervation of Muscle*, ed. S. Thesleff, pp. 117–49. New York: Academic. 351 pp.

Rand, R. P., Sengupta, S. 1972. Cardiolipin forms hexagonal structures with divalent cations. *Biochim. Biophys. Acta* 255:484–92

Reeber, A., Vincendon, G., Gombos, G., Bock, E. 1978. Synaptin, Na+, K+ ATPase, LDH, CNPase and cytochrome-c oxidase in adult rat brain synaptic vesicle fractions. *FEBS Lett.* 86:171–73

Reiss-Husson, F. 1967. Structure des phases liquide-cristallines de differents phospholipides, monoglycerides, sphingolipides, anhydres ou en presence d'eau. *J. Mol. Biol.* 25:363–82

Remler, M. P. 1973. A semi-quantitative theory of synaptic vesicle movements. *Biophys. J.* 13:104–17

Rotshenker, S., Erulkar, S. D., Rahamimoff, R. 1976. Reduction in the frequency of miniature end-plate potentials by nerve stimulation in low calcium solutions. *Brain Res.* 101:362–65

Satir, B. H., Oberg, S. G. 1977. Paramecium fusion rosettes: Possible function as Ca++ gates. *Science* 199:536–38

Satir, B., Schooley, C., Satir, P. 1973. Membrane fusion in a model system. Mucocyst secretion in *Tetrahymena. J. Cell Biol.* 56:153–76

Schenker, C., Mroz, E. A., Leeman, S. E. 1976. Release of substance P from isolated nerve endings. *Nature* 264:790–92

Schneider, F. H., Smith, A. D., Winkler, H. 1967. Secretion from the adrenal medulla: Biochemical evidence for exocytosis. *Br. J. Pharmacol. Chemother.* 31:94–104

Schober, R., Nitsch, C., Rinne, U., Morris, S. J. 1977. Calcium-induced displacement of membrane-associated particles upon aggregation of chromaffin granules. *Science* 195:495–97

Schubert, D., Klier, F. G. 1977. Storage and release of acetylcholine by a clonal cell line. *Proc. Natl. Acad. Sci. USA* 74:5184–88

Sheridan, M., Whittaker, V. P. 1964. Isolated synaptic vesicles: Morphology and acetylcholine content. *J. Physiol. London* 175:25P–26P

Shimoni, Y., Alnaes, E., Rahamimoff, R. 1977. Is hyperosmotic neurosecretion from motor nerve endings a calcium-dependent process? *Nature* 267:170–72

Silverstein, S. C., Steinman, R. M., Cohn, Z. A. 1977. Endocytosis. *Ann. Rev. Biochem.* 46:669–722

Simpson, L. L. 1974. Studies on the binding of botulinum toxin type A to the rat phrenic nerve-hemidiaphragm preparation. *Neuropharmacology* 13:683–91

Sperk, G., Baldessarini, R. J. 1977. Stabilizing effect of sucrose on leakage of cytoplasm from rat brain synaptosomes in saline media. *J. Neurochem.* 28:1403–5

Strong, P. N., Goerke, J., Oberg, S. G., Kelly, R. B. 1976. β-bungarotoxin, a presynaptic toxin with enzymatic activity. *Proc. Natl. Acad. Sci. USA* 73:178–82

Strong, P. N., Heuser, J. E., Kelly, R. B. 1977. Selective enzymatic hydrolysis of nerve terminal phospholipids by β-bungarotoxin: Biochemical and morphological studies. In *Cellular Neurobiology*, ed. Z. Hall, R. B. Kelly, C. F. Fox, pp. 227–49. New York: Allan R. Liss. 320 pp.

Strong, P. N., Kelly, R. B. 1977. Membranes undergoing phase transitions are preferentially hydrolyzed by β-bungarotoxin. *Biochim. Biophys. Acta* 469:231–35

Suszkiw, J. B. 1976. Acetylcholine translocation in synaptic vesicle ghosts *in vitro. J. Neurochem.* 27:853–57

Tanaka, R., Asaga, H., Takeda, M. 1976. Nucleoside triphosphate and cation requirement for dopamine uptakes by plain synaptic vesicles isolated from rat cerebrums. *Brain Res.* 115:273–83

Tapia, R., Meza-Ruiz, G. 1977. Inhibition by ruthenium red of the calcium-dependent release of [³H]GABA in synaptosomal fractions. *Brain Res.* 126:160–66

Therien, H. M., Mushynski, W. E. 1976. Isolation of synaptic junctional complexes of high structural integrity from rat brain. *J. Cell Biol.* 71:807–22

Toll, L., Gundersen, C. B. Jr., Howard, B. D. 1977. Energy utilization in the uptake of catecholamines by synaptic vesicles and adrenal chromaffin granules. *Brain Res.* 136:59–66

Trifaro, J. M. 1978. Contractile proteins in tissues originating in the neural crest. *Neuroscience* 3:1–24

Tsudzuki, T., Fujii, T., Tanaka, R. 1977. Synaptic vesicle fractions devoid of adenosine triphosphatase activity from bovine caudatolenticular nuclei and thalamus. *J. Biochem.* 82:709–17

Van Der Bosch, J., McConnell, H. M. 1975. Fusion of dipalmitoylphosphatidylcholine vesicle membranes by concanavalin A. *Proc. Natl. Acad. Sci. USA* 72:4409–13

VanDerKloot, W., Kita, H. 1973. The possible role of fixed membrane surface charges in acetylcholine release at the frog neuromuscular junction. *J. Membr. Biol.* 14:365–82

van Heyningen, S. 1977. Cholera toxin. *Biol. Rev. Cambridge Philos. Soc.* 52:509–49

Volsky, D., Loyter, A. 1977. Rearrangement of intramembranous particles and fusion promoted in chicken erythrocytes by intracellular Ca^{++}. *Biochim. Biophys. Acta* 471:243–59

Voyta, J. C., Slakey, L. L., Westhead, E. W. 1978. Accessibility of lysolecithin in catecholamine secretory vesicles to acyl CoA: lysolecithin acyl transferase. *Biochem. Biophys. Res. Commun.* 80: 413–17

Vyskocil, F., Illes, P. 1977. Non-quantal release of transmitter at mouse neuromuscular junction and its dependence on the activity of Na$^+$–K$^+$ ATP-ase. *Pfluegers Arch.* 370:295–97

Wagner, J. A., Carlson, S. S., Kelly, R. B. 1978. Chemical and physical characterization of cholinergic synaptic vesicles. *Biochemistry* 17:1199–1206

Wernig, A., Stover, H., Tonge, D. 1977. The labeling of motor end-plates in skeletal muscle of mice with ^{125}I-tetanus toxin. *Naunyn-Schmiedebergs Arch. Pharmakol.* 298:37–42

White, T. D. 1977. Direct detection of depolarisation-induced release of ATP from a synaptosomal preparation. *Nature* 267:67–68

White, T. D. 1978. Release of ATP from a synaptosomal preparation by elevated extracellular K$^+$ and by veratridine. *J. Neurochem.* 30:329–36

Whittaker, V. P. 1968. The morphology of fractions of rat forebrain synaptosomes separated on continuous sucrose density gradients. *Biochem. J.* 106:412–17

Whittaker, V. P. 1969. The synaptosome. In *Handbook of Neurochemistry*, ed. A. Lajtha, Vol. 2, pp. 327–64. New York: Plenum. 579 pp.

Whittaker, V. P., Dowdall, M. J., Dowe, G. H. C., Facino, R. M., Scotto, J. 1974. Proteins of cholinergic synaptic vesicles from the electric organ of Torpedo: Characterization of a low molecular weight acidic protein. *Brain Res.* 75:115–31

Whittaker, V. P., Essman, W. B., Dowe, G. H. C. 1972. The isolation of pure cholinergic synaptic vesicle from the electric organs of elasmobranch fish of the family torpedinidae. *Biochem. J.* 128: 833–46

Whittaker, V. P., Sheridan, M. N. 1965. The morphology and acetylcholine content of isolated cerebral cortical synaptic vesicles. *J. Neurochem.* 12:363–72

Wilson, W. S., Schulz, R. A., Cooper, J. R. 1973. The isolation of cholinergic synaptic vesicles from bovine superior cervical ganglion and estimation of their acetylcholine content. *J. Neurochem.* 20:659–67

Winkler, H., Schneider, F. H., Rufener, C., Nakane, P. K., Hortnagl, H. 1974. Membranes of adrenal medulla: Their role in exocytosis. *Adv. Cytopharmacol.* 2:127–40

Wonnacott, S., Marchbanks, R. M. 1976. Inhibition by botulinum toxin of depolarization evoked release of carbon-14 acetylcholine from synaptosomes in vitro. *Biochem. J.* 156:701–12

Yen, S. S., Klein, R. L., Chen-Yen, S. H. 1973. Highly purified splenic nerve vesicles: Early post-mortem effects on norepinephrine content and pools. *J. Neurocytol.* 2:1–12

Zengel, J. E., Magleby, K. L. 1977. Transmitter release during repetitive stimulation: Selective changes produced by Sr^{2+} and Ba^{2+}. *Science* 197:67–69

Zimmermann, H., Denston, C. R. 1977a. Recycling of synaptic vesicles in the cholinergic synapses of the Torpedo electric organ during induced transmitter release. *Neuroscience* 2:695–714

Zimmermann, H., Denston, C. R. 1977b. Separation of synaptic vesicles of different functional states from the cholinergic synapses of the Torpedo electric organ. *Neuroscience* 2:715–30

Zimmerman, J. M., Piffaretti, J. C. 1977. Interaction of tetanus toxin and toxoid with cultured neuroblastoma cells. *Naunyn-Schmiedebergs Arch. Pharmakol.* 296:271–77

Zimmermann, H., Whittaker, V. P. 1977. Morphological and biochemical heterogeneity of cholinergic synaptic vesicles. *Nature* 267:633–35.

Ann. Rev. Neurosci. 1979. 2:447–65
Copyright © 1979 by Annual Reviews Inc. All rights reserved

MODULATORY ACTIONS OF NEUROTRANSMITTERS

♦11530

Irving Kupfermann

Division of Neurobiology and Behavior and Department of Psychiatry,
College of Physicians and Surgeons of Columbia University
and The New York State Psychiatric Institute, New York, N.Y. 10032

INTRODUCTION

Increasing attention has been turned towards an analysis of a variety of nonconventional synaptic actions that are often termed modulatory. A recent workshop, sponsored by the Neurosciences Research Program, was devoted entirely to a detailed discussion of neural and behavioral modulation (Kandel, Krasne, Strumwasser and Truman, editors, manuscript in preparation). The concept of modulatory synaptic action is presently evolving and there is no universal agreement on an appropriate definition. The current review does not attempt to develop a logically consistent definition of modulation. Regardless of the definition, it appears clear that the concept can be understood best in contrast to a current model of the most widespread type of conventional synaptic action (see e.g. Kandel 1976). According to this model (Fig. 1), a transmitter agent released presynaptically interacts with a postsynaptic receptor-ionophore complex. This results in a rapid and brief opening of specific ion channels. The membrane potential of the cell has little or no effect on the magnitude of the conductance increase associated with the transmitter-sensitive channels, and hence these are termed voltage-independent or voltage-insensitive channels. The membrane potential change resulting from the alteration of ionic permeability during activation of the receptor-ionophore complex can affect a different set of ionic channels, whose conductance is voltage-dependent, and which are responsible for the action potential. The transmitter either excites the cell by moving the membrane potential toward threshold for an action potential, or inhibits the cell by stabilizing the membrane potential below threshold. One concept of neuromodulators is that, unlike conventional

447

0147-006X/79/0315-0447$01.00

transmitters, they do not simply excite or inhibit an electrically excitable cell, but rather are involved in altering the effects of other events occurring at the cell. The purest examples are those in which synaptic action at a cell results in no conductance or potential change, but instead alters the conductances or potentials produced either by endogenous pacemaker properties or by other synaptic inputs to the cell (Barker et al 1978b, Hidaka, Osa & Twarog 1967, Weiss, Cohen & Kupfermann 1978, Zieglgänsberger & Bayerl 1976). In many instances, an excitatory synaptic input has been called modulatory despite the fact that the input itself directly produces a conductance or potential change. One reason this type of action has been termed modulatory is that the direct action is relatively incidental to a more important function of altering the response of the cell to other inputs. A second and related concept of modulation can be traced back to a distinction that Florey (1967) made between the nature of transmitter action and that of hormone or neuromodulator action. It was pointed out that transmitter action is typically localized and limited to the subsynaptic region; it is brief, and does not involve intracellular events. By contrast, hormonal actions are not limited to the synaptic regions of the neuron; their effect is prolonged, and can involve intracellular actions. Although Florey limited his definition of modulators to blood-borne substances, it was natural to extend this distinction to include substances released from nerve terminals,

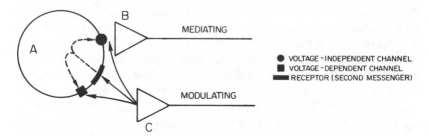

Figure 1   Model of synaptic action. Element A represents an electrically excitable element (neuronal dendrite, soma, presynaptic terminal, or muscle). Element B represents a conventional (mediating) synaptic terminal that releases a transmitter that interacts with a receptor-ionophore complex associated with a voltage-independent channel (●). Element C represents a modulatory terminal that also releases a transmitter. The modulatory transmitter can alter the effects of the mediating transmitter at the voltage-independent channels either by altering the receptor associated with that channel, or by altering the extracellular concentration of the mediating transmitter. Alternatively the modulatory transmitter can alter voltage-dependent channels (■) that control ions involved in spike activity, endogenous pacemaker properties of the cell, or, in the case of presynaptic terminals and muscle, transmitter release and contraction, respectively. The modulatory action can operate either extracellularly at the receptor associated with the voltage-dependent channel, or intracellularly by means of a specialized receptor (━) whose activation alters the level of an intracellular messenger.

but which function more like hormones than classical transmitters (Barker & Smith 1977, Bloom 1976, Libet 1970).

The present review covers some recent developments concerning a variety of nonconventional synaptic actions, from the point of view of describing how the actions could result in modulation of neural activity. First considered are examples in which one transmitter appears to modulate the effects of voltage-insensitive conductances. Two types of actions are considered: one in which the modulating substance appears to act by an intracellular route, typically involving an intracellular second messenger, and a second in which the substance acts by an extracellular route. The second topic considered is modulatory actions on voltage-sensitive processes. The modulatory effects on neuronal pacemaker properties are discussed, followed by a consideration of the possible role of cyclic nucleotides in voltage-sensitive processes. The third major topic considered is that of modulatory synaptic actions at presynaptic terminals. Since presynaptic terminals can receive axoaxonic synaptic input, the same synaptic processes that occur at cell bodies and dendrites can occur at terminals. This review does not attempt a comprehensive coverage of presynaptic control, but rather only considers presynaptic control in relation to axo-axonic inputs that produce relatively long-lasting effects. The last topic considered is the possible behavioral function of modulatory synaptic actions.

Since the topic of synaptic modulation from a broad perspective has not been reviewed in the Annual Reviews, the literature covered is not limited to publications of the last year. Emphasis, however, is placed on recent papers. Due to severe space limitations, no attempts are made to fully ascribe priority for various observations and ideas. Furthermore, certain topics and papers have been excluded for little reason other than to save space.

MODULATION OF VOLTAGE-INSENSITIVE CHANNELS

A variety of studies suggest that voltage-insensitive channels may be affected by mechanisms other than conventional transmitter actions of short duration. In some instances it has been suggested that the modulatory substance acts by an extracellular route, whereas in other cases the modulatory substance has a hormone-like action that involves an intracellular messenger, typically a cyclic nucleotide.

Intracellular Actions and Cyclic Nucleotides

SYMPATHETIC GANGLION   Research on sympathetic ganglia has provided an important source of data suggesting modulatory synaptic functions involving an intracellular messenger. Libet and co-workers (Libet 1978, Libet & Tosaka 1970) have suggested that in the sympathetic ganglion, dopamine may specifically amplify a slow cholinergic EPSP that occurs in the principal cells and that this amplification is mediated by cAMP (Libet, Kobayashi & Tanaka 1975). Although there is a disconcerting degree of variability from one vertebrate species to another, a basic pattern of synaptic organization has been worked out in the rabbit superior cervical ganglion (Libet 1978). Preganglionic nerve stimulation produces a sequence of excitatory and inhibitory synaptic potentials in the principal cells that comprise the output of the ganglion. An initial fast EPSP (fEPSP) appears to be monosynaptic. The fEPSP depends on nicotinic cholinergic receptors, and is a conventional synaptic potential, in that it is due to a brief increase in membrane conductance to specific ions. In addition to the fEPSP, the cells also exhibit two slow potentials. A slow EPSP (sEPSP) appears to be mediated by a monosynaptic cholinergic input, but unlike the fast EPSP, the receptors involved are muscarinic. Finally, the principal cells show a slow IPSP that appears to be due to the release of dopamine (and perhaps other catecholamines) from interneurons in the ganglion. In addition to a marked difference in time course, the slow synaptic potentials have other properties dramatically different from the fast potential. Unlike the fEPSP, the slow potentials are sensitive to temperature and metabolic inhibitors. Furthermore, the slow potentials are not associated with a net conductance increase, as measured by means of constant current intracellular pulses. Rather, they are associated with either no conductance change, or a decrease in conductance.

Based on the observation that preganglionic stimulation and dopamine enhance the synthesis of cAMP in the sympathetic ganglion of several different species, Greengard and co-workers (Greengard 1976) hypothesized that the sIPSP is mediated by an increase in intracellular cAMP. The following scheme was suggested: (a) dopamine enhances the synthesis of cAMP; (b) the cAMP activates a protein kinase, which (c) phosphorylates a membrane protein that controls ionic permeability. Since acetylcholine enhances the synthesis of cGMP in the ganglion (Weight, Petzold & Greengard 1974), it has also been suggested that the cholinergically elicited slow EPSP is mediated by cGMP. A key observation supporting the hypothesis that cyclic nucleotides mediate slow synaptic potentials was that extracellular application of derivatives of cGMP or cAMP (McAfee & Greengard 1972, Weight, Petzold & Greengard 1974) to the sympathetic ganglion produced depolarization or hyperpolarization, respectively, of cells in the ganglion. Several laboratories, however, have failed to replicate these results (Busis et al 1978, Dun, Kaibara & Karczmar 1977b, Libet 1978). It has been pointed out that the technique that was used to record slow potentials (sucrose gap) is subject to artifacts when solution changes are made (Libet 1978). Recent studies utilizing intracellular recording and injection of cAMP into sympathetic ganglion cells of frog, rat, and rabbit have found no evidence that cAMP produces a hyperpolarization (Busis et al 1978, Gallagher & Shinnick-Gallagher 1977, Kobayashi, Hashiguchi & Ushiyama 1978). Intracellular injection of cGMP into neurons of sympathetic ganglia of frog (Busis et al 1978) or rat (Gallagher & Shinnick-Gallagher 1977) and extracellular application of dibutyrl cyclic AMP (DbcGMP) in frog and rabbit (Busis et al 1978, Dun, Kaibara & Karczmar 1977a) have been reported to produce either no consistent changes of potential, or changes that do not resemble those produced by acetylcholine. On the other hand, one study in the rabbit sympathetic ganglion found extracellular application of low doses of cGMP associated with an intracellularly recorded depolarization that resembles the slow EPSP, in that it is not associated with an apparent increase of membrane conductance (Hashiguchi, Ushiyama & Kobayashi 1978). More detailed studies are needed to determine whether the potentials generated by cGMP, and the slow potentials elicited by preganglionic nerve stimulation are identical in their ionic and biochemical mechanisms. The recent negative findings involving direct application of cyclic nucleotides seriously challenge the hypothesis that these nucleotides play a role in the generation of slow potentials. Nevertheless the findings are not conclusive, since neither extracellular application nor intracellular injection of a cyclic nucleotide is entirely equivalent to a physiologically induced increase of intracellular levels. For reviews of the role of cyclic nucleotides in the generation of slow potentials in the vertebrate brain see Moore & Bloom (this volume) and Phillis (1977). For extensive general reviews of cyclic nucleotides and the nervous system see Daly (1977) and Nathanson (1977).

Although there is reason to question the role of cAMP in the sIPSP in the sympathetic ganglion, there is further evidence to support the suggestion of Libet, Kobayashi & Tanaka (1975) that cAMP may serve to modulate the size of the sEPSP. In four out of nine cells investigated in rabbit superior cervical ganglion, intracellular injection of cAMP enhanced the size of the sEPSP evoked by preganglionic stimulation (Kobayashi, Hashiguchi & Ushiyama 1978). Similar results were obtained in most cells tested, when DbcAMP was added to the superfusate. The enhancement of the sEPSP after injection of cAMP was not associated with any membrane potential changes in the postsynaptic cells. This suggests that the slow hyperpolarization seen with preganglionic stimulation or application of dopamine is not causally related to the associated enhancement of the sEPSP. It is significant that, although cAMP and dopamine enhance the sEPSP, the fast EPSP is not affected, which suggests that cAMP may be modulating a specific type of receptor. Application of noradrenalin to cerebellar Purkinje cells similarly has selective effects, enhancing the inhibition produced by GABA but not by glycine (Woodward et al 1978).

The mechanisms underlying the slow inhibitory synaptic potentials in the sympathetic ganglion and elsewhere have continued to be explored. Several studies have provided reasonably conclusive evidence that the sIPSP in the sympathetic ganglion is not the result of the activation of an electrogenic sodium pump associated with a oubain-sensitive Na-K ATPase. Although prolonged exposure to oubain depresses the sIPSP, this effect apparently occurs much later than inhibition of the pump (Libet, Tanaka & Tosaka 1977, Smith & Weight 1977). Thus, the depression of the sIPSP appears to be a nonspecific effect due to prolonged application of oubain.

There is still debate in the literature about whether the slow potentials in sympathetic ganglia are due to simple decreases in ionic conductance (Weight 1974) or whether additional mechanisms are involved (Hashiguchi, Ushiyama & Kobayashi 1978, Kuba & Koketsu 1976a). There is also some question about what transmitter mediates the sIPSP, particularly in the frog, in which acetylcholine as well as catecholamines can produce an sIPSP (Libet & Kobayashi 1974). Unfortunately the use of pharmacological manipulations to dissociate various synaptic potentials evoked by nerve stimulation makes it difficult to resolve these issues. Some clarification may result from recent studies of the experimentally advantageous autonomic ganglion of the heart of the mud puppy (Hartzell et al 1977). This ganglion exhibits a fast EPSP and slow IPSP that superficially resemble the potentials seen in sympathetic ganglia. The heart ganglion has catecholamine-containing cells and one might expect that the sIPSP would be mediated by a catecholamine, as appears to be the case for mammalian sympathetic ganglia. Nevertheless, in an elegant study by Hartzell et al (1977), it was convincingly shown that both the fEPSP and sIPSP result from cholinergic actions. During changes of membrane potential, the sIPSP exhibited anomalous behavior similar to that of slow potentials described in sympathetic ganglia. However, in the heart ganglion it proved possible to demonstrate that the anomalous behavior of the sIPSP is not the result of a conductance decrease, but rather of the voltage dependence of an otherwise conventional synaptic potential produced by an increase of conductance. These findings highlight the difficulty of drawing conclusions about conductance changes, based on changes in the size of voltage steps produced by constant current pulses.

SYNAPTIC ACTIVATION OF AN ELECTROGENIC PUMP   Although studies utilizing oubain to demonstrate the involvement of an electrogenic pump in the peripheral or central nervous system (for review see Phillis 1977) are difficult to interpret, in muscle cells it is much easier to perform needed control experiments. A thorough study by Clausen & Flatman (1977) on slow (soleus) muscle indicates that catecholamines, at physiological concentrations, activate a Na-K pump, which in turn appears to produce a hyperpolarization. DbcAMP mimicked the effect of catecholamines on Na^+ efflux, K^+ influx, ratio of intracellular K^+ to Na^+, and membrane potential. The DbcAMP appeared to activate a Na-K ATPase since the effects were all blocked by oubain, an inhibitor of the Na-K ATPase. Other studies of rat skeletal muscle indicate that the intact muscle contains a cAMP-sensitive Na-K ATPase, but that the cAMP sensitivity is lost in muscle homogenates (Rogus, Cheng & Zierler 1977). This reemphasizes that caution must be used in applying the findings obtained in tissue homogenates to intact tissue.

Extracellular Action

One means by which one transmitter may modulate the effect of another is to affect transmitter interaction with the receptor-ionophore complex. There is no proof that these interactions normally occur in the nervous system, but a number of observations suggest this.

Interactions between putative amino acid transmitters have been described in a number of instances. In the basic experimental paradigm for these studies, intracellular records are obtained from a neuron (or muscle) and the response to test ionotophoretic pulses of one substance are compared with and without a prepulse of a second substance applied to the same region of the cell. Utilizing this design, researchers have shown that aspartate enhances the response to pulses of glutamate (Crawford & McBurney 1977, Freeman 1976, McCreery & Carpenter 1978). In crayfish muscle, the enhancement can occur with a dose of aspartate that by itself produces little or no potential change in the postsynaptic cell. It has also been shown that in neurons in mouse CNS tissue culture, application of glycine can convert a GABA-induced hyperpolarizing response to a depolarizing response (Barker & Ransom 1978). In the latter case, it was suggested that the reversal of the test response might be due to an increase of chloride concentration following the conditioning application of transmitter, but it is difficult to conceive how this could occur. In the case of an IPSP exclusively due to an increase in the permeability to chloride, the increase of internal chloride would gradually shift the equilibrium potential in a depolarizing direction, until the cell reached resting potential. At that point, the cell would be at the chloride equilibrium potential and no further net chloride would enter. Thus, without invoking additional assumptions, it would appear that a pure chloride IPSP could not raise internal chloride sufficiently to result in depolarizing chloride-IPSPs. An alternative explanation of these results is that the test application activates depolarizing, as well as hyperpolarizing, conductances and that the application of the conditioning transmitter selectively desensitizes the receptors mediating the hyperpolarizing responses, or sensitizes the receptors mediating the depolarization.

In an elegant series of studies utilizing analysis of channel noise and miniature endplate currents at the crustacean neuromuscular junction, Crawford & McBurney (1977) have shown that the action of aspartate on glutamate-evoked potentials resembles, in detail, the action of anticholinesterases at vertebrate cholinergic synapses. These data suggest that aspartate increases the effective concentration of glutamate ions at the synapse, presumably by blocking their uptake. This notion is supported by the observation that removal of sodium, an ion necessary for uptake of amino acids, decreases the potentiating effect of aspartate on inhibitory responses evoked by application of glutamate to neurons in *Aplysia* (McCreery & Carpenter 1978). It has been suggested that at crayfish muscle, aspartate may reverse the process of desensitization evoked by glutamate (Dudel 1977). It was observed that glutamate responses were potentiated by aspartate only when the glutamate test pulses were prolonged to the point at which "desensitization" occurred, and the glutamate responses were depressed relative to the initial response. It is possible that these data might be explained by the previously discussed finding that aspartate may decrease the uptake of glutamate at crustacean muscle. Perhaps some part of desensitization to prolonged glutamate is due to stimulation of the uptake mechanism. Thus, aspartate might block uptake and would maximally enhance responses only when the uptake mechanism was fully active.

An additional mechanism that could explain transmitter interaction is that one transmitter can directly increase or decrease the sensitivity of the receptor-ionophore complex (Barker et al 1978b, Guidotti et al 1978, Shank & Freeman 1975, Zieglgänsberger & Bayerl 1976). Although this mechanism is unlikely in the case of the interaction of amino acids, there is some positive evidence for non–amino acid substances. For example, iontophoretic application of opiates can selectively depress sodium currents associated with excitatory synaptic actions evoked by glutamate or acetylcholine (Barker et al 1978b, Zieglgänsberger & Bayerl 1976). Opiates depress excitatory actions without substantially affecting the resting conductance or the voltage-sensitive sodium conductance associated with the action potential in the postsynaptic cell. An analysis of glutamate-evoked currents in mouse spinal neurons in tissue culture indicated that enkephalin depresses the glutamate response in a noncompetitive manner (Barker et al 1978b). The analysis suggested that the enkephalin does not alter the number of glutamate molecules required for a unitary conductance event, and that there is no apparent change of affinity of glutamate for its receptor. Thus, opiates may alter the coupling of the glutamate receptor to its ionophore, or may act on the ionophore directly. These experiments raise the possibility that the opiate receptor may be the ionic channel that is activated by glutamate. It is also possible that opiate effects may not be due to a direct action on conductance mechanisms, but instead may be an indirect effect mediated by an intracellular messenger.

The discovery that the nervous system contains endogenous opiate-like substances and opiate receptors suggests that other neuroactive drugs may also reflect the existence of similar endogenous substances. This is supported by the finding that the brain contains specific binding sites for the benzodiazepines (a class of antianxiety agents) and that the binding capacity of different benzodiazepines correlates highly with their effectiveness in modifying various behaviors (Squires & Braestrup 1977). In a cell-free system, the benzodiazepines appear to increase the affinity of GABA for its binding sites (Guidotti et al 1978). Guidotti et al (1978) therefore have suggested that the benzodiazepines may produce an allosteric alteration of the GABA binding site, which results in increased effectiveness of the inhibitory effects of GABA in the central nervous system. In a recent preliminary report Costa, Guidotti & Toffano (1978) provided evidence that a brain synaptic membrane fraction contains a thermostable protein that acts as an inhibitor of the high affinity receptor for GABA. The inhibition of this factor is competitively counteracted by benzodiazepines in a stereo-specific manner. Thus, the endogenous inhibitor may be the normal ligand for the receptors that bind benzodiazepines. Recent studies on cells in spinal cord tissue culture have shown directly that a benzodiazepine can selectively enhance inhibitory responses elicited by GABA (Choi, Farb & Fischbach 1977).

MODULATION OF VOLTAGE-SENSITIVE CHANNELS

Transmitter agents typically are thought to affect ion channels that are voltage independent. It is now known that in certain cases, transmitter-evoked opening or closing of ion channels is voltage dependent (Dudel 1974, Hartzell et al 1977, Kuba & Koketsu 1976b, Magleby & Stevens 1972), although the effects are generally small. As discussed below, substances considered to be neuromodulators now have been shown to produce rather marked alterations in voltage-sensitive channels.

Regulation of Pacemaker or Burst Activity

Certain neurons exhibit endogenous pacemaker or burst-generating properties. One important consequence of the effect of neuromodulators on voltage-sensitive channels is that the burst generating capacity of such neurons can be altered. Studies of Barker, Gainer, and colleagues on individual molluscan neurons first demonstrated that this can occur in response to vertebrate peptide hormones (e.g. vasopressin), as well as to unidentified substances found in the nervous system of mollusks (reviewed by Barker & Smith 1977). The peptide hormones, at remarkably low concentrations, could cause silent cells to go into a bursting mode, or could increase the burst activity of neurons already bursting. These actions were associated with an accentuation of a negative slope characteristic as determined by current-voltage plots. The negative slope characteristic appears to be a necessary component of burst activity in many neurons (for review see Gulrajani & Roberge 1978). In *Aplysia* there is evidence that a substance (or substances) released by neuroendocrine cells (bag cells) in the abdominal ganglion can exert a prolonged action on numerous neurons in the ganglion (Mayeri, Brownell, Branton and Simon, manuscripts submitted for publication). A great variety of actions have been described, including alterations of the pacemaker properties of several identified neurons. The product of the bag cells appears to be released into the ganglionic sheath, surrounding the target cells. Thus, this may represent a type of local hormone action.

The stomatogastric ganglion of the lobster (Selverston, Russell & Miller 1976) provides a convenient model for the study of the generation and modulation of burst activity. The isolated stomatogastric ganglion often exhibits poor burst activity. If, however, the ganglion is left connected to other ganglia, or if input nerves are stimulated, bursting is much more vigorous and reliable (Selverston, Russell & Miller 1976). Associated with this increased burst capability is the appearance of "plateau" formation in many neurons throughout the ganglion (Russell & Hartline 1978). The plateau formation refers to a prolonged depolarization occurring spontaneously or following a brief intracellular depolarizing pulse. The cells in the ganglion that generate the "gastric" rhythm appear to be interconnected in such a way that they could, in principle, generate burst activity even though individual cells have no endogenous bursting properties. Nevertheless, the rhythm appears to require the presence of particular membrane properties associated with bursting as well as appropriate interconnections between neurons. There is evidence that the burst properties of stomatogastric neurons is uncovered by one or more biogenic amines that are either blood borne or are released by activity of neurons that have cell bodies in other ganglia, but which terminate in the stomatogastric ganglion (Barker, Kushner & Hooper 1978a).

Just as modulator substances can uncover burst properties of neurons, they can also depress them. In spontaneously bursting cells in the abdominal ganglion of *Aplysia,* cholinergic input from the interneuron L10 produces a slow IPSP that suppresses the bursting (Pinsker & Kandel 1969). The slow IPSP responds very unusually to changes of membrane potential. When the cell is hyperpolarized, the IPSP reduces in size to the point where no potential is observed. However, it does not reverse, even with substantial hyperpolarization (Pinsker & Kandel 1969). Thus, it does not behave either like a simple conductance-increase or a conductance-decrease PSP. Earlier work suggested that this IPSP might be due to an electrogenic Na^+ pump (Pinsker & Kandel 1969) or to a conductance increase at a region electrically distant from the cell body (Kehoe & Ascher 1970, Kunze & Brown 1971). Recent evidence based on voltage clamp data suggests that the IPSP results from the turning off of a voltage-dependent slow inward current that underlies part of the depolarizing pacemaker mechanism of the cell (Wilson & Wachtel, manuscripts submitted for publication). Since the slow inward current is present only at relatively depolarized levels of membrane potential, the IPSP disappears when the cell is hyperpolarized. Interestingly, when tested with pulses of ACh, this mechanism is present only at membranes in the neuropile (Wilson & Wachtel, manuscript in preparation). The cell body shows a conventional conductance-increase IPSP. These findings help explain the discrepancies among earlier studies of the slow IPSP in *Aplysia* (Kehoe & Ascher 1970, Kunze & Brown 1971, Pinsker & Kandel 1969). These studies, as well as those showing activation of burst properties, indicate that modulatory actions do not merely function to excite or inhibit spike activity in bursting neurons, but instead directly turn on or off the burst generating capacity of the cells by affecting the voltage-sensitive channels that account for burst behavior.

Role of Cyclic Nucleotides at Voltage-Sensitive Channels

Several recent studies in *Aplysia* provide support for the notion that transmitter actions affecting voltage-sensitive channels are mediated by cAMP. The release of transmitter from presynaptic terminals of mechanoreceptor neurons in *Aplysia* is enhanced by intracellular injection of cAMP into the sensory neuron (Brunelli, Castellucci & Kandel 1976). Furthermore, cAMP mimics the effects produced by various experimental conditions including: (*a*) electrical stimulation of nerves, which produces presynaptic facilitation and (*b*) application of serotonin, which is believed to be the transmitter that mediates presynaptic facilitation in *Aplysia* (Kandel 1978). Finally, nerve stimulation and application of serotonin increase cAMP in the abdominal ganglion (see Kandel 1978, and later section on presynaptic facilitation, for further discussion of the possible mechanisms).

Adrenergic agents increase the force of contraction and enhance pacemaker activity of cardiac cells by a mechanism involving cAMP (Tsien 1977). The cAMP may regulate several different voltage-dependent mechanisms, as well as such non-voltage-sensitive phenomena as calcium sensitivity of contractile regulatory proteins (Reuter 1974, Tsien 1974, 1977). The data of Nawrath (1977) suggest that cGMP mediates part of the negative inotropic effect of acetylcholine on vertebrate heart. An analog of cGMP (8-bromo-cGMP) was found to shorten the cardiac action potential, and ionic flux measurements indicated that this was the result of a decrease of the Ca^{++} influx that occurs during the action potential. Data obtained from *Aplysia* suggest that pacemaker rhythms in neurons may also be mediated by cyclic nucleotides (Treistman & Levitan 1976).

Recent evidence in *Aplysia* indicates that the effect of serotonin on the force of contraction of buccal muscle also may be mediated by cAMP (Weiss et al 1976, 1978b). The buccal muscles are innervated by a serotonergic cell (the metacerebral cell, or MCC), whose activity enhances the force of contractions elicited by cholinergic motor neurons. Several lines of evidence indicate that the enhancement of muscle contraction is due, at least in part, to a direct effect of the MCC (and of serotonin) on excitation contraction coupling of the muscle. Nevertheless, activity of the MCC does not apparently produce any change in resting potential or conductance of muscle fibers. These data suggested that the modulatory effect of the MCC may involve an intracellular second messenger. Consistent with this idea is the observation that a phosphodiesterase inhibitor, RO 201724, potentiates the effect of the MCC on muscle contraction, and phosphodiesterase resistant analogs of cAMP mimic the potentiating effect of the MCC. Finally, firing of an individual MCC produces a measurable increase of the rate of accumulation of cAMP from a radio-labeled precursor. Serotonin also increases the excitability of the heart of *Aplysia* (Mayeri et al 1974) and enhances cAMP levels in cardiac tissue (Mandelbaum, Koester, Weiss & Kupfermann, unpublished observations).

MODULATORY SYNAPTIC ACTIONS AT PRESYNAPTIC TERMINALS

Presynaptic Facilitation

Recent evidence suggests that just as transmitters can produce long-lasting modulation of the excitability of neurons, they can produce long-lasting modulation of transmitter secretion from presynaptic terminals. Presynaptic facilitation has been extensively studied at the terminals of mechanosensory neurons in *Aplysia* (reviewed by Kandel 1978). Stimulation of nerves or of individual identified interneurons produces a long-lasting (several minutes or more) enhancement of transmitter release evoked by action potentials in the sensory neurons. Several observations indicate strongly that the presynaptic facilitation results from the action of serotonin (or a related amine) that is released by interneurons that terminate at or near the terminals of the sensory neurons. Evidence recently obtained by Klein and Kandel (discussed in Kandel 1978) suggests that the amine may produce a long-lasting enhancement of the inward calcium current elicited by the action potential that invades the sensory-neuron terminal. In other words, a modulatory action of an aminergic transmitter affects the voltage-sensitive calcium channels, perhaps in a manner similar to the action of adrenergic agents on cardiac muscle (Reuter 1974). This conclusion derives from intracellular studies of the cell body of the sensory neuron. The cell bodies of the neurons are located within the central ganglion, and are probably sufficiently close to the terminals to provide a reflection of their activity. Nerve stimulation produces a small slow potential change in the sensory neuron. As appears to be the case for the slow IPSP in the sympathetic ganglion (see previous section) the potential changes may not be causally related to the primary function of the synaptic input. In addition to the slow potentials, nerve stimulation produces a slight prolongation of the action potential. In order to accentuate inward currents, the sensory neurons were bathed with tetraethylammonium (TEA), which blocks the delayed outward current that normally repolarizes the membrane and terminates inward currents associated with the action potential. Under the influence of TEA, the action potential in sensory neurons became greatly prolonged, in large part because of an accentuated inward calcium current. Nerve stimulation now produced a dramatic enhancement of the duration of the action potential, and the time course of the enhancement ran parallel to the enhanced release of transmitter from the terminals of the sensory neurons. Enhancement of the action potential and of transmitter release of sensory neurons was also produced by bath application of serotonin, intracellular injection of cAMP, and extracellular application of phosphodiesterase inhibitors. These observations strengthen previous data suggesting that presynaptic facilitation is due to serotonergic stimulation of cAMP in terminals of sensory neurons (Brunelli, Castellucci & Kandel 1976). Shimahara & Tauc (1977) also suggested that serotonin may mediate heterosynaptic facilitation in *Aplysia* (at the giant cell) and on the basis of indirect data, suggested that this effect may be due to enhanced presynaptic influx of calcium, associated with an increase of cAMP. Work in vertebrate central neurons has also provided evidence for enhancement of inward currents, following intracellular injection of cAMP, and it was suggested that cAMP might facilitate calcium entry and regulate transmitter release (Krnjević & Van Meter 1976).

Presynaptic Inhibition

Conventional presynaptic inhibition is typically short lasting, but a long-lasting variety has also been described. For example, in the closer muscles of various species of crabs, a train of stimuli applied to the inhibitory axon results in presynaptic inhibition that persists for several minutes (Rath-mayer & Florey 1974). Thus, presynaptic inhibition may be mediated either by a conventional synaptic action or by a long-duration, modulatory action. Recent findings suggest that the presynaptic inhibitory action of opiates at sensory terminals that contain substance P (Jessell & Iversen 1978, Mac-donald & Nelson 1978) may be due to an action in which the opiates decrease the influx of calcium through voltage-sensitive calcium channels (Mudge, Fishbach & Leeman 1978). When the enkephalin was applied to the cell bodies of dorsal root ganglion cells in culture, it decreased the duration of the action potential, which is mediated, in part, by calcium influx. A variety of studies suggest that the presynaptic inhibitory action of morphine and many other substances at peripheral noradrenergic terminals is also due to a decrease of the availability of calcium at the terminal (reviewed by Starke 1977). Finally, decreased calcium influx has been sug-gested as an explanation of presynaptic inhibition produced by adrenaline in sympathetic ganglia of bullfrogs (Minota & Koketsu 1977). These data suggest that long-lasting presynaptic inhibition may be the mirror opposite of long-lasting presynaptic facilitation.

BEHAVIORAL FUNCTIONS OF MODULATORY SYNAPTIC ACTIONS

The behavioral function of modulatory synaptic actions has been studied in several invertebrate preparations. The modulatory synaptic action that occurs in the presynaptic terminals of sensory mechanoreceptive neurons in *Aplysia* clearly has been shown to execute behavioral sensitization or dishabituation of defensive withdrawal reflexes (see previous section). Be-havioral modulation of inking, another type of defensive response in *Ap-lysia,* also may involve modulatory synaptic action (Carew & Kandel 1977). Noxious stimuli induce slow excitatory synaptic potentials in motor neu-rons that innervate the ink gland of the animal. The synaptic potential is associated with a decrease of membrane conductance, and this results in an amplification of other excitatory input to the motor neurons. If a subthresh-old stimulus, which produces the decreased-conductance EPSP, is followed by another subthreshold stimulus, the PSPs can sum very effectively over a relatively long period of time, and excite the cells. A similar amplification of one synaptic potential by another associated with a decrease of mem-brane conductance has been described in the sympathetic ganglion (Schul-man & Weight 1976; see also Kandel & Tauc 1966, Siggins et al 1971, Weiss & Kupfermann 1976). This type of long-lasting amplification appears to underlie a form of behavioral sensitization in which inking in *Aplysia* will occur if an initially ineffective noxious stimulus is shortly followed by another normally ineffective stimulus.

Another example in *Aplysia* (Weiss, Cohen & Kupfermann 1978), in which there is evidence for a behavioral function for modulatory synaptic actions, comes from studies of the metacerebral cell (MCC). Central branches of this serotonergic cell produce long-lasting modulatory actions on motor neurons of the buccal mass, the organ that effects the biting response in gastropod mollusks. Peripheral branches of the MCC function to enhance the strength of contractions produced by activity of the buccal motor neurons. The central and peripheral effects of the MCC indicate that it is a neuron specialized for modulatory actions and that it serves as the final common path that executes various behavioral effects (specifically on biting) associated with a food-arousal state of the animal. The results obtained from chronic recordings of the MCC of *Aplysia* are consistent with this notion. The cell becomes active when animals are exposed to food and the level of activity correlates with the degree of arousal of the animal, as judged by the latency to bite and the strength of the biting responses (Kupfermann & Weiss, unpublished observations). In a number of mollusks other than *Aplysia,* the MCC also modulates various aspects of feeding behavior (Berry & Pentreath 1976, Gelperin & Chang 1976, Gillette & Davis 1977, Granzow & Kater 1977).

Studies in arthropods have also obtained evidence for arousal effects mediated by modulatory synaptic actions of biogenic amines. In the crayfish (Konishi & Kravitz 1978) and locust (Evans & O'Shea 1978, Hoyle & Dagan 1978), octopamine and other amines may have similar roles to that of serotonin in *Aplysia* and adrenergic agents in vertebrates. Modulatory synaptic actions of biogenic amines may also be related to food arousal effects in the stomatogastric ganglion of the lobster. This ganglion may be set into prolonged burst activity by modulatory synaptic input that is elicited by exposure of the animal to food stimuli (Barker et al 1978a, Morris & Maynard 1970, Russell & Hartline 1978). The burst activity functions to drive the stomach rhythmically (see Selverston, Russell & Miller 1976)

SUMMARY AND PERSPECTIVES

A relatively short time ago the individual neuron was viewed as functioning by means of the simple summation of brief inhibitory or excitatory events. The complexity of the nervous system was the outcome largely of the connections between neurons. Recent studies have uncovered a new set of phenomena that indicate that relatively complex information processing may occur at the level of the individual neuron. For example, rather than producing additive effects, synaptic inputs can produce multiplicative effects that serve to alter the gain of the system. In addition, synaptic inputs may be able to alter specific classes of other inputs selectively. One could envision that each class of synaptic input to a cell could be selectively depressed or enhanced by a corresponding modulatory input. Since the modulatory actions can be transmitted intracellularly via second messengers, an extensive array of presynaptic connections may be unnecessary. It remains to be determined which of the modulatory phenomena currently reviewed are functionally important and which are only pharmacological or experimental curiosities.

Are there any common attributes of the various synaptic phenomena that have been termed modulatory? The most common features of modulatory synaptic effects are long-duration of action and contingent action. Contingent action refers to the property that modulatory transmitters often have little or no effects in themselves, but instead they alter the effects of other events. Long-duration and contingent action endow modulatory effects with properties ideally suited to the control of behavioral modulations such as learning, motivational state, arousal, and sensitization. While there is no *necessary* connection between behavioral modulation and neural modulation, the available evidence from invertebrates suggests that there often is a connection. It is likely that examples of neuromodulation in vertebrates also will be tied to behavioral modulation.

ACKNOWLEDGMENTS

I thank Drs. T. Carew, V. Castellucci, E. Kandel, J. Schwartz, and K. Weiss for their useful comments on an earlier draft of this paper. Preparation of this review and research of the author were supported, in part, by NIH grant NS 12492.

Literature Cited

Barker, D. L., Kushner, P. D., Hooper, N. K. 1978a. Modulatory candidates in the crustacean stomatogastric nervous system: Synthesis of dopamine and octopamine. *Brain Res.* In press

Barker, J. L., Neale, J. H., Smith, J. G. Jr., Macdonald, R. L. 1978b. Opiate peptide modulation of amino acid responses suggests novel form of neuronal communication. *Science* 199:1451–53

Barker, J. L., Ransom, B. R. 1978. Amino acid pharmacology of mammalian central neurones grown in tissue culture. *J. Physiol. London* In press

Barker, J. L., Smith, T. G. Jr. 1977. Peptides as neurohormones. In *Approaches to the Cell Biology of Neurons,* ed. W. M. Cowan, J. A. Ferendelli. Neuroscience Symp. II:340–73. Bethesda, Md: Society for Neuroscience

Berry, M. S., Pentreath, V. W. 1976. Properties of a symmetric pair of serotonin-containing neurones in the cerebral ganglia of *Planorbis. J. Exp. Biol.* 65:361–80

Bloom, F. E. 1976. The role of cyclic nucleotides in central synaptic function. *Adv. Biochem. Psychopharmacol.* 15:273–82

Brunelli, M., Castellucci, V., Kandel, E. R. 1976. Synaptic facilitation and behavioral sensitization in *Aplysia:* Possible role of serotonin and cyclic AMP. *Science* 194:1178–80

Busis, N. A., Schulman, J. A., Smith, P. A., Weight, F. F. 1978 Do cyclic nucleotides mediate slow postsynaptic potentials in sympathetic ganglia? *Br. J. Pharmacol.* In press

Carew, T. J., Kandel, E. R. 1977. Inking in *Aplysia californica* III. Two different synaptic conductance mechanisms for triggering central program for inking. *J. Neurophysiol.* 40:721–34

Choi, D. W., Farb, D. H., Fischbach, G. D. 1977. Chlordiazepoxide selectively augments GABA action in spinal cord cell cultures. *Nature* 269:342–44

Clausen, T., Flatman, J. A. 1977. The effect of catecholamines on Na-K transport and membrane potential in rat soleus muscle. *J. Physiol. London* 270:383–414

Costa, E., Guidotti, A., Toffano, G. 1978 Molecular mechanisms mediating the action of diazepam on GABA receptors. *Br. J. Psychiatry* In press

Crawford, A. C., McBurney, R. N. 1977. The termination of transmitter action at the crustacean excitatory neuromuscular junction. *J. Physiol. London* 268: 711–29

Daly, J. 1977. *Cyclic nucleotides in the nervous system.* New York: Plenum. 401 pp.

Dudel, J. 1974. Nonlinear voltage dependence of excitatory synaptic current in crayfish muscle. *Pfluegers Arch.* 352: 227–41

Dudel, J. 1977. Aspartate and other inhibitors of excitatory synaptic transmission in crayfish muscle. *Pfluegers Arch.* 369:7–16

Dun, N. J., Kaibara, K., Karczmar, A. G. 1977a. Direct postsynaptic membrane effect of dibutyryl cyclic GMP on mammalian sympathetic neurons. *Neuropharmacology* 16:715–17

Dun, N. J., Kaibara, K., Karczmar, A. G. 1977b. Dopamine and adenosine 3',5'-monophosphate responses of single mammalian sympathetic neurons. *Science* 197:778–80

Evans, P. D., O'Shea, M. 1978. The identification of an octopamine neurone and the modulation of a myogenic rhythm in the locust. *J. Exp. Biol.* 73:235–60

Florey, E. 1967. Neurotransmitters and modulators in the animal kingdom. *Fed. Proc.* 26:1164–78

Freeman, A. R. 1976. Polyfunctional role of glutamic acid in excitatory synaptic transmission. *Prog. Neurobiol.* 6: 137–53

Gallagher, J. P., Shinnick-Gallagher, P. 1977. Cyclic nucleotides injected intracellularly into rat superior cervical ganglion cells. *Science* 198:851–52

Gelperin, A., Chang, J. J. 1976. Molluscan feeding motor program: Response to lip chemostimulation and modulation by indentified serotonergic interneurons. *Neurosci. Abstr.* 2:322

Gillette, R., Davis, W. J. 1977. The role of the metacerebral giant neuron in the feeding behavior of *Pleurobranchaea*. *J. Comp. Physiol.* 116:129–59

Granzow, B., Kater, S. B. 1977. Identified higher-order neurons controlling the feeding motor program of *Helisoma*. *Neuroscience* 2:1049–63

Greengard, P. 1976. Possible role for cyclic nucleotides and phosphorylated membrane proteins in postsynaptic actions of neurotransmitters. *Nature* 260: 101–8

Guidotti, A., Toffano, G., Grandison, L., Costa, E. 1978. Second messenger responses and regulation of high affinity receptor binding to study pharmacological modifications of GABAergic transmission. In press

Gulrajani, R. M., Roberge, F. A. 1978. Possible mechanisms underlying bursting pacemaker discharges in invertebrate neurons. *Fed. Proc.* 37:2146–52

Hartzell, H. C., Kuffler, S. W., Stickgold, R., Yoshikami, D. 1977. Synaptic excitation and inhibition resulting from direct action of acetylcholine on two types of chemoreceptors on individual amphibian parasympathetic neurones. *J. Physiol. London* 271:817–46

Hashiguchi, T., Ushiyama, N. S., Kobayashi, H. 1978. Does cyclic GMP mediate the slow excitatory synaptic potential in symphathetic ganglia? *Nature* 271: 267–68

Hidaka, T., Osa, T., Twarog, B. M. 1967. The action of 5-hydroxytryptamine on *Mytilus* smooth muscle. *J. Physiol. London* 192:869–77

Hoyle, G., Dagan, D. 1978. Physiological characteristics and reflex activation of DUM (octopaminergic) neurons of locust metathoracic ganglion. *J. Neurobiol.* 1:59–79

Jessell, T. M., Iversen, L. L. 1978. Opiate analgesics inhibit substance P release from rat trigeminal nucleus. *Nature* 268:549–51

Kandel, E. R. 1976. *Cellular Basis of Behavior. An introduction to behavioral neurobiology.* San Francisco: W. H. Freeman. 727 pp.

Kandel, E. R. 1978. *A Cell-biological Approach to Learning.* Rockville, Md: Society for Neuroscience. 90 pp.

Kandel, E. R., Tauc, L. 1966. Anomalous rectification in the metacerebral giant cells and its consequences for synaptic transmission. *J. Physiol. London* 183: 287–304

Kehoe, J., Ascher, P. 1970. Re-evaluation of the synaptic activation of an electrogenic sodium pump. *Nature* 225: 820–23

Kobayashi, H., Hashiguchi, T., Ushiyama, N. S. 1978. Postsynaptic modulation of excitatory processes in sympathetic ganglia by cyclic AMP. *Nature* 271:268–70

Konishi, S., Kravitz, E. A. 1978. The physiological properties of amine-containing neurones in the lobster nervous system. *J. Physiol. London* 279:215–29

Krnjević, K., Van Meter, W. G. 1976. Cyclic nucleotides in spinal cells. *Can. J. Physiol. Pharmacol.* 54:416–21

Kuba, K., Koketsu, K. 1976a. Analysis of the slow excitatory postsynaptic potential in bullfrog sympathetic ganglion cells. *Jpn. J. Physiol.* 26:651–69

Kuba, K., Koketsu, K. 1976b. The muscarinic effects of acetylcholine on the action potential of bullfrog sympathetic ganglion cells. *Jpn. J. Physiol.* 26: 703–16

Kunze, D. L., Brown, A. M. 1971. Internal potassium and chloride activities and the effects of acetylcholine on identifiable *Aplysia* neurones. *Nature New Biol.* 229:229–31

Libet, B. 1970. Generation of slow inhibitory and excitatory postsynaptic potentials. *Fed. Proc.* 29:1945–56

Libet, B. 1978. Dopaminergic synaptic processes in the superior cervical ganglion: Models for synaptic actions. In *The Neurobiology of Dopamine,* ed. Horn, A. S., Korf, J., Westerink, B. London: Academic. In press

Libet, B., Kobayashi, H. 1974. Adrenergic mediation of slow inhibitory postsynaptic potential in sympathetic ganglia of the frog. *J. Neurophysiol.* 37:805–14

Libet, B., Kobayashi, H., Tanaka, T. 1975. Synaptic coupling into the production and storage of a neuronal memory trace. *Nature* 258:155–57

Libet, B., Tanaka, T., Tosaka, T. 1977. Different sensitivities of acetylcholine-induced "after-hyperpolarization" compared to dopamine-induced hyperpolarization, to ouabain or to lithium-replacement of sodium, in rabbit sympathetic ganglia. *Life Sci.* 20:1863–70

Libet, B., Tosaka, T. 1970. Dopamine as a synaptic transmitter and modulator in sympathetic ganglia: A different mode of synaptic action. *Proc. Natl. Acad. Sci. USA* 67:667–73

Macdonald, R. L., Nelson, P. G. 1978. Specific-opiate-induced depression of transmitter release from dorsal root ganglion cells in culture. *Science* 199:1449–50

Magleby, K. L., Stevens, C. F. 1972. The effect of voltage on the time course of end-plate currents. *J. Physiol. London.* 223:151–71

Mayeri, E., Koester, J., Kupfermann, I., Liebeswar, G., Kandel, E. R. 1974.

Neural control of circulation in *Aplysia:* I. Motoneurons *J. Neurophysiol.* 37:458–75

McAfee, D. A., Greengard, P. 1972. Adenosine 3',5'-monophosphate: Electrophysiological evidence for a role in synaptic transmission. *Science* 178: 310–12

McCreery, M. J., Carpenter, D. O. 1978. Synergistic action of L-glutamate (GLU) and L-aspartate (ASP) in *Aplysia. Fed. Proc.* 37:582

Minota, S., Koketsu, K. 1977. Effects of adenaline on the action potential of sympathetic ganglion cells in bullfrogs. *Jpn. J. Physiol.* 27:353–66

Morris, J., Maynard, D. M. 1970. Recordings from the stomatogastric nervous system in intact lobsters. *Comp. Biochem. Physiol.* 33:969–74

Mudge, A., Fischbach, G. D., Leeman, S. 1978. Effects of enkephalin on cultured neurons: Inhibition of substance P release and decrease in action potential duration. In *Endorphins in Mental Health Research,* ed. Usdin, Bunney, Kline. London: Macmillan. In press

Nathanson, J. A. 1977. Cyclic nucleotides and nervous system functions. *Physiol. Revs.* 57:157–256

Nawrath, H. 1977. Does cyclic GMP mediate the negative inotropic effect of a acetylcholine in the heart? *Nature* 267:72–74

Phillis, J. W. 1977. The role of cyclic nucleotides in the CNS. *Can. J. Neurol. Sci.* 4:151–95

Pinsker, H., Kandel, E. R. 1969. Synaptic activation of an electrogenic sodium pump. *Science* 163:931–35

Rathmayer, W., Florey, E. 1974. Presynaptic inhibition of long duration at crab neuromuscular junctions. *Pfluegers Arch.* 348:77–81

Reuter, H. 1974. Localization of beta adrenergic receptors, and effects of noradrenaline and cyclic nucleotides on action potentials, ionic current and tensions in mammalian cardiac muscle. *J. Physiol. London* 242:429–51

Rogus, F., Cheng, L. C., Zierler, K. 1977. β-adrenergic effect on Na^+-K^+ transport in rat skeletal muscle. *Biochim. Biophys. Acta* 464:347–55

Russell, D. F., Hartline, D. K. 1978. Bursting neural networks: A re-examination. *Science* 200:453–56

Schulman, J. A., Weight, F. F. 1976. Synaptic transmission: Long-lasting potentiation by a postsynaptic mechanism. *Science* 194:1437–39

Selverston, A. I., Russell, D. F., Miller, J. P. 1976. The stomatogastric nervous system: Structure and function of a small neural network. *Prog. Neurobiol.* 7:215–90

Shank, R. P., Freeman, A. R. 1975. Cooperative interaction of glutamate and aspartate with receptors in the neuromuscular excitatory membrane in walking limbs of the lobster. *J. Neurobiol.* 6:289–302

Shimahara, T., Tauc, L. 1977. Cyclic AMP induced by serotonin modulates the activity of an identified synapse in *Aplysia* by facilitating the active permeability to calcium. *Brain Res.* 127:168–72

Siggins, G. R., Oliver, A. P., Hoffer, B. J., Bloom, F. E. 1971. Cyclic adenosine monophosphate and norepinephrine: Effects on transmembrane properties of cerebellar Purkinje cells. *Science* 171:192–94

Smith, P. A., Weight, F. F. 1977. Role of electrogenic sodium pump in slow synaptic inhibition is re-evaluated. *Nature* 267:68–70

Squires, R. F., Braestrup, C. 1977. Benzodiazepine receptors in rat brain. *Nature* 266:732–33

Starke, K. 1977. Regulation of noradrenaline release by presynaptic receptor systems. *Rev. Physiol. Biochem. Exp. Pharmacol.* 77:1–124

Treistman, S. N., Levitan, I. B. 1976. Alteration of electrical activity in molluscan neurones by cyclic nucleotides and peptide factors. *Nature* 261:62–64

Tsien, R. W. 1974. Effects of epinephrine on the pacemaker potassium current of cardiac Purkinje fibers. *J. Gen. Physiol.* 64:293–319

Tsien, R. W. 1977. Cyclic AMP and contractile activity in heart. In *Advances in Cyclic Nucleotide Research*, ed. P. Greengard, G. A. Robison, 8:363–420. New York: Raven. 592 pp.

Weight, F. F. 1974. Synaptic potentials resulting from conductance decreases. In *Synaptic Transmission and Neuronal Interaction*, ed. M. V. L. Bennett, pp. 141–52. New York: Raven. 401 pp.

Weight, F. F., Petzold, G., Greengard, P. 1974. Guanosine 3′,5′-monophosphate in sympathetic ganglia: Increase associated with synaptic transmission. *Science* 186:942–44

Weiss, K. R., Cohen, J. L., Kupfermann, I. 1978a. Modulatory control of buccal musculature by a serotonergic neuron (metacerebral cell) in *Aplysia. J. Neurophysiol.* 41:181–203

Weiss, K. R., Kupfermann, I. 1976. Homology of the giant serotonergic neurons (metacerebral cells) in *Aplysia* and pulmonate molluscs. *Brain Res.* 117:33–49

Weiss, K. R., Schonberg, M., Cohen, J., Mandelbaum, D., Kupfermann, I. 1976. Modulation of muscle contraction by a serotonergic neuron: Possible role of cAMP. *Neurosci. Abstr.* 2:338

Weiss, K. R., Schonberg, M., Mandelbaum, D. E., Kupfermann, I. 1978b. Activity of an individual serotonergic neurone in *Aplysia* enhances synthesis of cyclic adenosine monophosphate. *Nature.* 272:727–28

Woodward, D. J., Moises, H. C., Hoffer, B. J., Freedman, R. 1978. Norepinephrine modulation of Purkinje cell responses evoked by afferent stimulation and by microiontophoresis of amino acid transmitters. In *Iontophoresis and Transmitter Mechanisms in the Mammalian Central Nervous System*, ed. R. W. Ryall, J. S. Kelly. pp 441–43 Amsterdam: Elsevier 494 pp.

Zieglgänsberger, W., Bayerl, H. 1976. The mechanism of inhibition of neuronal activity by opiates in the spinal cord of cat. *Brain Res.* 115:111–28

Ann. Rev. Neurosci. 1979. 2:467–504
Copyright © 1979 by Annual Reviews Inc. All rights reserved

AXONAL TRANSPORT: ♦11531
COMPONENTS, MECHANISMS, AND SPECIFICITY

James H. Schwartz

Physiology and Neurology Departments, Division of Neurobiology
and Behavior, College of Physicians and Surgeons of Columbia University,
New York, New York 10032

INTRODUCTION

The neuron is a regionally differentiated cell. Axonal transport is the process by which materials, chiefly membranous and metabolically active, are moved rapidly from one region to another. The study of transport has recently attracted neurobiologists with widely different interests. Some are drawn to it because of their primary interest in neuronal physiology and biochemistry; these are eager to understand transport in order to comprehend the cellular dynamics of neuronal functioning. Others, interested in the organization of the brain, have used transport as a neuroanatomical tool.

Axonal transport occurs in two directions, from the cell body to terminals (anterograde) and from terminals to the cell body (retrograde). Once it is recognized that almost all macromolecular synthesis occurs in the perikaryon, the functional significance of anterograde transport becomes obvious. Essentially all subcellular organelles and membrane components in axons, dendrites, and nerve terminals must originate in the cell body, if not fully assembled, then at least in some precursor form. Transport also occurs in the retrograde direction, returning worn out materials from nerve terminals to the cell body for either degradation or restoration. Although its role in the pathogenesis of neurotropic viral diseases was suggested in the 1920s (see Baringer 1975), retrograde transport only recently has been shown to play an important part in the life cycle of the synaptic vesicle (see Heuser & Reese 1974, Holtzman 1977). Both anterograde and retrograde axonal transport have also been widely used during the past decade as the basis for

467

new methods of tracing neuronal pathways (Cowan & Cuénod 1975). Because of techniques made possible by the germinal observations of Grafstein (1971), Holtzman et al (1971), and Kristensson et al (1971), the 1970s have witnessed advances in the understanding of neuroanatomical organization that rival the Golden Age of Ramon y Cajal.

It would be useful at the start to distinguish axonal transport from the slow movement of materials originally described by Weiss & Hiscoe (1948). These two processes are mechanistically and functionally unrelated. Slow transport, currently named *axoplasmic flow,* ceases when the axon is separated from the pericaryon (Frizell et al 1975, McLean et al 1976, Lasek & Black 1977). Dependent upon the cell body, slow transport does not occur in the retrograde direction (for a review with considerable emphasis on slow transport, see Heslop 1975).

By weight, most of the protein in axons moves by slow transport (Komiya & Kurokawa 1978). Actin, myosin and myosin-like proteins, tubulin and the subunits of neurofilaments (Hoffman & Lasek 1975, Willard 1977, Lorenz & Willard 1978) are the chief constituents of the semisolid axonal column in all neurons; they move out of the cell body and down the axon at a rate of about 1 mm/day, being produced continuously at the pericaryon and undergoing internal dissolution at the nerve ending (Weiss 1972, Lasek & Black 1977). Also moving slowly with these cytoskeletal and structural components are the soluble enzymes concerned with intermediary metabolism (Partlow et al 1972). Most of these enzymes are common to all neurons, but several of them, which are specific to the synthesis of neurotransmitter substances, are an exception to this rule (choline acetyltransferase: Dahlström et al 1974, Davies et al 1977, Kása et al 1973, Saunders et al 1973; tyrosine hydroxylase and aromatic amino acid decarboxylase: Dairman et al 1973, Oesch et al 1973, Wooten & Coyle 1973). At least a portion of the tyrosine hydroxylase in axons has been reported to move rapidly, however (Jarrott & Geffen 1972, Brimijoin & Wiermaa 1977a).

In this review I am concerned primarily with fast axonal transport (for other recent reviews, see Grafstein 1975, 1977, Heslop 1975, Lubinska 1975). Whereas earlier reviews have emphasized elements common to axonal transport in all neurons, I focus here on its selectivity in specific neurons. First I consider general aspects of the molecular machinery of transport, and in particular its relationship to motility in non-neural cells. Secondly I review the relationship of axonal transport to the biogenesis and turnover of membranes. Of preeminent interest to biology is cellular diversity, and, as a class of cells, neurons are by far the most diverse. Although no one component or reaction can account for the numerous distinctions that exist between neurons, many of the differences reside in selective membrane components, both those on the cell surface and those forming compartments within the cell. The sequence of steps involved in axonal transport is a superb example of the organization of cellular diversity. To conclude I review studies illustrating that diversity depends on an assortment of molecules and processes with overlapping specificities, rather than on a single specific component.

General Properties of Fast Transport

Translocation by fast transport in warm-blooded animals occurs at a rate of about 400 mm/day and in all animals is an intrinsic or local axonal property independent of the pericaryon, as has been demonstrated in instances too numerous to cite comprehensively (but for examples, see Gross & Beidler 1975, Goldberg et al 1978, Ochs 1975a). Unlike slow transport, which supplies the bulk of the axoplasm, fast transport functions to distribute intracytoplasmic membranes, newly synthesized and assembled in the cell body, primarily to the neuron's terminals, but also to intraaxonal membrane systems (see for example, Ambron et al 1974, Bennett et al 1973, Cancalon & Beidler 1975, Di Giamberardino et al 1973, Droz et al 1975, Krygier-Brévart et al 1974, Lorenz & Willard 1978, Thompson et al 1976). It is important at the outset to emphasize the view that there is a striking similarity between nerve cells and secretory cells in general. A major share of a neuron's metabolism is invested in the elaboration of membranes. These are not destined for immediate incorporation into the plasmalemma, but, like membranes in secretory cells, first serve as intracytoplasmic compartments for the storage and transport of neurosecretory products, neurotransmitters, related packaging substances, and possibly constituents other than transmitters that are released during synaptic transmission. The total amounts of neuronal constituents that may be moving by fast transport, however, can be expected to vary with the type of neuron and, in neurons of the same kind, with their functional state. Thus, by weight, a large proportion of the protein synthesized in a neurosecretory neuron is in transit since the hormone is a protein (Norström & Sjöstrand 1971a). In contrast, only a relatively small fraction of the neuron's protein is transported along axons of dorsal root ganglion cells (see, for example, Ochs 1972a). There is some evidence that a neuron transports more protein when active than when at rest: stimulated neurosecretory neurons (Norstöm and Sjöstrand 1971b, 1972, Norström 1975, Yukitake et al 1977) and spinal motoneurons (Lux et al 1970) synthesize and transport more protein than do unstimulated neurons of the same kind.

How Transport Is Observed

There have been two approaches to the experimental observation of transport. The first can be termed destination analysis, in which the parameters of the process are inferred from the time when material from one region of the neuron arrives at another. Ligature was first used to produce an artificial destination, and has been the most widely used technique for studying transport (for reviews, see Dahlström 1971, Lubinska 1975). An elegant method of producing physiologically reversible ligatures by local axonal cooling has been developed by Brimijoin (1975) and used by him (Brimijoin & Wiermaa 1977a, b) and by others (Goldberg et al 1978, Hanson 1978) to study the rapid transport of neurotransmitters and related enzymes. Perhaps the most fruitful example of destination analysis was developed by Grafstein (1967 and reviewed, 1977). Here material synthesized in the retina after intra-ocular injection of radioactive precursors is monitored at its destination, the projections of the optic nerve (as recent examples, see Elam 1975, Heacock & Agranoff 1977, Hubel & Wiesel 1977, Willard et al 1974).

The second approach involves direct kinetic analysis. One of the most useful kinetic preparations is the cat sciatic nerve, introduced by Ochs and his co-workers (see Ochs 1972a). Transport, primarily of labeled proteins synthesized in dorsal root ganglion cells after injection of radioactive amino acids, is counted in uniform sequential segments cut along the nerve. Transport profiles showing the distribution of labeled protein with distance along the nerve are obtained from different animals at various times after the injection. This approach has been widely used for studying axonal transport in other vertebrates (Edström 1974, Frizell & Sjöstrand 1974, Gross & Beidler 1975, Kennedy et al 1972) and in invertebrates (Fernandez et al 1971, Gamache & Gamache 1974, Koike et al 1972, Heslop & Howes 1972). Another important method of kinetic analysis is cinematic observation. Direct microscopic analysis of the movement of unidentified large particles in living axons in culture was initiated by Matsumoto as early as 1920, and has recently been utilized to examine rates of movement of particles in a variety of nerves (for recent examples, see Breuer et al 1975, Cooper & Smith 1974, Forman et al 1977a, Leestma & Freeman 1977, Smith & Koles 1976). In another variation of kinetic analysis, β-radiation from ^{14}C-labeled protein moving along axons has been monitored directly in living peripheral nerve bundles of the frog (Snyder et al 1976, Takenaka et al 1978).

MECHANISM OF FAST AXONAL TRANSPORT

Molecular Components

Microtubules and contractile proteins (neural actins and myosins) are the favored candidates for the polymeric molecules that might mediate transport. Together they could provide the mechanical and biochemical properties required for intracellular movement: microtubules as skeletal elements and contractile proteins as generators of force. All biological polymers are directional. Microtubules (Snell et al 1974) and F-actin and myosin (Needham 1971) in particular have been shown to be polar. A transport mechanism involving polar molecules has the potential of operating in either the anterograde or retrograde direction if the polymer were oriented appropriately. Moreover, in the same segment of axon, the same mechanism could work in both directions if some of the polymer population had one polarity, and the rest the opposite polarity (Ochs 1975b, Kerkut 1975, Schwartz et al 1976).

Three kinds of filamentous structures can be distinguished in neruons: microtubules and microfilaments, also present in many other cell types, and neurofilaments, which are characteristic of neurons and glial cells. Morphologically, microtubules appear as hollow cylinders 20–30 nm in diameter, neurofilaments as rods 10 nm in diameter, and microfilaments as thin fibers 5–8 nm in diameter. While there is general agreement that microtubules are composed of polymerized tubulins and microfilaments chiefly of actin, the nature of the protein constituents of neurofilaments is at present highly controversial. Electronmicroscopic observation of axons and dendrites reveals microtubules and neurofilaments (which are not composed of contractile subunits) oriented longitudinally along the proximo-distal axis of the process (see for example, Byers 1974, Droz et al 1973, Lentz 1972, Smith et al 1975, Thompson et al 1976). Usually neurofilaments predominate; Brown et al (1977), for example, have shown that 10 nm filaments (incorrectly termed "microfilaments" by the authors) are twice as numerous as microtubules in the identified axon of R2, the giant neuron of the *Aplysia* abdominal ganglion, and this is our impression in a number of other identified *Aplysia* axons. By electron microscopy and biochemical analysis Gilbert (1975) has shown that most of the protein in the giant axon of *Myxicola* is in the form of neurofilaments; no microtubules could be detected.

The disposition of polymerized actin in axons is less certain because it is partially destroyed by treatment with osmium tetroxide during the usual fixation procedures (Maupin-Szamier & Pollard 1978). Although actin fibers are apparently well-preserved if protected with endogenous myosin or exogenous heavy meromyosin, preservation of ultrastructure is understandably poor when nervous tissue is treated with glycerol for the purpose of labeling with heavy meromyosin. With the technique of Ishikawa et al (1969), networks of decorated actin fibers were revealed throughout the neuron. In neuroblastoma cells they usually run parallel to the long axis of the axon (Chang & Goldman 1973). In rat brain, they run in bundles running longitudinally along dendrites, but also are disposed in transverse lattices (LeBeux & Willemot 1975). Mitochondria and membranes of the axoplasmic reticulum are attached to the decorated actin filaments or enmeshed in their network. Association of microfilaments with the external membrane was also seen. Because of these observations and because of the association of actin with the plasmalemma found in other cell types (see Bray 1978), it is likely that contractile proteins are anchored perpendicularly to the axolemmal surface, resulting in a circumferential as well as a longitudinal orientation. In normally fixed tissue, filamentous elements have occasionally been observed projecting laterally from microtubules and neurofilaments (Fernandez et al 1971, Smith et al 1975), and these sometimes branched, anastomosing structures are likely to be microfilaments and to be contractile in nature (LeBeux & Willemot 1975).

Indications for involvement of a polymeric macromolecule in the mechanism of transport have been provided by studies on the effects of temperature. There is a critical temperature, characteristic of the species, below which translocation stops abruptly (warm-blooded animals: cat sciatic nerve 11–13°, Ochs & Smith 1975; adrenergic axons in the rabbit sciatic nerve 13°, Cosens et al 1976; cold-blooded animals: frog 5–7°, Edström & Hanson 1973, Forman et al 1977b, Takenaka et al 1978; garfish olfactory nerve 6–7° C, Gross & Beidler 1973; gastropod molluscs 3–4° C, Heslop & Howes 1972, Goldberg et al 1978). A process that exhibits a critical temperature is likely to involve a limiting step that depends on a component that undergoes a change in state (molecular structure or phase). Depolymerization of microtubules in the cold has been thought to underlie the phenomenon of a critical temperature for transport (see, for example, Brimijoin 1975). It is therefore surprising that there is no careful comparative study correlating the critical temperatures of microtubular depolymerization and of fast transport.

Obviously this argument for microtubules is indirect, and could equally implicate other cellular components. Polymerization of actins from all tissues requires a critical concentration for nucleation and elongation; nonmuscle actins appear to have higher critical concentrations than does muscle actin and these critical concentrations are dramatically increased in the cold (Korn 1978). Functionally significant phase changes of lipids in membranes at critical temperatures are another possibility (see Hazel 1973). It must be noted that, at least in the cat sciatic nerve, the low temperatures at which transport is blocked do not crucially affect the supply of ATP or creatine phosphate needed for transport (Ochs & Smith 1975). Nor does the microviscosity of axoplasm change abruptly at 11° C as might be expected if cold blocked transport because of a marked increase in axoplasmic viscosity (Haak et al 1976).

The argument in favor of the participation of microtubules and the contractile proteins does not rest solely on temperature studies. There is more specific evidence for each of these macromolecular components.

MICROTUBULES   The fortunate existence of a class of natural products that cause depolymerization of microtubules, of which the alkaloids colchicine and vinblastin are the most familiar examples, has led to the chief experimental evidence that microtubules are involved in fast transport. In the vast majority of instances, application of these inhibitors of mitosis has been found to block transport in both vertebrate (most recent reviews: Samson 1976, Hanson & Edström 1978) and invertebrate axons (crustacea: Fernandez et al 1970, Fernandez et al 1971; insects: Schafer 1973; molluscs: Heslop & Howes 1972, Goldman et al 1976, Berry & Schwartz 1977; echinoderms: Gamache & Gamache 1974). Numerous though they are these studies do not provide compelling evidence for the direct involvement of microtubules, since their function may be cytoskeletal rather than mechanistic; depolymerization of microtubules may be to axonal transport as a broken femur is to walking. Furthermore it has frequently been noted that these depolymerizing drugs have targets other than microtubules. Strong indication that colchicine affects transport selectively by its action on microtubules, however, was provided by experiments with lumicolchicine, a photochemical derivative, which produces many of colchicine's effects other than those on microtubules, but does not cause depolymerization or interfere with axonal transport (see for example, Banks & Till 1975, Dahlström et al 1975b).

Striking electron micrographs showing the association of vesicles and other membranous organelles with microtubules provide morphological support for their participation in transport (LaVail & LaVail 1974, Smith et al 1975). Additional intriguing evidence that microtubules might be involved comes from experiments in which protein in dorsal root ganglia was labeled by injection of ^{3}H-amino acids. It was found that 3–5 times more protein was transported along the sensory axons in the sciatic nerve than into dorsal root axons toward the spinal cord (for recent studies, see Ochs 1972b, Barker et al 1976, Komiya & Kurokawa 1978). Barker et al (1976) suggested that the quantity transported might be related to the greater density of microtubules found in the spinal branch.

Gratifying support for a direct role of microtubules would be provided if the absence of transport could be demonstrated in the giant axon of *Myxicola,* which is devoid of microtubules (Gilbert 1975). Unfortunately this has not been accomplished. Although it has not been possible to demonstrate transport, both John Heslop and Raymond Lasek (personal communications) feel that their many unsuccessful attempts do not rule out the possibility convincingly. Their uncertainty is due to the technical difficulty of working with the giant axon, which is a syncitium formed from 5 pairs of cell bodies closely spaced at each of the 30 segments along the central cord of the marine worm.

Although strong correlation has usually been observed between the effects of antimitotic agents on microtubule assembly in vitro and their potency in inhibiting axonal transport (Banks & Till 1975, Friede & Ho 1977, Hammond & Smith 1977, Hanson & Edström 1977, Paulson & McClure 1975), the most severe challenge to the idea that microtubules are involved in the mechanism of transport is the number of instances in which blockage of transport has not been found to parallel disruption of microtubules. In the crayfish cord, colchicine (Fernandez et al 1970) and vinblastin (Fernandez et al 1971) do not appear to disrupt microtubules at concentrations sufficient to block transport. Transport of hypophyseal neurosecretory granules along axons in the median eminence is interrupted by administration of colchicine or vinblastin (Gainer et al 1977, Flament-Durand & Dustin 1972, Dustin et al 1975, Hindelang-Gertner et al 1976; Norström 1975). While Hindelang-Gertner et al (1976) found microtubules disrupted, Flament-Durand & Dustin (1972) failed to observe differences between microtubules in hypothalamic neurons in treated and untreated rats. Perhaps most influential are Byers's (1974) quantitative electron microscopic radioautographic studies on desheathed rabbit vagus nerves exposed to a moderate dose (7.5 mM) of colchicine. She observed a uniform decrease of microtubules in axons all along the nerve; transport of ^{3}H-proline-labeled protein was inhibited only in regions of axons in the proximal, but not in the distal, segments of the nerve. Since microtubules were lost but transport continued, Byers argued that there is no direct correlation between microtubules and transport.

All experiments of this kind have two critical weaknesses. First, they presume a strict one-to-one relationship between transport and the state of preservation of microtubules ultimately observed by electron microscopy. This of course is not necessary; only some microtubules may be involved in transport. Alternatively if all microtubules were involved, they might normally be operating below their maximum carrying capacity; any polymers remaining after treatment with colchicine might compensate by increasing their individual load. Secondly, drugs may have differential effects on different neurons (even of the same type) and on different regions of the same axon. Thus, the vagus is a complex nerve, and hypothalamic neurons may differ in their susceptibility to antimitotic drugs. For observations of this kind to be explicit in all respects, experiments should be done not only in a single identified axon, but in that portion of the axon in which transport is occurring when the drug is applied.

CONTRACTILE PROTEINS   Although plausible because actin and myosin are the prime biological movers, the idea that contractile proteins are involved in the mechanism of axonal transport lacks substantial evidence.

One reason that evidence is scanty is the lack of pharmacological agents with even the specificity that colchicine exhibits for microtubules. [The reported actions of cytochalasin B are conflicting (see Samson 1976) and it appears that the actions of this drug are not selective enough for interpretation.] Quite obviously the plausibility of the idea springs from the function of actomyosin in muscle contraction, now long understood (Needham 1971) and its role in motility in other cells, which has only recently become a field of widespread interest (see R. Goldman et al 1976, Korn 1978, Trifaro 1978).

The evidence on hand is of three kinds. First, actin and myosin are present in neurons: actin, at least, in substantial amounts (for general references: Trifaro 1978; for contractile proteins in specific axons: Hoffman & Lasek 1975, Lorenz & Willard 1978, Lubit et al 1978, Unsicker et al 1978, Willard 1977). Although necessary to the idea that they are involved in transport, the presence of contractile proteins in axons obviously does not constitute secure evidence.

The ionic requirements of transport, especially the role of Ca^{++}, provide the second kind of evidence for the possible participation of the contractile proteins. The ionic composition of solutions bathing nervous tissue containing cell bodies and axons of neurons engaged in transport has been manipulated in a variety of studies.

If contractile proteins are active in the mechanism of transport, Ca^{++} would be expected to play a direct and critical role. Although somewhat controversial, available data now indicate that Ca^{++}, as anticipated, is required for translocation of materials along the axon by fast transport. Banks et al (1973), found that exposure to a Ca^{++}-free solution containing EGTA prevented export of adrenergic storage vesicles from the pericaryon into the axon, but had no effect on the movement of the dense core vesicles already present in nonmyelinated axons in the cat hypogastric nerve. Similar effects of the lack of Ca^{++} were described for transport of labeled proteins in sensory afferents of frog dorsal root ganglion cells (Edström 1974, Dravid & Hammerschlag 1975, Hammerschlag et al 1977). The selective effect of Ca^{++}-deprivation on export provoked Hammerschlag et al (1975) to propose that although Ca^{++} is not needed for translocation of an organelle along axons, it is required in the initiation of transport, a process which, for example, might involve coupling of the organelle to be transported to the transport apparatus. There are a number of conditions under which export can be distinguished from translocation (Ambron et al 1975, and see below) but there is a more straightforward explanation for the different effects withdrawal of Ca^{++} seemed to bring about on cell body and axon: removal of external Ca^{++} from the bathing solution appears to be insufficient for depleting axonal Ca^{++}. In contrast to the observations already described with hypogastric and frog nerves, Ochs et al (1977) found that transport of protein is blocked within 150 min of bathing the isolated cat sciatic nerve in a Ca^{++}-free solution, if (and only if) the sheath has first been removed from the nerve.

In support of a role for Ca^{++} in transport, Hammerschlag et al (1976) found that addition of Co^{++}, an antagonist of Ca^{++}, depressed transport without affecting the synthesis of proteins or their glycosylation (Lavoie et al 1978). Knull & Wells (1975) and Neale & Barker (1977) observed that $^{45}Ca^{++}$, presumably bound to protein, is transported rapidly in neurons of chick, frog, goldfish, and *Aplysia*. This may not be pertinent to the role of contractile proteins in transport, however, since actin and myosin move by slow transport. Nevertheless, the transport of the regulatory proteins tropomyosin and troponin C, which bind Ca^{++}, have not been studied. It is likely that the success of tracing methods with cobalt salts injected directly as intraneuronal histochemical markers (Pitman et al 1972) results from the rapid movement of Co^{++} transported as an analogue of Ca^{++}.

There is full agreement that exposure to concentrations of Ca^{++} greater than normal blocks translocation (Edström 1974, Garcia et al 1974). Ochs et al (1977) suggested that elevated intra-axonal concentrations of Ca^{++} may depolymerize microtubules.

Changes in K^+ or Na^+, increased or decreased tonicity, and the addition of ouabain have been found either to be without effect or to be only slightly inhibitory (Edström 1975, Garcia et al 1974, Partlow et al 1972). These changes are difficult to interpret, however, since they are likely to be indirect. Moreover biochemical evidence that microtubules are associated with ouabain-sensitive Na^+, K^+-ATPase is most likely artifactual since the tubulin, even though purified by two cycles of polymerization from splenic nerve, still contained small amounts of phospholipid suggesting residual microsomal or plasma membrane contamination (Banks 1976). A Mg^{++}, Ca^{++}-ATPase extracted from cat sciatic nerve is a soluble component (Khan & Ochs 1974).

Edström & Mattsson (1975, 1976) provided indirect but nevertheless intriguing evidence for the involvement of contractile proteins in their studies with Zn^{++} and Cd^{++} and with other sulfhydryl blocking agents. When frog and rat sciatic nerves were treated with high concentrations of these agents, transport, not surprisingly, was inhibited. Remarkably, however, low concentrations of Zn^{++} and Cd^{++} caused an increase (up to 175%) in the amount (but not in the rate) of protein transported. Edström & Mattsson (1976) argued that stimulation might result from activation of an axonal ATPase, as previously described for myosin from muscle and for flagellar dynein from sea urchin spermatazoa.

The third, and possibly least direct, kind of evidence for the participation of contractile proteins in axonal transport is its dependence on temperature; rates of fast transport are quite compatible with a force-generating mechanism involving actomyosin. Transport is an active process with values of Q_{10} greater than 2 in neurons from both warm- and cold-blooded vertebrates (see, for example, Forman et al 1977b and Takenaka et al 1978), and in invertebrates (Heslop & Howes 1972, Goldberg et al 1978). Forman et al (1977b) found that values of observed transport velocities from a variety of vertebrates fell on a theoretical straight line when plotted semilogarithmically against temperature in the range from 10–30° C. Transport is likely to involve a similar temperature-sensitive process in invertebrates as well, since the velocities of serotonergic vesicles measured in an identified *Aplysia* axon (48 mm/day at 14° C; 130 mm/day at 23° C, Goldberg et al 1978) both fall precisely on the theoretical line drawn for the vertebrate data. Rates published from other invertebrate experiments are somewhat slower, but for experimental reasons were less accurate estimates (Heslop & Howes 1972, Ambron et al 1974, Goldman et al 1976, Goldberg et al 1976) or were complicated by diffusion (Koike et al 1972, Treistman & Schwartz 1977). The exceptionally slow rate obtained by Fernandez et al (1971) for the fast component in the crayfish nerve cord was probably the result of inadequate experimental design; shorter sampling intervals would have been needed to detect any rate faster than 10 mm/day at room temperature.

Rates of fast anterograde transport are compared with those of retrograde transport and with rates of other processes conjectured to involve contractile proteins in Table 1. Retrograde transport, as estimated from ligation experiments or as measured by microscopic cinematography, is about half as fast as in the anterograde direction. The rates of both fall within the range of maximal shortening speeds of unloaded vertebrate muscle. They are also in the same range as the velocities of amoeboid movement, but considerably faster than the movement of integral surface proteins in plasma membranes [for example, capping and the movement toward the cell body of red blood cells linked to concanavalin A bound to surface receptors in axons of dissociated sympathetic chick neurons (Koda & Partlow 1976)].

The plausible involvement of contractile proteins in the force-generating mechanism of transport awaits stronger support than that now available. We have no sufficiently specific drug for administration to the neuron externally. More convincing evidence might be provided by experiments with skinned axon preparations or with nonpermeable agents injected directly into neurons. A promising example is the cyclic peptide phalloidin, which promotes actin polymerization; Wehland et al (1977) injected the mushroom poison into cultured rat fibroblasts and found that it aggregated microfilaments and blocked locomotion.

Table 1 Rates of fast axonal transport compared with rates of other kinds of cells motility possibly involving contractile proteins

Motility	Temperature (°C)	Rates (μm/sec)	(mm/day)	References
Anterograde Fast Transport				
Labeled Proteins	38	4.74	410	Ochs 1972b, Ochs & Smith 1975
Acetylcholinesterase	38	4.44	384	Hanson 1978
Dopamine β-hydroxylase	37	3.47	300	Brimijoin 1975, Brimijoin & Wiermaa 1977b
Serotonergic Vesicles	23	1.51	130	Goldberg et al 1978
Retrograde Fast Transport				
Acetylcholinesterase	38	2.54	220	Ranish & Ochs 1972
Unidentified Particles	38	3.94	263	Calculated from the formula in Forman et al 1977b
Unidentified Particles	23	0.59	51	Calculated from the formula in Forman et al 1977b
Muscle Contraction				
(V_{max} without load)				
Skeletal				
Cat, rat, mouse	35–37	4–24	363–2076	Bárány 1967. Estimated from the maximal speed of shortening given in units of muscle lengths/sec. and assuming 1 μm to be the half-length of the sarcomere.
Frog sartorius	22	10	865	Bárány 1967.
Smooth				
Rabbit uterine	37	0.20	17.2	Bárány 1967.
Isolated frog cell	20	3.64	315	Fay et al 1976. Estimated from the maximal speed as above, but using the length of the isolated smooth muscle cell shown prior to stimulation.
Amoeboid Movement				
Amoeba	23	0.5–4.5	43–388	Anderson 1973
Lymphocytes in Culture	37	0.55	48	Anderson 1973
Transmembrane Movement				
Capping	37	0.05	4.3	Bretscher 1976
RBC-Con A Particles	37	0.014	1.2	Koda & Partlow 1976

Organelles that Might Participate in the Mechanism of Fast Transport

In the anterograde direction it is most reasonable to think of axonal transport as one step in the biogenesis of membranes, and, in the retrograde direction, as a step in the cycle of membrane turnover. Droz 1975, Holtzman 1977, Holtzman et al 1977, and Lasek & Hoffman 1976, have stressed this idea in conceptually important reviews, and have indicated the importance of the secretory and lysosomal paradigms.

Biochemical, histochemical, and electron microscopic analyses indicate that material moving along the axon by rapid transport is mainly particulate (McEwen & Grafstein 1968, Grafstein 1977), consisting primarily of membranous organelles. Thus components of membranes have been observed to be transported rapidly in the anterograde direction after incorporation and assembly in the cell body. Typically, in neurons engaged in chemical synaptic transmission, vesicles (or their precursors) involved in storage and release of neurotransmitters are the functionally most important constituents of fast axonal transport (Dahlström 1971, Banks & Mayor 1972, Dahlström et al 1975a, Howes et al 1974, Thompson et al 1976, Goldman et al 1976). In neurosecretory neurons, the chief product is the neurosecretory granule. The organelles transported in the retrograde direction are also membranous structures and these, which appear to be chiefly lysosomal, are usually detected by their exogenously derived contents. During the past decade, horseradish peroxidase (HRP) has been most frequently used as a histochemical marker because of the coupling that occurs at synapses between exocytosis and endocytosis (Holtzman et al 1971, Heuser & Reese 1974). Endocytotic uptake of HRP occurs to some extent in all regions of the neuron, however (Hansson 1973, Coleman et al 1976), and uptake into lysosomal elements can be greatly enhanced by injury (see, for example, Kristensson & Olsson 1976). Keefer (1978) lucidly discusses the various intracellular distributions of the reaction product and how they relate to the extent of damage done to the neuron.

Neurosecretory granules, synaptic vesicles, and the axoplasmic reticulum are derived from the endoplasmic reticulum and the Golgi apparatus; these will be considered first, followed by the axoplasmic reticulum and lysosomes.

THE ENDOPLASMIC RETICULUM AND GOLGI APPARATUS   Secretion can be approached from the vantage point either of the secretory product or of the membrane compartments through which that product passes. The magisterial review of Palade (1975) stresses the former point of view; those of Siekevitz (1972) and of Morré (1977), the latter. Neurosecretion and the release of transmitters are exocytotic processes; membranes destined for neurosecretory granules or for synaptic vesicles are synthesized in the endoplasmic reticulum and processed in the Golgi apparatus in the pericaryon. Ultimately they must reach nerve terminals. In neurons with somatic release sites (see for example, Hamori et al 1978) there is no need for developing an entirely new way of bringing vesicles to the active zone. In neurosecretory neurons, axons are short, and some motile mechanism must be differentiated to move granules and their contents, now not to the plasmalemma, but to the more distant nerve terminal. Axonal transport in conventional neurons would represent the extreme example of this differentiation. It is likely that secretion itself represents an elaboration of a more general process for synthesis and assembly of the external membrane. With synaptic transmission, because of the extreme regional differentiation of the neuron, the membrane that is most rapidly synthesized is not inserted in the local plasmalemma in the pericaryon, but rather is associated with intracytoplasmic membranes destined for specialized regions of distant axolemma and synaptolemma. There is extensive radioautographic evidence for these ideas (see Hendrickson 1972, Droz et al 1973, Droz et al 1975). Neither plasmalemma nor proximal axolemma were significantly labeled 3 hr after injection of ^{3}H-fucose in L10, an identified *Aplysia* interneuron, when membranous organelles in both cell body and axon were intensely labeled; only after 16–20 hr did the external membranes become labeled (Thompson et al 1976). Similar results were obtained with R2 and the giant cerebral neuron, two other identified *Aplysia* neurons, injected with ^{3}H-N-acetylgalactosamine. In all of these neurons, the relative specific activity (the ratio of % silver grains over an organelle to the % area of the micrographs occupied by that organelle) of axonal vesicles was 3–10 time greater than that of somatic vesicles (which, with Golgi membranes, were among the most intensely labeled components in the pericaryon). In contrast, when the plasmalemma did become significantly labeled, its relative specific activity exceeded that of the axolemma of the proximal axon (Thompson et al 1976, Ambron & Schwartz 1979), which suggests the possibility that the axolemma is largely renewed by lateral diffusion of membrane components from the plasmalemma. Subcellular fractionation studies also support the idea that intraaxoplasmic and synaptic plasma membranes are the ones most rapidly labeled by axonal transport (Di Giamberardino et al 1973, Krygier-Brévart el al 1974, Cancalon & Beidler 1975, Lorenz & Willard 1978, Ambron & Schwartz 1979).

The complex congeries of events that takes place in the rough and smooth endoplasmic reticula and in the Golgi apparatus during biogenesis of membranes is beyond the scope of this review. It is necessary, however, to indicate two important aspects of these membrane systems that are pertinent to axonal transport. First, these membrane systems exist together only in the pericaryon, and second, the series of events that takes place within them can be described as sequential, directional, or vectorial, and, like other instances of biological assembly, does not occur in random order. The absence of ribosomes in regions of the neuron other than the pericaryon (and the very proximal portion of dendrites) is sufficient to account for the fact that essentially all amino acid incorporation into protein is somatic. Lack of biosynthetic machinery, however, does not explain why the bulk of lipids are synthesized in the pericaryon, since both phospholipids (Larabee & Brinley 1968, Benes et al 1973, Gould 1976) and glycolipids (Barondes 1974, Fishman & Brady 1976, Rösner et al 1973) apparently can also be synthesized and inserted into membranes in axons and synapses. There is abundant evidence for the somatic synthesis of membrane lipids and their subsequent axonal transport (Abe et al 1973, Forman et al 1971, Grafstein et al 1975, Rösner et al 1973, Forman et al 1972, Forman & Ledeen 1972, McGregor et al 1973, Sherbany et al 1979). Since lipid is an essential constituent of membranes, it is likely that its insertion into nascent membranes is programmed coordinately with insertion of protein.

This idea would explain a number of similar observations that inhibitors of protein synthesis block dramatically the export of phospholipids (Abe et al 1973, Grafstein et al 1975), glycolipids (Forman et al 1971, Forman et al 1972, Sherbany et al 1979), and glycoproteins (Forman et al 1971, Edström & Mattsson 1973, Ambron et al 1975) from the cell body into the axon, under conditions in which synthesis of lipids or glycosylation of proteins in the cell body are affected only slightly or not at all. Blockage of export ultimately results in cessation of transport, not because the drugs affect translocation directly, but because organelles to be transported become unavailable in the axon (Brimijoin 1974, Ambron et al 1975, Goldberg et al 1978). These differential effects of the inhibitors on the synthesis and export of nonprotein membrane constituents can be understood if it is assumed that (a) the nascent membrane components cannot be exported on their own, but must be incorporated into completed organelles; (b) completion of organelles requires continued protein synthesis; and (c) nascent (labeled) components cannot be incorporated out of sequence, for example by exchange with similar components in already completed organelles.

The analogy between axonal transport and the movement of granules from the smooth endoplasmic reticulum and Golgi to the plasma membrane in secretion is strongly supported by the effects of antimitotic drugs on a variety of secretory cells. Lacy et al (1968) first noticed that colchicine blocked secretion of insulin from β cells in isolated rat pancreatic islets, and proposed that movement of secretory granules to the plasma membrane is mediated by microtubules. Subsequently, agents that depolymerize microtubules have been found to block secretion of substances released by exocytosis, but not of substances whose release does not involve vesicles. Thus secretion of lipoproteins (LeMarchand et al 1973, Stein et al 1974), collagen (Ehrlich et al 1974), albumin, and other plasma proteins (Redman et al 1975), and matrix deposition by chondrocytes (Moskalewski et al 1976b) and amylase (Patzelt et al 1977) were found to be inhibited, and the original observation on insulin confirmed (Moskalewski et al 1976a, Seybold et al 1975); colchicine did not block secretion of lecithin or cholesterol by liver cells, substances not released by a vesicular process (Stein et al 1974). Although the problem of pharmacological specificity is again obvious, in most of these studies the most likely nonspecific effects of the drugs were looked for and eliminated from consideration, including colchicine's anticholinergic action (Trifaro et al 1972), which might have been implicated because of the responsiveness of some of these glandular tissues to acetylcholine. As in hypothalamic neurosecretory cells (Flament-Durant & Dustin 1972, Flament-Durant et al 1975, Hindelang-Gertner et al 1976), accumulations of secretory granules, pileup of secretory products in Golgi elements, and dilatation of Golgi-associated vesicles were observed in the nonneural secretory tissues treated with antimitotic agents.

SMOOTH AXOPLASMIC RETICULUM   Droz (1975) proposed that the smooth endoplasmic reticulum forms a network of continuous channels, some of which mediate the rapid translocation of newly synthesized material from the pericaryon to the nerve ending; others carry used materials back to the cell body. This popular idea is supported by two kinds of evidence: first, the reticulum becomes labeled rapidly by ^{3}H-proteins originating in the cell body, as revealed by radioautography (Hendrickson 1972, Lentz 1972, Droz et al 1973, Bennett et al 1973, Droz et al 1974, Byers 1974, Droz et al 1975, Markov et al 1976, Thompson et al 1976, R. T. Ambron, L. J. Shkolnik and J. H. Schwartz, unpublished). In the few studies in which the radioautographs were analyzed properly by morphometric procedures (Hendrickson 1972, Thompson et al 1976, R. T. Ambron, L. J. Shkolnik and J. H. Schwartz, unpublished), 25–28% of the silver grains in axonal protein or glycoprotein were found over the axoplasmic reticulum, which occupied 3–6% of the total cross-sectional area; the relative specific activity of the reticulum was considerably higher than that of axoplasm or of some of the other membranous organelles, but was only 25–35% of the specific activity of vesicles.

The other kind of support for the participation of the reticulum in transport is the possibility that it is continuous from the pericaryon to the nerve terminal, and therefore could serve as a conduit. Droz et al (1975) and

Markov et al (1976) obtained evidence for its continuity in chicken and rat ciliary and in chicken spinal ganglia impregnated with uranylacetate and copper and lead citrates using high voltage electron microscopy. From axon terminals and from other parts of the tissue containing axons which were slightly compressed before fixation, Droz and his co-workers present micrographs of thick sections, which contain convincing images of anastomosing tubules running the length of the microscopic field, and emptying into subaxolemmal cisternal spaces. In addition, synaptic vesicles appeared to be connected with the tubules of the reticulum if the specimens were tilted. With radioautography, the reticulum was found to be intensely labeled.

Hendrickson & Cowan (1971), Lentz (1972), and Byers (1974) had previously noticed that the highest axonal grain density due to ^{3}H-protein originating from the pericaryon lay just beneath the axolemma. Since the reticulum and its cisternae appeared to be preferentially localized to that region and since the ^{3}H-proteins are presumably moving, this laminar distribution of silver grains has been taken as support for the participation of the reticulum in transport. A laminar distribution has not been invariably observed, however (Ochs 1972b, Thompson et al 1976). Furthermore, although silver grains from ^{3}H-serotonin being transported along an identified serotonergic axon were also found to be laminar, this distribution coincided with that of serotonergic vesicles (which were significantly labeled) and not with that of the smooth endoplasmic reticulum (which was not labeled) (J. E. Goldman et al 1976).

The existing evidence that membranes of the axoplasmic reticulum contain materials moved by fast transport is strong; the evidence that the reticulum mediates transport is weak. The splendid micrographs of Droz and his co-workers were not analyzed quantitatively, and attractive though the idea may be that there is a membranous conduit connecting the reticular systems of the pericaryon with nerve terminals, it has not been shown conclusively. Pressure injury was required to enhance the appearance of the reticulum along axons, and dramatic changes of internal membranes at sites of injury are a familiar phenomenon (see Till & Banks 1976). Furthermore in the small (less than twice the transverse diameter of a mitochondrion) but highly uniform axons of the olfactory nerve of the European pike, Kreutzberg & Gross (1977) found that 14% of axonal cross sections contained no axoplasmic reticulum, and in an additional 14%, its presence or absence could not be determined; they therefore concluded that the reticulum is not continuous in these axons.

It is impossible to extract useful information about the mechanism of transport from these radioautographic studies, even from those that were analyzed quantitatively, because of the uncertainty whether the silver grains were distributed over structures actually being moved. Thus, the possible irrelevance of the laminar distribution of silver grains that they observed was noted by Hendrickson & Cowan (1971), who stated explicitly that their radioautographs were not likely to contain mobile components since they were sampled from regions of the axons long since passed by the moving front. A most important insight gained from careful kinetic studies of transport is that most of the membranous material is left behind the moving front. In the garfish olfactory nerve, only 20% of the labeled protein that enters the axons reaches the terminals, the rest being deposited in the axon along the way (Gross & Beidler 1975, Cancalon & Beidler 1975). Because the radioactive material moving with the front differs from that deposited along the axon with respect to both protein composition (as revealed by SDS gel electrophoresis: Ambron & Schwartz 1979, Weiss et al 1978, Gross & Weiss 1977, Cancalon et al 1976) and subcellular distribution (as determined by differential centrifugation: Sabri & Ochs 1973, Cancalon & Beidler 1975, Weiss et al 1978), the population of moving organelles is likely to differ from the population of stationary ones. Unfortunately these have so far not been distinguished in the morphological studies available. In all of these studies, there is little doubt that the radioactive material had reached its position by fast transport; the unanswered question is whether, at the time of fixation, it had already been deposited in stationary or slowly-moving structures, or was still in the process of being transported. This question can only be answered by examining regions of an axon known to contain the moving front of material, and this probably could be best accomplished under conditions in which discrete pulses of radioactivity move along the axon, as achieved for example in the experiments of Gross & Beidler 1975, Weiss et al 1978, and Goldberg et al 1978.

LYSOSOMES   The final goal of anterograde transport is the nerve terminal, where the membranous organelles, synthesized and assembled in the pericaryon, are transformed. Some of the material is released and some locally deposited, but ultimately much is returned to the pericaryon within lysosomes. Exogenous proteins and other particles taken up at nerve endings by endocytosis (or by injury anywhere along the axon) eventually gain access to lysosomes. Formation and functioning of lysosomes is beyond the scope of this review (see Holtzman 1976, Silverstein et al 1977). The continuous process of recycling and return is mediated by retrograde transport, which, like anterograde transport, is likely to be a neuronal specialization of a process common to all cells. In rat peritoneal macrophages, phagosomes move away from the plasma membrane (where they are formed) by long saltatory movements. Bhisey & Freed (1971) showed that colchicine depolymerized microtubules and arrested these movements; the drug did not prevent the engulfment of heat-killed *E. coli,* nor did it alter the phagosome's content of acid phosphatase.

In all studies in which particles were large enough to be observed directly by light microscopy, movement in the retrograde direction predominated (Forman et al 1977a). In the few instances in which the moving particles were subsequently examined by electron microscopy, they were identified as two lysosomes traveling together, a lysosome together with a multivesicular body and a single mitochondrion (Breuer et al 1975). The majority of electron microscopic observations on the localization of HRP, tetanus toxin, nerve growth factor (NGF), and viruses, however, have been either at nerve terminals, close to the site of uptake, or at the distant cell body, the final destination. Most pertinent to transport is the nature of the axonal organelles containing the foreign particles, and presumed to be in motion. Lack of extensive information on the distribution of the material in transit is understandable, since most of the studies were performed with the intention of neuroanatomical tracing. In many instances it is inconvenient to search for axonal profiles between the terminals and cell bodies because they are difficult to identify and contain HRP only transiently (see, for example Repérant 1975). Nevertheless, HRP has been localized primarily to 100 nm axonal vesicles, and also to multivesicular bodies and to the tubules of the smooth axoplasmic reticulum (Hansson 1973, LaVail & LaVail 1974, 1975, Nauta et al 1975, Sherlock et al 1975, Kristensson & Olsson 1976, Brownson et al 1977). A similar localization was found for Herpes virus (Baringer 1975), ^{125}I-labeled tetanus toxin (Schwab & Thoenen 1977), ^{125}I-NGF coupled to HRP (Schwab 1977), and ^{125}I-tetanus toxin bound to colloidal gold particles (Schwab & Thoenen 1978).

As already discussed for the radioautographic studies in the preceding section, it is difficult to extract useful information on mechanism of transport from these localization studies, not only because of the lack of quantitation, but also because of the uncertainty whether the organelles containing the particles are actually in motion in the form in which they are observed in the electron micrographs. Both in the anterograde and in the retrograde direction, transport involves membranous organelles. How these membrane systems participate in the mechanism of transport is still an unanswered question.

Mechanistic Hypotheses for Fast Axonal Transport

The mechanism of fast axonal transport is not yet known. There are formally two types of hypothesis, one of which stems from Schmitt (1968), the other from Weiss (1970). Schmitt suggested that microtubules provide an essentially stationary and passive track on which specific organelles could move in a stepwise fashion. Translocation would be the result of a local (but sequential) energy-dependent reaction, either involving conformational changes occurring as the result of interaction of the organelle with a subunit of the track (rachette mechanism) or a sliding-filament mechanism. Samson

(1971) proposed that microtubules were "macroions," polyelectrolytes with branching polysaccharide units that attach to organelles. Through repeated cycles of attachment, contraction and expansion brought about by an in-tramolecular conformational change, and detachment, the organelle would be moved along. The energy of ATP would be needed at one point in the cycle. Ochs (1972a, 1975b) suggested the justifiably most popular mecha-nism in which organelles are attached to microtubules through actomyosin cross bridges or "sliding filaments," which provide the motive force by contracting in a manner analogous to muscle contraction.

Alternatively, Weiss (1970) postulated that materials are moved between tracks by "intratubular convection." Formally this hypothesis involves spe-cific channels within the axon through which components are translocated essentially by peristalsis. The motive force for the Schmitt hypothesis seems obvious, being analogous to the movement of mRNA across the ribosome, or the movement of contractile porteins, both of which make use of the energy of ATP to bring about conformational changes in highly ordered macromolecules. But the motive mechanism in the Weiss hypothesis is more difficult to specify, and might use the forces involved in lateral diffu-sion or capping, whose mechanisms are not yet understood, but which are thought to be linked in some way through the membrane to contractile proteins (Droz 1975, Marchisio et al 1975, see also Bourguignon & Singer 1977, Bray 1978). This seems unlikely, however. Transmembrane motile processes occur at a much slower rate than fast transport (Table 1). Their range of velocities is comparable to that of lateral diffusion (Bretscher 1976). Because the rates of these processes are so much slower, if capping and other "transmembrane" motile processes are mediated by contractile proteins, their loading or other geometrical conditions are likely to be rather different from those obtaining in fast axonal transport.

There are two interesting and recent elaborations of Weiss's hypothesis. First, Droz et al (1975) proposed that fast axonal transport occurs specifi-cally in the channels of the axoplasmic reticulum. The second is Gross's (1975) and Odell's (1976) proposal that orgenelles that are transported rapidly are moved in carrier streams of axoplasm; axoplasmic streaming (which is analogous to protoplasmic streaming as seen, for example, in protozoa and plant cells) would be faster in annular regions around axonal microtubules. In this "microstream concept," Gross (1975) specifies that axonal microtubules generate a vectorial shear force using the energy of ATP enzymatically. Odell (1976) provides a mathematical treatment of a postulated propagation of successive waves brought about by contraction-relaxation of contractile axonal proteins, producing fluid flow.

I favor Schmitt's hypothesis involving microtubules, contractile proteins, and formed vesicles rather than Droz's adaptation of Weiss's hypothesis, which involves continuous sheets of membrane. Visual evidence from microscopic cinematography showing discrete jumping particles is highly persuasive, and argues against a continuous peristaltic mechanism (but one must be reminded that the sun appears to rise and set daily). Kinetic studies also indicate that transport is an intermittent process. Studies in the giant cerebral cell, an identified serotonergic *Aplysia* neuron, are consistent with the theory that transport occurs only when serotonergic vesicles are associated with essentially stationary tracks. Association is intermittent, binding or dissociation being sensitive to local vesicle concentration (Goldberg et al 1976, Schwartz et al 1976, Goldberg et al 1978). Although easier to picture with the elements of Schmitt's hypothesis, our kinetic model would become formally equivalent to Gross's (1975) and Odell's (1976) microstream hypothesis if the feature of concentration dependence were incorporated.

Weiss's hypothesis is attractive chiefly because it would explain the widespread distribution of transported materials in axonal tubules and cisternae. It is possible however that some of this arises artifactually from injury (see Kristensson & Olsson 1976). In ligation experiments it is generally observed that some proportion of a rapidly transported component appears to be immobile (acetylcholine: Saunders et al 1973, Dahlström et al 1974; acetylcholinesterase: Ranish & Ochs 1972, Lubinska 1975, Tuček 1975, Fonnum et al 1976, Holmes et al 1977; norepinephrine and dopamine β-hydroxylase: Brimijoin & Wiermaa 1977a; various enzymes: Partlow et al 1972, Schmidt & McDougal 1978). A substantial amount of the stationary material may be explained by experimental artifact: 25–35% of the total norepinephrine in the sciatic nerve (Haggendal et al 1975), and much of the dopamine β-hydroxylase (Brimijoin 1977) and monoamine oxidase (Schmidt et al 1978) is contained in nerve endings in the perineurium, and must be tallied as immobile transmitter and enzyme. But some of the immobile components are truly intraaxonal (see for example, Brimijoin 1974, 1976, 1977). The relationship between mobile and stationary compartments within axons is not yet understood; it may be that the axoplasmic reticulum is a plastic structure, itself stationary as a network of tubules, but capable of transformation into vesicles which can participate in transport.

SPECIFICITY OF AXONAL TRANSPORT

Since all nervous tissue develops from the epithelium of the neural crest, it is not surprising that the various types of differentiated neurons have many properties and molecular components in common. Even though the separations generally achieved were quite crude, comparisons by polyacrylamide gel electrophoresis in sodium dodecyl sulfate of proteins being transported along different axon branches of the same neurons [with the exception of a single study with cat dorsal root ganglion cells (Anderson & McClure 1973)] have failed to reveal any differences (frog dorsal root ganglion cells, Barker et al 1976, Barker et al 1977; rat dorsal root, Bisby 1977, White & White 1977; rat optic collaterals to lateral geniculate nucleii and superior colliculi, Siegel & McClure 1975; axon branches of a single identified *Aplysia* neuron, Schwartz et al 1976). Moreover, Barker et al 1975, and Barker et al 1976, have provided evidence that similar molecular weight classes of proteins are also transported along axons of a variety of different sensory, motor, and sympathetic axons. Despite this indication of gross similarity, it is obvious that different substances are transported by different neurons. Grafstein (1975, 1977) and Wilson & Stone (1979) have reviewed transport of specific molecular components.

Specific Anterograde Transport

In the anterograde direction, neurosecretory cells characteristically transport neurosecretory granules and their contents, first usually a large precursor molecule and later, as the result of proteolytic processing within the granule, the hormone itself (neurohypophysis: Norström & Sjöstrand 1971a, Gainer et al 1977; insect neurosecretory neurons: Gabe 1972; *Aplysia:* Loh & Gainer 1975, Berry & Schwartz 1977). Adrenergic neurons transport norepinephrine and dopamine β-hydroxylase (rabbit sciatic nerve: Brimijoin 1975, Brimijoin & Helland 1976, Brimijoin 1977, Brimijoin & Wiermaa 1977b; projections of the locus coeruleus in the rat: Jones et al 1977, Levin & Stolk 1977). Dopaminergic neurons in the zona compacta of the substantia nigra transport dopamine (Fibiger et al 1973, Fibiger & McGeer 1974, McGeer et al 1975); serotonergic neurons transport serotonin (identified serotonergic *Aplysia* neurons: Goldman & Schwartz 1974, Goldman et al 1976; projections of the midbrain raphe: Halaris et al 1976); and cholinergic neurons, acetylcholine (rat sciatic nerve: Heiwall et al 1976; R2, an identified cholinergic *Aplysia* neuron: Koike et al 1972, Treistman & Schwartz 1977; chemoreceptor neurons of insect antennae of the type later shown to be cholinergic: Schafer 1973). Transport within specific storage vesicles has been demonstrated for catecholamines (Dahlström 1971, Banks & Mayor 1972) and for serotonin (Howes et al 1974, Goldman et al 1976), but has not been shown convincingly for acetylcholine (Koike et al 1972, Eisenstadt & Schwartz 1975, Treistman & Schwartz 1977) or for glutamate (Roberts et al 1973). Radioautographic localization of the biogenic amines with the electron microscope is made possible because their primary amino group allows fixation in place with aldehyde fixatives. Acetylcholine has no useful functional groups, and the amino acid transmitters are general metabolites whose distribution within the cell might be difficult to interpret.

Selectivity of transport within a neuron is likely to depend not on the transmitter but on the specificity of the storage vesicle. In identified serotonergic *Aplysia* neurons, serotonin introduced by intrasomatic pressure injection is rapidly transported in vesicular form; ^{3}H-serotonin injected into an identified cholinergic neuron is neither packaged nor transported (Goldman & Schwartz 1974, Goldman et al 1976, Treistman & Schwartz 1977). Specificity of transport is not absolute, however. Goldberg & Schwartz (1978) have probed its extent in the giant serotonergic neuron of the *Aplysia* cerebral ganglion (GCN) by introducing a variety of substances by pressure injection. ^{3}H-dopamine, ^{3}H-d, 1-octopamine, and ^{3}H-histamine are rapidly transported, but ^{3}H-5-hydroxytryptophan, ^{3}H-gamma amino butyric acid, and the metabolite of serotonin (probably β-glucuronyl-O-serotonin, Goldman & Schwartz 1977) are not. ^{3}H-choline is totally converted to betaine and phosphorylcholine in the serotonergic neuron, and these metabolites are also not transported. In order to be transported, an analogue of serotonin apparently must have an aromatic moiety and bear a positive charge at intracellular values of pH. A similar degree of specificity has been observed in dopaminergic neurons of the rat substantia nigra. McGeer et al (1975) injected 17 different substances stereotaxically; in addition to dopamine, only serotonin, norepinephrine, and octopamine gave any evidence of specific accumulation in the ipsilateral striatum; the other substances (which included histamine) did not. Leger et al (1977), injected ^{3}H-serotonin into the neostriatum of the rat and found that radioactivity (which they characterized as serotonin by column chromatography) had been transported to the substantia nigra within dopaminergic fibers in the retrograde direction; they suggested the existence of a transport system for monoamines common to all aminergic neurons. The molecular basis for this common feature is likely to be similar recognition sites in the membranes of the various aminergic storage vesicles.

Specific Retrograde Transport

Selectivity of retrograde transport depends on selective uptake into the neuron; this in turn is determined by specific membrane binding sites. Uptake of HRP and other "fluid phase markers," although enhanced by synaptic activity, nevertheless occurs relatively slowly since the enzyme is merely trapped as a solute in the extracellular fluid included in the endocytotic vesicle. Much greater amounts of a substance can be taken up by endocytosis if that substance binds to the plasma membrane (for a review of endocytosis, see Silverstein et al 1977). The presence or absence of appropriate receptors in an active form in the membrane of an axon terminal is decisive in determining whether a given substance can gain access to the neuron's lysosomal system, and, consequently, whether it will be transported in the retrograde direction in significant amounts. The best analyzed example of absorptive pinocytosis are those of tetanus toxin and NGF. Tetanus toxin binds to the sialogangliosides G_{D1b} and G_{T1}, which apparently are available in the plasma membranes of all non-dividing neurons (Mirsky et al 1978), and is transported in all (motor, sensory, adrenergic, and central) neurons (Price et al 1975, Erdmann et al 1975, Stöckel, Schwab & Thoenen 1975, Neale & Dimpfel 1976, Schwab et al 1977). Cholera toxin, with high affinity for G_{M1}, is also transported in adrenergic sensory and somatic motor neurons, as is wheat germ agglutinin, a lectin with high affinity for N-acetylglucosamine residues (Stoeckel, Schwab & Thoenen 1977). Transport of ^{125}I-labeled tetanus toxin was partially inhibited by exposing the tissue to G_{T1}, and transport of cholera toxin by G_{M1}. Transport of the wheat germ agglutinin and NGF were unaffected by prior treatment with the gangliosides.

There is some evidence that suggest that HRP is not simply a fluid-phase marker (Bunt et al 1976). The A form of the enzyme (with an isoelectric point of 4.5) is not transported from visual cortex to the lateral geniculate nucleus, but the B and the C forms (both with values of pI around 9) are transported. There are also differences among the carbohydrate compositions of these glycoproteins. If the difference in transport indicates that the B and C forms specifically interact with the membrane, then the assumption that HRP is a fluid-phase marker is not entirely valid. If the A form is toxic, however, its transport might not be expected. Discrimination of the mode of access is likely to be critical to interpretation of experiments in which uptake is coupled with synaptic activity. Farquhar (1978) found the distribution of native ferritin markedly different from that of cationic ferritin in dissociated neurons of the anterior pituitary in culture. Native ferritin, a fluid-phase marker, is taken up in relatively small amounts into endocytotic vesicles and is localized only in lysosomes. In contrast, cationic ferritin binds to the cell surface, and is taken up in large amounts by vesicles. In addition to lysosomes, cationic ferritin gains access to Golgi cisternae, forming granules and the rigid lamellae corresponding to the elements of GERL.

NGF is transported selectively in adrenergic and sensory neurons (Hendry, Stach & Herrup 1974, Hendry et al 1974, Stoeckel, Schwab & Thoenen 1975), presumably because of specific receptors that exclude insulin, a molecule distantly related to NGF as well as cytochrome C, ovalbumin, ferritin, and HRP, all labeled with iodine-125, and used as control substances; no transport occurs if NGF is selectively modified by treatment with N-bromosuccinimide, which destroys its biological activity (Stoeckel, Paravicini & Thoenen 1974, Stoeckel & Thoenen 1975). Even though NGF can be transported to nerve cell bodies no longer capable of responding to its action (Stoeckel, Schwab & Thoenen 1975), it is attractive to think that transport plays an important role during development of the nervous system (Paravicini, Stoeckel & Thoenen 1975, Hendry 1977, Otten et al 1977).

Another example of selective transport is that of antibody to dopamine β-hydroxylase, which is taken up specifically in adrenergic central neurons (Jacobowitz et al 1975, Zeigler et al 1976) and peripheral neurons (Fillenz et al 1976). Because of its specificity, and because of the relatively large amounts of antibody taken up, entry into the neuron is likely to be mediated by absorptive pinocytosis. Formation of an enzyme-antibody complex in the terminal plasmalemma with subsequent endocytosis is a plausible explanation, since some dopamine β-hydroxylase from adrenergic vesicles is likely to be incorporated transiently in the external membrane after synaptic release. Retrograde transport of active dopamine β-hydroxylase is difficult to demonstrate (Brimijoin & Helland 1976), but transport of substantial amounts of enzymatically inactive molecules can be shown by immunofluorescence (Nagatsu et al 1976). Presumably antibody taken up exogenously gains access to this lysosomal system for membrane retrieval, which is normally in operation.

The pathogenesis of virus disease is beyond the scope of this review. Nevertheless the viral capsid is an important determinant of host range since infection is initiated by the adsorption of virus particles to the host cell's membrane. Neurotropic properties of viruses, expecially Herpes and rabies virus, may be explained by specific binding to axon terminal membranes with subsequent enhanced access to the retrograde transport system (Baringer 1975).

The Significance of Specificity

Absolute specificity of any one biological process is not enountered, nor would it necessarily be desirable. At the molecular level, greater specificity is achieved by tighter binding, and the processes of interest require associations that are reversible to some degree. Sufficient specificity can be achieved simply by building a pathway with a series of moderately specific steps. Two examples are pertinent to transport. The first involves the possible significance of multiple potential transmitter substances within a single neuron. A neuron may be exposed to small amounts of foreign transmitter substances or their precursors. Thus the concentration of foreign transmitters in some identified molluscan neurons is said to be approximately 50 times lower than that of the appropriate native transmitter (see Burnstock 1976). Although the presence of this much foreign transmitter is almost certainly experimental artifact (see, for example, Osborne 1977), even if accurate, it would not lead to a deviation from Dale's principle (that a neuron releases only one transmitter from all of its terminals) if a selective process, even one with a fairly low degree of specificity, intervened between the soluble alien transmitters and synaptic release. Additional specific processes would provide even greater orthodoxy. Just such steps are in fact known to exist: entry of substrate into the neuron by high affinity uptake, conversion to transmitter by biosynthetic enzymes with fair degrees of specificity, and transport by virtue of selective packaging. On the other hand, if two neuroactive substances should arise through processing native precursors within a single neuron's secretory granules—for example, two different peptide hormones from one common polypeptide precursor—this would be an exception proving the rule.

Neurotropic viruses serve as an example of selectivity of retrograde transport, where lysosomes are involved. These viruses can multiply in cells other than neurons. At the concentration of virus physiologically available in the infected animal, they are effectively excluded from cells that do not have appropriate surface receptors. In neurons, however, a chain of specific processes facilitates the advent of the virion at the pericaryon where it can proliferate.

CONCLUSIONS

Fast axonal transport encompasses a complex set of cellular activities, involves many organelles, and is the result of numerous molecular components. Because of features similar to other phenomena involving intracellular motility—for example, amoeboid movement, movement of the spindle apparatus during mitosis, and the movement of granules during secretion—it is likely that contractile proteins are involved and that axonal transport is a specialization of some universal cellular process. The structural integrity of the axon is obviously pertinent to transport; consequently cytoskeletal elements should play some role in the transport process. It is generally agreed that all components that are rapidly transported move by virtue of their association with membranes; thus synthesis, assembly, and subsequent processing of membranes must be implicated. Movement implies work, and understandably energy metabolism must be involved.

Even though it can be considered a common cellular pathway in all neurons that have the task of moving materials from one region to another over relatively large distances, fast transport nevertheless reflects the highly differentiated makeup of the particular cell in which it is taking place. As a result, one type of neuron will not transport the same materials moved by another. Thus, in the anterograde direction, adrenergic neurons transport norepinephrine specifically, not solely because the transmitter happens to be present in large amounts in the axoplasm, but because the adrenergic vesicle, which is only present in adrenergic neurons and concentrates the norepinephrine, is rapidly transported. Similarly dopamine β-hydroxylase is transported by virtue of its association with the adrenergic vesicle. These neurons will not transport acetylcholine or any other substance chemically unrelated to their proper neurotransmitter because the adrenergic vesicle will not take up acetylcholine. Similar specificity applies to retrograde transport. Specific viruses, toxins, and hormones are rapidly transported in the retrograde direction from nerve terminals of some neurons but not others. It seems likely that their transport is contingent on their uptake and subsequent incorporation into lysosomes in nerve terminals, which are then rapidly transported. Uptake into the neuron appears to involve binding to specific receptors in the terminal's membrane.

Thus specificity of transport can be thought of as residing in a set of steps, none of which are absolutely specific, but which together lead to neurons functioning with great individuality. Selectivity is determined by the specificity of several portals of access to a common cellular process; obvious examples are uptake of the transported substance into the neuron and uptake into specific vesicle carriers. Perhaps more important are the sites of entry to the transport process of the specific transported organelles themselves. The structural components of vesicles and other membranous organelles are synthesized in the reticular and Golgi systems of the pericaryon. Synthesis and assembly of membranes into organelles is known to be compartmentalized and directional, and it is likely that the last steps in assembly lead directly into the transport process.

The evidence reviewed permits us to identify subcellular organelles that participate in fast axonal transport with a reasonable degree of assurance; convergence of information from cell biology and from biochemistry allows the formulation of plausible relationships between these organelles and likely molecular components. Future experiments must securely identify the molecular species that play direct causal roles, not only those which operate in the force-generating mechanism, but also those that endow fast transport with its specific features.

ACKNOWLEDGEMENT

I am grateful to my colleagues and students for their many valuable comments and suggestions, and to a number of authors who kindly sent me reprints and information about their work prior to publication. I thank my student, Leonard J. Cleary, for library research on retrograde transport, Peter J. Schwartz for his help with the literature cited, and Helen M. Jordan for her unfailing secretarial help. Work in this laboratory is supported by the National Institutes of Health (NS 12066), the National Science Foundation (BNS 77-16505), and the Sloan Foundation.

Literature Cited

Abe, T., Haga, T., Kurokawa, M. 1973. Rapid transport of phosphotidylcholine occurring simultaneously with protein transport in the frog sciatic nerve. *Biochem. J.* 136:731–40

Ambron, R. T., Goldman, J. E., Schwartz, J. H. 1974. Axonal transport of newly-synthesized glycoproteins in a single identified neuron of *Aplysia californica*. *J. Cell Biol.* 61:655–75

Ambron, R. T., Goldman, J. E., Schwartz, J. H. 1975. Effect of inhibiting protein synthesis on axonal transport of membrane glycoproteins in an identified neuron of *Aplysia*. *Brain Res.* 94:307–23

Ambron, R. T., Schwartz, J. H. 1979. Regional aspects of neuronal glycoprotein synthesis. In *Complex Carbohydrates of Nervous Tissue*, ed. R. U. Margolis, R. K. Margolis. New York: Plenum. In press.

Anderson, J. D. 1973. Amoeboid movement. In *Comparative Animal Physiology*, ed. C. L. Prosser, pp. 799–808. Philadelphia: Saunders.

Anderson, L. L., McClure, W. O. 1973. Differential transport of protein in axons: comparison between the sciatic nerve and dorsal columns of cats. *Proc. Natl. Acad. Sci. USA* 70:1521–25

Banks, P. 1976. ATP hydrolase activity associated with microtubules reassembled from bovine splenic nerve—a cautionary tale. *J. Neurochem.* 27:1465–71

Banks, P., Mayor, D. 1972. Intra-axonal transport in noradrenergic neurons in the sympathetic nervous system. *Biochem. Soc. Symp.* 36:133–49

Banks, P., Mayor, D., Mraz, P. 1973. Metabolic aspects of the synthesis and intra-axonal transport of noradrenaline storage vesicles. *J. Physiol. London* 229:383–94

Banks, P., Till, R. 1975. A correlation between the effects of anti-mitotic drugs on microtubule assembly in vitro and the inhibition of axonal transport in noradrenergic neurones. *J. Physiol. London* 252:283–94

Bárány, M. 1967. ATPase activity of myosin correlated with speed of muscle shortening. *J. Gen. Physiol.* 50:197–218

Baringer, J. R. 1975. Herpes simplex virus infection of nervous tissue in animals and man. *Prog. Med. Virol.* 20:1–26

Barker, J. L., Hoffman, P. M., Gainer, H., Lasek, R. J. 1975. Rapid transport of proteins in the sonic motor system of the toadfish. *Brain Res.* 97:291–301

Barker, J. L., Neale, J. H., Bonner, W. M. 1977. Slab gel analysis of rapidly transported proteins in the isolated frog nervous system. *Brain Res.* 124:191–96

Barker, J. L., Neale, J. H., Gainer, H. 1976. Rapidly transported proteins in sensory, motor and sympathetic nerves of the isolated frog nervous system. *Brain Res.* 105:497–515

Barondes, S. H. 1974. Synaptic macromolecules: identification and metabolism, *Ann. Rev. Biochem.* 43:147–68

Benes, F., Higgins, J. A., Barrett, R. J. 1973. Ultrastructural localization of phospholipid synthesis in the rat trigeminal nerve during myelination. *J. Cell Biol.* 57:613–29

Bennett, G., DiGiamberardino, L., Koenig, H. L., Droz, B. 1973. Axonal migration of protein and glycoprotein to nerve endings. II. Radio-autographic analysis of the renewal of glycoproteins in nerve endings of chicken ciliary ganglion after intracerebral injection of [³H]fucose and [³]glucosamine. *Brain Res.* 60:129–46

Berry, R. W., Schwartz, A. W. 1977. Axonal transport and axonal processing of low molecular weight proteins from the abdominal ganglion of *Aplysia*. *Brain Res.* 129:75–90

Bhisey, A. N., Freed, J. J. 1971. Altered movement of endosomes in colchicine-treated cultured macrophages. *Exp. Cell Res.* 64:430–38

Bisby, M. A. 1977. Similar polypeptide composition of fast-transported proteins in rat motor and sensory axons. *J. Neurobiol.* 8:303–14

Bourguignon, L. Y. W., Singer, S. J. 1977. Transmembrane interactions and the mechanism of capping of surface receptors by their specific ligands. *Proc. Natl. Acad. Sci. USA* 74:5031–35

Bray, D. 1978. Membrane movements and microfilaments. *Nature* 273:265–66

Bretscher, M. 1976. Directed lipid flow in cell membranes. *Nature* 260:21–23

Breuer, A. C., Christian, C. N., Henkart, M., Nelson, P. G. 1975. Computer analysis of organelle translocation in primary neuronal cultures and continuous cell lines. *J. Cell Biol.* 65:562–76

Brimijoin, S. 1974. Local changes in subcellular distribution of dopamine-β-hydroxylase (EC 1.14.2.1) after blockage of axonal transport. *J. Neurochem.* 22:347–53

Brimijoin, S. 1975. Stop-Flow: a new technique for measuring axonal transport, and its application to the transport of dopamine-β-hydroxylase. *J. Neurobiol.* 6:379–94

Brimijoin, S. 1976. Cyclohexamide alters axonal transport and subcellular distribution of dopamine-β-hydroxylase. *J. Neurochem.* 26:35–40

Brimijoin, S. 1977. A histofluorescence study of events accompanying accumulation and migration of norepinephrine within locally cooled nerves. *J. Neurobiol.* 8:251–63

Brimijoin, S., Helland, L. 1976. Rapid retrograde transport of dopamine-β-hydroxylase as examined by the stop-flow technique. *Brain Res.* 102:217–28

Brimijoin, S., Wiermaa, M. J. 1977a. Rapid axonal transport of tyrosine hydroxylase in rabbit sciatic nerves. *Brain Res.* 121:77–96

Brimijoin S., Wiermaa, M. J. 1977b. Direct comparison of the rapid axonal transport of norepinephrine and dopamine-β-hydroxylase activity. *J. Neurobiol.* 8:239–50

Brown, A. V. W., Brown, G. L., Hopkins, J., Madden, F. T. 1977. The distribution of tubulin in the chief ganglia, connectives and giant cell of *Aplysia californica.* *Exp. Cell Res.* 107:63–70

Brownson, R. H., Uusitalo, R., Palkama, A. 1977. Intraaxonal transport of horseradish peroxidase in the sympathetic nervous system. *Brain Res.* 120:407–22

Bunt, A. H., Haschke, R. H., Lund, R. D., Calkins, D. F. 1976. Factors affecting retrograde axonal transport of horseradish peroxidase in the visual system. *Brain Res.* 102:152–55

Burnstock, G. 1976. Do some nerve cells release more than one transmitter? *Neuroscience* 1:239–48

Byers, M. R. 1974. Structural correlates of rapid axonal transport: evidence that microtubules may not be directly involved. *Brain Res.* 75:97–113

Cancalon, P., Beidler, L. M. 1975. Distribution along the axon and into various subcellular fractions of molecules labeled with (^{3}H)leucine and rapidly transported in the garfish olfactory nerve. *Brain Res.* 89:225–44

Cancalon, P., Elam, J. S., Beidler, L. M. 1976. SDS gel electrophoresis of rapidly transported proteins in garfish olfactory nerve. *J. Neurochem.* 27:687–93

Chang, C.-M., Goldman, R. D. 1973. The localization of actin-like fibers in cultured neuroblastoma cells as revealed by heavy meromyosin binding. *J. Cell Biol.* 57:867–74

Coleman, D. R., Scalia, F., Cabrales, E. 1976. Light and electron microscopic observations on the anterograde transport of horseradish peroxidase in the optic pathway in the mouse and rat. *Brain Res.* 102:156–63

Cooper, P. D., Smith R. S. 1974. The movement of optically detectable organelles in myelinated axons of *Xenopus Laevis.* *J. Physiol. London* 242:77–97

Cosens, B., Thacker, D., Brimijoin S. 1976. Temperature-dependence of rapid axonal transport in sympathetic nerves of the rabbit. *J. Neurobiol.* 7:339–54

Cowan, W. M., Cuénod, M. eds. 1975. *The Use of Axonal Transport for Studies of Neuronal Conductivity.* Amsterdam: Elsevier.

Dahlström, A. 1971. Axoplasmic transport (with particular respect to adrenergic neurons). *Philos. Trans. R. Soc. London Ser. B* 261:325–58

Dahlström, A., Heiwall, P.-O., Häggendal, Heiwall, P.-O., Saunders, N. R. 1974. Proximodistal transport of acetylcholine in peripheral cholinergic neurons. In *Dynamics of Degeneration and Growth in Neurons,* ed. K. Fuxe, L. Olson, Y. Zotterman, pp. 275–89. Oxford and New York: Pergamon.

Dahlström, A., Heiwall, P.-O., Häggendahl, J., Saunders, N. R. 1975a. Effect of antimitotic drugs on the intraaxonal transport of neurotransmitters in rat adrenergic and cholinergic nerves. *Ann. NY Acad. Sci.* 253:507–16

Dahlström, A., Heiwall, P. O., Larsson, P. A. 1975b. Comparison between the effect of colchicine and lumicolchicine on axonal transport in rat motor neurons. *J. Neural Transm.* 37:305–11

Dairman, W., Geffen, L., Marchelle, M. 1973. Axoplasmic transport of aromatic L-amino acid decarboxylase (EC 4.1.1.26) and dopamine-β-hydroxylase (EC 1.14.2.1) in rat sciatic nerve. *J. Neurochem.* 20:1617–23

Davies, L. P., Whittaker, V. P., Zimmermann, H. 1977. Axonal transport in the electromotor nerves of *Torpedo Marmorata.* *Exp. Brain Res.* 30:493–510

Di Giamberardino, L., Bennett, G., Koenig, H. L., Droz, B. 1973. Axonal migration of protein and glycoprotein to nerve endings. III. Cell fraction analysis of chicken ciliary ganglion after intracerebral injection of labeled precursors of proteins and glycoproteins. *Brain Res.* 60:147–59

Dravid, A. R., Hammerschlag, R. 1975. Axoplasmic transport of proteins in vitro in primary afferent neurons of frog spinal cord: effect of Ca^{++}-free incubation condition. *J. Neurochem.* 24:711–18

Droz, B. 1975. Synthetic machinery and axoplasmic transport: maintenance of neuronal connectivity. In *The Basic Neurosciences,* ed. D. B. Tower, vol. 1, pp. 111–27. New York: Raven.

Droz, B., Bruner, J., Gerschenfeld, H. 1974. Transport axonal de protéines marquées dans le motoneurone géant de l'écrevisse. *Biochimie* 56:1613–20

Droz, B., Koenig, H. L., Di Giamberardino, L. 1973. Axonal migration of protein and glycoprotein to nerve endings. I. Radioautographic analysis of the renewal of protein in nerve endings of chicken ciliary ganglion after intracerebral injection of [^{3}H]lysine. *Brain Res.* 60:93–127

Droz, B., Rambourg, A., Koenig, H. L. 1975. The smooth endoplasmic reticulum: structure and role in the renewal of axonal membrane and synaptic vesicles by fast axonal transport. *Brain Res.* 93: 1–13

Dustin, P., Hubert, J. P., Flament-Durand, J. 1975. Action of colchicine on axonal flow and pituicytes in the hypothalamopituitary system of the rat. *Ann. NY Acad. Sci.* 253:670–84

Edström, A. 1974. Effects of Ca^{++} and Mg^{++} on rapid axonal transport of proteins in vitro in frog sciatic nerves. *J. Cell Biol.* 61:812–18

Edström, A. 1975. Ionic requirements for rapid axonal transport in vitro in frog sciatic nerves. *Acta Physiol. Scand.* 93:104–12

Edström, A., Hanson, M. 1973. Temperature effects on fast axonal transport of proteins in vitro in frog sciatic nerves. *Brain Res.* 58:345–54

Edström, A., Mattsson, H. 1973. Electrophoretic characterization of leucine-, glucosamine- and fucose-labelled proteins rapidly transported in frog sciatic nerve. *J. Neurochem.* 21:1499–1507

Edström, A., Mattsson, H. 1975. Small amounts of zinc stimulate rapid axonal transport in vitro. *Brain Res.* 86:162–67

Edström, A., Mattsson, H. 1976. Inhibition and stimulation of rapid axonal transport in vitro by sulphydryl blockers. *Brain Res.* 108:381–95

Ehrlich, H. P., Ross, R., Bernstein, P. 1974. Effects of antimicrotubular agents on the secretion of collagen. A biochemical and morphological study. *J. Cell Biol.* 62:390–405

Eisenstadt, M. L., Schwartz, J. H. 1975. Metabolism of acetylcholine in the nervous system of *Aplysia californica.* III. Studies of an identified cholinergic neuron. *J. Gen. Physiol.* 65:293–313

Elam, J. S. 1975. Association of proteins undergoing slow axonal transport with goldfish visual system myelin. *Brain Res.* 97:303–15

Erdmann, G., Wiegand, H., Wellhoner, H. H. 1975. Intraaxonal and extraaxonal transport of ^{125}I-tetanus toxin in early local tetanus. *Naunyn-Schmiedebergs Arch. Pharmakol.* 290:357–73

Farquhar, M. G. 1978. Recovery of surface membrane in anterior pituitary cells. Variations in traffic detected with anionic and cationic ferritin. *J. Cell Biol.* 77:35–42

Fay, F. S., Cooke, P. H., Canaday, P. G. 1976. Contractile properties of isolated smooth muscle cells. In *Physiology of Smooth Muscle,* ed. E. Bülbring, M. F. Shuba, pp. 249–64. New York: Raven.

Fernandez, H. L., Burton, P. R., Samson, F. E. 1971. Axoplasmic transport in the crayfish nerve cord. The role of fibrillar constituents of neurons. *J. Cell Biol.* 51:176–92

Fernandez, H. L., Huneeus, F. C., Davison, P. F. 1970. Studies on the mechanism of axoplasmic transport in the crayfish cord. *J. Neurobiol.* 1:395–407

Fibiger, H. C., McGeer, E. G. 1974. Accumulation and axoplasmic transport of dopamine but not of amino acids by axons of the nigro-neostriatal projection. *Brain Res.* 72:366–69

Fibiger, H. C., McGeer, E. G., Atmadja S. 1973. Axoplasmic transport of dopamine in nigro-striatal neurons. *J. Neurochem.* 21:373–85

Fillenz, M., Gagnon, C., Stoeckel, K., Thoenen, H. 1976. Selective uptake and retrograde axonal transport of dopamine-β-hydroxylase antibodies in peripheral adrenergic neurons. *Brain Res.* 114:293–303

Fishman, P. H., Brady, R. O. 1976. Biosynthesis and function of gangliosides. *Science* 194:905–15

Flament-Durand, J., Couck, A. M., Dustin P. 1975. Studies on the transport of secretory granules in the magnocellular hypothalamic neurons of the rat. II. Action of vincristine on axonal flow and neurotubules in the paraventricular and supraoptic nuclei. *Cell Tissue Res.* 164:1–9

Flament-Durand, J., Dustin, P. 1972. Studies on the transport of secretory granules in the magnocellular hypothalamic neurons. I. Action of colchicine on axonal flow and neurotubules in the paraventricular nuclei. *Z. Zellforsch. Mikrosk. Anat.* 130:440–54

Fonnum, F., Frizell, M., Sjöstrand, J. 1976. Redistribution of choline acetyltransferase and acetylcholinesterase in an isolated nerve segment of the rabbit vagus nerve. *J. Neurochem.* 26:427–29

Forman, D. S., Grafstein, B., McEwen, B. S. 1972. Rapid axonal transport of [³H]fucosyl glycoproteins in the goldfish optic system. *Brain Res.* 48:327–42

Forman, D. S., Ledeen, R. W. 1972. Axonal transport of gangliosides in the goldfish optic nerve. *Science* 177:630–33

Forman, D. S., McEwen, B. S., Grafstein, B. 1971. Rapid transport of radioactivity in goldfish optic nerve following injections of labeled glucosamine. *Brain Res.* 28:119–30

Forman, D. S., Padjen, A. L., Siggins, G. R. 1977a. Axonal transport of organelles visualized by light microscopy: cinemicrographic and computer analysis. *Brain Res.* 136:197–213

Forman, D. S., Padjen, A. L., Siggins, G. R. 1977b. Effect of temperature on the rapid retrograde transport of microscopically visible intra-axonal organelles. *Brain Res.* 136:215–26

Friede, R. L., Ho, K. C. 1977. The relation of axonal transport of mitochondria with microtubules and other axoplasmic organelles. *J. Physiol. London* 265:507–19

Frizell, M., McLean, W. G., Sjöstrand, J. 1975. Slow axonal transport of proteins: blockage by interruption of contact between cell body and axon. *Brain Res.* 86:67–73

Frizell, M., Sjöstrand, J. 1974. The axonal transport of (³H)fucose labelled glycoproteins in normal and regenerating peripheral nerves. *Brain Res.* 78:109–23

Gabe, M. 1972. Histochemical data on the evolution of protocephalic neurosecretions in pterygot insects during axonal transport. *Acta Histochem.* 43:168–83

Gainer, H., Sarne, Y., Brownstein, M. J. 1977. Biosynthesis and axonal transport of rate neurohypophysial proteins and peptides. *J. Cell Biol.* 73:366–81

Gamache, F. W. Jr., Gamache, J. F. 1974. Changes in axonal transport in neurones of *Asterias Vulgaris* and *Asterias Forbesei* produced by colchicine and dimethyl sulfoxide. *Cell Tissue Res.* 152:423–35

Garcia, A. G., Kirpekar, S. M., Prat, J. C., Wakade, A. R. 1974. Metabolic and ionic requirements for the axoplasmic transport of dopamine β-hydroxylase. *J. Physiol. London* 241:809–21

Gilbert, D. S. 1975. Axoplasm architecture and physical properties as seen in the *Myxicola* giant axon. *J. Physiol. London* 253:257–301

Goldberg, D. J., Goldman, J. E., Schwartz, J. H. 1976. Alterations in amounts and rates of serotonin transported in an axon of the giant cerebral neuron of *Aplysia californica. J. Physiol. London* 259:473–90

Goldberg, D. J., Schwartz, J. H. 1978. Axonal transport of foreign transmitters in an identified neuron of *Aplysia. Neurosci. Abstr.* 4:194

Goldberg, D. J., Sherbany, A. A., Schwartz, J. H. 1978. Kinetic properties of normal and perturbed axonal transport of serotonin in a single identified axon. *J. Physiol. London* 281:559–79

Goldman, J. E., Kim, K. S., Schwartz, J. H. 1976. Axonal transport of ³H-serotonin in an identified neuron of *Aplysia californica. J. Cell Biol.* 70:304–18

Goldman, J. E., Schwartz, J. H. 1974. Cellular specificity of serotonin storage and axonal transport in identified neurons of *Aplysia californica. J. Physiol. London* 242:61–76

Goldman, J. E., Schwartz, J. H. 1977. Metabolism of ³H-serotonin in the marine mollusc, *Aplysia californica. Brain Res.* 136:77–88

Goldman, R., Pollard, T., Rosenbaum, J., eds. 1976. *Cell Motility.* 3 vols. Cold Spring Harbor, NY: Cold Spring Harbor Laboratory. 1373 pp

Gould, R. M. 1976. Inositol lipid synthesis in axons unmyelinated fibers of peripheral nerve. *Brain Res.* 117:169–74

Grafstein, B. 1967. Transport of protein by goldfish optic nerve fibers. *Science* 157:196–98

Grafstein, B. 1971. Transneuronal transfer of radioactivity in the central nervous system. *Science* 172:177–79

Grafstein, B. 1975. Principles of anterograde axonal transport in relation to studies of neuronal connectivity. In *The Use of Axonal Transport for Studies of Neuronal Connectivity,* ed. W. M. Cowan, M. Cuénod, pp. 47–67. Amsterdam: Elsevier

Grafstein, B. 1977. Axonal transport: the intracellular traffic of the neuron. *Handbook of Physiology, The Nervous System. I. Cellular Biology of Neurons,* ed. E. R. Kandel, pp. 691–717. Baltimore: Williams and Wilkins.

Grafstein, B., Miller, J. A., Ledeen, R. W., Haley, J., Specht, S. C. 1975. Axonal transport of phospholipid in goldfish optic system. *Exp. Neurol.* 46:261–81

Gross, G. W. 1975. The microstream concept of axoplasmic and dendritic transport. *Adv. Neurol.* 12:283–96

Gross, G. W., Beidler, L. M. 1973. Fast axonal transport in the C-fibers of the garfish olfactory nerve. *J. Neurobiol.* 4:413–28

Gross, G. W., Beidler, L. M. 1975. A quantitative analysis of isotope concentration profiles and rapid transport velocities in the C-fibers of the garfish olfactory nerve. *J. Neurobiol.* 6:213–32

Gross, G. W., Weiss, D. G. 1977. Subcellular fractionation of rapidly transported axonal material in olfactory nerve: evidence for a size-dependent molecule separation during transport. *Neurosci. Lett.* 5:15–20

Haak, R. A., Kleinhans, F. W., Ochs, S. 1976. The viscosity of mammalian nerve axoplasm measured by electron spin resonance. *J. Physiol. London* 263:115–37

Haggendal, J., Dahlström, A., Larsson, P. A. 1975. Rapid transport of noradrenaline in adrenergic axons of rat sciatic nerve distal to a crush. *Acta Physiol. Scand.* 94:386–92

Halaris, A. E., Jones, B. E., Mocre, R. Y. 1976. Axonal transport in serotonin neurons of the midbrain raphe. *Brain Res.* 107:555–74

Hammerschlag, R., Bakhit, C., Chiu, A. Y., Dravid, A. R. 1977. Role of calcium in the initiation of fast axonal transport of protein: effects of divalent cations. *J. Neurobiol.* 8:439–51

Hammerschlag, R., Chiu, A. Y., Dravid, A. R. 1976. Inhibition of fast axonal transport of (^{3}H)protein by cobalt ions. *Brain Res.* 114:353–58

Hammerschlag, R., Dravid, A. R., Chiu, A. Y. 1975. Mechanism of axonal transport: a proposed role for calcium ions. *Science* 188:273–75

Hammond, G. R., Smith, R. S. 1977. Inhibition of the rapid movement of optically detectable axonal particles colchicine and vinblastine. *Brain Res.* 128:227–42

Hamori, J., Pasik, T., Pasik, P. 1978. Electron-microscopic identification of axonal initial segments belonging to interneurons in the dorsal lateral geniculate nucleus of the monkey. *Neuroscience* 3:403–12

Hanson, M. 1978. A new method to study fast axonal transport in vivo. *Brain Res.* 153:121–26

Hanson, M., Edström, A. 1977. Fast axonal transport: effect of antimitotic drugs and inhibitors of energy metabolism on the rate and amount of transported protein in frog sciatic nerves. *J. Neurobiol.* 8:97–108

Hanson, M., Edström, A. 1978. Mitosis inhibitors and axonal transport. *Int. Rev. Cytol. Suppl.* 7:373–402

Hansson, H.-A. 1973. Uptake and intracellular bidirectional transport of horseradish peroxidase in retinal ganglion cells. *Exp. Eye Res.* 16:377–88

Hazel, J. R. 1973. The regulation of cellular function by temperature-induced alterations in membrane composition. *Effects of Temperature on Ectothermic Organisms,* ed. W. Wieser, pp. 55–67. New York, Heidelberg, Berlin: Springer.

Heacock, A. M. Agranoff, B. W. 1977. Reutilization of precursor following axonal transport of [^{3}H]proline-labeled protein. *Brain Res.* 122:243–54

Heiwall, P.-O., Saunders, N. R., Dahlström, A., Haggendal, J. 1976. The effect of local application of vinblastine or colchicine on acetylcholine accumulation rat sciatic nerve. *Acta Physiol. Scand.* 96:478–85

Hendrickson, A. E. 1972. Electron microscopic distribution of axoplasmic transport. *J. Comp. Neurol.* 144:381–97

Hendrickson, A. E., Cowan, W. M. 1971. Changes in the rate of axoplasmic transport during postnatal development of the rabbit's optic nerve tract. *Exp. Neurol.* 30:403–22

Hendry, I. A. 1977. The effect of the retrograde axonal transport of nerve growth factor on the morphology of adrenergic neurones. *Brain Res.* 134:213–23

Hendry, I. A., Stach, R., Herrup, K. 1974. Characteristics of the retrograde axonal transport system for nerve growth factor in the sympathetic nervous system. *Brain Res.* 82:117–28

Hendry, I. A., Stockel, K., Thoenen, H., Iversen, L. L. 1974. The retrograde axonal transport of nerve growth factor. *Brain Res.* 68:103–21

Heslop, J. P. 1975. Axonal flow and fast transport in nerves. *Adv. Comp. Physiol. Biochem.* 6:75–163

Heslop, J. P., Howes, E. A. 1972. Temperature and inhibitor effects on fast axonal transport in a molluscan nerve. *J. Neurochem.* 19:1709–16

Heuser, J. E., Reese, T. S. 1974. Morphology of synaptic vesicle discharge and reformation at the frog neuromuscular junction. In *Synaptic Transmission and Neuronal Interaction,* ed. M.V.L Bennett, pp. 59–77. New York: Raven.

Hindelang-Gertner, C., Stoekel, M.-E., Porte, A., Stutinsky, F. 1976. Colchicine effects on neurosecretory neurons and other hypothalamic and hypophyseal cells, with special reference to changes in the cytoplasmic membranes. *Cell Tissue Res.* 170:17–41

Hoffman, P. N., Lasek, R. J. 1975. The slow component of axonal transport. Identification of major structural polypeptides of the axon and their generality among mammlian neurons. *J. Cell Biol.* 66:351–66

Holmes, M. J., Turner, C. J., Fried, J. A., Cooper, E., Diamond, J. 1977. Neuronal transport in salamander nerves and its blockade by colchicine. *Brain Res.* 136:31–43

Holtzman, E. 1976. *Lysosomes: A Survey.* New York: Springer. 298 pp.

Holtzman, E. 1977. The origin and fate of secretory packages, especially synaptic vesicles. *Neuroscience* 2:327–55

Holtzman, E., Freeman, A. R., Kashner, L. A. 1971. Stimulation-dependent alterations in peroxidase uptake at lobster neuromuscular junctions. *Science* 173:733–36

Holtzman, E., Schacher, S., Evans, J., Teichberg, S. 1977. Origin and fate of the membranes of secretion granules and synaptic vesicles: membrane circulation in neurons, gland cells and retinal photoreceptors. In *The Synthesis, Assembly and Turnover of Cell Surface Components,* ed. G. Poste, G. L. Nicolson, pp. 165–246. Amsterdam: Elsevier/North Holland Biomedical Press

Howes, E. A., McLaughlin, B. J., Heslop, J. P. 1974. The autoradiographical association of fast transported material with dense core vesicles in the central nervous system of *Anodonta cygnea* (L.) *Cell Tissue Res.* 153:545–58

Hubel, D. H., Wiesel, T. N. 1977. Ferrier Lecture. Functional architecture of macaque monkey visual cortex. *Proc. R. Soc. London Ser. B* 198:1–59

Ishikawa, H., Bischoff, R., Holtzer, H. 1969. Formation of arrow head complexes with heavy meromyosin in a variety of cell types. *J. Cell Biol.* 43:312–28

Jacobowitz, D. M., Zeigler, M. G., Thomas, J. A. 1975. In vivo uptake of antibody to dopamine-β-hydroxylase into sympathetic elements. *Brain Res.* 91:165–70

Jarrott, B., Geffen, L. B. 1972. Rapid axoplasmic transport of tyrosine hydroxylase in relation to other cytoplasmic constituents. *Proc. Natl. Acad. Sci. USA.* 69:3440–42

Jones, B. E., Halaris, A. E., McIlhany, M., Moore, R. Y. 1977. Ascending projec-

tions of the *locus coeruleus* in the rat. I. Axonal transport in central noradrenaline neurons. *Brain Res.* 127:1–21

Kása, P., Mann, S. P., Karcsv, S., Tóth, L., Jordan, S. 1973. Transport of choline acetyltransferase and acetylcholinesterase in rat sciatic nerve: a biochemical and electron histochemical study. *J. Neurochem.* 21:431–36

Keefer, D. A. 1978. Horseradish peroxidase as a retrogradely-transported, detailed dendritic marker. *Brain Res.* 140:15–32

Kennedy, R. D., Fink, B. R., Byers, M. R. 1972. The effect of halothane on rapid axonal transport in the rabbit vagus. *Anesthesiology* 36:433–43

Kerkut, G. A. 1975. Axoplasmic transport. *Comp. Biochem. Physiol. A.* 51:701–4

Khan, M. A., Ochs, S. 1974. Magnesium or calcium activated ATPase in mammalian nerve. *Brain Res.* 81:413–26

Knull, H. R., Wells, W. W. 1975. Axonal transport of cations in the chick optic system. *Brain Res.* 100:121–24

Koda, L. Y., Partlow, L. M. 1976. Membrane marker movement on sympathetic axons in tissue culture. *J. Neurobiol.* 7:157–72

Koike, H., Eisenstadt, M., Schwartz, J. H. 1972. Axonal transport of newly synthesized acetylcholine in an identified neuron of *Aplysia. Brain Res.* 37:152–59

Komiya, Y., Kurokawa, M. 1978. Asymmetry of protein transport in two branches of bifurcating axons. *Brain Res.* 139:354–58

Korn, E. D. 1978. Biochemistry of actomyosin-dependent cell motility (A review). *Proc. Natl. Acad. Sci. USA* 75:588–99

Kreutzberg, G. W., Gross, G. W. 1977. General morphology and axonal ultrastructure of the olfactory nerve of the pike, *Esox lucius. Cell Tissue Res.* 181:443–57

Kristensson, K., Olsson, Y. 1976. Retrograde transport of horseradish peroxidase in transected axons. 3. Entry into injured axons and subsequent localization in perikaryon. *Brain Res.* 115:201–13

Kristensson, K., Olsson, Y., Sjöstrand, J. 1971. Axonal uptake and retrograde transport of exogenous proteins in the hypoglossal nerve. *Brain Res.* 32:399–406

Krygier-Brévart, V., Weiss, D. G., Mehl, E., Schubert, P., Kreutzberg, G. W. 1974. Maintenance of synaptic membranes by the fast axonal flow. *Brain Res.* 77:97–110

Lacy, P. E., Howell, S. L., Young, D. A.,

Fink, C. J. 1968. New hypothesis of insulin secretion. *Nature* 219:1177–79

Larabee, M. D., Brinley, F. J. Jr. 1968. Incorporation of labelled phosphate into phospholipids in squid giant axons. *J. Neurochem.* 15:533–45

Lasek, R. J., Black, M. M. 1977. How do axons stop growing? Some clues from the metabolism of the proteins in the slow component of axonal transport. In *Mechanisms, Regulation and Special Functions of Protein Synthesis in the Brain,* ed. S. Roberts, A. Lajtha, W. H. Gispen, pp. 161–69. Amsterdam: Elsevier/North Holland Biomedical Press

Lasek, R. J., Hoffman, P. N. 1976. The neuronal cytoskeleton, axonal transport and axonal growth. In *Cell Motility,* ed. R. Goldman, T. Pollard, J. Rosenbaum, pp. 1021–49. Cold Spring Harbor: Cold Spring Harbor Laboratory.

LaVail, J. H., LaVail, M. M. 1974. The retrograde intraaxonal transport of horseradish peroxidase in the chick visual system: a light and electron microscopic study. *J. Comp. Neurol.* 157:303–57

LaVail, M. M., LaVail, J. H. 1975. Retrograde intraaxonal transport of horseradish peroxidase in retinal ganglion cells of the chick. *Brain Res.* 85:273–80

Lavoie, P.-A., Hammerschlag, R., Tjan, A. 1978. Cobalt ions inhibit fast axonal transport of [³H]-glycoproteins but not glycosylation. *Brain Res.* 149:535–40

LeBeux, Y. J., Willemot, J. 1975. An ultrastructural study of the microfilaments in rat brain by means of heavy meromyosin labeling. I. The perikaryon, the dendrites and the axon. *Cell Tissue Res.* 160:1–36

Leestma, J. E., Freeman, S. S. 1977. Computer-assisted analysis of particulate axoplasmic flow in organized CNS tissue cultures. *J. Neurobiol.* 8:453–67

Leger, L., Pujol, J. F., Bobillier, P., Jouvet M. 1977. Retrograde axoplasmic transport of serotonin in central mono-aminergic neurons. *C. R. Acad. Sci. Ser. D.* 285:1179–82

LeMarchand, Y., Singh, A., Assimacopoulos-Jeannet, F., Orci, L., Rouiller, C., Jeanrenaud, B. 1973. A role for the microtubular system in the release of very low density lipoproteins by perfused mouse livers. *J. Biol. Chem.* 248: 6862–70

Lentz, T. L. 1972. Distribution of Leucine-³H during axoplasmic transport within regenerating neurons as determined by electron-microscope radioautography. *J. Cell Biol.* 52:719–32

Levin, B. E., Stolk, J. M. 1977. Axoplasmic transport of norepinephrine in the locus coeruleus-hypothalamic system in the rat. *Brain Res.* 120:303–15

Loh, Y. P., Gainer, H. 1975. Low molecular weight specific proteins in identified molluscan neurons. II. Processing, turnover and transport. *Brain Res.* 92:193–205

Lorenz, T., Willard, M. 1978. Subcellular fractionation of intra-axonally transport polypeptides in the rabbit visual system. 1978. *Proc. Natl. Acad. Sci. USA* 75:505–9

Lubinska, L. 1975. On axoplasmic flow. *Int. Rev. Neurobiol.* 17:241–96

Lubit, B. W., Sherbany, A. A., Schwartz, J. H. 1978. Immunohistochemical localization of actin in identified neurons of *Aplysia. Neurosci. Abstr.* 4:332

Lux, H. D., Schubert, P., Kreutzberg, G. W., Globus, A. 1970. Excitation and axonal flow: autoradiographic study on motoneurons intracellularly injected with a ³H-amino acid. *Exp. Brain Res.* 10:197–204

Marchisio, P. C., Cremo, F., Sjöstrand, J. 1975. Axonal transport in embryonic neurons. The possibility of a proximodistal axolemmal transfer of glycoproteins. *Brain Res.* 85:281–85

Markov, D., Rambourg, A., Droz, B. 1976. Smooth endoplasmic reticulum and fast axonal transport of glycoproteins, an electron microscope radioautographic study of thick sections after heavy metals impregnation. *J. Micros. Paris* 25:57–60

Matsumoto, T. 1920. The granules, vacuoles, and mitochondria in sympathetic nerve-fibers cultivated in vitro. *Johns Hopkins Hosp. Bull.* 31:91–93

Maupin-Szamier, P., Pollard, T. D. 1978. Actin filament destruction by osmium tetroxide. *J. Cell Biol.* 77:837–52

McEwen, B., Grafstein, B. 1968. Fast and slow components in axonal transport of protein. *J. Cell Biol.* 38:494–508

McGeer, E. G., Searl, K., Fibiger, H. C. 1975. Chemical specificity of dopamine transport in the nigro-neostriatal projection. *J. Neurochem.* 24:283–88

McGregor, A., Jeffrey, P. L., Klingman, J. D., Austin, L. 1973. Axoplasmic flow of cholesterol in chicken sciatic nerve. *Brain Res.* 63:466–69

McLean, W. G., Frizell, M., Sjöstrand, J. 1976. Slow axonal transport of labeled proteins in sensory fibers of rabbit vagus nerve. *J. Neurochem.* 26:1213–16

Mirsky, R., Wendon, L. M. B., Black, P., Stolkin, C., Bray, D. 1978. Tetanus toxin: a cell surface marker for neurones in culture. *Brain Res.* 148:251–59

Morré, D. J. 1977. The Golgi apparatus and membrane biogenesis. In *The Synthesis, Assembly and Turnover of Cell Surface Components,* ed. G. Poste, G. L. Nicolson, pp. 1–83. Amsterdam: Elsevier North Holland Biomedical Press

Moskalewski, S., Thyberg, J., Friberg, V. 1976a. In vitro influence of colchicine on the Golgi complex in A- and B- cells of guinea pig pancreatic islets. *J. Ultrastruct. Res.* 54:304–17

Moskalewski, S., Thyberg, J., Lohmander, S., Friberg, V. 1976b. Influence of colchicine and vinblastin on the Golgi complex and matrix deposition in chondrocyte aggregation. *Exp. Cell Res.* 95: 440–54

Nagatsu, I., Kondo, Y., Kato, T., Nagatsu, I. 1976. Retrograde axoplasmic transport of inactive dopamine β-hydroxylase in sciatic nerves. *Brain Res.* 116:277–85

Nauta, H. J. W., Kaiserman-Abramof, I. R., Lasek, R. J. 1975. Electron microscopic observations of horseradish peroxidase transported from the caudoputamen to the substantia nigra in the rat: possible involvement of the agranular reticulum. *Brain Res.* 85:373–84

Neale, J. H., Barker, J. L. 1977. Bidirectional axonal transport of $^{45}Ca^{++}$; studies in isolated frog sensory, motor and sympathetic neurons, *Aplysia* cerebral ganglion and the goldfish visual system. *Brain Res.* 129:45–59

Neale, J. H., Dimpfel, W. 1976. Movement of labeled macromolecules to the goldfish optic tectum following intraocular injection of ^{125}I-labelled tetanus toxin. *Exp. Neurol.* 53:355–62

Needham, D. M. 1971. *Machina Carnis.* Cambridge Univ. Press. 782 pp.

Norström, A. 1975. Axonal transport and turnover of neurohypophyseal proteins in the rat. *Ann. NY Acad. Sci.* 248: 46–63

Norström, A., Sjöstrand, J. 1971a. Axonal transport of proteins in the hypothalamo-neurohypophysial system of the rat. *J. Neurochem.* 18:29–39

Norström, A., Sjöstrand, J. 1971b. Effect of haemorrhage on the rapid axonal transport of neurohypophyseal proteins of the rat. *J. Neurochem.* 18:2017–26

Norström, A., Sjöstrand, J. 1972. Effect of suckling and parturition on axonal transport and turnover of neurohypophysial proteins of the rat. *J. Endocrinol.* 52:107–17

Ochs, S. 1972a. Fast transport of materials in mammalian nerve fibers. *Science* 176: 252–60

Ochs, S. 1972b. Rate of fast axoplasmic transport in mammalian nerve fibers. *J. Physiol. London* 227:627–45

Ochs, S. 1975a. Retention and redistribution of proteins in mammalian nerve fibres by axoplasmic transport. *J. Physiol. London* 253:459–75

Ochs. 1975b. Axoplasmic transport. In *The Nervous System, v. I: The Basic Neurosciences,* ed. D. B. Tower, pp. 137–46. New York: Raven.

Ochs, S., Smith, C. 1975. Low temperature slowing and cold-block of fast axoplasmic transport in mammalian nerves in vitro. *J. Neurobiol.* 6:85–102

Ochs, S., Worth, R. M., Chan, S. Y. 1977. Calcium requirement for axoplasmic transport in mammalian nerve. *Nature* 270:748–50

Odell, G. M. 1976. A new mathematical continuum theory of axoplasmic transport. *J. Theor. Biol.* 60:223–27

Oesch, F., Otten, U., Thoenen, H. 1973. Relationship between the rate of axoplasmic transport and subcellular distribution of enzymes involved in the synthesis of norepinephrine. *J. Neurochem.* 20:1691–1706

Osborne, N. N. 1977. Do snail neurones contain more than one neurotransmitter? *Nature* 270:622–23

Otten, U., Schwab, M., Gagnon, C., Thoenen, H. 1977. Selective induction of tyrosine hydroxylase and dopamine β-hydroxylase by nerve growth factor. Comparison between adrenal medulla and sympathetic ganglia of adult and newborn rats. *Brain Res.* 133:291–303

Palade, G. 1975. Intracellular aspects of the process of protein synthesis. *Science* 189:347–58

Paravicini, U., Stoeckel, K., Thoenen, H. 1975. Biological importance of retrograde axonal transport of nerve growth factor in adrenergic neurons. *Brain Res.* 84:279–91

Partlow, L. M., Ross, C. D., Motwani, R., McDougal, D. B. Jr. 1972. Transport of axonal enzymes in surviving segments of frog sciatic nerve. *J. Gen. Physiol.* 60:388–405

Patzelt, C., Brown, D., Jeanrenaud, B. 1977. Inhibitory effect of colchicine on amylase secretion by rat parotid gland. *J. Cell Biol.* 73:578–93

Paulson, J. C., McClure, W. O. 1975. Microtubules and axoplasmic transport. Inhibition of transport by podophyllotoxin: an interaction with microtubular protein. *J. Cell Biol.* 67:461–67

Pitman, R. M., Tweedle, C. D., Cohen, M. J. 1972. Branching of central neurons: intracellular cobalt injection for light and electron microscopy. *Science* 176: 412–14

Price, D. L., Griffin, J., Young, A., Peck, K., Stocks, A. 1975. Tetanus toxin: direct evidence for retrograde intraaxonal transport. *Science* 188:945–47

Ranish, N., Ochs, S. 1972. Fast axoplasmic transport of acetylcholinesterase in mammalian nerve fibres. *J. Neurochem.* 19:2641–49

Redman, C. M., Banerjee, D., Howell, K., Palade, G. E. 1975. Colchicine inhibition of plasma protein release from rat hepatocytes. *J. Cell Biol.* 66:42–59

Repérant, J. 1975. The orthograde transport of horseradish peroxidase in the visual system. *Brain Res.* 85:307–12

Roberts, P. J., Keen, P., Mitchell, J. F. 1973. The distribution and axonal transport of free amino acids and related compounds in the dorsal sensory neuron of the rat, as determined by the dansyl reaction. *J. Neurochem.* 21:199–209

Rösner, H., Weigandt, H., Rahmann, H. 1973. Sialic acid incorporation into gangliosides and glycoproteins of the fish brain. *J. Neurochem.* 21:655–65

Sabri, M. I., Ochs, S. 1973. Characterization of fast and slow transported proteins in dorsal root and sciatic nerve of cat. *J. Neurobiol.* 4:145–65

Samson, F. E. Jr. 1971. Mechanism of axoplasmic transport. *J. Neurobiol.* 2: 347–60

Samson, F. E. 1976. Pharmacology of drugs that affect intracellular movement. *Ann. Rev. Pharmacol. Toxicol.* 16:143–59

Saunders, N. R., Dziegielewska, K., Haggendal, C. J., Dahlström, A. B. 1973. Slow accumulation of choline acetyltransferase in crushed sciatic nerves of the rat. *J. Neurobiol.* 4:95–103

Schafer, R. 1973. Acetylcholine: fast axoplasmic transport in insect chemoreceptor fibers. *Science* 180:315–17

Schmidt, R. E., McDougal, D. B. Jr. 1978. Axonal transport of selected particle-specific enzymes in rat sciatic nerve in vivo and its response to injury. *J. Neurochem.* 30:527–35

Schmidt, R. E., Ross, C. D., McDougal, D. B. Jr. 1978. Effects of sympathectomy on axoplasmic transport of selected enzymes including MAO and other mitochondrial enzymes. *J. Neurochem.* 30: 537–41

Schmitt, F. O. 1968. Fibrous proteins—neuronal organelles. *Proc. Natl. Acad. Sci. USA* 60:1092–1101

Schwab, M. E. 1977. Ultrastructural localization of nerve growth factor-horseradish peroxidase (NGF-HVP) coupling product after retrograde axonal transport in adrenergic neurons. *Brain Res.* 130: 190–96

Schwab, M., Agid, Y., Glowinski, J., Thoenen, H. 1977. Retrograde axonal transport of ^{125}I tetanus toxin as a tool for tracing fiber connections in the central nervous system: connections of the rostral part of the rat neostriatum. *Brain Res.* 126:211–24

Schwab, M., Thoenen, H. 1977. Selective trans-synaptic migration of tetanus toxin after retrograde axonal transport in peripheral sympathetic nerves: a comparison with nerve growth factor. *Brain Res.* 122:459–74

Schwab, M. E., Thoenen, H. 1978. Selective binding, uptake, and retrograde transport of tetanus toxin by nerve terminals in the rat iris. An electron microscopic study using colloidal gold as a tracer. *J. Cell Biol.* 77:1–13

Schwartz, J. H., Goldman, J. E., Ambron, R. T., Goldberg, D. J. 1976. Axonal transport of vesicles carrying ^3H-serotonin in the metacerebral neuron of *Aplysia californica*. *Cold Spring Harbor Symp. Quant. Biol.* 40:83–92

Seybold, J., Bieger, W., Kern, H. F. 1975. Studies on intracellular transport of secretory proteins in the rat exocrine pancreas. II. Inhibition by antimicrotubular agents. *Virchows Arch. A* 368:309–27

Sherbany, A. A., Ambron, R. T., Schwartz, J. H. 1979. Membrane glycolipids: regional synthesis and axonal transport in a single identified neuron of *Aplysia californica*. *Science*. In press

Sherlock, D. A., Field, P. M., Raisman, G. 1975. Retrograde transport of horseradish peroxidase in the magnocellular neurosecretory system of the rat. *Brain Res.* 88:403–14

Siegel, L. G., McClure, W. O. 1975. Fractionation of protein carried by axoplasmic transport. II. Comparison in the rat of proteins carried to the optic relay nuclei. *Neurobiology* 5:167–77

Siekevitz, P. 1972. Biological membranes: the dynamics of their organization. *Ann. Rev. Physiol.* 34:117–40

Silverstein, S. C., Steinman, R. M., Cohn, Z. A. 1977. Endocytosis. *Ann. Rev. Biochem.* 46:669–722

Smith, D. C., Järlfors, U., Cameron, B. F. 1975. Morphological evidence for the participation of microtubules in axonal transport. *Ann. NY Acad. Sci.* 253:470–506

Smith, R. S., Koles, Z. J. 1976. Mean velocity of optically detected intraaxonal particles measured by a cross-correlation method. *Can. J. Physiol. Pharmacol.* 54:859–69

Snell, W. J., Dentler, W. L., Haimo, L. T., Binder, L. I., Rosenbaum, J. L. 1974. Assembly of chick brain tubulin onto isolated basal bodies of *Chlamydomonas reinhardi. Science* 185:357–60

Snyder, R. E., Reynolds, R. A., Smith, R. S., Kandal, W. S. 1976. Application of a multiwire proportional chamber to the detection of axoplasmic transport. *Can. J. Physiol. Pharmacol.* 54:238–44

Stein, O., Sanger, L., Stein, Y. 1974. Colchicine-induced inhibition of lipoprotein and protein secretion into the serum and lack of interference with secretion of biliary phospholipids and cholesterol by rat liver in vivo. *J. Cell Biol.* 62:90–103

Stöckel, K., Paravicini, U., Thoenen, H. 1974. Specificity of the retrograde axonal transport of nerve growth factor. *Brain Res.* 76:413–21

Stöckel, K., Schwab, M., Thoenen, H. 1975. Comparison between the retrograde axonal transport of nerve growth factor and tetanus toxin in motor, sensory and adrenergic neurons. *Brain Res.* 99:1–16

Stoeckel, K., Schwab, M., Thoenen, H. 1975. Specificity of retrograde transport of nerve growth factor (NGF) in sensory neurons: a biochemical and morphological study. *Brain Res.* 89:1–14

Stoeckel, K., Schwab, M., Thoenen, H. 1977. Role of gangliosides in the uptake and retrograde axonal transport of cholera and tetanus toxin as compared to nerve growth factor and wheat germ agglutinin. *Brain Res.* 132:273–85

Stoeckel, K., Thoenen, H. 1975. Retrograde axonal transport of nerve growth factor: specificity and biological importance. *Brain Res.* 85:337–41

Takenaka, T., Horie, H., Sugita, T. 1978. New technique for measuring dynamic axonal transport and its application to temperature effects. *J. Neurobiol.* 9:317–24

Thompson, E. B., Schwartz, J. H., Kandel, E. R. 1976. A radioautographic analysis in the light and electron microscope of identified *Aplysia* neurons and their processes after intrasomatic injection of L-^{3}H-fucose. *Brain Res.* 112:251–81

Till, R., Banks, P. 1976. Pharmacological and ultrastructural studies on the electron dense cores of the vesicles that accumulate in noradrenergic axons constricted in vitro. *Neuroscience* 1:49–55

Treistman, S. N., Schwartz, J. H. 1977. Metabolism of acetylcholine in the nervous system of *Aplysia californica*. IV. Studies of an indentified cholinergic axon. *J. Gen. Physiol.* 69:725–41

Trifaro, J. M. 1978. Contractile proteins in tissues originating in the neural crest. *Neuroscience* 3:1–24

Trifaro, J. M., Collier, B., Lastowecka, A., Stern, D. 1972. Inhibition by colchicine and by vinblastin of acetylcholine-induced catecholamine release from the adrenal gland: an anticholinergic action, not an effect upon microtubules. *Mol. Pharmacol.* 8:264–67

Tuček, S. 1975. Transport of choline acetyltransferase and acetylcholinesterase in the central stump and isolated segments of a peripheral nerve. *Brain Res.* 86:259–70

Unsicker, K., Drenckhahn, D., Gröschel-Stewart, J., Schumacher, V., Griesser, G. H. 1978. Immunohistochemical evidence of myosin in peripheral nerves and spinal cord of the rat. *Neuroscience* 3:301–6

Wehland, J., Osborn, M., Weber, K. 1977. Phalloidin-induced actin polymerization in the cytoplasm of cultured cells interferes with cell locomotion and growth. *Proc. Natl. Acad. Sci. USA* 74:561–17

Weiss, P. A. 1970. Neuronal dynamics and neuroplasmic flow. In *The Neurosciences, Second Study Program*, ed. F. O. Schmitt, G. C. Quarton, T. Melnechuk, pp. 840–50. New York: The Rockefeller Univ. Press

Weiss, P. A. 1972. Neuronal dynamics and axonal flow. V. The semisolid state of the moving axonal column. *Proc. Natl. Acad. Sci. USA* 69:620–23

Weiss, P., Hiscoe, H. 1948. Experiments on the mechanism of nerve growth. *J. Exp. Zool.* 107:315–95

Weiss, D. G., Krygier-Brévart, V., Gross, G. W., Kreutzberg, G. W. 1978. Rapid axoplasmic transport in the olfactory nerve of the pike: II. Analysis of transported proteins by SDS gel electrophoresis. *Brain Res.* 139:77–87

White, F. P., White, S. R. 1977. Characterization of proteins transported at different rates by axoplasmic flow in the dorsal root afferents of rats. *J. Neurobiol.* 8:315–24

Willard, M. 1977. The identification of two intra-axonally transported polypeptides resembling myosin in some respects in the rabbit visual system. *J. Cell Biol.* 75:1–11

Willard, M., Cowan, W. M., Vagelos, P. R. 1974. The polypeptide composition of intra-axonally transported proteins: Evidence for four transport velocities. *Proc. Natl. Acad. Sci. USA.* 71:2183–87

Wilson, D. L., Stone, G. C. 1979. Axoplasmic Transport of Proteins. *Ann. Rev. Biophys. Bioeng.* 8:27–45

Wooten, G. F., Coyle, J. T. 1973. Axonal transport of catecholamine synthesizing and metabolizing enzymes. *J. Neurochem.* 20:1361–71

Yukitake, Y., Taniguchi, Y., Kurosumi, K. 1977. Ultrastructural studies on the secretory cycle of the neurosecretory cells and the formation of herring bodies in the paraventricular nucleus of the rat. *Cell Tissue Res.* 177:1–8

Zeigler, M. G., Thomas, J. A., Jacobowitz, D. M. 1976. Retrograde axonal transport of antibody to dopamine-β-hydroxylase. *Brain Res.* 104:390–95

Ann. Rev. Neurosci. 1979. 2:505–18

THE BIOLOGY OF AFFECTIVE DISORDERS

♦11532

Edward J. Sachar[1] and Miron Baron

Department of Psychiatry, Columbia University College of Physicians &
Surgeons, and the New York State Psychiatric Institute, New York, N.Y. 10032

INTRODUCTION

Biological research on depression was hindered for many years by a lack
of precision and uniformity among researchers in defining clinical depressive syndromes. As a prominent Viennese investigator pointed out in 1917
"Even in descriptive psychiatry the definition of melancholia is uncertain;
it takes on various clinical forms—some of them suggesting somatic rather
than psychogenic affections—that do not seem definitely to warrant reduction to a unity" (Freud 1917). In the past decade, however, rigorous and
reliable systems for clinically classifying affective disorders have been developed (Katz & Hirschfeld 1978, Feighner et al 1972, Spitzer, Endicott &
Robins 1975), and from the welter of conditions involving unhappiness,
misery, grief, disappointment, despair, etc have emerged certain depressive
syndromes that seem clearly to be "somatic affections."

This review will focus on two of these syndromes—the major depressive
illnesses of the unipolar (recurrent depressions) and bipolar (manic depressive) types. These disorders have been labelled by various terms in the past
by different investigators—e.g. "endogenous," "psychotic," "primary,"
"manic-depressive," etc. In the forthcoming revision of the American Psychiatric Association's Diagnostic and Statistical Manual for Mental Disorders (DSM III), they will be termed Major Affective Disorders.

[1]Supported in part by U.S.P.H.S. grant MH-29841.

0147-006X/79/0315-0505$01.00

CLINICAL FEATURES OF MAJOR AFFECTIVE DISORDERS

Unipolar Depressions

The clinical features of this type of depression can be briefly summarized. The average age of onset of the first episode is about 40 and women are affected 2–3 times more than men. The episode of depression is characterized by a pervasive unhappy mood, and/or a generalized loss of interests and of the pleasure response. The lack of pleasure includes both the ability to anticipate pleasure and to experience pleasure. In addition, the diagnosis requires several of the following symptoms: disturbed sleep (usually insomnia), diminished appetite, loss of energy, decreased sex drive, psychomotor agitation (restlessness) or retardation (slowing down of thoughts and actions), difficulty in concentration, and thoughts of self-reproach, suicide, and pessimism that may reach delusional proportions. Decreased salivation and constipation are also common features. The diagnosis also involves exclusion criteria—no signs of schizophrenia, organic brain disease, etc. The untreated episode lasts an average of 7 months (Angst 1978).

When the syndrome is carefully defined in this manner, a number of interesting features of the disorder come into focus, which suggest a biological factor: (*a*) A number of the characteristic clinical features suggest hypothalamic dysfunction: disturbances of mood, appetite, sexual drive, sleep, and autonomic activity. (*b*) In at least 60% of the episodes, no significant psychosocial precipitant can be found (Endicott & Spitzer 1978). Even in those in which a psychological precipitant is found, the condition, once established, typically appears autonomous i.e. unresponsive to psychotherapy or environmental manipulation, while somatic treatments are of proven effectiveness. (*c*) The disturbance is recurrent. Angst (1978), for example, followed 159 patients for an average of 19 years, and found that the median number of depressive episodes over this period was 4.

Bipolar (Manic Depressive) Illness

Patients with this disorder suffer both depressive and manic episodes. Men and women are approximately equally affected, and the age of onset is about a decade younger than unipolar depression. The depressive episodes are very similar clinically to those seen in unipolar depressive illness, with a few minor statistical differences. Hypersomnia is more likely to be seen in bipolar depression than in unipolar depression, while psychomotor agitation is relatively uncommon in bipolar depression.

The cardinal symptoms of a manic episode are quite striking, and may seem to be at the opposite pole from the depressive episode. The diagnosis requires an elevated, expansive or irritable mood lasting at least a week. In addition, several of the following symptoms must be present: overactivity, overtalkativeness, increased energy, pressure of ideas and thoughts, grandiosity, distractibility, decreased need for sleep, reckless involvements (buying sprees, sexual indiscretions, foolish investments). In severe cases, the patients are delusional.

With careful clinical definition, once again, several features of the illness suggest biological factors: (a) The syndrome of mania also suggests hypothalamic involvement, with disturbances in mood, energy, appetite, sleep, and sexual function. (b) Significant psychosocial precipitants cannot be found in the majority of episodes. (c) It is a recurrent illness, even more so than unipolar disease. In Angst's series of 95 cases observed for an average of 23 years, the median number of episodes of mania and depression combined was 9, or about twice the frequency of unipolar episodes (Angst 1978).

GENETICS OF BIPOLAR AND UNIPOLAR AFFECTIVE DISORDERS

A genetic factor in affective disorders is suggested by the numerous studies showing high rates of the illness in first-degree relatives (parents, sibs, and children) of patients with affective illness, as compared with the general population (Gershon et al 1976). As an even more crucial test of different genetic contributions, there is a consistently higher concordance rate (both suffering the illness) for affective disorder in monozygotic (MZ) than in dizygotic (DZ) twins. The overall concordance rate for MZ twin pairs is 69%; the rate for DZ twins is 13% (Gershon et al 1976).

Of course, families and twins share common environments as well as genes, but adoption studies offer a method of separating biological from rearing influences. Higher rates of affective illness in the biological parents of ill adoptees than in their adoptive parents or parents of control adoptees have recently been reported (Mendlewicz & Rainer 1977). Furthermore, MZ twins reared apart demonstrate a concordance rate similar to that of MZ twins reared together (Price 1968). The presence of many discordant MZ twin pairs indicates, however, that nongenetic contributing factors may influence the vulnerability to affective disorder. Nevertheless, it can be considered established that genetic factors play a significant role in these disorders. Current research has focused on the precise nature of the genetic transmission, which remains obscure.

More specific genetic strategies have included attempts to fit classical models of inheritance to data on patterns of illness in families. A single recessive gene model was rejected. A single dominant autosomal gene with low penetrance (incomplete expression of the genotype due to other genetic and/or nongenetic influences) and a polygenic (multiple genes) model have been proposed, but were found to be indistinguishable from each other (Gershon et al 1976). Thus, affective illness appeared unlike the classic Mendelian inherited genetic disorders. Furthermore, a genetic model would need to take into account the possibility that affective disorders may have biologic and genetic heterogeneity even though clinically they may appear similar. For example, it would be necessary to determine whether unipolar and bipolar illness were two separate illnesses or mild and severe forms of the same genetic illness, and whether unipolar depressions might involve more than one genetic entity.

Strategies that can be devised to solve these problems include: (a) trying to sort out homogeneous subsets from among the affective disorders on the basis of clinical, pharmacologic, or biologic characteristics; (b) formulation of genetic models for traits transmitted without clear Mendelian ratios, in which individuals nevertheless can be termed affected or unaffected (all-or-none traits), and which have multiple phenotypic (clinical) manifestations: the multiple threshold model; (c) studying linkage (the tendency of genes on the same chromosome to segregate together), to known chromosomal markers. Data and hypotheses relevant to these strategies are discussed below.

Bipolar (BP)—Unipolar (UP) Dichotomy

Leonhard's hypothesis (Leonhard, Korf & Schulz 1962) that bipolar (episodes of mania and depression) and unipolar (recurrent depression) affective disorders are distinct entities has had a major impact on the applicability of the strategies described to genetic research in affective disorders. A higher probability of illness in relatives of bipolar individuals than in relatives of unipolars, and high concordance rate for clinical polarity (i.e. bipolar vs. unipolar illness) between the ill index subjects and their co-twins or other first-degree relatives (Gershon et al 1976) seemed at first to support the notion that these were two distinct illnesses genetically. However, numerous studies recently reviewed (Gershon 1978) of attempts to find biological differences between bipolar and unipolar patients have consistently failed to produce replicable results. Among the biological measures studied have been platelet monoamine oxidase (MAO) and erythrocyte catechol-o-methyltransferase (COMT), spinal fluid 5-hydroxyindoleacetic acid (HIAA) and homovanillic acid (HVA), and urinary excretion of methoxyhydroxyphenylglycol (MHPG), visual evoked EEG responses, cortisol secretion, prophylactic response to lithium carbonate, and erythrocyte/plasma lithium ratio. Only the increased probability of bipolar depressions to respond to lithium, and to switch into mania with L-dopa or antidepressant treatment, seems to distinguish this group. Since distinguishing biologic and pharmacologic characteristics are generally considered a likely corollary to genetic traits, the hypothesis that the *clinical* distinction between unipolar and bipolar reflects an underlying *genetic* distinction cannot be validly deduced from these data. This conclusion is consistent with other clinical genetic data, as discussed below, and has led to a multiple-threshold model of inheritance.

Multiple-Threshold Model of Inheritance

Viewing unipolar and bipolar depressive illness as mild and severe forms of the same inherited disorder permits application of a newly developed multiple threshold genetic model. Threshold models of vulnerability to illness have been used to determine modes of transmission of semicontinuous (all-or-none) traits that are related to an underlying graded vulnerability to illness. In these models it is assumed that there is an underlying graded attribute related to the etiology of the disorder and referred to as an individual's vulnerability or liability. The liability curve is taken to have a normal distribution and the point beyond which all individuals are expected to manifest the disorder is termed the threshold. The curve of distribution for relatives of affected persons has a similar form as the general population distribution, but is characterized by a higher mean liability (see Figure 1).

The development of this genetic continuum model has coincided with the development of the spectrum model in clinical psychiatry, which suggests that for certain mental illnesses such as schizophrenia and affective disorders there is a spectrum ranging from mild to severe forms. Recently, models involving multiple thresholds, which are suitable for application to psychiatric spectrum disorders, have been developed (Reich, James & Mor-

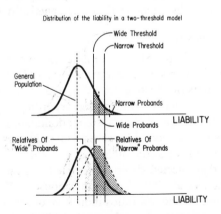

Distribution of the liability in a two-threshold model

Figure 1   Schematic representation of genetic vulnerability (liability) to an illness in a two threshold model. "Wide" refers to the milder form and "narrow" to the severe form of illness. It is proposed that unipolar depressions are the wide form, and bipolar (manic-depressive) illness the narrow form of a single disorder.

ris 1972). Using these models, ill individuals are subdivided into "narrow" and "wide" categories. The narrow (more extreme or severe) form of the illness has a higher liability threshold and is more deviant from the general population mean than the other. Narrow and wider illness forms represent points on a genetic continuum. The two-threshold hypothesis would predict differing patterns of manifestation of the illnesses in families of patients, depending on whether the patient had the severe or mild form of the disease. Compared to families of a patient with the mild (wide) form of the disorder, the severe (narrow form) case would be expected to have a larger absolute number of ill family members, and within the ill group, a larger percentage of those with the severe form as opposed to the mild form.

Recent reviews of clinical genetic data (Gershon et al 1976, Baron, submitted) indicated that most reliable family studies are generally consistent with a two-threshold model wherein bipolar and unipolar affective illnesses represent the narrow (severe) and wide (mild) illness forms, respectively. This concept of shared but graded liability in affective disorders has recently been tested for multifactorial (polygenic) and single major gene transmissions, using the unipolar form as the lower threshold, and the bipolar as the higher (heavier genetic loading) form (Gershon, Baron & Leckman 1975). Although larger populations will have to be studied in order to specify a mode of inheritance, the results clearly indicated that the same underlying genetic abnormality could be manifested clinically as either bipolar or unipolar illness. This conclusion is also supported by twin data which note that 19% of all reported monozygotic twin pairs concordant for major affective illness had one bipolar and one unipolar twin (Perris 1974). Thus, bipolar and unipolar disorders need not be genetically distinct.

Linkage Hypotheses

As discussed above, simple Mendelian inheritance does not fit the overall data on families of patients with affective disorder. The possibility remains, however, that subsets of affective illness, nested within the wider range of affective disorders, are indeed transmitted according to simple Mendelian ratios. A sex-linked (to the X chromosome) dominant mode of inheritance of bipolar illness was suggested by several investigators, based on the low rates of father-son transmission, high rate of mother-son transmission, and the observation that affected sisters of female cases outnumbered affected brothers (Gershon et al 1976, Crowe & Smouse 1977). However, other workers have failed to find sex ratios indicative of sex linkage (Gershon et al 1976), and an alternative hypothesis, that bipolar illness is transmitted as an autosomal dominant trait whose expression is merely dependent upon sex effect rather than on linkage to the X chromosome, has recently been proposed (Leckman & Gershon 1978).

Allied to the hypothesis of X-chromosome transmission is the hypothesis of linkage to specific X-chromosome markers, like color blindness. Bipolar affective disorder has been reported to be closely linked to the known X-chromosome markers of protan and deutan color blindness and the Xg blood group in selected pedigrees (Mendlewicz & Fleiss 1974). These findings, however, have recently been criticized (Gershon & Bunney 1976) on several grounds, and further studies will be necessary to establish this hypothesis.

Autosomal linkage studies of various blood cators in families with affective disorders—the histocompatibility (HLA) antigen system, the ABO blood group, alpha-haptoglobin and complement C_3 have also not yielded a consistent association of autosomal genetic markers with affective illness (Gershon et al 1977).

In summary, the multiple threshold model of inheritance offers the best analytic method available at present for subsuming the known spectrum of clinical and biological characteristics of bipolar and unipolar affective disorders under a single genetic diathesis. It may also elucidate modes of transmission and clarify the interaction of genetic and nongenetic factors. Attempts at delineating genetic types of affective illness on the basis of linkage to known genetic markers have not demonstrated consistent results. The search for a biological measure of individual vulnerability to affective disorder continues.

PHARMACOLOGY AND BIOCHEMISTRY

The major biological research effort in affective disorders was stimulated by the success of somatic, particularly pharmacologic therapies. A number of somatic treatments have proven to be effective in the treatment of major unipolar and bipolar affective disorders. These are: (a) electroconvulsive therapy; (b) antidepressant drugs of the monoamine oxidase inhibitor and tricyclic classes; (c) lithium; and (d) neuroleptic or antischizophrenic agents (phenothiazines and butyrophenones).

Of all these modalities of treatment, electroconvulsive therapy (ECT) has been used the longest, almost 4 decades, and continues to prove the most effective. It produces remission or marked improvement in about 90% of patients with well-defined endogenomorphic depressions. The critical therapeutic factor is the induction of a generalized electrophysiological seizure; approximately 4 to 12 such seizures (average 8), induced at intervals over a 2 to 4 week period suffices to achieve the therapeutic outcome. Several such treatments over a shorter period can terminate a manic episode.

Both monoamine oxidase inhibitors and tricyclics are reported to produce remission or marked improvement in about 70% of depressed patients, although the success rate of tricyclics may well achieve 85% in high dose regimens, or in studies where blood drug levels are monitored and dose adjusted to reach an optimal concentration (Glassman & Perel 1978). Bipolar depressives have a tendency to switch into mania during treatment with both classes of antidepressants (Bunney 1978). While some patients may begin to improve immediately, usually there is a lag of 1 to 3 weeks before the symptoms begin to improve rapidly.

Lithium does not appear particularly effective as an antidepressant in unipolar cases, but may be effective in bipolar cases. It is quite effective in terminating manic episodes, with the improvement beginning as soon as therapeutic blood levels are achieved. Maintenance lithium therapy has a significant prophylactic effect in preventing recurrence or attenuating of both manic and depressive episodes, especially the former (Gerbino, Oleshánsky & Gershon 1978).

Finally, neuroleptics have been shown to be as effective as antidepressants in the treatment of depressions, particularly of the agitated type (Overall et al 1964, Klein 1967, Raskin et al 1970), and they are also quite effective in the treatment of manic episodes (Gerbino, Oleshánsky & Gershon 1978).

As we shall see, the neurochemical hypotheses of depressive illness have been derived largely from the studies of the neurochemical effects of the MAO inhibitors and the tricyclics, and only more recently, and less persuasively, have the hypotheses tried to take into account the therapeutic actions of ECT, lithium, and neuroleptics. The effectiveness of the latter in depression, in fact, has been virtually ignored by the hypothesizers.

The vast supporting data for the monoamine hypotheses of depressive illness have been frequently and extensively reviewed in numerous papers (Schildkraut 1978, Murphy, Campbell & Costa 1978, Baldessarini 1975, Van Praag 1977), are well known to most neuroscientists, and therefore only the bare highlights are discussed here, and only those references are included that are not covered in the review articles. Particularly recommended is the recent volume of review papers in this field by numerous leading investigators (Lipton, DiMascio & Killam 1978) and the critical review by Baldessarini in 1975.

In its simplest terms the monoamine hypothesis states that depressive illness is associated with a functional decrease in noradrenergic and/or serotonergic activity in the hypothalamus and related limbic structures; as a corollary, the therapeutic effect of the antidepressants supposedly occurs by functionally increasing the activity of noradrenalin and/or serotonin in these regions. Mania was initially believed to result from an excess of noradrenergic activity either by itself or in association with a serotonergic deficit, but this hypothesis has been weaker, and is currently held with less confidence.

The highlights of the supporting data include the early observation that reserpine precipitated some sort of depressive syndrome in perhaps as many as 15% of the people who received it. It is interesting that it is difficult now to review those earlier reports to determine how many cases would fit the modern criteria for major effective disorder. However, since reserpine depletes the brain of serotonin and noradrenalin primarily by releasing these neurotransmitters intraneuronally for catabolism by mitochondrial MAO, an effect prevented by pretreatment with MAO inhibitors, these observations formed the beginnings of the monoamine hypotheses.

The discovery that the tricyclic antidepressants blocked the active reuptake of released serotonin and noradrenalin by the presynaptic neurone, thereby preventing catabolism by MAO, further supported the view that the therapeutic effect was mediated by increasing the functional availability of these neurotransmitters for synaptic transmission.

The use of precursors of serotonin and noradrenalin in depressed patients seemed to provide more evidence in the support of the hypothesis: the serotonin precursor 1-tryptophan potentiated the therapeutic effect of MAO inhibitors, and more recently, 5-hydroxy tryptophan (an immediate precursor) also has been reported to have antidepressant action. While L-dopa (a noradrenalin precursor) was not particularly effective as an antidepressant, in one study it precipitated manic or hypomanic episodes in a high proportion of bipolar depressives—lending support to the idea of noradrenalin excess in mania.

Amphetamine, which releases catecholamines from nerve endings, elevated mood in normals, and some depressives, again arguing for an important role of noradrenalin in mood regulation. This fitted well with the research showing that the brain tracts associated with self-stimulation (the "reward" or "pleasure" pathways) were made up of catecholaminergic neurones (Stein 1978). The pervasive loss of the pleasure response in depressives could be thought to relate to disturbances in this system.

In one report, the serotonin synthesis inhibitor parachlorophenylalanine (PCPA), precipitated abrupt relapses in depressed patients recovering after tricyclic therapy—again implicating serotonin. Triiodothyronine was found to potentiate the antidepressant effect of moderate doses of tricyclics, presumably by sensitization of brain catecholamine receptors.

All these psychopharmacologic observations provided strong circumstantial evidence for the biogenic amine hypothesis. However, it remained difficult to account for many clinical phenomena. For example, reuptake blockade by tricyclics occurs rapidly, but clinical response takes 2 to 3 weeks. (The time course for MAO inhibition with MAO inhibitor drugs, however, is more consistent with clinical response.) Tricyclics vary widely in their relative abilities to block serotonin or noradrenalin reuptake, yet their clinical efficacies in depressed patients are about the same, particularly if doses are adjusted to blood level. Finally, recently discovered tricyclics, such as mianserin, appear to be clinically efficacious, yet have no effect on reuptake of serotonin or noradrenalin (Ghose, Coppen & Turner 1976, Van der Burg et al 1970, Kafoe, DeRidder & Leonard 1976). Thus, there is a recent effort to study other neurochemical effects of the drugs, for example, on receptor sensitivity. An example of this new strategy is evident in the recent report that the tricyclics desmethylimipramine and chlorimipramine, which differ markedly in their relative acute effects on serotonin and nora-drenalin reuptake, both increase serotonin receptor sensitivity equally when administered chronically (DeMontigny & Aghajanian 1977).

However, a major problem in such research is the fact that the various antidepressant drugs are roughly equipotent clinically. Research on the neuroleptics was greatly aided by the 100-fold difference among them in clinical potency; it was thus possible to sort out among their myriad neuro-chemical effects one effect—relative potency of dopamine blockade—which correlated well with relative clinical potency. A super-potent antidepres-sant, effective in doses 50 to 100 times smaller than present drugs, has yet to be discovered, but would obviously be very useful to the researcher.

Another problem for the monoamine hypothesis is the failure, until recently, to demonstrate any significant effect of ECT on brain monoamine systems in animals (Ebert et al 1973). A recent report of increase in cate-cholamine receptor sensitivity may be the first breakthrough in this area (Modigh 1976), but the fact that ECT is also effective in mania would still require explanation.

It has similarly been difficult to account neurochemically for lithium ion's therapeutic and prophylactic effects in both mania and depression. Effects on passage of other ions across membranes, on the uptake of serotonin, and the conversion of tryptophan to serotonin have been among the major foci of recent work, but it remains speculative which, if any, of these effects are related to the clinical effects (Gerbino, Oleshánsky & Gershon 1978). It is worth keeping in mind, then, that it is risky to assume a fundamental connection between a therapeutic drug's effect and the etiology of the disorder—many drugs work in hypertension by affecting the final common manifestation of the disease, rather than its multiple etiologies.

Biochemical studies on depressed patients, some of which have been referred to in our discussion of genetic investigations, also have been used to provide evidence for a deficit in monoamine metabolism. Two biochemi-cal measures for which there is a substantial literature are urinary MHPG and spinal fluid 5H1AA. MHPG is a glycol metabolite of noradrenalin, and the more recent animal studies suggest that 30 to 50% of the MHPG in the urine is derived from the degradation of noradrenalin from the brain. Several investigators have reported a mean reduction in urinary MHPG in primary unipolar and bipolar depressives, compared to normals. While a subgroup in each study seems particularly low, it remains controversial whether there is a bimodal distribution.

Several investigators have also reported a decrease in the serotonin metabolite 5H1AA in spinal fluid, either as absolute concentration, or in the rate of accumulation after blockade of efflux from CSF with probenecid. Again, a subgroup seems particularly low.

These observations have led to the hypothesis that there are two biochemical subtypes of depression—a "low noradrenalin" and a "low serotonin" group. It was suggested that these groups might respond differentially to tricyclic drugs that differentially block noradrenalin or serotonin reuptake. Studies that support this view (Goodwin, Cowdry & Webster 1978) have, however, used doses of drugs now recognized as being suboptimal. Furthermore, it is increasingly questioned whether reuptake blockade, which has been used to differentiate "serotonergic" and "noradrenergic" drugs, is the most important mechanism of the therapeutic action of tricyclics.

In any case, a major problem in such biochemical research is the likelihood that the brain monoamine dysfunction is restricted to a small region of the hypothalamus and limbic system, and even if the urinary and spinal fluid measurement accurately reflected monoaminergic state of the brain, they might overlook small but significant regional disturbances.

In summary, the earlier thermostat version of the moanoamine hypothesis—too little or too much of a particular neurotransmitter—is in the process of revision. The efficacy of antischizophrenic agents, lithium, and ECT in both mania and depression would suggest a different type of regulatory disorder, perhaps one involving a disturbance in the balance between several biochemically different neuronal systems in the brain. This view would be consistent with the involvement of brain serotonergic and noradrenergic pathways at a critical point in the chemical pathology of affective disorders and in the clinical response to therapeutic agents, but would not exclude the additional involvement of other transmitters.

NEUROENDOCRINE FUNCTION

There are two reasons to expect disturbances in neuroendocrine function in severe depression: (a) The existence of many clinical signs consistent with hypothalamic dysfunction would suggest that hypothalamic modulation of neuroendocrine activity might also be affected, and (b) the neurotransmitter systems implicated in aspects of depression—serotonin and noradrenalin—also play a major role in neoroendocrine regulation.

An extensive literature has documented a hypersecretion of cortisol in severe depressions, secondary to excessive ACTH secretion (see reviews by Sachar 1975, Carroll 1978). Plasma cortisol concentration is elevated as is the urinary excretion of corticosteroids. The 24-hour pattern of episodic secretion is also characteristically abnormal. Whereas the normal subject essentially ceases to secrete cortisol in the evening and early morning hours, the depressed patient actively secretes cortisol throughout this period, leading to a flattening of the circadian pattern. The hypersecretion is resistant to feedback suppression, particularly in the evening, by the synthetic corticosteroid dexamethasone. The hypersecretion almost certainly cannot be a reflection of a simple stress response—it persists during sleep and is unrelated to sleep disturbance, it is not affected by antianxiety agents and occurs in apathetic as well as anxious patients. Moreover, other psychiatric disorders marked by anxiety do not show this characteristic abnormality in the circadian pattern nor the resistance to dexamethasone suppression. With clinical recovery, the cortisol secretory pattern returns to normal.

It is tempting to speculate that the cortisol secretory abnormality reflects a disturbance in a neurotransmitter regulatory system acting on the hypothalamic neuroendocrine cells that release Corticotropin Releasing Hormone (CRH) which stimulates ACTH. Several neurotransmitters are involved in the complex regulation of CRH, including a noradrenergic system inhibitory to CRH and ACTH secretion. A functional decrease in noradrenergic activity could account for the cortisol hypersecretion in depression.

Another characteristic endocrine abnormality in depression is a reduced growth hormone (HGH) response to hypoglycemia, induced by a standard dose of insulin. This HGH response to hypoglycemia can also be blunted by blockers and depletors of brain noradrenalin and serotonin, once again suggesting that the neuroendocrine abnormality in depressive illness may be related to central monoaminergic dysfunction (Sachar 1975, Carroll 1978).

A blunted TSH response to infusion of standard doses of Thyrotropin Releasing Hormone (TRH) has also been widely reported in depression. This response is presumably mediated at the pituitary level, and its relation to hypothalamic abnormality is obscure, particularly since other measures of thyroid function are normal in depressed patients. It is possible that the blunted TSH response is secondary to elevated cortisol levels in depression, since exogenous corticosteroid administration will also inhibit the TSH response (Prange 1975).

SUMMARY

The clinical features, genetic data, psychopharmacological studies, hormonal abnormalities, and biochemical observations all serve to define major depressive illness as an inherited neurochemical disorder affecting the hypothalamus, and probably involving monoamine pathways. While the precise nature of the defect and its mode of transmission remain obscure, the rapid development of this field in less than two decades permits optimism that major depressive illness will be among the first "functional" psychiatric disorders to have its chemical pathology elucidated.

Literature Cited

Angst, J. 1978. The course of affective disorders. In *Handbook of Biological Psychiatry*, ed. H. M. van Praag, O. J. Rafaelsen, M. Lader, E. J. Sachar. New York: Dekker. In press

Baldessarini, R. J. 1975. Biogenic amine hypotheses in affective disorders. In *Nature and Treatment of Depression*, ed. F. Flach, S. Draghi, pp. 347–86. New York: Wiley. 422 pp.

Baron, M. Genetic models of the transmission of bipolar and unipolar affective disorders; the multiple threshold concept. (Submitted for publication.)

Bunney, W. E. 1978. Psychopharmacology of the switch process in affective illness. In *Psychopharmacology: A Generation of Progress*, ed. M. A. Lipton, A. DiMascio, K. F. Killam, pp. 1249–59. New York: Raven. 1731 pp.

Carroll, B. J. 1978. Neuroendocrine function in psychiatric disorders. See Bunney 1978, pp. 487–97

Crowe, R. R., Smouse, P. E. 1977. The genetic implications of age-dependent penetrance in manic-depressive illness. *J. Psychiatr. Res.* 13:273–85

deMontigny, C., Aghajanian, G. K. 1977. Sensitization of postsynaptic serotonin receptors by chronic pretreatment with tricyclic antidepressants. *Neurosci. Abstr.* III: 248

Ebert, M., Baldessarini, R., Lipinski, J., Berv, K. 1973. Effects of electroconvulsive seizures on amine metabolism in rat brain. *Arch. Gen. Psychiatry* 29:397–401

Endicott, J., Spitzer, R. 1978. Use of RDC and SADS in the diagnosis of affective disorders. *Am. J. Psychiatry.* In press

Feighner, J. P., Robins, E., Guze, S. B., Woodruff, R. A. Jr., Winokur, G., Munos, R. 1972. Diagnostic use in psychiatric research. *Arch. Gen. Psychiatry* 22:57–63

Freud, S. 1956. Mourning and melancholia [1917]. In *Collected Papers*, Vol. IV. London: Hogarth, pp. 152–70

Gerbino, L., Oleshansky, M., Gershon, S. 1978. Clinical use and mode of action of lithium. See Bunney 1978, pp. 1261–76

Gershon, E. 1978. The search for genetic markers in affective disorders. See Bunney 1978, pp. 1197–1212

Gershon, E. S., Baron, M., Leckman, J. F. 1975. Genetic models of the transmission of affective disorders. *J. Psychiatr. Res.* 12:301–17

Gershon, E. S., Bunney, W. E. Jr. 1976. The question of X-linkage in bipolar manic-depressive illness. *J. Psychiatr. Res.* 13:99–117

Gershon, E. S., Bunney, W. E., Jr., Leckman, J., Van Eerdewegh, M., DeBauche, B. 1976. The inheritance of affective disorders: A review of data and of hypotheses. *Behav. Genet.* 6:227–61

Gershon, E. S., Targum, S. D., Kessler, L. R., Mazure, C. M., Bunney, W. E. Jr. 1977. Genetic studies and biologic strategies in the affective disorders. *Prog. Med. Genet.* 2:101–64

Ghose, K., Coppen, A., Turner, P. 1976. Autonomic actions and interactions of mianserin hydrochloride and amytriptyline in patients with depressive illness. *Psychopharmacology* 49:201–4

Glassman, A. H., Perel, J. M. 1978. Tricyclic blood levels and clinical outcome. See Bunney 1978, pp. 917–22

Goodwin, F. K., Cowdry, R. W., Webster, M. H. 1978. Predictors of drug response in the affective disorders. See Bunney 1978, pp. 1277–88

Kafoe, W. F., DeRidder, J. J., Leonard, B. E. 1976. The effect of a tetracyclic antidepressant compound, Org. GB94, on the turnover of biogenic amines in rat brain. *Biochem. Pharmacol.* 25: 2455–60

Katz, M. M., Hirschfeld, R. M. 1978. Phenomenology and classification of depression. See Bunney 1978, pp. 1185–95

Klein, D. F. 1967. Importance of psychiatric diagnosis in prediction of clinical drug effects. *Arch. Gen. Psychiatry* 16:118–26

Leckman, J. F., Gershon, E. S. 1978. Autosomal models of sex effect in bipolar-related major affective illness. In press

Leonhard, K., Korf, I., Schulz, H. 1962. Die Temperamente in den Familien der monopolaren und bipolaren phasischen Psychosen. *Psychiatr. Neurol.* 143:416

Lipton, M., DiMascio, A., Killam, K. F., eds. 1978. *Psychopharmacology: A Generation of Progress.* New York: Raven. 1731 pp.

Mendlewicz, J., Fleiss, J. L. 1974. Linkage studies with X-chromosome markers in bipolar (manic-depressive) and unipolar (depressive) illness. *Biol. Psychiatry* 9:261–94

Mendlewicz, J., Rainer, J. D. 1977. Adoption study in manic-depressive illness. *Nature* 268:327–29

Modigh, K. 1976. Long term effects of electroconvulsive shock on synthesis, turnover, and uptake of brain monoamines. *Psychopharmacology* 49:179–85

Murphy, D. L., Campbell, I., Costa, J. L. 1978. Current status of the indoleamine hypothesis of the affective disorders. See Bunney 1978, pp. 1235–47

Overall, J. E., Hollister, L. E., Meyer, F., Kimbell I. Jr., Shelton, J. 1964. Imipramine and thioridazine in depressed and schizophrenic patients. *J. Am. Med. Assoc.* 189:605–8

Perris, C. 1974. The genetics of affective disorders. In *Biological Psychiatry* ed. J. Mendels, pp. 385–415. New York: Wiley. 527 pp.

Prange, A. J. 1975. Patterns of pituitary responses to thyrotropin releasing hormone in depressed patients. In *Phenomenology and Treatment of Depression,* ed. W. E. Fann, pp. 1–15. New York: Spectrum

Price, J. 1968. The genetics of depressive behavior. *Br. J. Psychiatry.* Special Publ. No. 2, pp. 37–45

Raskin, A., Schulterbrandt, J. G., Reatig, N., McKeon, J. J. 1970. Differential response to chlorpromazine, imipramine, and placebo. *Arch. Gen. Psychiatry* 23:164–73

Reich, T., James, J. W., Morris, C. A. 1972. The use of multiple thresholds in determining the mode of transmission of semi-continuous traits. *Ann. Hum. Genet.* 36:163

Sachar, E. J. 1975. Neuroendocrine abnormalities in depressive illness. In *Topics in Psychoendocrinology,* ed. E. J. Sachar, pp. 135–56. New York: Grune & Stratton. 182 pp.

Schildkraut, J. 1978. Current status of the catecholamine hypothesis of affective disorders. See Bunney 1978, pp. 1223–34

Spitzer, R. L., Endicott, J. E., Robins, E. 1975. *Research Diagnostic Criteria.* New York: N.Y. State Psychiatric Institute

Stein, L. 1978. Reward transmitters: Catecholamines and opioid peptides. See Bunney 1978, pp. 569–81

Van der Burg, W. J., Bonta, I. L., Delobbelle, J., et al. 1970. A novel type of substituted piperazine with high antiserotonin potency. *J. Med. Chem.* 13:35

van Praag, H. M. 1977. Indoleamines in depression. In *Neuroregulators and Psychiatric Disorders,* ed. E. Usdein, D. Hamburg, J. Barchas, pp. 163–76. N.Y.: Oxford Univ. Press. 627 pp.

AUTHOR INDEX

SUBJECT INDEX

CUMULATIVE INDEXES

CONTRIBUTING AUTHORS, VOLUMES 1–2

CHAPTER TITLES, VOLUMES 1–2